The Receptors

Volume 32

Series Editor

Giuseppe Di Giovanni
Department of Physiology & Biochemistry
Faculty of Medicine and Surgery
University of Malta
Msida, Malta

The Receptors book Series, founded in the 1980's, is a broad-based and well-respected series on all aspects of receptor neurophysiology. The series presents published volumes that comprehensively review neural receptors for a specific hormone or neurotransmitter by invited leading specialists. Particular attention is paid to in-depth studies of receptors' role in health and neuropathological processes. Recent volumes in the series cover chemical, physical, modeling, biological, pharmacological, anatomical aspects and drug discovery regarding different receptors. All books in this series have, with a rigorous editing, a strong reference value and provide essential up-to-date resources for neuroscience researchers, lecturers, students and pharmaceutical research.

More information about this series at http://www.springer.com/series/7668

Bruno P. Guiard • Giuseppe Di Giovanni
Editors

$5\text{-}HT_{2A}$ Receptors in the Central Nervous System

Humana Press

Editors
Bruno P. Guiard
Faculté de Pharmacie
Université Paris Sud
Université Paris-Saclay
Chatenay-Malabry, France

Centre de Recherches sur la Cognition
Animale (CRCA)
Centre de Biologie Intégrative (CBI)
Université de Toulouse; CNRS, UPS
Toulouse, France

Giuseppe Di Giovanni
Department of Physiology
and Biochemistry
University of Malta
Msida MSD, Malta

The Receptors
ISBN 978-3-319-70472-2 ISBN 978-3-319-70474-6 (eBook)
https://doi.org/10.1007/978-3-319-70474-6

Library of Congress Control Number: 2017964095

Printed on acid-free paper

This Humana Press imprint is published by Springer Nature
The registered company is Springer International Publishing AG
The registered company address is: Gewerbestrasse 11, 6330 Cham, Switzerland

Preface

It was a great honour and pleasure for us to prepare this book entitled *5-HT_{2A} Receptors in the Central Nervous System*. The serotonin 5-HT_{2A} receptor, cloned in 1994, is principally coupled to $G_{q/11}$ protein and it is expressed in different brain regions involved in cognition, perception, sensorimotor gating and mood. A major flaw in the study of the 5-HT_{2A} receptor has relied on the lack of selective tools for mapping its distribution in the brain and examining its specific contribution to physiological and pathological processes. Nevertheless, by combining pharmacological approaches with 5-HT_2 receptor ligands or antibodies with genetic tools (e.g., constitutive or tissue-specific 5-HT_{2A} receptor knockout mice), it has been possible to unveil a complex and fascinating organization of this receptor at both cellular and subcellular levels. Interestingly, in the last few years, our knowledge on the 5-HT_{2A} receptor has undergone a revolution owing to the discovery of many peculiar pharmacological properties. For example, the 5-HT_{2A} receptors have been shown to interact with various G protein coupled receptors to form stable homo- or heteromeric complexes, such as D2, mGlu2 and CB1 receptors likely responsible for changes in binding and coupling properties. There is another mechanism by which 5-HT_{2A} receptor can regulate its signalling. Although it is well documented that agonist stimulation leads to internalization of the 5-HT_{2A} receptors through a βarrestins2-dependent mechanism, recent evidence suggests that βarrestins can also facilitate G protein-independent signalling by functioning as adaptor proteins, notably with the 5-HT_{2A} receptors. The first chapters of this book are therefore aimed at reviewing the recent discovery of the pharmacological properties of the 5-HT_{2A} receptors since their better understanding may open avenues for the design of new therapeutic compounds and thereby to improve the treatment of a number of psychiatric and neurological disorders. The next section of chapters reviews our understanding of the role(s) of the 5-HT_{2A} receptors in several brain functions including sleep, memory, emotion and food intake. This compelling evidence reviewed here raises the possibility that the 5-HT_{2A} receptor might be a relevant and promising target for the treatment of pathologies such as schizophrenia, mood and eating disorders, pain, epilepsy or Parkinson's disease. The final section of chapters further supports this evidence by illustrating how the 5-HT_{2A} receptor directly or indirectly controls neuronal excitability and brain plasticity through its interactions with monoaminergic, GABAergic and glutamatergic neurons but also with neurotrophic factors. The scope of the book is vast, going from the 5-HT_{2A} receptor genome to its therapeutic applications based on clinical and preclinical observations. We have embarked on this unique editorial enterprise with the aim of providing the most recent achievements, a result of the efforts of an international group of scholars. We have also prepared this book for a wide audience (students, researchers, practitioners and caregivers) offering them a valuable and integrated insight into the mechanisms of action of the 5-HT_{2A} receptors. We hope that the contents of this book will further inspire and stimulate discussions and new interdisciplinary research on the 5-HT_{2A} receptor.

We would like to thank all the authors for their contribution and outstanding review manuscripts in their area of expertise. A particular acknowledgement should be given to Springer editors, Ms. Portia Wong, the Developmental Editor, who helped us to manage this project and Ms. Simina Calin, the Neuroscience Editor, for coordinating the entire project.

Msida, Malta Bruno Guiard
Toulouse, France Giuseppe Di Giovanni
September, 2017

Contents

Contributors

Albert Adell Institute of Biomedicine and Biotechnology of Cantabria, IBBTEC (CSIC-University of Cantabria), Santander, Spain

Rosa María Anaya-Jiménez Facultad de Ciencias de la Salud, Universidad Anáhuac México Norte, México City, México

D.N. Arvanitis Institut des Maladies Métaboliques et Cardiovasculaires, INSERM UMR1048, Université Paul Sabatier, Toulouse, France

Ewa Błasiak Faculty of Biochemistry, Biophysics and Biotechnology, Jagiellonian University, Kraków, Poland

Laura M. Bohn Departments of Molecular Therapeutics and Neuroscience, The Scripps Research Institute, Jupiter, FL, USA

Cristiano Bombardi Department of Veterinary Medical Sciences, University of Bologna, Ozzano dell'Emilia, BO, Italy

Anna Castañé CSIC-Institut d'Investigacions Biomèdiques de Barcelona-CIBERSAM-IDIBAPS, Barcelona, Spain

Christine Courteix INSERM U1107, NEURO-DOL, Clermont-Ferrand, France

Laboratoire de Physiologie, UFR Pharmacie, Université Clermont Auvergne, Clermont-Ferrand, France

Vincenzo Crunelli Neuroscience Division, School of Bioscience, Cardiff University, Cardiff, UK

Department of Physiology and Biochemistry, University of Malta, Msida, Malta

Kristína Csatlósová Institute of Molecular Physiology and Genetics, Slovak Academy of Sciences, Bratislava, Slovakia

Philippe De Deurwaerdère Université de Bordeaux, Bordeaux, France

Centre National de la Recherche Scientifique, UMR 5287, Bordeaux, France

Eliyahu Dremencov Institute of Molecular Physiology and Genetics, Slovak Academy of Sciences, Bratislava, Slovakia

Institute of Experimental Endocrinology, Biomedical Research Center, Slovak Academy of Sciences, Bratislava, Slovakia

Guillaume Drutel Université de Bordeaux, Bordeaux, France

Institut National de la Santé et de la Recherche Médicale, U 1215, Bordeaux, France

Amandine Dupuis Inserm U1107, NEURO-DOL, Clermont-Ferrand, France

Laboratoire de Physiologie, UFR Pharmacie, Université Clermont Auvergne, Clermont-Ferrand, France

Barbora Ďurišová Institute of Molecular Physiology and Genetics, Slovak Academy of Sciences, Bratislava, Slovakia

Marta Dziedzicka-Wasylewska Faculty of Biochemistry, Biophysics and Biotechnology, Jagiellonian University, Kraków, Poland

Institute of Pharmacology, Polish Academy of Sciences, Kraków, Poland

Giuseppe Di Giovanni Neuroscience Division, School of Bioscience, Cardiff University, Cardiff, UK

Department of Physiology and Biochemistry, University of Malta, Msida, Malta

Javier González-Maeso Department of Physiology and Biophysics, Virginia Commonwealth University, School of Medicine, Richmond, VA, USA

Philip Gorwood CMME (Hôpital Sainte-Anne) and Université Paris-Descartes, Paris, France

Centre de Psychiatrie et Neurosciences (INSERM UMR894), INSERM, Paris, France

Bruno P. Guiard Faculté de Pharmacie, Université Paris Sud, Université Paris-Saclay, France

Centre de Recherches sur la Cognition Animale (CRCA), Centre de Biologie Intégrative (CBI), Université de Toulouse; CNRS, UPS, Toulouse, France

Matthias M. Herth Center for Integrated Molecular Brain Imaging, Copenhagen University Hospital, Rigshospitalet, Copenhagen, Denmark

Department of Drug Design and Pharmacology, Faculty of Health and Medicine, University of Copenhagen, Copenhagen, Denmark

Department of Clinical Physiology, Nuclear Medicine and PET, Copenhagen University Hospital, Copenhagen, Denmark

Minal Jaggar Department of Biological Sciences, Tata Institute of Fundamental Research, Mumbai, Maharashtra, India

M. Kermorgant Institut des Maladies Métaboliques et Cardiovasculaires, INSERM UMR1048, Université Paul Sabatier, Toulouse, France

Gitte M. Knudsen Center for Integrated Molecular Brain Imaging, Copenhagen University Hospital, Rigshospitalet, Copenhagen, Denmark

Ľubica Lacinová Institute of Molecular Physiology and Genetics, Slovak Academy of Sciences, Bratislava, Slovakia

Laurence Lanfumey Centre de Psychiatrie et Neurosciences (INSERM UMR894), Paris, France

Sylwia Łukasiewicz Faculty of Biochemistry, Biophysics and Biotechnology, Jagiellonian University, Kraków, Poland

Pierre-Yves Martin Inserm U1107, NEURO-DOL, Clermont-Ferrand, France

Laboratoire de Physiologie, UFR Pharmacie, Université Clermont Auvergne, Clermont-Ferrand, France

Alfredo Meneses Department of Pharmacobiology, CINVESTAV, México City, México

Cristina Miguelez Department of Pharmacology, Faculty of Medicine and Nursing, University of the Basque Country (UPV/EHU), Leioa, Spain

Jaime M. Monti Department of Pharmacology and Therapeutics, School of Medicine Clinics Hospital, University of the Republic, Montevideo, Uruguay

Lucia Moravčíková Institute of Molecular Physiology and Genetics, Slovak Academy of Sciences, Bratislava, Slovakia

José L. Moreno Department of Physiology and Biophysics, Virginia Commonwealth University, School of Medicine, Richmond, VA, USA

Teresa Morera-Herreras Department of Pharmacology, Faculty of Medicine and Nursing, University of the Basque Country (UPV/EHU), Leioa, Spain

L. Moulédous Research Center on Animal Cognition, Center for Integrative Biology, Université Paul Sabatier, Toulouse Cedex, France

Rossana Nieto-Vera Facultad de Ciencias de la Salud, Universidad Anáhuac México Norte, México City, México

Katarína Ondáčová Institute of Molecular Physiology and Genetics, Slovak Academy of Sciences, Bratislava, Slovakia

Seithikurippu R. Pandi Perumal Somnogen Canada Inc, Toronto, ON, Canada

Michaela Pavlovičová Institute of Molecular Physiology and Genetics, Slovak Academy of Sciences, Bratislava, Slovakia

A. Pavy-Le Traon Institut des Maladies Métaboliques et Cardiovasculaires, INSERM UMR1048, Université Paul Sabatier, Toulouse, France

Department of Neurology, Institute for Neurosciences, University Hospital of Toulouse, Toulouse, France

Nicolas Ramoz Centre de Psychiatrie et Neurosciences (INSERM UMR894), Paris, France

P. Roullet Research Center on Animal Cognition, Center for Integrative Biology, Université Paul Sabatier, Toulouse Cedex, France

Cullen L. Schmid Departments of Molecular Therapeutics and Neuroscience, The Scripps Research Institute, Jupiter, FL, USA

J.M. Senard Institut des Maladies Métaboliques et Cardiovasculaires, INSERM UMR1048, Université Paul Sabatier, Toulouse, France

Department of Clinical Pharmacology, University Hospital of Toulouse, Toulouse, France

Benoit Sion INSERM U1107, NEURO-DOL, Clermont-Ferrand, France

Laboratoire de Physiologie, UFR Pharmacie, Université Clermont Auvergne, Clermont-Ferrand, France

Pablo Torterolo Department of Physiology, School of Medicine, University of the Republic, Montevideo, Uruguay

Vidita A. Vaidya Department of Biological Sciences, Tata Institute of Fundamental Research, Mumbai, Maharashtra, India

Marcello Venzi AstraZeneca Translational Science Centre at Karolinska Institutet, Stockholm, Sweden

Odile Viltart Centre de Psychiatrie et Neurosciences (INSERM UMR894), Paris, France

D. Warren Spence Sleep and Altertness Clinic, University Health Network, Toronto, ON, Canada

Part I
Properties and Distribution of the 5-HT_{2A}R

Pharmacological Analysis in Favour of a Physiological Role for the Constitutive Activity of 5-HT$_{2A}$ Receptors in Learning

Philippe De Deurwaerdère, Guillaume Drutel, and Giuseppe Di Giovanni

Abstract The Serotonin2A (5-hydroxytryptamin, 5-HT$_{2A}$) receptor is one of the numerous seven transmembrane G protein coupled receptors for serotonin (5-HT) originally described as displaying a low affinity for its endogenous ligand. It is densely expressed in the cortex and the hippocampus of rodents, primates and humans brain. A role of 5-HT$_{2A}$ receptors in learning and memory has been proposed for years. In some behavioural tasks in rodents, 5-HT$_{2A}$ receptors would display a constitutive activity, a spontaneous activity of the receptor occurring without the presence of the endogenous ligand and silenced by inverse agonists. Nonetheless, the demonstration of the existence of such a subtle activity in living organisms relies on specific criteria and on clear-cut pharmacological evaluation. While it has been claimed that 5-HT$_{2A}$ receptor constitutive activity participates in the conditioned eyeblink response in rabbits, such an activity would not be systematically observed in other models of learning and conditioning such as the conditioned avoidance response in rats. Here, we propose a thorough pharmacological analysis of the available data arguing in favour of the involvement of constitutive activity of 5-HT$_{2A}$ receptors, mostly in learning tasks and discuss the functional significance of such an activity.

Keywords Pharmacology of 5-HT$_2$ receptors • MDL11,939 • Selectivity • Pavlovian conditioning • Conditioned avoidance response • Autoshaping learning task • Inverse agonism • Intracellular signaling pathways • Serotonin

---

P. De Deurwaerdère (✉)
Université de Bordeaux, Bordeaux, France

Centre National de la Recherche Scientifique UMR 5287, Bordeaux, France
e-mail: deurwaer@u-bordeaux.fr

G. Drutel
Université de Bordeaux, Bordeaux, France

Institut National de la Santé et de la Recherche Médicale, U 1215, Bordeaux, France

G. Di Giovanni
Neuroscience Division, School of Bioscience, Cardiff University, Cardiff, UK

Department of Physiology and Biochemistry, University of Malta, Msida, Malta

B.P. Guiard, G. Di Giovanni (eds.), *5-HT$_{2A}$ Receptors in the Central Nervous System*, The Receptors 32, https://doi.org/10.1007/978-3-319-70474-6_1

Abbreviations

5,7-DHT	5,7-dihydroxytryptamin
5-HT	Serotonin
5-HT_{2A} receptor	5-hydroxytryptamine2A receptor
5-HT_{2C} receptor	5-hydroxytryptamine2C receptor
BOL	d-bromolysergic acid diethylamide
CAR	Conditioned avoidance response
CHO	Chinese hamster ovary
DA	dopamine
DOI	(±)-1(2,5-dimethoxy-4-iodophenyl)-2-aminopropane hydrochloride
DOM	d,l-2,5-dimethoxy-4-methylamphetamine
GPCR	G-Protein coupled receptor
HEK-293	Human embryonic kidney 293
IP	Inositol phosphate
LSD	d-lysergic acid diethylamide
MDA	d,l-methylenedioxyamphetamine
MDMA	d,l-methylenedioxymethamphetamine
PLA2	Phospholipase A2
PLC	Phospholipase C
PLD	Phospholipase D
SERT	Serotonin transporter

Introduction

The occurrence of a constitutive activity has been proposed for several seven-transmembrane receptors *in vivo* [1–3]. In a neurobiological point of view, this notion is puzzling regarding the way we usually teach and conceive neurotransmission. Indeed, it conceptually limits the role of the neurotransmitter released in the synaptic cleft and stimulation of its receptor. The constitutive activity would imply that the activity of the cell bearing the receptor triggers the ability of the receptor to be stimulated on its own and more so if a neurotransmitter is present. The demonstration of the existence of an endogenous constitutive activity for a receptor is a complex pharmacological tale *in vivo*, often leading to strong presumptions rather than absolute certainty.

Serotonin2A (5-HT_{2A}) receptors participate in numerous biological functions peripherally and centrally. They were originally characterized as a 5-HT receptor displaying a low affinity for its endogenous ligand serotonin (5-HT) (5-HT_2 receptor subtype) [4–6]. Typically, such a receptor could intervene to mediate a phasic enhancement of 5-HT tone associated with an enhancement of 5-HT extracellular levels [7]. In heterologous expression systems, its high expression is associated with a low constitutive activity, a spontaneous activity occurring without the presence of an agonist [8, 9]. This agonist-independent activity of the receptor obtained under certain conditions allowed pharmacologists to show that many drugs that were thought to act as neutral antagonists were actually inverse agonists that silenced the enhanced constitutive activity of 5-HT_{2A} receptors [10]. This activity in heterologous expression systems remains lower compared to that displayed by other receptors, such as the 5-HT_{2C} receptor, questioning the physiological relevance of such an activity in living organisms. Nonetheless, some data support the existence of such an activity *in vivo* in learning [11], and other data indirectly give some credit to this possibility.

In this chapter, after briefly presenting the interest in 5-HT_{2A} receptor pharmacology, we will review the pharmacological criteria before presenting the data in favour of the existence of a constitutive activity of 5-HT_{2A} receptors *in vitro*. Thereafter, we will discuss the evidence that suggests that 5-HT_{2A} receptors can adopt a constitutive activity mostly in learning tasks in rodents.

Physiology and Pathophysiology of 5-HT_{2A} Receptors

The 5-HT_{2A} receptor subtype was discovered several years ago and was named "D" followed by 5-HT_2 receptor subtype. Its peculiarity, with respect to the so-called 5-HT_1 receptors, was related to its low affinity for 5-HT, ranging in several hundred nM range to μM. After the discovery of at least 14 receptor subtypes for 5-HT, 5-HT_{2A} receptors remain the 5-HT receptor subtype with the lowest affinity for 5-HT [4, 5].

Using numerous radioligands, including ^{3}H-ketanserin and ^{3}H-MDL100907, it was consistently reported that 5-HT_{2A} receptors were mainly and densely expressed in cortical regions as well as in the hippocampus and amygdala [12–16]. The labelling is modest in the basal ganglia, particularly faint in the caudate nucleus and putamen of human and subhuman primates except in the striosomes [12, 14–16]. The distribution has been confirmed using in situ hybridization [16, 17]. The anatomical description of 5-HT_{2A} receptors in the brain has been confirmed in living human brains in PET studies, using various radiolabelled compounds [18, 19]. 5-HT_{2A} receptors are also expressed in peripheral tissues, including platelets, stomach, and gut, on smooth muscle cells [20].

The use of antibodies directed against 5-HT_{2A} receptors or in situ hybridization permitted to identify the presence of the receptor on pyramidal cells and on interneurons in layer V of the cortex [21]. They primarily mediate excitation in both neuronal types [22, 23]. In the basal ganglia, they would be expressed by striatonigral GABAergic neurons, some cholinergic interneurons, and some DA neurons in the ventral tegmental area and substantia nigra pars compacta [24, 25]. The demonstration that 5-HT_{2A} receptors are present presynaptically on striatal DA terminals is a matter of debate. In the hippocampus, 5-HT_{2A} receptors are expressed and concentrated in the apical dendrites of the pyramidal cells [26]. Moreover strong evidence suggests that presynaptic 5-HT_{2A} receptors at thalamocortical synapses play an essential role in associative learning [27].

5-HT_{2A} receptors have been implicated in the actions of hallucinogenic drugs [20, 28, 29] and in cognitive function [21, 30–32]. Historically, this involvement gave support to a role of 5-HT and 5-HT receptors in psychosis and schizophrenia [33], and this hypothesis has become tangible over the years [34–36]. Meltzer et al. [37] proposed that the therapeutic benefit of atypical antipsychotic drugs compared to typical antipsychotics for treating positive and negative symptoms of schizophrenia was due to a higher ratio affinity to 5-HT_{2A} receptors as opposed to dopamine (DA) D2 receptors [37]. To date, the pharmacology of 5-HT_{2A} receptors could result in treatments of various neuropsychiatric diseases, including schizophrenia [38], drug abuse [39], depression [40], anxiety [41] and psychosis in Parkinson's disease [42].

Constitutive Activity of 5-HT_{2A} Receptors: Pharmacology and *In Vitro* Consideration

Constitutive Activity and the Inverse Agonists

The constitutive activity of seven-transmembrane receptors is well characterized in heterologous recombinant systems *in vitro* [8, 9, 43–47]. It corresponds to the ability of a given receptor to spontaneously activate and regulate cellular signaling systems in the absence of occupancy by a ligand. This property has been considered as an artificial property, as it depends on the density of the receptor expressed at the cell surface and the total absence of the endogenous ligand, two conditions that can be easily controlled *in vitro* [9, 48]. A large number of seven-transmembrane receptors exhibit constitutive activity *in vitro*, and there is often a strong and sometimes linear relationship between the magnitude of the constitutive activity of a given receptor and its expression in various heterologous recombinant systems *in vitro* [3]. The constitutive activity is also dependent on the quantities of G-proteins or other interacting proteins expressed in these cells, and this is another parameter that can be managed *in vitro*.

The constitutive activity of a receptor is intimately linked to the notion of "inverse agonism". While agonists were still able to trigger intracellular systems transduction pathways in a situation where a given receptor presumably displayed a constitutive activity, other drugs that were considered antagonists were able to silence the constitutive activity of a receptor *in vitro* (Fig. 1). Logically, these drugs were called "inverse agonist". These drugs are the solely, direct pharmacological tool that can determine the existence of a constitutive activity in a given receptor. For several receptors, including the 5-HT_{2A} receptors [10, 49], most drugs labeled "antagonists", such as ritanserin, behaved as inverse agonists in various heterologous recombinant systems *in vitro* (Fig. 1). As a classical pharmacological response, the effect of an inverse agonist should be diminished in case of receptor occupancy by an antagonist. An antagonist or neutral antagonist is defined as a drug that has no negative or positive intrinsic activity capable of occluding the binding of both the agonist and the inverse agonist, thereby preventing their intracellular effects (Fig. 1).

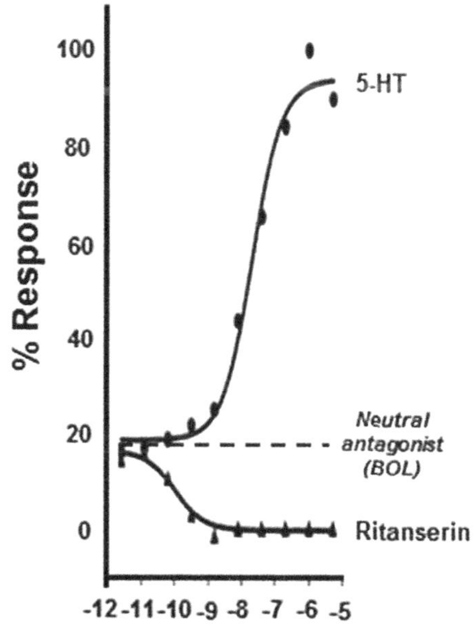

Fig. 1 Constitutive activity of 5-HT_{2A} receptors. Representative concentration response experiments for the human 5-HT_{2A} receptor. The y-axis is the percentage of response on phosphatidyl inositol production defined as Full Agonist Response (5-HT) or Full Inverse Agonist Response (ritanserin); the x-axis is the negative logarithm of drug concentration. The curve with the *dashed line* reports the putative effect of a neutral antagonist such as we could expect from d-bromolysergic acid diethylamide (BOL). EC50 values (mean and standard deviation) are 29 ± 12 nM for 5-HT and 0.67 ± 0.3 nM for ritanserin. Adapted from [10]

In searching for the physiological or pathophysiological existence of a constitutive activity of a given receptor, the neutral antagonist that is selective for this receptor is probably the most precious of the pharmacological tools. Indeed, it determines whether a pharmacological response attributed to an agonist or an inverse agonist is dependent on the receptor itself. Most of the time, neutral antagonists are rare [46]. In addition to the functional attribution of full agonists, partial agonists, neutral antagonists, partial inverse agonists and full inverse agonists toward one intracellular signaling system, it is now accepted that several G-Protein Coupled Receptors (GPCR) may couple various intracellular signaling pathways via the interaction with different G proteins or other G protein-independent pathways [45]. Some agonists are more capable of stimulating one signaling pathway than others. This "bias agonism" corresponds to "agonist-directed trafficking of receptor stimulus" or "functional agonist selectivity" [44]. This property, which is always directly addressed *in vitro*, would illustrate the ability of the receptor to adopt different conformations. It implies that the active form of a receptor, classically termed R*, has several active forms due to isomerization, each of these active forms being related to one specific signaling pathway.

The property of several receptors to trigger different intracellular systems relocates in most cases the neutral antagonist to a "protean" ligand, a drug that behaves

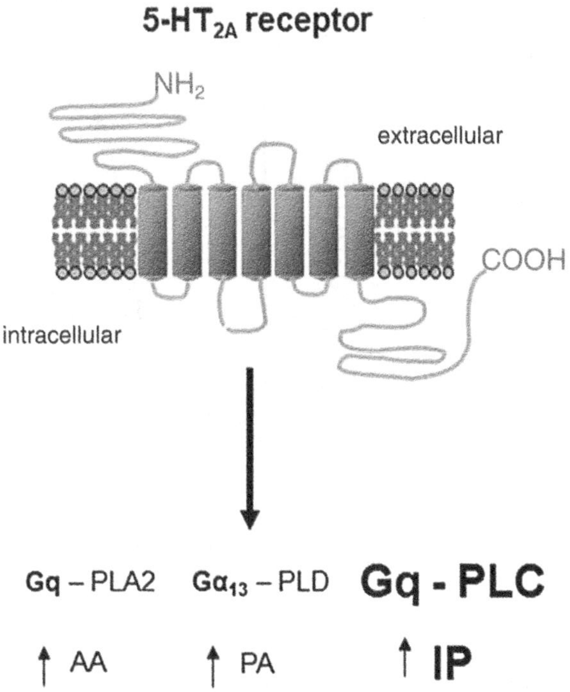

Fig. 2 Some cellular partnerships of 5-HT_{2A} receptors. It shows the main intracellular signaling pathways associated with 5-HT_{2A} receptor transduction, the phospholipase C pathway being presumably the most important. It can trigger heterogeneous responses in different intracellular pathways. *IP* inositol triphosphate, *AA* arachidonic acid, *PLC* phospholipase C, *PLA2* phospholipase A2

differently toward diverse signaling pathways [46]. Several drugs defined as antagonists displayed either weak agonist activity, no activity, or inverse agonist activity, depending on the signaling system considered. This has been nicely illustrated in terms of the selective ligands for the 5-HT_{2C} receptor or the H_3 receptor [1, 50, 51]. The pharmacological properties of the H3 ligands are tissue-dependent and species-dependent [50]. Thus, a drug behaving as a neutral antagonist in one system might behave as weak agonist or inverse agonist in another one. This is a major challenge when producing selective pharmacological compounds in translational research.

To summarize, the pharmacological properties of drugs have to be determined and their selectivity established in heterologous recombinant systems *in vitro*. This would be indicative of their behavior with respect to one or several signaling pathways; however, it would not necessarily correspond to definite proof of their pharmacological properties in living organisms.

Constitutive Activity of 5-HT_{2A} Receptors In Vitro

The 5-HT_{2A} receptor is coupled to G-protein and stimulates phosphoinositide-specific phospholipase C (PLC) with a consequent increase in inositol triphosphate (IP) [5, 52, 53] (Fig. 2). Most data obtained in heterologous recombinant systems *in vitro* regarding 5-HT_{2A} receptors were conducted by measuring IP hydrolysis. The ability of 5-HT_{2A} receptor to spontaneously activate PLC is extremely low *in vitro*

[54, 55]. It is about 10 times lower compared to the 5-HT_{2C} receptor [55, 56]. Constitutive activity of 5-HT_{2A} receptors can be obtained either by mutation [29, 54, 56, 57] or by overexpressing G proteins [10]. Constitutive 5-HT_{2A} receptor activity can be amplified using specific receptor selection and amplification technology (R-SAT) assays [10, 29].

This low agonist-independent activity of native 5-HT_{2A} receptors *in vitro* has two consequences. First, the pharmacological profile of various tools is not well defined because in some cases, the spontaneous activity of the native 5-HT_{2A} receptors is very low [54]. In case of basal activity obtained by mutated 5-HT_{2A} receptors and/or specific assays as mentioned above, drugs such as ritanserin, ketanserin, M100907 and many atypical antipsychotic drugs, behaved as inverse agonists [10, 54]. Ritanserin is a full 5-HT_{2A} receptor inverse agonist and was considered as the reference compound for this pharmacological class [10] (Fig. 1). The second consequence is that it is difficult to imagine that 5-HT_{2A} receptors, with this low propensity to spontaneously activate the PLC signaling pathway *in vitro*, may display a constitutive activity *in vivo*.

It is noteworthy that the 5-HT_{2A} receptor has been shown to activate phospholipase D (PLD) and phospholipase A2 (PLA2) by interacting with additional G-proteins (Fig. 2). The 5-HT_{2A} receptor activation closes potassium channels, producing neuronal depolarization [22, 58]. It also interacts with β-arrestin *in vivo*, but this interaction seems different in HEK-293 cells [59]. This is an important consideration because it could regulate the density of 5-HT_{2A} receptors at the cell surface and could directly affect the desensitization process induced by agonists [59, 60]. The interaction of GPCR with PSD-95/Disc Large/Zona Occludens-1 (PDZ) domain containing proteins are involved in cell specific functions such as signaling and trafficking [61]. It has been demonstrated that the PDZ protein, synapse-associated protein 97 (SAP97), interacts with both 5-HT_{2A} and 5-HT_{2C} receptors [62]. This interaction antagonizes the endocytosis of 5-HT_{2A} receptors [63]. Another PDZ protein PDZK1, alternatively called Na(+)/H(+) exchange regulatory cofactor 3 (NHERF3), interacts with 5HT_{2A} receptor and induces suppression of 5-HT_{2A} receptor internalization. Second generation of antipsychotics downregulate 5-HT_{2A} receptor-mediated signaling. Thus, the impairment of the interaction of PDZ domain containing proteins with 5-HT_{2A} receptors may result in a down-regulation of 5-HT_{2A} receptors surface expression associated with decreased 5-HT_{2A} receptor signaling. The cellular background in which the 5-HT_{2A} receptor is expressed appears to determine the regulation properties of the receptor. Depending on the cell system, this can lead to opposite consequences of 5-HT_{2A} receptor expression after agonist exposure [64]. Moreover, 5-HT_{2A} receptors form heterodimers and can be associated with glutamate mGluR2 receptors, D2 receptors, CB1 cannabinoid receptors or 5-HT_{2C} receptors with possible pathophysiological consequences in cognition [65–68]. Thus, the location of the receptor in neurons, the presence of specific proteins interacting with the receptor intracellularly and on the plasmamembrane, and the locally available materials (G proteins/β-arrestin) are important factors that differ from heterologous expression systems.

The pharmacology of 5-HT_{2A} receptor in native tissues *in vitro* has always been different from the other 5-HT receptors [6]. There is a higher potency of agonists that displace ^{3}H-agonist (usually DOI) binding compared to ^{3}H-antagonists (usually ketanserin). Conversely, some compounds like d-lysergic acid diethylamide (LSD), which can display weak partial agonist activity, and several antagonists are equipotent in displacing 5-HT$_{2A}$ receptor binding of ^{3}H-agonists or ^{3}H-antagonists [6]. This highlights drastic differences in the binding affinity between the inactive state, maintained by antagonists (inverse agonists), and the active states of the 5-HT$_{2A}$ receptor. Meanwhile, the binding of an agonist *in vivo* would be conditioned by the state of activity of 5-HT$_{2A}$ receptor (active versus inactive form). Combined together, all these data show how complex the interaction of several ligands with 5-HT$_{2A}$ receptors could be. For instance, LSD activates PLA2 instead of activating PLC or enhances dopamine D2 receptor protomer recognition and the signaling of D$_2$–5-HT$_{2A}$ receptor complexes [69].

Based on these considerations, the difficulties in establishing the pharmacological behavior of 5-HT$_{2A}$ receptor ligands in heterologous cell systems are a major obstacle when studying the existence of a constitutive activity of 5-HT$_{2A}$ receptor *in vivo*.

Constitutive Activity of 5-HT$_{2A}$ Receptors *In Vivo*

The living organism implies a functional 5-HT system that maintains basal extracellular levels in virtually all organs and responds to phasic stimulation. The lower affinity to 5-HT of 5-HT$_{2A}$ receptors compared to other 5-HT receptors, notably the 5-HT$_{2C}$ receptors, favors the hypothesis that the 5-HT$_{2A}$ receptor typically mediates phasic responses associated with an enhancement of 5-HT extracellular levels [7]. In various biological functions, acute pharmacological blockade of 5-HT$_{2A}$ receptors does not alter basal activity, implying that the receptor does not exert tonic controls upon neurobiological networks and organs acutely [70–72].

The doubts of the existence of constitutive activity *in vivo* come from studies with antagonists. Indeed, when various antagonists display different degrees of responses irrespective of changes in extracellular levels of the endogenous ligand, this could suggest the existence of a constitutive activity. This pattern was encountered in the conditioned eyeblink response in rabbits regarding 5-HT$_{2A}$ receptor antagonists [11].

The Conditioned Eye Blink Response and the 5-HT$_{2A}$ Receptors

The conditioned eyeblink response in rabbits is classically used as a Pavlovian model of learning. Precisely, the rabbit's nictitating membrane response can be conditioned to the presentation of a tone stimulus. The efficacy of the conditioned response is compared to the effect of an unconditioned stimulus, such as an air puff or shock, while the rate of acquisition of the conditioned responses progressively increases over a week of conditioning [11, 31, 73]. The rate of acquisition of the

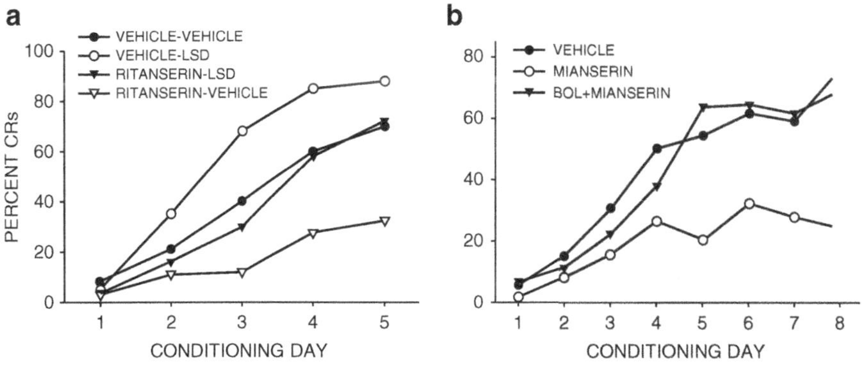

Fig. 3 (**a**) Antagonism by ritanserin (1 μmol/kg) of the enhancement of CR acquisition produced by LSD (0.030 μmol/kg). Ritanserin was injected subcutaneously 60 min prior, and LSD was injected intravenously 20 min prior to each acquisition session. Acquisition of the nictitating membrane response was measured during the pairing of a tone conditioned stimulus and air puff unconditioned stimulus. Data are taken from Welsh et al. [74]. (**b**) Antagonism by BOL (5.8 μmol/kg) of the retardant effects of mianserin (10 μmol/kg) on acquisition of the nictitating membrane response. Mianserin was injected 1 h and BOL 20 min prior to each conditioning session by use of the pairing of a tone CS and air puff US. All injections were subcutaneous. Data are taken from [75]

conditioned eye blink responses was shown to be enhanced by some hallucinogenic drugs. This model is of interest when studying the effect of hallucinogenic drugs because their dose range to effectively produces hallucinations in humans is approximately similar in the conditioned eyeblink response in rabbits [11, 74]. Various direct agonists at 5-HT_{2A} receptors, including LSD (Fig. 3a), (±)-1(2,5-dimethoxy-4-iodophenyl)-2-aminopropane hydrochloride (DOI), and d,l-2,5-dimethoxy-4-methylamphetamine (DOM) enhance associative learning in rabbits [11, 74]. Similarly, the weak agonists and 5-HT releasers, d,l-methylenedioxymethamphetamine (MDMA) and d,l-methylenedioxyamphetamine (MDA) also increased the rate of acquisition of the conditioned response [11, 76, 77]. Conversely, the non-hallucinogenic drug lisuride, which has a strong affinity for 5-HT_{2A} receptors but induces cortical effects that are not produced by LSD (Table 1) [92], did not modify the conditioned response [74]. Nevertheless, the effects of hallucinogenic drugs are possibly related to the activation of 5-HT_{2A} receptors because ritanserin blocked the facilitatory effects of LSD [74] (Fig. 3a).

While some 5-HT_{2A} receptor blocking agents did not individually alter the acquisition of the conditioned eyeblink response (BOL, ketanserin, LY53,857), others, like ritanserin, retarded the acquisition of the response by themselves [73, 75, 93] (Fig. 3b). This was a surprising result, and the authors further characterized the negative efficacy of these drugs. Of note, both ketanserin and ritanserin behaved as inverse agonists in R-SAT assays *in vitro* [10, 54], while in the conditioned eye blink response, ritanserin behaved as an inverse agonist and ketanserin as a neutral antagonist. Since the eye blink response is an integrative and complex response, it is modulated by numerous neurotransmitter systems and receptors [94, 95]. The selectivity of the 5-HT_2 receptor compounds that were used has to be addressed.

Table 1 Affinities of various 5-HT_2 receptor ligands towards 5-HT_{2C}, 5-HT_{2B} and 5-HT_{2A} receptors and other sites

	5-HT_2 receptor subtype Ki (nM)		Other sites (<300 nM)	References	
	5-HT_{2A}	5-HT_{2B}	5-HT_{2C}		
5-HT_2 agonists					
5-HT	16	13.5	5.7		[78]
d-LSD	2.9	4.9	23	5-HT_{1A}, 5-HT_5, 5-HT_6, 5-HT_7, D2, D3	[78–80]
DOI	0.65	53	5.37		[78]
DOM	21	nd	42	nd	[6, 81]
MDMA	5100	500		SERT (600 nM), DAT and NET; α2, M3, H1	[82]
MDA	5300	nd	2932	SERT, DAT and NET;	[81, 83]
Lisuride	5	1.4	20	5-HT_{1A}; 5-HT_6, 5-HT_7; α1-2; D1-4; 5-HT_{1B}; 5-HT_{1D}	[6, 80]
m-CPP	55	40	14	5-HT_{1B}, 5-HT_3, α1-2, SERT	[78, 84]
TFMPP	251	23	63	5-HT_{1B}, α1-2, SERT	[78, 80]
Quipazine	125	85	53	5-HT_3	[78, 85, 86]
5-HT_2 antagonists					
5-HT_{2A} antagonists					
MDL 100907	1.8	1000	88	H1, α1	[71, 78]
MDL11,939	26	3311	263	5-HT_{1B} (70)	[78]
5-$HT_{2B/2C}$ antagonists					
SB 200646	6606	nd	138		[87]
LY215840	22	2.7	3.7	5-HT_{1A}, 5-HT_7	[88]
5-HT_2 antagonists					
Ritanserin	4.6	2.1	6.6	5-HT_{1D}, 5-HT_6, 5-HT_7, H1, D2, D3, D4	[78]
Mianserin	18	12	5.5	5-HT_{1D}, 5-HT_6, 5-HT_7, 5-HT_3, α1-2, H1, NET	[78]
Ketanserin	8.1	741	62	Tetrabenazine site, α1-2, H1	[78, 89]
LY 53857	1.7	6.9	4.2	5-HT_{1A}	[6, 80, 90]

Table 1 (continued)

	5-HT_2 receptor subtype Ki (nM)		Other sites (<300 nM)	References	
BOL	5.5	nd	51	5-HT_6, 5-HT_7, D2, H2,	[6, 91]
Mesulergine	46	3.5	1.8	5-HT_{1A}, 5-HT_7, D2	[78]
Cyproheptadine	8.7	5.9	2.1	5-HT_{1A}, 5-HT_3, 5-HT_5, 5-HT_6, 5-HT_7, M1-5, D2, D3, H1, H4, NET	[6, 80, 85]

Values correspond to Ki (nM). References correspond to studies reporting the Ki of 5-HT_2 receptor ligands on 5-HT_2 receptor subtypes and/or on other binding sites. The other binding sites have been completed with the averaged data from the Ki Database (http://pdsp.med.unc.edu/kidb.php). Binding affinities below 300 nM have been considered except in the case of LSD or lisuride (many sites are already bound at 30 nM) and MDMA and MDA (μM affinity toward 5-HT_{2A} receptors). *SERT* serotonin transporter, *DAT* DA transporter, *NET* noradrenaline transporter, *5-$HT_{1\text{-}7}$* 5-HT receptor subtypes [5], *α1-2* adrenergic receptor subtypes, *M1-5* muscarinic receptor subtypes, *D1-D4* DA receptor subtypes, *H1-3* histaminergic receptor subtypes

Table 1 reports the affinity of the drugs that have been tested so far in the conditioned eyeblink response acquisition as well as in other models of learning that are discussed below. None of these drugs, except MDL11,939, M100907 and to some extent, SB200,646, are selective for 5-HT_{2A} and/or 5-HT_{2C} receptors (see also [96]). In particular, the agonists have poor selectivity with the possible exception of DOI, which preferentially binds 5-HT_2 receptors [6, 78, 97].

Role of Endogenous 5-HT in the Conditioned Eye Blink Response

One important criterion to establish when determining the existence of a constitutive activity of 5-HT_{2A} receptors in the conditioned eye blink response is to understand the putative role of endogenous 5-HT in this response. Based on the effects of the direct and indirect agonists, it seems that endogenous 5-HT does not play a role in this learning task. LSD and DOI are compounds known to inhibit 5-HT neuron firing rate and 5-HT release in various rat brain areas [98, 99]. On the contrary, MDMA and MDA enhance extracellular levels of 5-HT by reversing the function of the 5-HT transporter (SERT) [100], raising questions about a putative role of endogenous 5-HT in this response. Lisuride is also known to reduce 5-HT nerve activity [101], and the fact that this compound did not modify the conditioned eyeblink response suggests that the decrease in 5-HT extracellular levels is not a prerequisite in the action of LSD or DOI. Similarly, the 5-HT_{1A} agonist, 8-OHDPAT, which is known to reduce 5-HT neuron discharge and 5-HT extracellular levels [101], did not alter the rate of acquisition of the conditioned response [74]. Finally, destruction of 5-HT neurons by 5,7-dihydroxytryptamin (5,7-DHT) reduced 5-HT tissue content by more than 85% in the hippocampus and the cortex without affecting 5-HT_{2A} receptor density, the conditioned eye blink response per se, or the facilitatory effect of LSD on learning [94]. It appears that the endogenous 5-HT tone does not impact this learning response.

These data suggest that the efficacy of the ligands on the behavioral response is related to their peculiar interaction at 5-HT_{2A} receptors and/or other targets (Table 1). Consequently, the efficacy of some 5-HT_{2A} receptor blocking agents could be related to their specific interaction at the 5-HT_{2A} receptor rather than the blockade of the 5-HT endogenous tone on 5-HT_{2A} receptors during learning.

Direct Pharmacological Evidence for the Constitutive Activity of 5-HT_{2A} Receptor in the Conditioned Eyeblink Response

The retardation of the acquisition of the conditioned response has been observed with ritanserin, mianserin, MDL11,939, and pizotifen [11]. In line with the possibility that these effects are related to their direct interaction at 5-HT_{2A} receptors, the magnitude of the effects differs between these 5-HT_{2A} receptor ligands, ritanserin proving to be the most effective compound. In most cases, the effects of these agents are unique to the conditioned responses, as they do not alter basal response or modify the conditioned stimulus. The conditioned response differed with respect to the intensity of the conditioned stimulus, with a maximal response obtained at 80 db. Interestingly, the effects of 5-HT_{2A} receptor blocking agents varied according to the intensity. Ritanserin reduced the conditioned responses at several intensities, ranging from 70 db to 90 db, and MDL11,939 and LY53,857, the latter being normally ineffective, attenuated the conditioned response for 70 db and 80 db intensities [73]. On the one hand, these findings bring awareness of the complexity of the effects elicited by 5-HT_{2A} receptor blocking agents in the conditioned eye blink response. Indeed, the results obtained with LY53,857 could be interpreted as the very weak 5-HT_{2A} receptor inverse agonist property of this compound or attributed to the involvement of other 5-HT receptors (Table 1) at specific intensities of the conditioned stimulus. On the other hand, these findings are essentials as they showed that the conditioned eye blink response is phasic in nature. They correspond to the enhanced and attenuated effects on the rate of acquisition of the conditioned response by agonists and inverse agonists respectively as a function of the intensity of the stimulus [73].

The last arguments come from the demonstration that d-bromolysergic acid diethylamide (BOL), ineffective by itself, reduced the retardation of the acquisition induced by mianserin [75, 93] (Fig. 3b). The obstruction of 5-HT_{2A} receptors by BOL prevented the binding of the inverse agonist mianserin to 5-HT_{2A} receptors, thereby preventing its attenuating effect on learning. In addition, the lesion of 5-HT neurons by 5,7-DHT did not alter the retardation of acquisition of the conditioned response induced by MDL11,939 [94].

The authors have further demonstrated that the chronic administration of the 5-HT_{2A} receptor blocking agent, MDL11,939, enhanced the binding of 5-HT_{2A} receptors in the cortex and the hippocampus. Such an effect is expected from an inverse agonist [102]. The enhancement of 5-HT_{2A} receptor expression was also associated with a higher responsiveness of DOI in eliciting head bobs and an enhancement of learning [103, 104]. Conversely, the administration of agonists reduced the binding of 5-HT_{2A} receptors [103, 104].

The Pharmacological Limits: An Open Discussion

This impressive set of experiments in rabbits justifies the proposal that 5-HT_{2A} receptors adopt a constitutive activity in the conditioned eye blink responses. Nonetheless, the proposal is extremely fragile, as several interpretations do not resist to a deeper pharmacological analysis. The fact that 5-HT_{2A} receptor blocking agents differ in their ability to alter learning could be related to their non-selective pharmacological profile (Table 1). Among the drugs that inhibited the learning tasks, all, except MDL11,939, displayed a substantial affinity toward 5-HT_{2C}

receptors at least (Table 1). The contribution of 5-HT_{2C} receptors in this task has not been directly addressed, and their interference with the task is unknown. In fact, ketanserin has been described as an inverse agonist *in vitro* and could behave like this in other models *in vivo* (see below). Moreover, the reversal of the effects of mianserin by BOL is inconclusive because both drugs are really non-selective, BOL having been used as a neutral 5-HT_{2C} receptor antagonist to suggest the existence of the constitutive activity of native 5-HT_{2C} receptors in choroid plexus [105]. Because of its non-selective profile, BOL could shut down 5-HT_{2A} receptor-dependent mechanisms via indirect actions on various receptors. Indeed, BOL displays a similar binding profile on DA receptors comparable to LSD. Furthermore, BOL decreased the binding of 5-HT_{2A} receptors upon its chronic administration in rabbits, behaving like the other agonists [104], and would display an inverse agonist profile on cortical H2 receptors [91]. BOL is really difficult to manage and should not be considered to determine whether a pharmacological response of a ligand depends on 5-HT_{2A} receptor stimulation. As far as we know in these studies in rabbits, key experiments to propose the influence of the constitutive activity of 5-HT_{2A} receptors in learning are lacking. Notably, some experiments should have addressed the ability of ketanserin (considered here as a neutral antagonist) to reverse the attenuating learning effects of MDL11,939 and ritanserin or the enhancing learning effects of DOI.

MDL11,939 Under the Spotlight

Overall, the stronger arguments that support the existence of the constitutive activity of 5-HT_{2A} receptors in the conditioned eye blink response are (1) the retardation of learning induced by MDL11,939 and (2) the persistence of MDL11,939's effect in 5-HT neuron-lesioned rabbits. MDL11,939 has been less used compared to the selective 5-HT_{2A} receptor antagonist, M100907, and one may wonder if this drug has been shown to impair learning in other paradigms. In a paradigm of light-induced locomotor activity, a model to study sensory-motor activation, ritanserin reduced locomotor activity whereas MDL11,939 or the 5-HT_{2C} receptor antagonist SER082 were inactive [106]. Interestingly, in a conditioned defeat paradigm, corresponding to a learned, social defeat model in Syrian hamsters, MDL11,939 dose-dependently impaired the acquisition of the conditioned defeat but not its expression [107]. Furthermore, focal injection of MDL11,939 into the basolateral nucleus of the amygdala dose-dependently impaired acquisition of the conditioned defeat response. In contrast, the 5-HT_{2A} receptor agonist, TCB-2, had an opposite effect [108]. In a paradigm of trace fear conditioning memory in C57BL/6 J mice, systemic administration of one dose of MDL11,939 (0.5 mg/kg, i.p.) delayed the acquisition of extinction of fear memory [109]. In the context of memory retrieval in a paradigm of the object recognition task in rats, Bekinschtein et al. [110] reported that focal administration of one dose of MDL11,939 into the medial prefrontal cortex affected retrieval of an object in a spontaneous novelty preference task of context memory, while sparing single-item recognition memory [110].

While none of these studies addressed the possibility that MDL11,939's effects were due to the silencing of constitutive activity of 5-HT_{2A} receptors, all these studies report attenuating role of 5-HT_{2A} receptor blockade in the learning process. As mentioned above, chronic administration of MDL11,939 enhanced 5-HT_{2A} receptor binding in rabbits, but this effect has not been reported in mice [111]. Thus, the inverse agonist property of MDL11,939 would be more specific to rabbits than to other rodent species.

Constitutive Activity of 5-HT_{2A} Receptors in Other Models of Learning

The Conditioned Avoidance Response

Conditioned avoidance response (CAR) is also a classical Pavlovian response mostly studied in rats. In addition to its importance in learning, this paradigm is also used to evaluate the potential antipsychotic activity of a drug [112]. In one of his reviews, Harvey [11] reported the parallel between the results obtained in the CAR and the eyeblink response in rabbits [11]. The CAR was enhanced by non-selective agonists that stimulated 5-HT_{2A} receptors, unaltered by the non-selective blocking agents, spiperone, ketanserin, cinanserin and mianserin, and impaired by the non-selective blocking agents, ritanserin or cyproheptadine. Nonetheless, it has also been reported that the CAR was unaltered by mianserin, ritanserin and the selective 5-HT_{2A} receptor blocking agent, M100907 [113–116]. Mianserin, ritanserin, and M100907 potentiated the disrupting effects of the DA antagonist, raclopride, in this paradigm [114–116]. M100907 also potentiated the disrupting effect of the DA antagonist, haloperidol [113]. The 5-HT_{2A} antagonists in these studies potentiated the effects of subthreshold doses of the DA antagonists in the CAR.

It appears that the results obtained in the CAR diverge from the responses studied in rabbits on the 5-HT system. In fact, 8-OHDPAT impaired the acquisition of the response [117], suggesting the participation of the endogenous 5-HT tone in this response. In addition, there is no clear evidence that 5-HT_2 receptor inverse agonists disrupt the responses on their own in this model. Finally, this learning task is also sensitive to 5-HT_{2C} receptors since the selective 5-HT_{2C} agonists, WAY 163909 or CP809,101, disrupted the acquisition of the conditioned response [96, 118, 119].

The Autoshaping Learning Task

The autoshaping learning task is an operant system that produces a conditioned response in rodents. It combines the procedures of pavlovian conditioning consisting of the pairing of a light lever with the delivery of food. It also combines instrumental conditioning, which corresponds to the delivery of food upon pressing a lever [88, 120]. The results of this task are very interesting although extremely

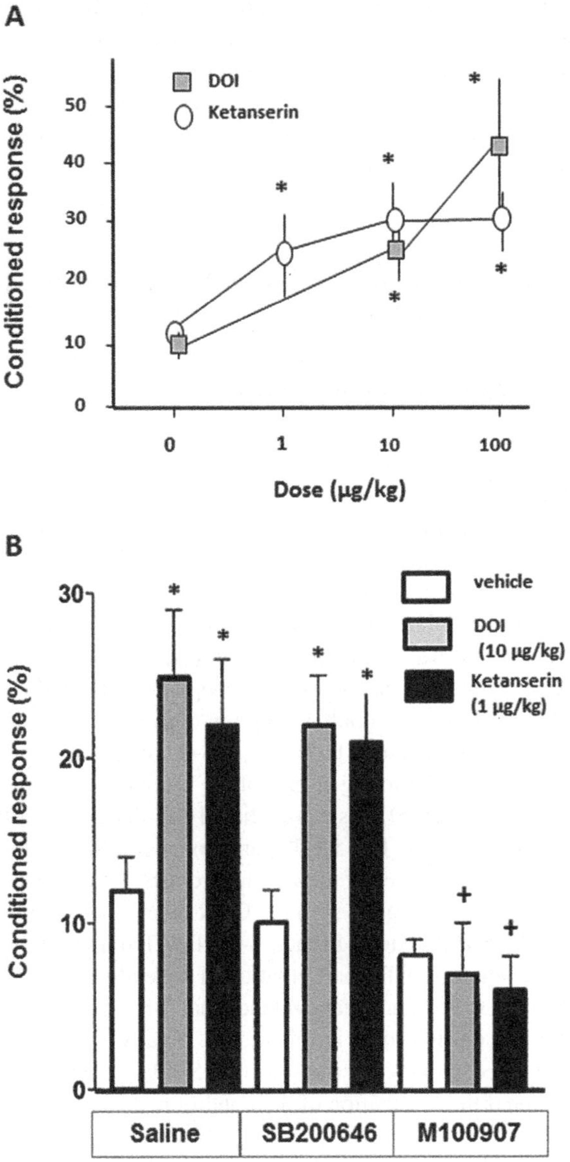

Fig. 4 (**a**) Effects of post-training acute injection (ip) of ketanserin or DOI on conditioned response of autoshaping task in fasted animals. Data are plotted according to percentage of conditioned responses. All rats received the injection immediately after the first training session. Values represent the mean ± S.E.M. of eight different animals. *$p < 0.05$ versus vehicle-injected controls Dunnett's test. This has been adapted from the study of [121]. (**b**) The effect of acute posttraining administration (i.p.) of SB200,646 (2 mg/kg) or MDL100907 (1 mg/kg) on the responses induced by DOI (0.1 mg/kg) and Ketanserin (0.1 mg/kg) in an autoshaping learning task in fasted animals. Data are plotted as percentage of conditioned responses. All rats received an injection immediately after the first training session, and data correspond to a session carried out 24 h later. Top bar values represent the mean, and vertical lines denote the S.E.M. of mean of eight different animals. *p < 0:05 versus vehicle-treated rats + $p < 0.05$ versus initial drug treatments. Data are adapted from the study of [88]

confusing regarding the role of the 5-HT system. The non-selective 5-$HT_{1B/2A/2B/2C}$ agonists, mCPP and TFMPP, as well as the non-selective 5-$HT_{2C/2A/D2}$ receptor antagonists, mesulergine (0.4 mg/kg) and 1-naphtyl piperazine (1-NP), impaired memory consolidation in a dose-dependent manner. Conversely, DOI (0.01, 0.1 mg/kg) dose-dependently improved memory consolidation (Fig. 4a). Strikingly, low doses of ketanserin (1, 10 and 100 μg/kg) also enhanced memory consolidation with the maximal effect almost obtained with the lower dose (Fig. 4a). Ritanserin (0.1, 0.2 and 0.4 mg/kg) induced an inverted U-shape curve with a significant enhancement of learning observed at 0.2 mg/kg only [121, 122]. Interestingly, three drugs, the most selective in the panel, did not modify memory consolidation: the selective 5-HT_{2A} receptor antagonist, M100907, and the preferential 5-$HT_{2B/2C}$ receptor antagonists, SB200,646 and LY215,840. It is also important to note that monoamine depletion induced by two daily injections of parachloamphetamine over 10 days or 5-HT depletion induced by parachlorophenylalanine did not alter task performance [88, 123].

This complex pharmacological picture could hide the involvement of 5-HT_{2C} and 5-HT_{2A} receptors in the consolidation of memory along with the involvement of 5-$HT_{1B/1D}$ autoreceptors, which were shown to participate in the effect of TFMPP [121]. First of all, SB200,646 and LY215,840 blocked the disrupting effects of both mCPP and mesulergine and attenuated those induced by TFMPP [88]. These findings strongly suggest that 5-HT_{2C} receptors were involved in the memory impairment induced by both the agonists and the antagonist. Mesulergine has been shown to behave as an inverse agonist *in vitro* and *in vivo* [2, 45, 124] and its effects could be related to the blockade of a constitutive activity of 5-HT_{2C} receptors. Recently, it has been reported in a similar task that drugs displaying high inverse agonist property at 5-HT_{2C} receptors including SB 206553 and mianserin were more potent to inhibit memory consolidation compared to other ligands [125]. Furthermore, a possible constitutive activity of endogenous 5-HT_{2C} receptors has been identified *in vivo* in the control of subcortical DA release as well as in the control of orofacial activity and muscle spasms consequent to spinal cord injury in rats [2, 51, 126–128]. Ritanserin can also behave as a 5-HT_{2C} receptor inverse agonist, and this property could participate in its curious dose-dependent effect with respect to that observed with ketanserin. The pharmacological concern in the memory experiments is the dosage of mCPP (10 mg/kg), TFMPP (10 mg/kg), and mesulergine (0.4 mg/kg) to impair memory. At these doses, the piperazines mCPP and TFMPP might also act as 5-HT releasers [129], in addition to numerous other potential targets, including 5-HT_{2A} receptors, while mesulergine could already block D_2 receptors at this dose [124] and 5-HT_{2A} receptors, consequently (Table 1, see below). As a matter of fact, M100907 (1 mg/kg) did reduce the attenuation of TFMPP, mCPP, and mesulergine on memory consolidation. M100907 even unmasked an excitatory effect of mCPP on learning [121]. This could suggest that M100907 loses its 5-HT_{2A} receptor selectivity toward 5-$HT_{2B/2C}$ receptors at this dose. Nonetheless, these results can suggest that the inhibitory responses of TFMPP, mCPP, and mesulergine are related to their direct action at 5-HT_{2A} receptors or their combined action at both 5-HT_{2A} and 5-HT_{2C} receptors.

In line with the latter possibility, the author reported that M100,907 blocked the excitatory effect induced by DOI and ketanserin on memory consolidation. Conversely, SB200,646 and LY215,840 were unable to alter the learning effects of DOI and ketanserin (Fig. 4b) [88]. These results indicated that 5-HT_{2A} receptors, but not 5-$HT_{2B/2C}$ receptors were involved in the responses induced by DOI and ketanserin (Fig. 4b). Furthermore, it strongly suggested that the ability of SB200,646 and LY215,840 to reverse the inhibitory effects of mCPP, TFMPP and mesulergine was preferentially due to their 5-$HT_{2B/2C}$ antagonistic properties.

These data are compatible with the role of 5-HT_{2A} receptors in an autoshaping task, involving a constitutive activity of 5-HT_{2A} receptors. Ketanserin and ritanserin would act as inverse agonists at 5-HT_{2A} receptors and should prevent the enhancing effects of DOI. The effects of DOI were blocked by ritanserin (0.1 mg/kg) and mesulergine (0.2 mg/kg). Nonetheless, ketanserin (1 µg/kg) did not block the effect of DOI (0.01 mg/kg). Rather, the enhancing effects of both compounds were additive [122]. It is regrettable that the lowest dose of ketanserin was used in this interaction. This low dose of ketanserin, as did higher doses of ritanserin or mesulergine, was sufficient to impair the effects of the weak partial 5-HT_{2A} receptor agonists, such as mCPP, but was probably insufficient to fully occlude 5-HT_{2A} receptors to alter the effect of DOI.

Apart the pharmacological weaknesses underlined above, this set of experiments and studies support the existence of a constitutive activity of 5-HT_{2A} receptors in the autoshaping learning task as well. It also reveals an interaction between several 5-HT receptors and 5-HT_{2A}-5-HT_{2C} receptors, which participate in the effects of non-selective compounds.

The finding that the agonist, DOI, and the inverse agonist, ketanserin, similarly concur to enhance memory consolidation could be puzzling, as they act as inverses in the function of 5-HT_{2A} receptors. Regardless, it is possibly related to the existence of at least two populations of 5-HT_{2A} receptors that exert opposite effects on learning, one displaying a constitutive activity. In fact, this is likely the situation of 5-HT_{2C} receptors. The injection of 5-HT_{2C} agonists and inverse agonists promotes purposeless oral movements in rats, the effects of both pharmacological classes being suppressed by a 5-HT_{2C} receptor antagonist [128]. The origin of the effects elicited by 5-HT_{2C} receptor agonists and inverse agonists is unlikely related to the same site because only the effects of the agonists are sensitive to a lesion of DA neurons [128]. Even when the effects triggered by 5-HT_{2C} receptor agonists and inverse agonists are opposite in vivo, as those reported in the control of subcortical DA release [51], it refers to distinct populations of 5-HT_{2C} receptors [2, 96, 130]. Regarding 5-HT_{2A} receptors, they are expressed by pyramidal cells as well as by interneurons in the cortex and in the hippocampus. The autoshaping learning task also involves procedural learning, containing corticostriatal mechanisms [120]. Thus, it is possible that multiple 5-HT_{2A} receptor-dependent regulations participate in memory consolidation.

The Physiological Meaning of the Constitutive Activity of 5-HT$_{2A}$ Receptors

Despite some pharmacological weaknesses, the experiments in rabbits and rats presented above bring up arguments to suggest that the constitutive activity of 5-HT$_{2A}$ receptors participates in some learning abilities. Physiologically, it is likely that such an activity precedes and convoys an increase in 5-HT tone in order to facilitate learning. This activity would represent one aspect of the phasic influence of a 5-HT$_{2A}$ receptor upon neurobiological networks, i.e., the transient presentation of the high affinity state of 5-HT$_{2A}$ receptors imposed by a change in activity of the cell expressing the receptor. In subcortical areas, 5-HT$_{2A}$ receptors have been shown to participate in a state-dependent facilitatory control of striatal DA release [131]. A state-dependent control means that some factors, which are not clearly identified, trigger the facilitatory control exerted by 5-HT$_{2A}$ receptors; this idea has been clearly demonstrated in the mechanism of action of MDMA. At that time, the authors thought that the increase in 5-HT induced by MDMA played a major role in this effect [7, 72, 132–134]. Although it is true, it has been reported that amphetamine, which poorly impacts 5-HT release, and haloperidol, which tends to decrease 5-HT release, triggered a similar 5-HT$_{2A}$ receptor state-dependent facilitatory on subcortical DA release [135, 136]. It is likely that this control still depends on endogenous 5-HT because the DA effects of amphetamine and haloperidol were reduced by 5-HT$_{1A}$ agonists in a manner that is comparable to 5-HT$_{2A}$ antagonists [136–138]. Nevertheless, it suggests that the phasic influence of 5-HT$_{2A}$ receptors is related to a specific receptor state rather than the endogenous tone of 5-HT. In this context, the constitutive activity would represent a phasic and strong activation of the 5-HT$_{2A}$ receptor, impacting the cell itself even before the phasic and expected enhancement of 5-HT release. In paradigms where 5-HT release does not play a major role, the constitutive activity of 5-HT$_{2A}$ receptors could be detected by inverse agonists and not antagonists.

We do not have clear evidence that such a mechanism could occur in some neuropsychiatric diseases. Although some antipsychotics are 5-HT$_{2A}$ receptor inverse agonists at 5-HT$_{2A}$ receptors, it does not imply that their action is related to silencing the constitutive activity of 5-HT$_{2A}$ receptors. It is interesting to note that 5-HT$_{2A}$ receptors could play a role in psychosis induced by L-DOPA treatments in Parkinson's disease. Primavanserin (ACP-103) is a 5-HT$_{2A}$ receptor inverse agonist that attenuated L-DOPA-induced psychosis in patients in phase III trials [42, 139, 140]. L-DOPA tends to acutely and chronically decrease 5-HT extracellular levels in various rat brain regions, including the cortex and the hippocampus [141, 142], suggesting that the activation of 5-HT$_{2A}$ receptors is not associated with an increase in 5-HT release. It is postulated that L-DOPA exerts various effects, including effects on DA in these regions, triggering a high affinity state of 5-HT$_{2A}$ receptors in frontocortical and/or hippocampal regions. Additional data are warranted to determine the extent to which the constitutive activity of 5-HT$_{2A}$ receptors is a risk in the development of L-DOPA-induced psychosis.

Conclusion

We have offered a thorough pharmacological examination of the data underscoring the existence of the constitutive activity of 5-HT_{2A} receptors *in vivo*, mostly in learning tasks. While the data are few, they give convincing evidence that the phenomenon exists. The link between *in vitro* data, to determine the pharmacological profile of 5-HT_{2A} agents, and *in vivo* data is particularly difficult in the case of 5-HT_{2A} receptors due to the weak constitutive activity of native 5-HT_{2A} receptors *in vitro*. The data *in vivo* indicate that the phasic influence of 5-HT_{2A} receptors does not necessarily depend on an increase in endogenous 5-HT release. The constitutive activity of 5-HT_{2A} receptors could correspond to a possible lag time between the activation of the 5-HT_{2A} receptor and the binding of 5-HT *in vivo*. In any case, the canonical definition of the constitutive activity of GPCR is based on intracellular signaling pathways *in vitro* and implies that intracellular signaling pathways are oppositely modulated by agonists and inverse agonists at a given GPCR. This canonical definition is not clearly met *in vivo* due to the existence of several populations of receptors possibly interfering with the parameter. Additional data are warranted to determine whether the constitutive activity of 5-HT_{2A} receptor may participate in the development of neuropsychiatric disorders.

Acknowledgments The authors are indebted to the EU COST Action CM1103 "Structure-based drug design for diagnosis and treatment of neurological diseases: dissecting and modulating complex function in the monoaminergic systems of the brain" for supporting their international collaboration.

References

1. Morisset S et al (2000) High constitutive activity of native H3 receptors regulates histamine neurons in brain. Nature 408(6814):860–864
2. Navailles S et al (2013a) Serotonin2C receptor constitutive activity: in vivo direct and indirect evidence and functional significance. Cent Nerv Syst Agents Med Chem 13(2):98–107
3. Seifert R, Wenzel-Seifert K (2002) Constitutive activity of G-protein-coupled receptors: cause of disease and common property of wild-type receptors. Naunyn Schmiedeberg's Arch Pharmacol 366(5):381–416
4. Hoyer D et al (1994) International Union of Pharmacology classification of receptors for 5-hydroxytryptamine (serotonin). Pharmacol Rev 46(2):157–203
5. Hoyer D, Hannon JP, Martin GR (2002) Molecular, pharmacological and functional diversity of 5-HT receptors. Pharmacol Biochem Behav 71(4):533–554
6. Zifa E, Fillion G (1992) 5-Hydroxytryptamine receptors. Pharmacol Rev 44(3):401–458
7. Schmidt CJ et al (1995) The role of 5-HT2A receptors in antipsychotic activity. Life Sci 56(25):2209–2222
8. Costa T, Herz A (1989) Antagonists with negative intrinsic activity at delta opioid receptors coupled to GTP-binding proteins. Proc Natl Acad Sci U S A 86(19):7321–7325
9. Milligan G, Bond RA, Lee M (1995) Inverse agonism: pharmacological curiosity or potential therapeutic strategy? Trends Pharmacol Sci 16(1):10–13

10. Weiner DM et al (2001) 5-hydroxytryptamine2A receptor inverse agonists as antipsychotics. J Pharmacol Exp Ther 299(1):268–276
11. Harvey JA (2003) Role of the serotonin 5-HT(2A) receptor in learning. Learn Mem 10(5):355–362
12. Hoyer D et al (1986) Serotonin receptors in the human brain. II. Characterization and autoradiographic localization of 5-HT1C and 5-HT2 recognition sites. Brain Res 376(1):97–107
13. Lopez-Gimenez JF et al (1997) Selective visualization of rat brain 5-HT2A receptors by autoradiography with [3H]MDL 100,907. Naunyn Schmiedeberg's Arch Pharmacol 356(4): 446–454
14. Lopez-Gimenez JF et al (1999) Human striosomes are enriched in 5-HT2A receptors: autoradiographical visualization with [3H]MDL100,907,[125I](+/−)DOI and [3H]ketanserin. Eur J Neurosci 11(10):3761–3765
15. Lopez-Gimenez JF et al (1998) [3H]MDL 100,907 labels 5-HT2A serotonin receptors selectively in primate brain. Neuropharmacology 37(9):1147–1158
16. Lopez-Gimenez JF et al (2001) Mapping of 5-HT2A receptors and their mRNA in monkey brain: [3H]MDL100,907 autoradiography and in situ hybridization studies. J Comp Neurol 429(4):571–589
17. Wright DE et al (1995) Comparative localization of serotonin1A, 1C, and 2 receptor subtype mRNAs in rat brain. J Comp Neurol 351(3):357–373
18. Blin J et al (1993) Loss of brain 5-HT2 receptors in Alzheimer's disease. In vivo assessment with positron emission tomography and [18F]setoperone. Brain 116(Pt 3):497–510
19. Erritzoe D et al (2010) A nonlinear relationship between cerebral serotonin transporter and 5-HT(2A) receptor binding: an in vivo molecular imaging study in humans. J Neurosci 30(9):3391–3397
20. Leysen JE (2004) 5-HT2 receptors. Curr Drug Targets CNS Neurol Disord 3(1):11–26
21. Jakab RL, Goldman-Rakic PS (1998) 5-Hydroxytryptamine2A serotonin receptors in the primate cerebral cortex: possible site of action of hallucinogenic and antipsychotic drugs in pyramidal cell apical dendrites. Proc Natl Acad Sci U S A 95(2):735–740
22. Aghajanian GK, Marek GJ (1997) Serotonin induces excitatory postsynaptic potentials in apical dendrites of neocortical pyramidal cells. Neuropharmacology 36(4–5):589–599
23. Tanaka E, North RA (1993) Actions of 5-hydroxytryptamine on neurons of the rat cingulate cortex. J Neurophysiol 69(5):1749–1757
24. Doherty MD, Pickel VM (2000) Ultrastructural localization of the serotonin 2A receptor in dopaminergic neurons in the ventral tegmental area. Brain Res 864(2):176–185
25. Nocjar C, Roth BL, Pehek EA (2002) Localization of 5-HT(2A) receptors on dopamine cells in subnuclei of the midbrain A10 cell group. Neuroscience 111(1):163–176
26. Bombardi C, Di Giovanni G (2013) Functional anatomy of 5-HT2A receptors in the amygdala and hippocampal complex: relevance to memory functions. Exp Brain Res 230(4):427–439
27. Barre A, Berthoux C, De Bundel D, Valjent E, Bockaert J, Marin P, Becamel C (2016) Presynaptic serotonin 2A receptors modulate thalamocortical plasticity and associative learning. Proceedings of the National Academy of Sciences of the United States of America 113:E1382–E1391
28. Roth BL et al (1998) Serotonin 5-HT2A receptors: molecular biology and mechanisms of regulation. Crit Rev Neurobiol 12(4):319–338
29. Shapiro DA et al (2002) Evidence for a model of agonist-induced activation of 5-hydroxytryptamine 2A serotonin receptors that involves the disruption of a strong ionic interaction between helices 3 and 6. J Biol Chem 277(13):11441–11449
30. Boulougouris V, Glennon JC, Robbins TW (2008) Dissociable effects of selective 5-HT2A and 5-HT2C receptor antagonists on serial spatial reversal learning in rats. Neuropsychopharmacology 33(8):2007–2019
31. Harvey JA (1996) Serotonergic regulation of associative learning. Behav Brain Res 73(1–2):47–50

32. Winstanley CA et al (2004) 5-HT2A and 5-HT2C receptor antagonists have opposing effects on a measure of impulsivity: interactions with global 5-HT depletion. Psychopharmacology 176(3–4):376–385
33. Woolley DW, Shaw E (1954) A biochemical and pharmacological suggestion about certain mental disorders. Proc Natl Acad Sci U S A 40(4):228–231
34. Abi-Saab W et al (2002) Ritanserin antagonism of m-chlorophenylpiperazine effects in neuroleptic-free schizophrenics patients: support for serotonin-2 receptor modulation of schizophrenia symptoms. Psychopharmacology 162(1):55–62
35. Iqbal N, van Praag HM (1995) The role of serotonin in schizophrenia. Eur Neuropsychopharmacol 5(Suppl):11–23
36. Krystal JH et al (1993) m-Chlorophenylpiperazine effects in neuroleptic-free schizophrenic patients. Evidence implicating serotonergic systems in the positive symptoms of schizophrenia. Arch Gen Psychiatry 50(8):624–635
37. Meltzer HY, Matsubara S, Lee JC (1989) Classification of typical and atypical antipsychotic drugs on the basis of dopamine D-1, D-2 and serotonin2 pKi values. J Pharmacol Exp Ther 251(1):238–246
38. Meltzer HY et al (2003) Serotonin receptors: their key role in drugs to treat schizophrenia. Prog Neuro-Psychopharmacol Biol Psychiatry 27(7):1159–1172
39. Howell LL, Cunningham KA (2015) Serotonin 5-HT2 receptor interactions with dopamine function: implications for therapeutics in cocaine use disorder. Pharmacol Rev 67(1):176–197
40. Guiard BP, Di Giovanni G (2015) Central serotonin-2A (5-HT2A) receptor dysfunction in depression and epilepsy: the missing link? Front Pharmacol 6:46
41. Weisstaub NV et al (2006) Cortical 5-HT2A receptor signaling modulates anxiety-like behaviors in mice. Science 313(5786):536–540
42. Cummings J et al (2014) Pimavanserin for patients with Parkinson's disease psychosis: a randomised, placebo-controlled phase 3 trial. Lancet 383(9916):533–540
43. Berg KA et al (2008) A conservative, single-amino acid substitution in the second cytoplasmic domain of the human Serotonin2C receptor alters both ligand-dependent and -independent receptor signaling. J Pharmacol Exp Ther 324(3):1084–1092
44. Berg KA et al (2005) Physiological relevance of constitutive activity of 5-HT2A and 5-HT2C receptors. Trends Pharmacol Sci 26(12):625–630
45. Berg KA et al (1998) Pleiotropic behavior of 5-HT2A and 5-HT2C receptor agonists. Ann N Y Acad Sci 861:104–110
46. Kenakin T (2001) Inverse, protean, and ligand-selective agonism: matters of receptor conformation. FASEB J 15(3):598–611
47. Kenakin T (2004) Efficacy as a vector: the relative prevalence and paucity of inverse agonism. Mol Pharmacol 65(1):2–11
48. Milligan G, Bond RA (1997) Inverse agonism and the regulation of receptor number. Trends Pharmacol Sci 18(12):468–474
49. Gray JA, Roth BL (2001) Paradoxical trafficking and regulation of 5-HT(2A) receptors by agonists and antagonists. Brain Res Bull 56(5):441–451
50. Arrang JM, Morisset S, Gbahou F (2007) Constitutive activity of the histamine H3 receptor. Trends Pharmacol Sci 28(7):350–357
51. De Deurwaerdere P et al (2004) Constitutive activity of the serotonin2C receptor inhibits in vivo dopamine release in the rat striatum and nucleus accumbens. J Neurosci 24(13): 3235–3241
52. Hannon J, Hoyer D (2008) Molecular biology of 5-HT receptors. Behav Brain Res 195(1): 198–213
53. Raymond JR et al (2001) Multiplicity of mechanisms of serotonin receptor signal transduction. Pharmacol Ther 92(2–3):179–212
54. Egan CT, Herrick-Davis K, Teitler M (1998) Creation of a constitutively activated state of the 5-hydroxytryptamine2A receptor by site-directed mutagenesis: inverse agonist activity of antipsychotic drugs. J Pharmacol Exp Ther 286(1):85–90

55. Grotewiel MS, Sanders-Bush E (1999) Differences in agonist-independent activity of 5-Ht2A and 5-HT2c receptors revealed by heterologous expression. Naunyn Schmiedeberg's Arch Pharmacol 359(1):21–27
56. Teitler M, Herrick-Davis K, Purohit A (2002) Constitutive activity of G-protein coupled receptors: emphasis on serotonin receptors. Curr Top Med Chem 2(6):529–538
57. Muntasir HA et al (2006) Inverse agonist activity of sarpogrelate, a selective 5-HT2A-receptor antagonist, at the constitutively active human 5-HT2A receptor. J Pharmacol Sci 102(2):189–195
58. Barnes NM, Sharp T (1999) A review of central 5-HT receptors and their function. Neuropharmacology 38(8):1083–1152
59. Gray JA et al (2003) The interaction of a constitutively active arrestin with the arrestin-insensitive 5-HT(2A) receptor induces agonist-independent internalization. Mol Pharmacol 63(5):961–972
60. Karaki S et al (2014) Quantitative phosphoproteomics unravels biased phosphorylation of serotonin 2A receptor at Ser280 by hallucinogenic versus nonhallucinogenic agonists. Mol Cell Proteomics 13(5):1273–1285
61. Romero G, von Zastrow M, Friedman PA (2011) Role of PDZ proteins in regulating trafficking, signaling, and function of GPCRs: means, motif, and opportunity. Adv Pharmacol 62:279–314
62. Becamel C et al (2004) The serotonin 5-HT2A and 5-HT2C receptors interact with specific sets of PDZ proteins. J Biol Chem 279(19):20257–20266
63. Dunn HA, Ferguson SS (2015) PDZ protein regulation of G protein-coupled receptor trafficking and signaling pathways. Mol Pharmacol 88(4):624–639
64. Grotewiel MS, Sanders-Bush E (1994) Regulation of serotonin2A receptors in heterologous expression systems. J Neurochem 63(4):1255–1260
65. Albizu L et al (2011) Functional crosstalk and heteromerization of serotonin 5-HT2A and dopamine D2 receptors. Neuropharmacology 61(4):770–777
66. Gonzalez-Maeso J et al (2008) Identification of a serotonin/glutamate receptor complex implicated in psychosis. Nature 452(7183):93–97
67. Teitler M, Herrick-Davis K (2014) Determining the oligomer number of native GPCR using florescence correlation spectroscopy and drug-induced inactivation-reactivation. Curr Pharm Biotechnol 15(10):927–937
68. Vinals X et al (2015) Cognitive impairment induced by delta9-tetrahydrocannabinol occurs through heteromers between cannabinoid CB1 and serotonin 5-HT2A receptors. PLoS Biol 13(7):e1002194
69. Borroto-Escuela DO et al (2014) Hallucinogenic 5-HT2AR agonists LSD and DOI enhance dopamine D2R protomer recognition and signaling of D2-5-HT2A heteroreceptor complexes. Biochem Biophys Res Commun 443(1):278–284
70. Di Giovanni G et al (1999) Selective blockade of serotonin-2C/2B receptors enhances mesolimbic and mesostriatal dopaminergic function: a combined in vivo electrophysiological and microdialysis study. Neuroscience 91(2):587–597
71. Kehne JH et al (1996) Effects of the selective 5-HT2A receptor antagonist MDL 100,907 on MDMA-induced locomotor stimulation in rats. Neuropsychopharmacology 15(2):116–124
72. Palfreyman MG et al (1993) Electrophysiological, biochemical and behavioral evidence for 5-HT2 and 5-HT3 mediated control of dopaminergic function. Psychopharmacology 112 (1 Suppl):S60–S67
73. Welsh SE, Romano AG, Harvey JA (1998b) Effects of serotonin 5-HT(2A/2C) antagonists on associative learning in the rabbit. Psychopharmacology 137(2):157–163
74. Welsh SE et al (1998a) Effects of LSD, ritanserin, 8-OH-DPAT, and lisuride on classical conditioning in the rabbit. Pharmacol Biochem Behav 59(2):469–475
75. Romano AG, Hood H, Harvey JA (2000) Dissociable effects of the 5-HT(2) antagonist mianserin on associative learning and performance in the rabbit. Pharmacol Biochem Behav 67(1):103–110

76. Romano AG, Harvey JA (1993) Enhanced learning following a single, acute dose of MDA. Pharmacol Biochem Behav 44(4):965–969
77. Romano AG, Harvey JA (1994) MDMA enhances associative and nonassociative learning in the rabbit. Pharmacol Biochem Behav 47(2):289–293
78. Knight AR et al (2004) Pharmacological characterisation of the agonist radioligand binding site of 5-HT(2A), 5-HT(2B) and 5-HT(2C) receptors. Naunyn Schmiedeberg's Arch Pharmacol 370(2):114–123
79. Engel G et al (1986) Identity of inhibitory presynaptic 5-hydroxytryptamine (5-HT) autoreceptors in the rat brain cortex with 5-HT1B binding sites. Naunyn Schmiedeberg's Arch Pharmacol 332(1):1–7
80. Hoyer D, Schoeffter P (1991) 5-HT receptors: subtypes and second messengers. J Recept Res 11(1–4):197–214
81. Titeler M, Lyon RA, Glennon RA (1988) Radioligand binding evidence implicates the brain 5-HT2 receptor as a site of action for LSD and phenylisopropylamine hallucinogens. Psychopharmacology 94(2):213–216
82. Battaglia G et al (1988) Pharmacologic profile of MDMA (3,4-methylenedioxymethampheta mine) at various brain recognition sites. Eur J Pharmacol 149(1–2):159–163
83. Lyon RA, Glennon RA, Titeler M (1986) 3,4-Methylenedioxymethamphetamine (MDMA): stereoselective interactions at brain 5-HT1 and 5-HT2 receptors. Psychopharmacology 88(4):525–526
84. Kahn RS, Wetzler S (1991) m-Chlorophenylpiperazine as a probe of serotonin function. Biol Psychiatry 30(11):1139–1166
85. Hoyer D, Neijt HC (1988) Identification of serotonin 5-HT3 recognition sites in membranes of N1E-115 neuroblastoma cells by radioligand binding. Mol Pharmacol 33(3):303–309
86. Kilpatrick GJ, Jones BJ, Tyers MB (1989) Binding of the 5-HT3 ligand, [3H]GR65630, to rat area postrema, vagus nerve and the brains of several species. Eur J Pharmacol 159(2):157–164
87. Forbes IT et al (1993) N-(1-methyl-5-indolyl)-N′-(3-pyridyl)urea hydrochloride: the first selective 5-HT1C receptor antagonist. J Med Chem 36(8):1104–1107
88. Meneses A (2002) Involvement of 5-HT(2A/2B/2C) receptors on memory formation: simple agonism, antagonism, or inverse agonism? Cell Mol Neurobiol 22(5–6):675–688
89. Leysen JE et al (1988) Identification of nonserotonergic [3H]ketanserin binding sites associated with nerve terminals in rat brain and with platelets; relation with release of biogenic amine metabolites induced by ketanserin- and tetrabenazine-like drugs. J Pharmacol Exp Ther 244(1):310–321
90. Wainscott DB et al (1996) Pharmacologic characterization of the human 5-hydroxytryptamine2B receptor: evidence for species differences. J Pharmacol Exp Ther 276(2): 720–727
91. Green JP et al (1977) Antagonism of histamine-activated adenylate cyclase in brain by D-lysergic acid diethylamide. Proc Natl Acad Sci U S A 74(12):5697–5701
92. Gonzalez-Maeso J et al (2007) Hallucinogens recruit specific cortical 5-HT(2A) receptor-mediated signaling pathways to affect behavior. Neuron 53(3):439–452
93. Harvey JA et al (1999) Effect of 5-HT2 receptor antagonists on a cranial nerve reflex in the rabbit: evidence for inverse agonism. Psychopharmacology 141(2):162–168
94. Romano AG et al (2006) Effect of serotonin depletion on 5-HT2A-mediated learning in the rabbit: evidence for constitutive activity of the 5-HT2A receptor in vivo. Psychopharmacology 184(2):173–181
95. Winsky L, Harvey JA (1992) 6-Hydroxydopamine induced impairment of Pavlovian conditioning in the rabbit. Neurochem Res 17(5):415–422
96. Di Giovanni G, De Deurwaerdere P (2016) New therapeutic opportunities for 5-HT2C receptor ligands in neuropsychiatric disorders. Pharmacol Ther 157:125–162
97. van Wijngaarden I, Tulp MT, Soudijn W (1990) The concept of selectivity in 5-HT receptor research. Eur J Pharmacol 188(6):301–312
98. Garratt JC et al (1991) Inhibition of 5-hydroxytryptamine neuronal activity by the 5-HT agonist, DOI. Eur J Pharmacol 199(3):349–355

99. Hery F, Ternaux JP (1981) Regulation of release processes in central serotoninergic neurons. J Physiol Paris 77(2–3):287–301
100. Wichems CH, Hollingsworth CK, Bennett BA (1995) Release of serotonin induced by 3,4-methylenedioxymethamphetamine (MDMA) and other substituted amphetamines in cultured fetal raphe neurons: further evidence for calcium-independent mechanisms of release. Brain Res 695(1):10–18
101. Carboni E, Di Chiara G (1989) Serotonin release estimated by transcortical dialysis in freely-moving rats. Neuroscience 32(3):637–645
102. Strange PG (2002) Mechanisms of inverse agonism at G-protein-coupled receptors. Trends Pharmacol Sci 23(2):89–95
103. Dave KD, Harvey JA, Aloyo VJ (2007) The time-course for up- and down-regulation of the cortical 5-hydroxytryptamine (5-HT)2A receptor density predicts 5-HT2A receptor-mediated behavior in the rabbit. J Pharmacol Exp Ther 323(1):327–335
104. Harvey JA et al (2004) Selective remodeling of rabbit frontal cortex: relationship between 5-HT2A receptor density and associative learning. Psychopharmacology 172(4):435–442
105. Barker EL et al (1994) Constitutively active 5-hydroxytryptamine2C receptors reveal novel inverse agonist activity of receptor ligands. J Biol Chem 269(16):11687–11690
106. Amato D et al (2015) Neuropharmacology of light-induced locomotor activation. Neuropharmacology 95:243–251
107. Harvey ML, Swallows CL, Cooper MA (2012) A double dissociation in the effects of 5-HT2A and 5-HT2C receptors on the acquisition and expression of conditioned defeat in Syrian hamsters. Behav Neurosci 126(4):530–537
108. Clinard CT et al (2015) Activation of 5-HT2a receptors in the basolateral amygdala promotes defeat-induced anxiety and the acquisition of conditioned defeat in Syrian hamsters. Neuropharmacology 90:102–112
109. Zhang G et al (2013) Stimulation of serotonin 2A receptors facilitates consolidation and extinction of fear memory in C57BL/6J mice. Neuropharmacology 64:403–413
110. Bekinschtein P et al (2013) Role of medial prefrontal cortex serotonin 2A receptors in the control of retrieval of recognition memory in rats. J Neurosci 33(40):15716–15725
111. Dougherty JP, Aloyo VJ (2011) Pharmacological and behavioral characterization of the 5-HT2A receptor in C57BL/6N mice. Psychopharmacology 215(3):581–593
112. Wadenberg ML (2010) Conditioned avoidance response in the development of new antipsychotics. Curr Pharm Des 16(3):358–370
113. Wadenberg MG et al (2001) Antagonism at 5-HT(2A) receptors potentiates the effect of haloperidol in a conditioned avoidance response task in rats. Pharmacol Biochem Behav 68(3):363–370
114. Wadenberg ML et al (1998) Enhancement of antipsychoticlike properties of raclopride in rats using the selective serotonin2A receptor antagonist MDL 100,907. Biol Psychiatry 44(6):508–515
115. Wadenberg ML et al (1996) Enhancement of antipsychotic-like properties of the dopamine D2 receptor antagonist, raclopride, by the additional treatment with the 5-HT2 receptor blocking agent, ritanserin, in the rat. Eur Neuropsychopharmacol 6(4):305–310
116. Wiker C et al (2005) Adjunctive treatment with mianserin enhances effects of raclopride on cortical dopamine output and, in parallel, its antipsychotic-like effect. Neuropsychiatr Dis Treat 1(4):356–372
117. Wadenberg ML, Ahlenius S (1988) Suppression of conditioned avoidance by 8-OH-DPAT in the rat. J Neural Transm 74(3):195–198
118. Marquis KL et al (2007) WAY-163909 [(7bR,10aR)-1,2,3,4,8,9,10,10a-octahydro-7bH-cyclopenta-[b][1,4]diazepino[6,7,1hi]indole]: a novel 5-hydroxytryptamine 2C receptor-selective agonist with preclinical antipsychotic-like activity. J Pharmacol Exp Ther 320(1):486–496
119. Siuciak JA et al (2007) CP-809,101, a selective 5-HT2C agonist, shows activity in animal models of antipsychotic activity. Neuropharmacology 52(2):279–290
120. Meneses A (2003) A pharmacological analysis of an associative learning task: 5-HT(1) to 5-HT(7) receptor subtypes function on a pavlovian/instrumental autoshaped memory. Learn Mem 10(5):363–372

121. Meneses A, Terron JA, Hong E (1997) Effects of the 5-HT receptor antagonists GR127935 (5-HT1B/1D) and MDL100907 (5-HT2A) in the consolidation of learning. Behav Brain Res 89(1–2):217–223
122. Meneses A, Hong E (1997) Role of 5-HT1B, 5-HT2A and 5-HT2C receptors in learning. Behav Brain Res 87(1):105–110
123. Meneses A, Hong E (1994) Mechanism of action of 8-OH-DPAT on learning and memory. Pharmacol Biochem Behav 49(4):1083–1086
124. Navailles S, De Deurwaerdère P (2010) The constitutive activity of 5-HT2C receptors as an additional modality of interaction of the serotonergic system in motor control. In: Di Giovanni G, Di Matteo V, Esposito E (eds) 5-HT2C receptors in the pathophysiology of CNS disease, The Receptors series: Springer. Humana Press, Totowa, pp 187–214
125. Walker EA, Foley JJ (2010) Acquisition session length modulates consolidation effects produced by 5-HT2C ligands in a mouse autoshaping-operant procedure. Behav Pharmacol 21(2):83–89
126. Leggio GM et al (2009) In vivo evidence that constitutive activity of serotonin2C receptors in the medial prefrontal cortex participates in the control of dopamine release in the rat nucleus accumbens: differential effects of inverse agonist versus antagonist. J Neurochem 111(2):614–623
127. Murray KC et al (2010) Recovery of motoneuron and locomotor function after spinal cord injury depends on constitutive activity in 5-HT2C receptors. Nat Med 16(6):694–700
128. Navailles S et al (2013b) Serotonin2C ligands exhibiting full negative and positive intrinsic activity elicit purposeless oral movements in rats: distinct effects of agonists and inverse agonists in a rat model of Parkinson's disease. Int J Neuropsychopharmacol 16(3):593–606
129. Eriksson E et al (1999) Effects of mCPP on the extracellular concentrations of serotonin and dopamine in rat brain. Neuropsychopharmacology 20(3):287–296
130. Navailles S et al (2006) Region-dependent regulation of mesoaccumbens dopamine neurons in vivo by the constitutive activity of central serotonin2C receptors. J Neurochem 99(4):1311–1319
131. De Deurwaerdere P, Di Giovanni G (2017) Serotonergic modulation of the activity of mesencephalic dopaminergic systems: therapeutic implications. Prog Neurobiol 151:175–236
132. Gudelsky GA, Yamamoto BK, Nash JF (1994) Potentiation of 3,4-methylenedioxymethamp hetamine-induced dopamine release and serotonin neurotoxicity by 5-HT2 receptor agonists. Eur J Pharmacol 264(3):325–330
133. Schmidt CJ et al (1992) 5-HT2 receptors exert a state-dependent regulation of dopaminergic function: studies with MDL 100,907 and the amphetamine analogue, 3,4-methylenedioxyme thamphetamine. Eur J Pharmacol 223(1):65–74
134. Schmidt CJ, Sullivan CK, Fadayel GM (1994) Blockade of striatal 5-hydroxytryptamine2 receptors reduces the increase in extracellular concentrations of dopamine produced by the amphetamine analogue 3,4-methylenedioxymethamphetamine. J Neurochem 62(4): 1382–1389
135. Ichikawa J, Meltzer HY (1995) DOI, a 5-HT2A/2C receptor agonist, potentiates amphetamine-induced dopamine release in rat striatum. Brain Res 698(1–2):204–208
136. Lucas G et al (2000) The effect of serotonergic agents on haloperidol-induced striatal dopamine release in vivo: opposite role of 5-HT(2A) and 5-HT(2C) receptor subtypes and significance of the haloperidol dose used. Neuropharmacology 39(6):1053–1063
137. Ichikawa J et al (1995) R(+)-8-OH-DPAT, a 5-HT1A receptor agonist, inhibits amphetamine-induced dopamine release in rat striatum and nucleus accumbens. Eur J Pharmacol 287(2): 179–184
138. Porras G et al (2002) 5-HT2A and 5-HT2C/2B receptor subtypes modulate dopamine release induced in vivo by amphetamine and morphine in both the rat nucleus accumbens and striatum. Neuropsychopharmacology 26(3):311–324
139. Meltzer HY et al (2010) Pimavanserin, a serotonin(2A) receptor inverse agonist, for the treatment of parkinson's disease psychosis. Neuropsychopharmacology 35(4):881–892

140. Weiner DM et al (2003) Psychosis of Parkinson's disease: serotonin 2A receptor inverse agonists as potential therapeutics. Curr Opin Investig Drugs 4(7):815–819
141. Navailles S et al (2010) High-frequency stimulation of the subthalamic nucleus and L-3,4-dihydroxyphenylalanine inhibit in vivo serotonin release in the prefrontal cortex and hippocampus in a rat model of Parkinson's disease. J Neurosci 30(6):2356–2364
142. Navailles S et al (2011) Chronic L-DOPA therapy alters central serotonergic function and L-DOPA-induced dopamine release in a region-dependent manner in a rat model of Parkinson's disease. Neurobiol Dis 41(2):585–590

βArrestins: Ligand-Directed Regulators of 5-HT$_{2A}$ Receptor Trafficking and Signaling Events

Cullen L. Schmid and Laura M. Bohn

Abstract βArrestins are scaffolding and regulatory proteins that both modify and mediate G protein coupled receptor responsiveness by desensitizing G protein signaling pathways and facilitating receptor internalization and alternate downstream signaling cascades. This chapter details the studies that demonstrate a role for βarrestin2 in regulating 5-HT$_{2A}$ receptor responsiveness in vitro and in vivo. The studies presented herein demonstrate that while the 5-HT$_{2A}$ receptor is capable of being regulated through interactions with GRKs and βarrestins, interactions with other proteins can facilitate receptor desensitization and internalization through non-βarrestin-mediated mechanisms. Moreover, the pathways utilized for each of these events are determined by both the complement of intracellular proteins expressed in residence with the 5-HT$_{2A}$ receptor and the agonist acting at the receptor.

Keywords βarrestins • G protein receptor kinases • Functional selectivity • Internalization • G protein coupling • MAP kinase

Classical Regulation of GPCRs by βarrestins

Both the agonist activated signaling cascades and the regulatory mechanisms that interact with the receptor to control the extent and duration of the response determines 5-HT$_{2A}$ receptor responsiveness. The canonical model of GPCR regulation posits that the agonist-bound receptor is regulated through its interactions with the serine/threonine GPCR kinases (GRKs) and the intracellular regulatory proteins, βarrestins (Fig. 1). Upon agonist activation, GPCRs are phosphorylated by GRKs, which initiate the desensitization of G protein-mediated signaling by promoting the recruitment of βarrestins. Once recruited, βarrestins can further inhibit G

C.L. Schmid (✉) • L.M. Bohn
Departments of Molecular Medicine and Neuroscience, The Scripps Research Institute, Jupiter, FL, USA
e-mail: cschmid@scripps.edu

B.P. Guiard, G. Di Giovanni (eds.), *5-HT$_{2A}$ Receptors in the Central Nervous System*, The Receptors 32, https://doi.org/10.1007/978-3-319-70474-6_2

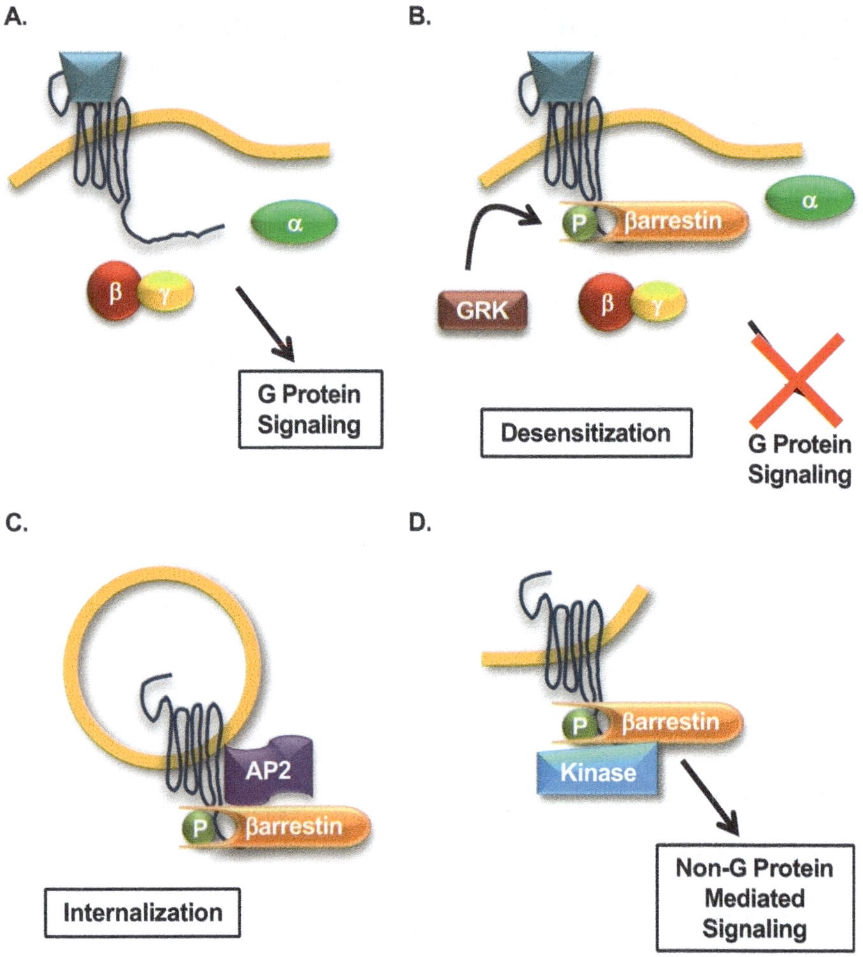

Fig. 1 Canonical model of GPCR regulation. (**a**) Agonist binding to a GPCR catalyzes the dissociation of the Gα subunit from the Gβγ heterodimer, initiating downstream signaling events. (**b**) Rapidly following agonist binding to a GPCR, the receptor is phosphorylated by GRKs, followed by βarrestin binding. βArrestin binding to a receptor desensitizes the GPCR by preventing further coupling to G proteins. (**c**) βarrestins also facilitate GPCR internalization by acting as adaptor proteins between the receptor and clathrin and AP2, proteins involved in endocytosis. (**d**) βArrestins also initiate signaling cascades by scaffolding components of non-G protein-mediated cascades to GPCRs

protein-coupling, initiate the internalization of receptors and scaffold additional signaling molecules to GPCRs, thus facilitating G protein-independent signaling cascades [1].

GRKs can be divided into three main families: the visual GRKs (GRK1 and 7), the GRK2 subfamily of GRKs (GRK2 and 3) and the GRK4 subfamily (GRK4, 5 and 6). The visual GRKs are mainly expressed in the retina and regulate visual

GPCRs. The GRK2 subfamily is ubiquitously expressed, while, of the GRK4 subfamily, only GRKs 5 and 6 are ubiquitously expressed. GRK4 is highly expressed in the testes, with some expression in the kidneys, uterus and brain, where expression is limited to the purkinje cells of the cerebellum [2–4]. While data on the cell-type specific expression of GRKs is limited, most mammalian cells express multiple isoforms of GRKs.

The phosphorylation of the agonist-bound GPCR by GRKs causes βarrestins to translocate to the receptor. βArrestin recruitment to, and interaction with, GPCRs has been visualized by confocal microscopy with fluorescently tagged βarrestins [5] and quantified by co-immunoprecipitation, bioluminescence/fluorescence resonance energy transfer, gene reporter and enzyme fragment complementation assays [6–10]. There are four arrestin isoforms: the visual arrestins (arrestin1 and arrestin4) and the βarrestins (βarrestin1/arrestin2 and βarrestin2/arrestin3). The two βarrestins are expressed in almost every mammalian cell and bind to hundreds of different GPCRs as well as non-GPCR signaling molecules, including trafficking proteins, Src family kinases, MAP kinases and E3 ubiquitin ligases [11]. βarrestin1 and βarrestin2 are very similar in terms of sequence and seem to compensate for one another, as evidenced by the fact that βarrestin1 and βarrestin2- knockout (KO) mice display no gross abnormalities [12–14]. However, there are differences between the two isoforms. For one, βarrestin2 has a nuclear export sequence while βarrestin1 does not, which results in overexpressed βarrestin1 accumulating in the nucleus and the cytoplasm, while βarrestin2 is localized to the cytoplasm [15]. Moreover, some GPCRs appear to be preferentially regulated by one or the other [16, 17] and there are some differences in the binding of each isoform to non-GPCR partners as well [18]. The differential interactions between βarrestins and their binding partners and the resulting differential functions seems to be strongly dictated by the cellular environment, such as the expression of a particular GRK [19].

βArrestin interactions with GPCRs serve to desensitize further signaling by sterically blocking further interactions between receptors and their cognate G proteins (Fig. 1b) [20]. In vitro, the role of βarrestins in the desensitization of many GPCRs has been shown by over-expressing βarrestins, thus enhancing receptor desensitization, or by disrupting GPCR interactions with βarrestins, thereby diminishing the desensitization of the signaling pathways [21–23]. For example, the over-expression of either βarrestin1 or βarrestin2 increases the desensitization of β_2-adrenergic receptor coupling to $G\alpha_s$ [24]. In contrast, agonist-induced β_2-adrenergic receptor signaling is enhanced in human embryonic kidney (HEK-293) cells in which both βarrestins have been silenced by small interfering RNAs (siRNA), or in mouse embryonic fibroblasts (MEF) generated from genetically modified mice lacking both βarrestins (βarr1/2-KO) [16, 25]. This negative regulatory role for βarrestins has also been demonstrated in vivo through the use of mice lacking βarrestin2 (βarr2-KO), where responses are enhanced in the absence of βarrestin2 [14].

βArrestin interactions with GPCRs also facilitate the internalization of receptors into intracellular vesicles (Fig. 1c). βArrestins promote GPCR internalization by acting as adaptors that link agonist-bound receptors to elements of clathrin-coated pits. Both βarrestin1 and βarrestin2 have been shown to directly interact with clathrin and the β_2-adaptin subunit of the adaptor protein 2 (AP2) complex [26–29]. The βarrestin-mediated recruitment of these proteins then targets receptors for endocytosis. Interfering with GPCR/βarrestin interactions has been shown to impair agonist-mediated trafficking for a number of GPCRs and in a number of different cells types [30]. The internalization of a GPCR then aids in the determination of receptor fate, as it can direct receptors to endosomes for recycling or to lysosomes for degradation [31], thus implicating βarrestins in receptor resensitization and down-regulation.

Different GPCRs interact with βarrestins to varying degrees: some exhibit weak interactions which lead to their transient internalization and recycling to the plasma membrane (Class A GPCRs), while others display stronger interactions with βarrestins, due to highly conserved serine/threonine phosphorylation site clusters in the C-terminal tail of the receptor, and more sustained internalization into endosomes (Class B GPCRs) [17, 32]. Those Class B receptors which have stronger interactions with βarrestins promote sustained G protein signaling due to the formation of GPCR/G protein/βarrestin super-complexes [33], suggesting that in some instances, βarrestins can actually promote, rather than desensitize, G protein signaling. For instance, a constitutively active form of βarrestin1 enhances the sustained G protein signaling of the Class B parathyroid hormone type 1 and vasopressin type 2 receptors. In contrast, the constitutively active βarrestin1 enhanced the desensitization of $G_{\alpha s}$ signaling for the Class A β_2-adrenergic receptor [34].

βArrestins can also mediate GPCR signaling by scaffolding elements of signal transduction cascades to receptors (Fig. 1d). This was first demonstrated for the β_2-adrenergic receptor, wherein agonist stimulation was shown to recruit Src to the receptor, but only when βarrestin1 was also expressed [35]. Since then, βarrestins have been shown to be integral members of in vitro receptor signaling scaffolds for a number of kinases, including ERK1/2, JNK, p38 and Akt [36–38]. Moreover, by binding multiple components of a signaling cascade simultaneously, βarrestins can increase the efficiency of the signaling between successive kinases. For example, when expressed in COS-7 cells, βarrestin2 forms a complex with the MAP kinase kinase kinase Ask1, the MAP kinase kinase MKK4 and JNK3 and over-expression of βarrestin2 increases Ask1-dependent phosphorylation of JNK3 [39, 40]. Moreover, while both βarrestins and G proteins can mediate signaling through the same downstream effectors, the two pathways have different time-courses of activation. For example, the AT_{1A} receptor activates ERK1/2 via both pathways: a rapid G protein-mediated pathway and a slower and more persistent βarrestin2-dependent pathway [41]. βArrestin2 has also been co-immunoprecipitated out of mouse brain with ERK1/2, Akt, Src and JNK3 kinases [39, 42–44], demonstrating the formation of these complexes in vivo. The physiological consequences of GPCR signaling through βarrestins has been correlated to decreased drug responsiveness in the βarr2-KO mice for a number of different receptors [14]. βarrestin-mediated signaling can be modulated by other proteins that interact with a GPCR: in the case of the 5-HT_{2C} receptor, calmodulin binding to the C-terminus of the receptor promotes recruitment of βarrestin2 and the inhibition of calmodulin

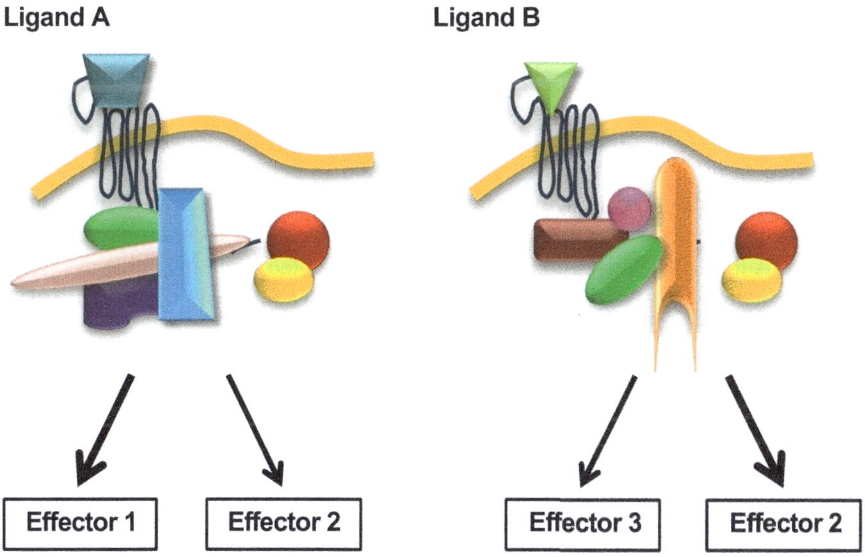

Fig. 2 Ligand directed signaling at GPCRs. Different agonists (*ligand A* and *ligand B*) acting at the same GPCR selectively recruit a subset of the signaling proteins (*depicted by different colored shapes*) expressed in close proximity to the receptor complex, which leads to the activation of downstream signaling cascades (*effector 1* and *effector 3*). Some proteins are recruited in a conserved manner to activate the same signaling cascades (*effector 2*), though the degree of activation of these pathways may differ between agonists (*depicted by arrows of different thickness*)

(whether by dominant negatives, mutation of the binding site or RNA interference) prevents βarrestin2-mediated signaling to ERK1/2 [45].

Functional Selectivity and βarrestin Bias

Traditionally, the conventional understanding of receptor pharmacology has been that an agonist fully activates all signal transduction pathways to which a GPCR is coupled, while partial agonists induce sub-maximal activation of these same pathways. Antagonists do not shift any of the responses away from basal levels, yet block further signaling, and inverse agonists reduce the basal activities of these pathways. However, these concepts of receptor pharmacology are too simple to conceptualize the full range of pharmacological profiles that are experimentally observed. For example, agonists at a particular GPCR can display full efficacy in certain signaling assays, while only partially activating or having no activity at others [46–48]. Current receptor pharmacology is based on the understanding that receptors can exist in multiple, ligand-specific conformational states which allow GPCRs to preferentially and differentially engage a subset of the multiple signaling pathways to which they are coupled [49, 50]. This concept that ligands differentially activate downstream signaling pathways has been referred to as "biased agonism" or "functional selectivity" (Fig. 2) [51–54]. The cellular environment can also influence the signaling that occurs downstream of GPCR activation by determining the complement of intracellular proteins available to couple to a receptor. In this way, the same ligand can induce differential signaling cascades at a particular receptor expressed in different cell types [55]. In addition to determining the downstream signaling cascades that are activated, ligands can also dictate the extent and nature of βarrestin interactions with GPCRs. For example, the AT_{1A} receptor agonist Sar^1, Ile^4, Ile^8-AngII selectively stimulates βarrestin2-mediated signaling pathways without inducing any detectable coupling to G proteins [48], thus being characterized as a ligand that is biased towards βarrestin2.

βarrestin2-Mediated Regulation of the 5-HT_{2A} Receptor

βArrestin2 Interactions with the 5-HT_{2A} Receptor

The Class A 5-HT_{2A} receptor is co-expressed and capable of interacting with βarrestins. Biochemical studies have demonstrated that this interaction can take place: purified βarrestin1 and βarrestin2 interact with a fusion protein that encodes the third intracellular loop of the receptor [56]. Moreover, the receptor and βarrestins are co-expressed endogenously, as demonstrated by dual label fluorescence confocal microscopy of pyramidal neurons from the rat frontal cortex [56]. These studies indicate that the regulatory protein may play a major role in regulating 5-HT_{2A} receptor responsiveness to ligands; however, the aforementioned study identified some neurons in which the 5-HT_{2A} receptor was expressed in the absence of βarrestins, suggesting that regulation of the 5-HT_{2A} receptor may not always follow the canonical model.

The 5-HT_{2A} receptor is among those GPCRs that can interact with both βarrestins. This interaction has been confirmed by co-immunopreciptation and confocal microscopy in HEK-293 cells following serotonin treatment [57–60] and quantified by the DiscoveRx PathHunter enzyme complementation assay in U2OS cells [60]. These qualitative and quantitative studies have confirmed that serotonin induces robust translocation of βarrestin2 to the 5-HT_{2A} receptor in vitro. In contrast, confocal studies suggest that serotonin only induces marginal translocation of βarrestin1 to the plasma membrane of HEK-293 cells expressing the 5-HT_{2A} receptor [57]. This could indicate that the 5-HT_{2A} receptor has a higher affinity for βarrestin2 than βarrestin1, although the assay is neither ratiometric nor quantitative and may simply reflect properties of the transfected βarrestin constructs. In addition to serotonin, other 5-HT_{2A} receptor agonists, such as quipazine [57] and the psychedelic tryptamines, also induce βarrestin2 translocation to the 5-HT_{2A} receptor [61]. Pretreatment with 5-HT_{2A} receptor antagonists such as clozapine will also block serotonin-mediated translocation of βarrestin2 [60].

5-HT_{2A} receptor interactions with βarrestins have been observed in vivo. Both βarrestins and the 5-HT_{2A} receptor are highly expressed in the rodent frontal cortex, [24, 56, 62–65], and as stated above, are co-expressed in many, but not all pyramidal neurons in the rat frontal cortex [56]. In addition, βarrestin2 co-immunoprecipitates with the 5-HT_{2A} receptor when it is isolated from the frontal cortex of mice treated with 100 mg/kg of 5-hydroxytryptamine, the metabolic precursor to serotonin [66]. Collectively, these studies indicate that 5-HT_{2A} receptors interact with βarrestins and suggest a role for the protein in regulating receptor responsiveness.

Desensitization of the 5-HT_{2A} Receptor

The mechanism by which the 5-HT_{2A} receptor is desensitized following ligand binding, let alone the role that βarrestins play in the process, is not well understood. Cell culture studies have demonstrated that exposure to agonist results in the desensitization of 5-HT_{2A} receptor-mediated phosphatidyl inositol (PI) hydrolysis [67, 68]. The Roth laboratory systematically mutated all of the serine and threonine residues in the cytoplasmic domains of the 5-HT_{2A} receptor and assessed the agonist-mediated desensitization of the IP_3 pathway. They show that the mutation of two serine residues, serine 421 in the C-terminal tail and serine 188 in the second intracellular loop, results in significant inhibition of quipazine-induced desensitization [69]. Serine 280 in the third intracellular loop also plays a role in the desensitization of the 5-HT_{2A} receptor, in an agonist dependent manner. The hallucinogenic compounds DOI (1-[2,5-dimethoxy-4-iodophenyl]-2-aminopropane) and LSD induce phosphorylation of serine 280 while the non-hallucinogenic compounds lisuride and ergotamine do not. Moreover, these agonists induce differing degrees of 5-HT_{2A} receptor desensitization: pretreatment of HEK-293 cells or cortical neurons with lisuride or ergotamine reduces serotonin-stimulated PI hydrolysis and ERK1/2 phosphorylation, while pretreatment with LSD and DOI does not [70]. These findings suggest that the phosphorylation of the 5-HT_{2A} receptor can impact agonist-induced desensitization.

The mechanism by which the 5-HT_{2A} receptor is desensitized varies for different cell types. In rat C6 glioma cells, which endogenously express the 5-HT_{2A} receptor, quipazine and serotonin-induced desensitization of PI-hydrolysis is attenuated by a dominant negative βarrestin1 (βarr$1_{319\text{-}418}$), that encodes only the C-terminal tail of the protein and not the GPCR binding domain [71–73]. In contrast, the βarr$1_{319\text{-}418}$ dominant negative has no effect on quipazine-induced desensitization in HEK-293 cells transiently transfected with the 5-HT_{2A} receptor [71]. The expression of a dominant negative GRK2 (GRK2_{K22R}), which lacks kinase activity [74], also blocks serotonin-mediated desensitization in C6 glioma cells [72], but not in CHO cells stably expressing the 5-HT_{2A} receptor [75]. These differences could reflect differences in cell lines or incomplete inhibition by the dominant negatives. An in vivo study in cocaine-withdrawn rats correlates the increased PLCβ activity of prefrontal cortical 5-HT$_{2A}$ receptors to reduced phosphorylation of the receptors and reduced GRK 5 levels [76]. Although inconclusive, these studies suggest that the 5-HT$_{2A}$ receptor can be classically desensitized via interactions with GRKs and βarrestins.

While βarrestins may play a role in the desensitization of the 5-HT$_{2A}$ receptor, other second messenger-dependent kinases, such as PKC can also dampen receptor signaling. The 5-HT$_{2A}$ receptor contains five putative PKC phosphorylation sites in its intracellular domains [77]. Berg et al. [75] demonstrated that the PKC inhibitors staurosporine and bisindolylmaleimide significantly inhibit serotonin-induced desensitization of PI hydrolysis in CHO cells stably transfected with the 5-HT$_{2A}$ receptor. However, PKC-mediated desensitization is also cell-type specific, as inhibition of PKC fails to alter serotonin-mediated desensitization of the IP$_3$ pathway in Chinese hamster lung fibroblasts stably expressing the 5-HT$_{2A}$ receptor [77].

Another serine/threonine kinase, p90 kDa ribosomal S6 family of serine/threonine kinases (RSK), has also been implicated in the agonist-mediated desensitization of the 5-HT$_{2A}$ receptor. Studies from the Roth laboratory have shown that RSK2 co-immunoprecipitates with the 5-HT$_{2A}$ receptor in HEK-293 cells, C6 glioma cells and rat cortical homogenates and is co-expressed with the 5-HT$_{2A}$ receptor in neurons in the rat frontal cortex [78, 79]. Moreover, purified and activated RSK2 directly phosphorylates serine 314 of the third intracellular loop of the 5-HT$_{2A}$ receptor in vitro [79]. The activation of intracellular Ca^{2+} release, PI hydrolysis and ERK1/2 phosphorylation are all potentiated following treatment with a panel of 5-HT$_{2A}$ receptor agonists in RSK2-KO MEFs or in WT MEFs expressing a RSK2-insensitive 5-HT$_{2A}$ receptor mutant (5-HT$_{2A}$ receptor$_{S314A}$) [78–80]. The enhanced responses observed in the absence of 5-HT$_{2A}$ receptor interactions with RSK2 suggest that RSK2 may act to dampen 5-HT$_{2A}$ receptor signaling in vitro. Taken together, these data demonstrate that there are multiple pathways by which the 5-HT$_{2A}$ receptor can be desensitized and that both the cell-type and the agonist may determine which pathway is utilized.

Internalization of the 5-HT$_{2A}$ Receptor

Agonist stimulation leads to the internalization of the 5-HT$_{2A}$ receptor through a clathrin-mediated pathway *in vitro*. Treatment of NIH-3 T3 cells with quipazine induces trafficking of the 5-HT$_{2A}$ receptor from the cell surface to intracellular vesicles that co-express clathrin [81, 82]. Moreover, pretreatment with concanavalin A or phenylarsine oxide, two chemical inhibitors of endocytosis, inhibits the quipazine-induced internalization of the 5-HT$_{2A}$ receptor in HEK-293 cells [71]. Finally, a dominant negative to dynamin (dynamin$_{K44A}$) also completely attenuates serotonin-mediated 5-HT$_{2A}$ receptor internalization in HEK-293 cells [57].

Although βarrestins facilitate the clathrin-mediated endocytosis of GPCRs, their involvement in the trafficking of the 5-HT$_{2A}$ receptor seems to be complex. For instance, expression of $βarr1_{319-418}$ or a similar dominant negative for βarrestin2 ($βarr2_{284-409}$) [83], has no effect on quipazine-induced internalization of the 5-HT_{2A} receptor as assessed by confocal microscopy [57]. The $βarr1_{319-418}$ mutant also does not affect serotonin-mediated internalization of the receptor, as quantified by a cell-surface biotinylation assay [57], suggesting that βarrestins are not involved in the internalization process in HEK-293 cells. However, another study from the Roth laboratory has demonstrated that the transfection of a constitutively active βarrestin1 mutant ($βarr1_{R169E}$), which binds to GPCRs regardless of the phosphorylation state of the receptor [84–86], results in the constitutive internalization of the 5-HT_{2A} receptor in HEK-293 cells, as determined by confocal microscopy. The 5-HT_{2A} receptor also co-immunoprecipitates with the $βarr1_{R169E}$ mutant in the absence of agonist [87], suggesting that the internalization of the 5-HT_{2A} receptor is sensitive to the expression of βarrestins in vitro. Although the dominant negative βarrestin mutants are thought to compete in receptor/clathrin-coated pit assembly, they may not fully inhibit the function of the endogenous βarrestins, which could explain these differential findings. Confocal and biotinylation studies with mouse embryonic fibroblasts with and without both βarrestin1 and βarrestin2 (βarr1/2-KO) indicate that serotonin-induced internalization of the 5-HT_{2A} receptor is dependent on βarrestins [88]. In contrast, these same studies demonstrate that the 5-HT_2 receptor agonist DOI is able to induce internalization in a βarrestin-independent manner.

5-HT_{2A} receptor internalization via a βarrestin2-mediated mechanism has also been observed in vivo. Several immunohistochemical studies have demonstrated that the 5-HT_{2A} receptor is found to be internalized in cortical neurons from untreated rats [62, 89], suggesting that the receptor is constitutively internalized. Moreover, βarrestins have been shown to co-localize with the 5-HT_{2A} receptor within intracellular vesicles of rat cortical neurons [56]. In primary cortical neurons cultured from postnatal day 1 mice, confocal microscopy shows that the endogenously expressed 5-HT_{2A} receptor is again found within the intracellular region of untreated neurons. Live-cell staining of an HA-tagged 5-HT_{2A} receptor that was transfected into primary cortical neurons reveals that the 5-HT_{2A} receptor is constitutively trafficked, as the fluorescently labeled receptor was internalized following antibody labeling [88]. However, in primary neurons cultured from βarr2-KO mice, the 5-HT_{2A} receptor is more prominently on the cellular membrane. Transfection of βarrestin2-YFP into the βarr2-KO neurons rescues the internalization of the receptor into endocytic vesicles. Moreover, the 5-HT_{2A} receptor colocalizes with βarrestin2-YFP in those endocytic vesicles [88].

The findings that the 5-HT_{2A} receptor can still internalize in the presence of dominant negative βarrestins or in βarr1/2-KO MEFs suggests that 5-HT_{2A} receptors can also be internalized through βarrestin-independent mechanisms. One such mechanism may involve interactions with caveolin-1, another multifunctional scaffolding protein involved in the targeting and internalization of GPCRs [90, 91]. The 5-HT_{2A} receptor co-immunoprecipitates with caveolin-1 in HEK-293 cells, C6 glioma cells and synaptic membranes prepared from rat frontal cortex [92]. Caveolin-1 also co-localizes with the 5-HT_{2A} receptor at the plasma membrane of HEK-293 cells [92].

Caveolins can facilitate clathrin-independent, but dynamin-dependent endocytosis of receptors [91], and caveolin-1 may promote 5-HT_{2A} receptor internalization, as serotonin induces the co-localization of the two proteins in intracellular vesicles of HEK-293 cells [92]. In addition to internalization, caveolins have been shown to target GPCRs to lipid microdomains of the plasma membrane, and both the 5-HT_{2A} receptor and PLC can localize to these caveolin-containing regions [93, 94], suggesting that caveolin-1 may facilitate downstream signaling events. Likewise, the over-expression of caveolin-1 increases 5-HT_{2A} receptor interactions with $G\alpha_q$ and siRNA knockdown of caveolin-1 attenuates serotonin-mediated increases in intracellular Ca^{2+} [92]. Therefore, caveolin-1 interactions with the 5-HT_{2A} receptor may also serve to scaffold the receptor with $G\alpha_q$ and/or PLC in lipid rafts, thereby facilitating agonist-mediated signaling.

It is possible that the 5-HT_{2A} receptor is internalized by both clathrin-mediated, βarrestin-facilitated endocytosis pathways, and caveolin-1-dependent mechanisms. The cholecystokinin receptor is a GPCR that also can be internalized through multiple endocytic pathways. While the majority of cholecystokinin receptors are normally internalized via βarrestin-mediated/clathrin-coated pits, nearly all of the receptors are still internalized in CHO cells through a caveolin-dependent mechanism when the clathrin-mediated pathway is inhibited [95]. Similar to the cholecystokinin receptor, the βarrestin-mediated pathway may be the preferred mechanism of 5-HT_{2A} receptor internalization. However, some agonists at the 5-HT_{2A} receptor may preferentially utilize the caveolin pathway, or may engage this alternate endocytic route only under conditions in which the main pathway is inhibited.

Another protein that impacts 5-HT_{2A} receptor trafficking is PSD-95. The 5-HT_{2A} receptor contains a PDZ-binding domain within its C-terminus, which directly interacts with the PDZ-domain of PSD-95. The two proteins also co-localize on the cell surface of HEK-293 cells and primary cortical neurons, as determined by confocal microscopy and co-immunoprecipitation studies [96, 97]. The interaction between the two proteins appears to localize the receptor to the plasma membrane, as over-expression of PSD-95 attenuates serotonin-mediated 5-HT_{2A} receptor internalization in HEK-293 cells, an effect that can be blocked by mutation of the receptor PDZ-binding domain [96]. Further, inhibition of 5-HT_{2A} receptor/PSD-95 interactions through mutation of the receptor PDZ-binding domain or the use of PSD-95-KO mice abrogates normal dendritic targeting of the receptor and sequesters receptors in the soma of primary cortical neurons [97, 98]. There appears to be an interplay between PSD-95 and βarrestin2 regulation of the 5-HT_{2A} receptor as well. In vehicle treated mice, PSD-95 is pre-associated with cortical 5-HT_{2A} receptors, as determined by co-immunoprecipitation studies. Following treatment with serotonin's metabolic precursor (5-HTP), PSD-95 disassociates from the 5-HT_{2A} receptor. In βarrestin2-KO mice, however, PSD-95 does not disassociate from the 5-HT_{2A} receptor [66]. Interestingly, overexpression of PSD-95 was shown to suppress βarrestin2 recruitment to the corticotrophin-releasing factor receptor [99], further suggesting a relationship between the two regulatory proteins. Interactions with PSD-95 also impact 5-HT_{2A} receptor signal transduction, as the over-expression of PSD-95 augments serotonin-induced IP accumulation in HEK-293 cells, while inhibiting the interaction by mutation of the receptor PDZ-binding domain reduces serotonin-mediated PI hydrolysis [96]. Moreover, DOI-induced ERK1/2 phosphorylation is attenuated in the cortex of PSD-95-KO mice [98]. Collectively, these studies suggest that PSD-95 is involved in properly targeting the receptor to the apical dendrites of cortical neurons, which appears to be integral for the activation of some 5-HT_{2A} receptor-mediated signaling cascades.

Like many GPCRs, the 5-HT_{2A} receptor can be down-regulated following exposure to agonists, both in vitro and in vivo. For instance, daily administration of LSD or DOI for 7 days significantly decreases [^{3}H]-ketanserin binding in the rat cortex [100, 101]. However, in vitro studies have shown that down-regulation can differ depending upon cell-type or agonist, as exposure to serotonin for 8 h decreases [^{125}I]-LSD binding in P11 rat pituitary tumor cells [67], while exposure to DOI for up to 24 h had no effect on [^{3}H]-ketanserin binding in NIH-3 T3 cells [68]. Chronic treatment with 5-HT_{2A} receptor antagonists, such as clozapine, can also down-regulate the receptor in the rat frontal cortex [102, 103]. Though receptor endocytosis can play a role in the down-regulation of some GPCRs by directing them towards degradation pathways [104], the mechanism of 5-HT_{2A} receptor down-regulation by both agonists and antagonists and the involvement of receptor trafficking is not defined.

Internalization of GPCRs can also lead to the de-phosphorylation and recycling of receptors for trafficking back to the plasma membrane, a process termed resensitization. The 5-HT_{2A} receptor is resensitized following agonist treatment in vitro [75]. The $\beta arr1_{319\text{-}418}$ and $dynamin_{K44A}$ dominant negatives significantly inhibit the resensitization of 5-HT_{2A} receptor-mediated PI hydrolysis after quipazine-induced desensitization in C6 glioma cells [71]. In contrast, both dominant negatives actually potentiate the resensitization of the 5-HT_{2A} receptor in HEK-293 cells [71, 87]. These studies demonstrate that requirement of receptor internalization and the involvement of βarrestins in the resensitization process again appears to be cell-type specific. Furthermore, the potentiation of 5-HT_{2A} receptor resensitization in HEK-293 cells under conditions in which internalization is blocked, suggests the existence of a mechanism in which the 5-HT_{2A} receptor can be resensitized on the cell-surface, a phenomenon that has also been shown to occur previously for the β_2-adrenergic receptor [105].

βArrestins as Facilitators of 5-HT_{2A} Receptor Signaling

As stated above, βarrestins can facilitate G protein-independent signaling by functioning as adaptor proteins, to promote the stable association of signaling proteins with GPCRs. This has also proven to be the case for the 5-HT_{2A} receptor. In MEF cells expressing the 5-HT_{2A} receptor, serotonin induces ERK1/2 phosphorylation through $G\alpha_q$/PLC-mediated and βarrestin-dependent pathways [88]. This is evident, as knocking out both βarrestins or pretreating the cells with the PLC inhibitor U73122 only partially decreases serotonin-induced ERK1/2 activation. However pretreating the βarr1/2-KO MEFs with the PLC inhibitor completely abrogates ERK1/2 activation. βarrestin2-facilitated signaling to ERK1/2 is also observed in the frontal cortex of mice treated with 5-HTP, but is reduced in βarr2-KO mice [88]. The stimulation of this βarrestin2-mediated signaling pathway is agonist dependent, as DOI does not induce βarrestin-mediated ERK1/2 phosphorylation in either the MEFs or in mouse brain, while the $G\alpha_q$/PLC-mediated signaling is preserved.

In neuronal cultures, the 5-HT_{2A} receptor agonist α-methylserotonin also induces ERK1/2 activation through a βarrestin-dependent mechanism, as siRNA mediated knockdown of either βarrestin blocked the phosphorylation of ERK1/2 [106]. Interestingly, the authors demonstrate that treatment with a cell-permeable dynamin inhibitory peptide blocks the α-methylserotonin-mediated activation of ERK1/2 [106]. This would suggest that there may be a correlation between an agonist's dependence upon βarrestins for receptor trafficking and their utilization of βarrestins in the phosphorylation of ERK1/2. In the MEF studies by Schmid et al. [88], this correlation is also apparent, wherein, serotonin mediates receptor internalization and ERK1/2 phosphorylation through βarrestin-mediated pathways, while DOI does not require βarrestins for the internalization of the 5-HT_{2A} receptor or the activation of ERK1/2. Internalization has been shown to be a prerequisite for the βarrestin-mediated activation of downstream signaling for other GPCRs, wherein the blockade of endocytosis inhibits βarrestin-mediated signaling [37, 107, 108]. Although additional studies are necessary to demonstrate causation, these data support the hypothesis that βarrestin-mediated endocytosis of the 5-HT_{2A} receptor and βarrestin-mediated signaling may be interrelated events.

βarrestin2 has also been shown to mediate 5-HT_{2A} receptor activation of the kinase Akt in primary neurons and in mouse brain, again in an agonist dependent manner [66]. In primary cortical cultures and in the mouse frontal cortex, serotonin (or its metabolic precursor 5-HTP), but neither of the *N*-methyl tryptamines, *N*-methyl serotonin and 5-methoxy-dimethyltryptamine (5-MeO-DMT), activates Akt. Serotonin is unable to stimulate Akt phosphorylation in βarr2-KO neurons and transfection of βarr2 into the knockout neurons rescues the signaling response. Co-immunoprecipitation studies from the frontal cortex of mice treated with 5-HTP reveal that βarrestin2 is serving to scaffold a signaling complex that involves Src and Akt to 5-HT_{2A} receptors. This complex does not form following treatment with 5-MeO-DMT. Furthermore, pretreatment with either the Src inhibitor, PP2 or the PI3 kinase inhibitor, LY294002, also blocks serotonin-induced Akt phosphorylation in mouse cortical neurons. Therefore, βarrestin2 serves to scaffold members of the kinase cascade to the receptor, thereby facilitating serotonin-induced activation of Akt.

βarrestin2 and the Head Twitch Response

5-HT_{2A} receptor activation in rodents is manifested as a rapid, discrete shaking of the head, termed the head twitch response. The head twitch response was first described as a method for assessing the central actions of serotonin in vivo [109]. Systemic injection of 5-HTP induces the head twitch response in rodents [66, 88, 109]. This response is presumably due to the resulting increase in brain serotonin levels rather than the direct actions of 5-HTP *per se*, as it can be significantly attenuated by pretreatment with decarboxylase inhibitors, which prevent serotonin synthesis, or potentiated by pretreatment with MAO-A inhibitors, which block the main degradation pathway [109]. Moreover, the time-course of the response closely correlates to the rise and fall of serotonin levels in the brain [109].

Extensive pharmacological studies have demonstrated that the 5-HT_{2A} receptor is the target for the head twitch response in rodents: agonists with strong affinity for the 5-HT_{2A} receptor all induce the head twitch response [110–113]; blockade of the 5-HT_{2A} receptor by selective antagonists is sufficient to prevent the agonist-induced head twitch responses [88, 114–117]; 5-HT_{2A} receptor knockout mice do not display head twitches following treatment with a range of hallucinogenic drugs, including LSD, DMT or DOI [112, 118, 119]. Moreover, several experiments have also demonstrated that the head twitch response is due to the activation of 5-HT_{2A} receptors expressed specifically in the rodent frontal cortex. Centrally expressed receptors were implicated by the fact that the systemic injection of serotonin, which is not brain penetrant, does not induce the head twitch response [109, 110], yet head twitches are induced by the direct injection of serotonin into the intracerebroventricular (i.c.v.) space [66, 120, 121]. The direct bilateral administration of the 5-HT_{2A} receptor agonists m-chloro-phenylpiperazine (m-CPP) and DOI into the rat medial prefrontal cortex activates the head twitch response [117]. Moreover, the Gingrich laboratory selectively restored 5-HT_{2A} receptor expression to cortical glutamatergic (primarily pyramidal) neurons of 5-HT_{2A} receptor-KO mice by crossing them with a second line of mice expressing cre-recombinase under the control of the *Emx1* promoter. The resulting cortical 5-HT_{2A} receptor expression was sufficient to rescue the LSD- and DOI-mediated head twitch responses in mice [112]. Collectively, these studies implicate the head twitch response as an in vivo model of selective 5-HT_{2A} receptor activation in the mouse frontal cortex.

The serotonin induced head twitch response, whether due to systemic injection of 5-HTP or intracerebroventricular injection of serotonin, is decreased in the βarr2-KO mice [66, 88]. Decreased behavioral responses to agonist in the βarr2-KO mice have been correlated to βarrestin2 serving a pro-signaling role in the pathways underlying the physiological response [122]. The decreased head twitch response to serotonin has similarly been correlated to the βarrestin2-mediated signaling to Akt that occurs through a PI3 kinase and Src dependent pathway in the mouse frontal cortex, as intracerebroventricular injection with either the Src inhibitor PP2, the PI3 kinase inhibitor LY294002 or Akt inhibitor VIII blocks serotonin-induced head twitches in vivo [66]. In parallel to the signaling profiles that showed that the

N-methyltryptamines do not activate the βarrestin2-mediated signaling cascade in the mouse cortex, the head twitch response that is induced by the *N*-methyltryptamines is not reduced in the βarr2-KO mice, nor is it sensitive to pretreatment with the Src, PI3 kinase or Akt inhibitors [66].

In stark contrast to serotonin, DOI and the *N*-methyltryptamines induce a similar or a more pronounced head twitch response in the WT and βarr2-KO mice, respectively [66, 88]. These findings are consistent with the in vivo and in vitro studies demonstrating that these agonists do not activate βarrestin-mediated signaling downstream of the 5-HT_{2A} receptor. While the mechanisms by which they activate the head twitch response remains to be determined, they presumably are composed of G protein-dependent/βarrestin-independent signaling pathways. This is supported by a study demonstrating that $G\alpha_q$-KO mice exhibit decreased DOI-induced head twitches compared to their WT littermates [123]. Furthermore, the enhanced responses to the *N*-methyltryptamines suggest that βarrestin2 may be negatively regulating the signaling that underlies the response, as other cases in which enhanced physiological responses observed in the βarr2-KO have been correlated to βarrestins playing desensitizing roles [12, 14, 124–127]. However, the direct involvement of βarrestin2 in the desensitization of 5-HT_{2A} receptor-mediated signaling that underlies the head twitch response remains to be determined.

The head twitch response profiles to serotonin and the *N*-methyltryptamines suggest that βarrestin2 both facilitates and dampens 5-HT_{2A} receptor signaling in the frontal cortex, depending upon the agonist bound to the receptor. These data indicate that the agonist not only dictates whether or not βarrestin2 is recruited to a receptor, but also determines the functional consequence of the interaction: serotonin recruits βarrestin2 to facilitate signaling, while *N*-methyltryptamine-induced recruitment of βarrestin2 may act to desensitize the 5-HT_{2A} receptor. The chemokine receptor CCR7 agonists, CCL19 and CCL21, provide in vitro precedence for the agonist-directed divergence in βarrestin2 function at a single receptor. Both agonists recruit βarrestins and stimulate ERK1/2 phosphorylation through a βarrestin2-dependent pathway, yet only CCL19 induces βarrestin-dependent receptor desensitization and internalization [128–131]. Previously, βarrestins have been implicated in both facilitating and desensitizing the same GPCR in vivo; however the regulatory role has appeared to be tissue or region specific. While βarrestin2 acts to desensitize the μ opioid receptor expressed in brain regions associated with antinociception [12, 132], data suggest that it may facilitate signaling in neurons in the gastrointestinal tract which are involved in the development of opioid-induced constipation [133]. The head twitch data indicate that βarrestins are both negatively and positively regulating a particular receptor expressed within the same neuronal population. Furthermore, serotonin and the *N*-methyltryptamines serve as in vivo examples of ligands, which specifically target distinct actions of βarrestins to either stimulate or dampen specific signaling cascades. GRKs may serve as cofactors which regulate the functional consequences of βarrestin interactions with GPCRs, as interactions with GRK2 and 3 have been shown to promote βarrestin-mediated desensitization of receptors while interactions with GRK5 and 6 facilitate βarrestin-mediated signaling cascades [131, 134, 135]. From these studies, we might infer that serotonin and the *N*-methyltryptamines may recruit specific GRKs to the 5-HT_{2A} receptor, which may impact the selective engagement of βarrestin2 to either facilitate or desensitize signal transduction. Therefore, GRK interactions with the 5-HT_{2A} receptor may represent another target through which 5-HT_{2A} receptor signaling could be modulated in vivo.

βarrestin2 and Atypical Antipsychotics

Clozapine and other antipsychotic drugs induce their antipsychotic effects largely due to their activity at 5-HT_{2A} receptors expressed in mesocortical pathways [136]. Clozapine is classically considered a potent inverse agonist at the 5-HT_{2A} receptor with respect to $G\alpha_q$ signaling [137]. This classification is complicated by the fact that chronic administration of the drug causes downregulation of the 5-HT_{2A} receptor in brain [138, 139] and the finding that it causes internalization in vitro and in vivo [57, 82, 140, 141]. The Roth laboratory has shown that 5-HT_{2A} receptor internalization induced by the 5-HT_{2A} receptor antagonist clozapine is unaffected by the $\beta arr1_{319\text{-}418}$ dominant negative in HEK-293 cells [57]. Moreover, clozapine maintains its ability to induce 5-HT_{2A} receptor internalization in βarr1/2-KO MEFs [60]. Interestingly, similar to serotonin, clozapine has also been shown to promote Akt phosphorylation in cultured neurons and in the rodent prefrontal cortex [60, 142–144]. However, unlike serotonin, clozapine induces the phosphorylation of Akt independent of βarrestin2 [60]. These results demonstrate that serotonin and clozapine use differential mechanisms to internalize the 5-HT_{2A} receptor and induce 5-HT_{2A} receptor-mediated Akt phosphorylation.

βArrestin2 also has no effect on clozapine's actions in vivo. In mice, the antipsychotic activity of drugs is modeled by their ability to inhibit dizocilpine (MK-801) or phencyclidine (PCP) hyperlocomotion [145], an effect that is due to their actions at 5-HT_{2A} receptors [146, 147]. Moreover, highly selective 5-HT_{2A} receptor antagonists induce hypolocomotion in mice, similar to that observed with clozapine [148]. Clozapine's ability to suppress MK-801 or PCP-induced hyperlocomotion is unaffected in βarr2-KO mice [60]. Interestingly, the inhibition of Akt by Akti-1/2 blocks clozapine-induced suppression of MK-801 induced hyperlocomotion, suggesting that the βarr2-independent activation of Akt may be important for the antipsychotic effect of the drug [60].

Conclusions

The cellular and animal studies presented in this chapter demonstrate that the interaction between the 5-HT_{2A} receptor and βarrestin2 is a critical point of divergence in agonist-directed 5-HT_{2A} receptor signaling. Moreover, the mechanisms underlying 5-HT_{2A} receptor regulation and signaling are cell-type dependent, which emphasizes the necessity to study the receptor in its endogenous environment. In addition to being dictated by the cellular environment, the regulation of the 5-HT_{2A} receptor is a function of the agonist bound to the receptor, with the agonist dictating the functional implications of βarrestin2 recruitment: whether it be to assist in the activation of additional signal transduction pathways by scaffolding them to the receptor or to inhibit further coupling to G proteins.

References

1. Gurevich VV, Gurevich EV, Cleghorn WM (2008) Arrestins as multi-functional signaling adaptors. Handb Exp Pharmacol 186:15–37. https://doi.org/10.1007/978-3-540-72843-6_2
2. Black JB, Premont RT, Daaka Y (2016) Feedback regulation of G protein-coupled receptor signaling by GRKs and arrestins. Semin Cell Dev Biol 50:95–104. https://doi.org/10.1016/j.semcdb.2015.12.015
3. Gurevich EV, Tesmer JJ, Mushegian A, Gurevich VV (2012) G protein-coupled receptor kinases: more than just kinases and not only for GPCRs. Pharmacol Ther 133(1):40–69. https://doi.org/10.1016/j.pharmthera.2011.08.001
4. Premont RT, Gainetdinov RR (2007) Physiological roles of G protein-coupled receptor kinases and arrestins. Annu Rev Physiol 69:511–534. https://doi.org/10.1146/annurev.physiol.69.022405.154731
5. Barak LS, Ferguson SS, Zhang J, Caron MG (1997) A beta-arrestin/green fluorescent protein biosensor for detecting G protein-coupled receptor activation. J Biol Chem 272(44):27497–27500
6. Drake MT, Violin JD, Whalen EJ, Wisler JW, Shenoy SK, Lefkowitz RJ (2008) Beta-arrestin-biased agonism at the beta2-adrenergic receptor. J Biol Chem 283(9):5669–5676
7. Groer CE, Tidgewell K, Moyer RA, Harding WW, Rothman RB, Prisinzano TE, Bohn LM (2007) An opioid agonist that does not induce micro-opioid receptor—arrestin interactions or receptor internalization. Mol Pharmacol 71(2):549–557
8. Hamdan FF, Audet M, Garneau P, Pelletier J, Bouvier M (2005) High-throughput screening of G protein-coupled receptor antagonists using a bioluminescence resonance energy transfer 1-based beta-arrestin2 recruitment assay. J Biomol Screen 10(5):463–475
9. van Der Lee MM, Bras M, van Koppen CJ, Zaman GJ (2008) beta-Arrestin recruitment assay for the identification of agonists of the sphingosine 1-phosphate receptor EDG1. J Biomol Screen 13(10):986–998. https://doi.org/10.1177/1087057108326144. [pii]: 1087057108326144
10. von Degenfeld G, Wehrman TS, Hammer MM, Blau HM (2007) A universal technology for monitoring G-protein-coupled receptor activation in vitro and noninvasively in live animals. FASEB J 21(14):3819–3826
11. Gurevich VV, Gurevich EV (2015a) Analyzing the roles of multi-functional proteins in cells: the case of arrestins and GRKs. Crit Rev Biochem Mol Biol 50(5):440–452. https://doi.org/10.3109/10409238.2015.1067185
12. Bohn LM, Lefkowitz RJ, Gainetdinov RR, Peppel K, Caron MG, Lin FT (1999) Enhanced morphine analgesia in mice lacking beta-arrestin 2. Science 286(5449):2495–2498
13. Conner DA, Mathier MA, Mortensen RM, Christe M, Vatner SF, Seidman CE, Seidman JG (1997) beta-Arrestin1 knockout mice appear normal but demonstrate altered cardiac responses to beta-adrenergic stimulation. Circ Res 81(6):1021–1026

14. Schmid CL, Bohn LM (2009) Physiological and pharmacological implications of beta-arrestin regulation. Pharmacol Ther 121(3):285–293. https://doi.org/10.1016/j.pharmthera.2008.11.005
15. Scott MG, Benmerah A, Muntaner O, Marullo S (2002) Recruitment of activated G protein-coupled receptors to pre-existing clathrin-coated pits in living cells. J Biol Chem 277(5):3552–3559. https://doi.org/10.1074/jbc.M106586200
16. Kohout TA, Lin FS, Perry SJ, Conner DA, Lefkowitz RJ (2001) beta-Arrestin 1 and 2 differentially regulate heptahelical receptor signaling and trafficking. Proc Natl Acad Sci U S A 98(4):1601–1606
17. Oakley RH, Laporte SA, Holt JA, Caron MG, Barak LS (2000) Differential affinities of visual arrestin, beta arrestin1, and beta arrestin2 for G protein-coupled receptors delineate two major classes of receptors. J Biol Chem 275(22):17201–17210. https://doi.org/10.1074/jbc.M910348199. [pii]: M910348199
18. Xiao K, McClatchy DB, Shukla AK, Zhao Y, Chen M, Shenoy SK, Yates JR 3rd, Lefkowitz RJ (2007) Functional specialization of beta-arrestin interactions revealed by proteomic analysis. Proc Natl Acad Sci U S A 104(29):12011–12016. https://doi.org/10.1073/pnas.0704849104
19. Bychkov E, Zurkovsky L, Garret MB, Ahmed MR, Gurevich EV (2012) Distinct cellular and subcellular distributions of G protein-coupled receptor kinase and arrestin isoforms in the striatum. PLoS One 7(11):e48912. https://doi.org/10.1371/journal.pone.0048912
20. Gurevich VV, Gurevich EV (2006) The structural basis of arrestin-mediated regulation of G-protein-coupled receptors. Pharmacol Ther 110(3):465–502. https://doi.org/10.1016/j.pharmthera.2005.09.008. [pii]: S0163-7258(05)00211-1
21. Lohse MJ, Andexinger S, Pitcher J, Trukawinski S, Codina J, Faure JP, Caron MG, Lefkowitz RJ (1992) Receptor-specific desensitization with purified proteins. Kinase dependence and receptor specificity of beta-arrestin and arrestin in the beta 2-adrenergic receptor and rhodopsin systems. J Biol Chem 267(12):8558–8564
22. Mundell SJ, Loudon RP, Benovic JL (1999) Characterization of G protein-coupled receptor regulation in antisense mRNA-expressing cells with reduced arrestin levels. Biochemistry 38(27):8723–8732. https://doi.org/10.1021/bi990361v. [pii]: bi990361v
23. Paing MM, Stutts AB, Kohout TA, Lefkowitz RJ, Trejo J (2002) beta-Arrestins regulate protease-activated receptor-1 desensitization but not internalization or down-regulation. J Biol Chem 277(2):1292–1300. https://doi.org/10.1074/jbc.M109160200. [pii]: M109160200
24. Attramadal H, Arriza JL, Aoki C, Dawson TM, Codina J, Kwatra MM, Snyder SH, Caron MG, Lefkowitz RJ (1992) Beta-arrestin2, a novel member of the arrestin/beta-arrestin gene family. J Biol Chem 267(25):17882–17890
25. Ahn S, Nelson CD, Garrison TR, Miller WE, Lefkowitz RJ (2003) Desensitization, internalization, and signaling functions of beta-arrestins demonstrated by RNA interference. Proc Natl Acad Sci U S A 100(4):1740–1744
26. Ferguson SS, Downey WE 3rd, Colapietro AM, Barak LS, Menard L, Caron MG (1996) Role of beta-arrestin in mediating agonist-promoted G protein-coupled receptor internalization. Science 271(5247):363–366
27. Goodman OB Jr, Krupnick JG, Santini F, Gurevich VV, Penn RB, Gagnon AW, Keen JH, Benovic JL (1996) Beta-arrestin acts as a clathrin adaptor in endocytosis of the beta2-adrenergic receptor. Nature 383(6599):447–450
28. Kim YM, Benovic JL (2002) Differential roles of arrestin-2 interaction with clathrin and adaptor protein 2 in G protein-coupled receptor trafficking. J Biol Chem 277(34):30760–30768. https://doi.org/10.1074/jbc.M204528200. [pii]: M204528200
29. Laporte SA, Oakley RH, Zhang J, Holt JA, Ferguson SS, Caron MG, Barak LS (1999) The beta2-adrenergic receptor/betaarrestin complex recruits the clathrin adaptor AP-2 during endocytosis. Proc Natl Acad Sci U S A 96(7):3712–3717
30. Gurevich VV, Gurevich EV (2015b) Arrestins: critical players in trafficking of many GPCRs. Prog Mol Biol Transl Sci 132:1–14. https://doi.org/10.1016/bs.pmbts.2015.02.010

31. Shenoy SK, Lefkowitz RJ (2003) Trafficking patterns of beta-arrestin and G protein-coupled receptors determined by the kinetics of beta-arrestin deubiquitination. J Biol Chem 278(16):14498–14506. https://doi.org/10.1074/jbc.M209626200. [pii]: M209626200
32. Oakley RH, Laporte SA, Holt JA, Barak LS, Caron MG (1999) Association of beta-arrestin with G protein-coupled receptors during clathrin-mediated endocytosis dictates the profile of receptor resensitization. J Biol Chem 274(45):32248–32257
33. Thomsen AR, Plouffe B, Cahill TJ 3rd, Shukla AK, Tarrasch JT, Dosey AM, Kahsai AW, Strachan RT, Pani B, Mahoney JP, Huang L, Berton B, Heydenreich FM, Sunahara RK, Skiniotis G, Bouvier M, Lefkowitz RJ (2016) GPCR-G protein-beta-arrestin super-complex mediates sustained G protein signaling. Cell 166(4):907–919. https://doi.org/10.1016/j.cell.2016.07.004
34. Violin JD, DiPilato LM, Yildirim N, Elston TC, Zhang J, Lefkowitz RJ (2008) beta2-adrenergic receptor signaling and desensitization elucidated by quantitative modeling of real time cAMP dynamics. J Biol Chem 283(5):2949–2961. https://doi.org/10.1074/jbc.M707009200
35. Luttrell LM, Ferguson SS, Daaka Y, Miller WE, Maudsley S, Della Rocca GJ, Lin F, Kawakatsu H, Owada K, Luttrell DK, Caron MG, Lefkowitz RJ (1999) Beta-arrestin-dependent formation of beta2 adrenergic receptor-Src protein kinase complexes. Science 283(5402):655–661
36. Cianfrocca R, Rosano L, Spinella F, Di Castro V, Natali PG, Bagnato A (2010) Beta-arrestin-1 mediates the endothelin-1-induced activation of Akt and integrin-linked kinase. Can J Physiol Pharmacol 88(8):796–801. https://doi.org/10.1139/y10-052. [pii]: y10-052
37. Daaka Y, Luttrell LM, Ahn S, Della Rocca GJ, Ferguson SS, Caron MG, Lefkowitz RJ (1998) Essential role for G protein-coupled receptor endocytosis in the activation of mitogen-activated protein kinase. J Biol Chem 273(2):685–688
38. Sun Y, Cheng Z, Ma L, Pei G (2002) Beta-arrestin2 is critically involved in CXCR4-mediated chemotaxis, and this is mediated by its enhancement of p38 MAPK activation. J Biol Chem 277(51):49212–49219. https://doi.org/10.1074/jbc.M207294200. [pii]: M207294200
39. McDonald PH, Chow CW, Miller WE, Laporte SA, Field ME, Lin FT, Davis RJ, Lefkowitz RJ (2000) Beta-arrestin 2: a receptor-regulated MAPK scaffold for the activation of JNK3. Science 290(5496):1574–1577
40. Miller WE, McDonald PH, Cai SF, Field ME, Davis RJ, Lefkowitz RJ (2001) Identification of a motif in the carboxyl terminus of beta -arrestin2 responsible for activation of JNK3. J Biol Chem 276(30):27770–27777. https://doi.org/10.1074/jbc.M102264200. [pii]: M102264200
41. Ahn S, Shenoy SK, Wei H, Lefkowitz RJ (2004) Differential kinetic and spatial patterns of beta-arrestin and G protein-mediated ERK activation by the angiotensin II receptor. J Biol Chem 279(34):35518–35525. https://doi.org/10.1074/jbc.M405878200. [pii]: M405878200
42. Beaulieu JM, Sotnikova TD, Marion S, Lefkowitz RJ, Gainetdinov RR, Caron MG (2005) An Akt/beta-arrestin 2/PP2A signaling complex mediates dopaminergic neurotransmission and behavior. Cell 122(2):261–273. https://doi.org/10.1016/j.cell.2005.05.012
43. Luan B, Zhao J, Wu H, Duan B, Shu G, Wang X, Li D, Jia W, Kang J, Pei G (2009) Deficiency of a beta-arrestin-2 signal complex contributes to insulin resistance. Nature 457(7233):1146–1149. https://doi.org/10.1038/nature07617. [pii]: nature07617
44. Urs NM, Daigle TL, Caron MG (2011) A dopamine D1 receptor-dependent beta-arrestin signaling complex potentially regulates morphine-induced psychomotor activation but not reward in mice. Neuropsychopharmacology 36(3):551–558. https://doi.org/10.1038/npp.2010.186. [pii]: npp2010186
45. Labasque M, Reiter E, Becamel C, Bockaert J, Marin P (2008) Physical interaction of calmodulin with the 5-hydroxytryptamine2C receptor C-terminus is essential for G protein-independent, arrestin-dependent receptor signaling. Mol Biol Cell 19(11):4640–4650. https://doi.org/10.1091/mbc.E08-04-0422
46. Azzi M, Charest PG, Angers S, Rousseau G, Kohout T, Bouvier M, Pineyro G (2003) Beta-arrestin-mediated activation of MAPK by inverse agonists reveals distinct active conformations for G protein-coupled receptors. Proc Natl Acad Sci U S A 100(20):11406–11411. https://doi.org/10.1073/pnas.1936664100. [pii]: 1936664100

47. Holloway AC, Qian H, Pipolo L, Ziogas J, Miura S, Karnik S, Southwell BR, Lew MJ, Thomas WG (2002) Side-chain substitutions within angiotensin II reveal different requirements for signaling, internalization, and phosphorylation of type 1A angiotensin receptors. Mol Pharmacol 61(4):768–777
48. Wei H, Ahn S, Shenoy SK, Karnik SS, Hunyady L, Luttrell LM, Lefkowitz RJ (2003) Independent beta-arrestin 2 and G protein-mediated pathways for angiotensin II activation of extracellular signal-regulated kinases 1 and 2. Proc Natl Acad Sci U S A 100(19):10782–10787. https://doi.org/10.1073/pnas.1834556100. [pii]: 1834556100
49. Kenakin T (2011) Functional selectivity and biased receptor signaling. J Pharmacol Exp Ther 336(2):296–302. https://doi.org/10.1124/jpet.110.173948
50. Shukla AK, Singh G, Ghosh E (2014) Emerging structural insights into biased GPCR signaling. Trends Biochem Sci 39(12):594–602. https://doi.org/10.1016/j.tibs.2014.10.001
51. Kenakin T (2005) New concepts in drug discovery: collateral efficacy and permissive antagonism. Nat Rev Drug Discov 4(11):919–927. https://doi.org/10.1038/nrd1875. [pii]: nrd1875
52. Kenakin T (2007) Collateral efficacy in drug discovery: taking advantage of the good (allosteric) nature of 7TM receptors. Trends Pharmacol Sci 28(8):407–415. https://doi.org/10.1016/j.tips.2007.06.009. [pii]: S0165-6147(07)00155-1
53. Rajagopal S, Rajagopal K, Lefkowitz RJ (2010) Teaching old receptors new tricks: biasing seven-transmembrane receptors. Nat Rev Drug Discov 9(5):373–386. https://doi.org/10.1038/nrd3024
54. Urban JD, Clarke WP, von Zastrow M, Nichols DE, Kobilka B, Weinstein H, Javitch JA, Roth BL, Christopoulos A, Sexton PM, Miller KJ, Spedding M, Mailman RB (2007) Functional selectivity and classical concepts of quantitative pharmacology. J Pharmacol Exp Ther 320(1):1–13. https://doi.org/10.1124/jpet.106.104463
55. Bohn LM (2009) Selectivity for G protein or arrestin-mediated signaling. In: Neve K (ed) Functional selectivity of G protein-coupled receptor ligands, 1st edn. Humana Press, Totowa, pp 71–85
56. Gelber EI, Kroeze WK, Willins DL, Gray JA, Sinar CA, Hyde EG, Gurevich V, Benovic J, Roth BL (1999) Structure and function of the third intracellular loop of the 5-hydroxytryptamine2A receptor: the third intracellular loop is alpha-helical and binds purified arrestins. J Neurochem 72(5):2206–2214
57. Bhatnagar A, Willins DL, Gray JA, Woods J, Benovic JL, Roth BL (2001) The dynamin-dependent, arrestin-independent internalization of 5-hydroxytryptamine 2A (5-HT2A) serotonin receptors reveals differential sorting of arrestins and 5-HT2A receptors during endocytosis. J Biol Chem 276(11):8269–8277. https://doi.org/10.1074/jbc.M006968200
58. Bhattacharya A, Sankar S, Panicker MM (2010) Differences in the C-terminus contribute to variations in trafficking between rat and human 5-HT(2A) receptor isoforms: identification of a primate-specific tripeptide ASK motif that confers GRK-2 and beta arrestin-2 interactions. J Neurochem 112(3):723–732. https://doi.org/10.1111/j.1471-4159.2009.06493.x. [pii]: JNC6493
59. Bohn LM, Schmid CL (2010) Serotonin receptor signaling and regulation via beta-arrestins. Crit Rev Biochem Mol Biol 45(6):555–566. https://doi.org/10.3109/10409238.2010.516741
60. Schmid CL, Streicher JM, Meltzer HY, Bohn LM (2014) Clozapine acts as an agonist at serotonin 2A receptors to counter MK-801-induced behaviors through a betaarrestin2-independent activation of Akt. Neuropsychopharmacology 39(8):1902–1913. https://doi.org/10.1038/npp.2014.38
61. Blough BE, Landavazo A, Decker AM, Partilla JS, Baumann MH, Rothman RB (2014) Interaction of psychoactive tryptamines with biogenic amine transporters and serotonin receptor subtypes. Psychopharmacology 231(21):4135–4144. https://doi.org/10.1007/s00213-014-3557-7

62. Cornea-Hebert V, Watkins KC, Roth BL, Kroeze WK, Gaudreau P, Leclerc N, Descarries L (2002) Similar ultrastructural distribution of the 5-HT(2A) serotonin receptor and microtubule-associated protein MAP1A in cortical dendrites of adult rat. Neuroscience 113(1):23–35
63. Gurevich EV, Benovic JL, Gurevich VV (2002) Arrestin2 and arrestin3 are differentially expressed in the rat brain during postnatal development. Neuroscience 109(3):421–436. [pii]: S0306452201005115
64. Miner LA, Backstrom JR, Sanders-Bush E, Sesack SR (2003) Ultrastructural localization of serotonin2A receptors in the middle layers of the rat prelimbic prefrontal cortex. Neuroscience 116(1):107–117
65. Willins DL, Deutch AY, Roth BL (1997) Serotonin 5-HT2A receptors are expressed on pyramidal cells and interneurons in the rat cortex. Synapse 27(1):79–82. https://doi.org/10.1002/(SICI)1098-2396(199709)27:1<79::AID-SYN8>3.0.CO;2-A
66. Schmid CL, Bohn LM (2010) Serotonin, but not N-methyltryptamines, activates the serotonin 2A receptor via a ss-arrestin2/Src/Akt signaling complex in vivo. J Neurosci 30(40):13513–13524. https://doi.org/10.1523/JNEUROSCI.1665-10.2010
67. Ivins KJ, Molinoff PB (1991) Desensitization and down-regulation of 5-HT2 receptors in P11 cells. J Pharmacol Exp Ther 259(1):423–429
68. Roth BL, Palvimaki EP, Berry S, Khan N, Sachs N, Uluer A, Choudhary MS (1995) 5-Hydroxytryptamine2A (5-HT2A) receptor desensitization can occur without down-regulation. J Pharmacol Exp Ther 275(3):1638–1646
69. Gray JA, Compton-Toth BA, Roth BL (2003b) Identification of two serine residues essential for agonist-induced 5-HT2A receptor desensitization. Biochemistry 42(36):10853–10862. https://doi.org/10.1021/bi035061z
70. Karaki S, Becamel C, Murat S, Mannoury la Cour C, Millan MJ, Prezeau L, Bockaert J, Marin P, Vandermoere F (2014) Quantitative phosphoproteomics unravels biased phosphorylation of serotonin 2A receptor at Ser280 by hallucinogenic versus nonhallucinogenic agonists. Mol Cell Proteomics 13:1273–1285
71. Gray JA, Sheffler DJ, Bhatnagar A, Woods JA, Hufeisen SJ, Benovic JL, Roth BL (2001) Cell-type specific effects of endocytosis inhibitors on 5-hydroxytryptamine(2A) receptor desensitization and resensitization reveal an arrestin-, GRK2-, and GRK5-independent mode of regulation in human embryonic kidney 293 cells. Mol Pharmacol 60(5):1020–1030
72. Hanley NR, Hensler JG (2002) Mechanisms of ligand-induced desensitization of the 5-hydroxytryptamine(2A) receptor. J Pharmacol Exp Ther 300(2):468–477
73. Krupnick JG, Santini F, Gagnon AW, Keen JH, Benovic JL (1997) Modulation of the arrestin-clathrin interaction in cells. Characterization of beta-arrestin dominant-negative mutants. J Biol Chem 272(51):32507–32512
74. Kong G, Penn R, Benovic JL (1994) A beta-adrenergic receptor kinase dominant negative mutant attenuates desensitization of the beta 2-adrenergic receptor. J Biol Chem 269(18):13084–13087
75. Berg KA, Stout BD, Maayani S, Clarke WP (2001) Differences in rapid desensitization of 5-hydroxytryptamine2A and 5-hydroxytryptamine2C receptor-mediated phospholipase C activation. J Pharmacol Exp Ther 299(2):593–602
76. Franklin JM, Carrasco GA (2015) Cocaine potentiates multiple 5-HT2A receptor signaling pathways and is associated with decreased phosphorylation of 5-HT2A receptors in vivo. J Mol Neurosci 55(3):770–777. https://doi.org/10.1007/s12031-014-0419-z
77. Vouret-Craviari V, Auberger P, Pouyssegur J, Van Obberghen-Schilling E (1995) Distinct mechanisms regulate 5-HT2 and thrombin receptor desensitization. J Biol Chem 270(9):4813–4821
78. Sheffler DJ, Kroeze WK, Garcia BG, Deutch AY, Hufeisen SJ, Leahy P, Bruning JC, Roth BL (2006) p90 ribosomal S6 kinase 2 exerts a tonic brake on G protein-coupled receptor signaling. Proc Natl Acad Sci U S A 103(12):4717–4722. https://doi.org/10.1073/pnas.0600585103. [pii]: 0600585103

79. Strachan RT, Sheffler DJ, Willard B, Kinter M, Kiselar JG, Roth BL (2009) Ribosomal S6 kinase 2 directly phosphorylates the 5-hydroxytryptamine 2A (5-HT2A) serotonin receptor, thereby modulating 5-HT2A signaling. J Biol Chem 284(9):5557–5573. https://doi.org/10.1074/jbc.M805705200. [pii]: M805705200
80. Strachan RT, Sciaky N, Cronan MR, Kroeze WK, Roth BL (2010) Genetic deletion of p90 ribosomal S6 kinase 2 alters patterns of 5-hydroxytryptamine 2A serotonin receptor functional selectivity. Mol Pharmacol 77(3):327–338. https://doi.org/10.1124/mol.109.061440. [pii]: mol.109.061440
81. Berry SA, Shah MC, Khan N, Roth BL (1996) Rapid agonist-induced internalization of the 5-hydroxytryptamine2A receptor occurs via the endosome pathway in vitro. Mol Pharmacol 50(2):306–313
82. Willins DL, Alsayegh L, Berry SA, Backstrom JR, Sanders-Bush E, Friedman L, Khan N, Roth BL (1998) Serotonergic antagonist effects on trafficking of serotonin 5-HT2A receptors in vitro and in vivo. Ann N Y Acad Sci 861:121–127
83. Orsini MJ, Benovic JL (1998) Characterization of dominant negative arrestins that inhibit beta2-adrenergic receptor internalization by distinct mechanisms. J Biol Chem 273(51):34616–34622
84. Gray-Keller MP, Detwiler PB, Benovic JL, Gurevich VV (1997) Arrestin with a single amino acid substitution quenches light-activated rhodopsin in a phosphorylation-independent fashion. Biochemistry 36(23):7058–7063. https://doi.org/10.1021/bi963110k. [pii]: bi963110k
85. Gurevich VV, Dion SB, Onorato JJ, Ptasienski J, Kim CM, Sterne-Marr R, Hosey MM, Benovic JL (1995) Arrestin interactions with G protein-coupled receptors. Direct binding studies of wild type and mutant arrestins with rhodopsin, beta 2-adrenergic, and m2 muscarinic cholinergic receptors. J Biol Chem 270(2):720–731
86. Kovoor A, Celver J, Abdryashitov RI, Chavkin C, Gurevich VV (1999) Targeted construction of phosphorylation-independent beta-arrestin mutants with constitutive activity in cells. J Biol Chem 274(11):6831–6834
87. Gray JA, Bhatnagar A, Gurevich VV, Roth BL (2003a) The interaction of a constitutively active arrestin with the arrestin-insensitive 5-HT(2A) receptor induces agonist-independent internalization. Mol Pharmacol 63(5):961–972
88. Schmid CL, Raehal KM, Bohn LM (2008) Agonist-directed signaling of the serotonin 2A receptor depends on beta-arrestin-2 interactions in vivo. Proc Natl Acad Sci U S A 105(3):1079–1084. https://doi.org/10.1073/pnas.0708862105
89. Cornea-Hebert V, Riad M, Wu C, Singh SK, Descarries L (1999) Cellular and subcellular distribution of the serotonin 5-HT2A receptor in the central nervous system of adult rat. J Comp Neurol 409(2):187–209. https://doi.org/10.1002/(SICI)1096-9861(19990628)409:2<187::AID-CNE2>3.0.CO;2-P
90. Cohen AW, Hnasko R, Schubert W, Lisanti MP (2004) Role of caveolae and caveolins in health and disease. Physiol Rev 84(4):1341–1379. https://doi.org/10.1152/physrev.00046.2003. [pii]: 84/4/1341
91. Le Roy C, Wrana JL (2005) Clathrin- and non-clathrin-mediated endocytic regulation of cell signalling. Nat Rev Mol Cell Biol 6(2):112–126. https://doi.org/10.1038/nrm1571. [pii]: nrm1571
92. Bhatnagar A, Sheffler DJ, Kroeze WK, Compton-Toth B, Roth BL (2004) Caveolin-1 interacts with 5-HT2A serotonin receptors and profoundly modulates the signaling of selected Galphaq-coupled protein receptors. J Biol Chem 279(33):34614–34623. https://doi.org/10.1074/jbc.M404673200. [pii]: M404673200
93. Dreja K, Voldstedlund M, Vinten J, Tranum-Jensen J, Hellstrand P, Sward K (2002) Cholesterol depletion disrupts caveolae and differentially impairs agonist-induced arterial contraction. Arterioscler Thromb Vasc Biol 22(8):1267–1272
94. Weerth SH, Holtzclaw LA, Russell JT (2007) Signaling proteins in raft-like microdomains are essential for Ca2+ wave propagation in glial cells. Cell Calcium 41(2):155–167. https://doi.org/10.1016/j.ceca.2006.06.006. [pii]: S0143-4160(06)00124-2

95. Roettger BF, Rentsch RU, Pinon D, Holicky E, Hadac E, Larkin JM, Miller LJ (1995) Dual pathways of internalization of the cholecystokinin receptor. J Cell Biol 128(6):1029–1041
96. Xia Z, Gray JA, Compton-Toth BA, Roth BL (2003a) A direct interaction of PSD-95 with 5-HT2A serotonin receptors regulates receptor trafficking and signal transduction. J Biol Chem 278(24):21901–21908. https://doi.org/10.1074/jbc.M301905200
97. Xia Z, Hufeisen SJ, Gray JA, Roth BL (2003b) The PDZ-binding domain is essential for the dendritic targeting of 5-HT2A serotonin receptors in cortical pyramidal neurons in vitro. Neuroscience 122(4):907–920
98. Abbas AI, Yadav PN, Yao WD, Arbuckle MI, Grant SG, Caron MG, Roth BL (2009) PSD-95 is essential for hallucinogen and atypical antipsychotic drug actions at serotonin receptors. J Neurosci 29(22):7124–7136. https://doi.org/10.1523/JNEUROSCI.1090-09.2009
99. Dunn HA, Chahal HS, Caetano FA, Holmes KD, Yuan GY, Parikh R, Heit B, Ferguson SS (2016) PSD-95 regulates CRFR1 localization, trafficking and beta-arrestin2 recruitment. Cell Signal 28(5):531–540. https://doi.org/10.1016/j.cellsig.2016.02.013
100. Buckholtz NS, Freedman DX, Middaugh LD (1985) Daily LSD administration selectively decreases serotonin2 receptor binding in rat brain. Eur J Pharmacol 109(3):421–425
101. Buckholtz NS, Zhou DF, Freedman DX (1988) Serotonin2 agonist administration down-regulates rat brain serotonin2 receptors. Life Sci 42(24):2439–2445
102. Gray JA, Roth BL (2001) Paradoxical trafficking and regulation of 5-HT(2A) receptors by agonists and antagonists. Brain Res Bull 56(5):441–451
103. Reynolds GP, Garrett NJ, Rupniak N, Jenner P, Marsden CD (1983) Chronic clozapine treatment of rats down-regulates cortical 5-HT2 receptors. Eur J Pharmacol 89(3–4):325–326
104. Gagnon AW, Kallal L, Benovic JL (1998) Role of clathrin-mediated endocytosis in agonist-induced down-regulation of the beta2-adrenergic receptor. J Biol Chem 273(12):6976–6981
105. Jockers R, Angers S, Da Silva A, Benaroch P, Strosberg AD, Bouvier M, Marullo S (1999) Beta(2)-adrenergic receptor down-regulation. Evidence for a pathway that does not require endocytosis. J Biol Chem 274(41):28900–28908
106. Yuen EY, Jiang Q, Chen P, Feng J, Yan Z (2008) Activation of 5-HT2A/C receptors counteracts 5-HT1A regulation of n-methyl-D-aspartate receptor channels in pyramidal neurons of prefrontal cortex. J Biol Chem 283(25):17194–17204. https://doi.org/10.1074/jbc.M801713200. [pii]: M801713200
107. Lin FT, Daaka Y, Lefkowitz RJ (1998) beta-arrestins regulate mitogenic signaling and clathrin-mediated endocytosis of the insulin-like growth factor I receptor. J Biol Chem 273(48):31640–31643
108. Luttrell LM, Daaka Y, Della Rocca GJ, Lefkowitz RJ (1997) G protein-coupled receptors mediate two functionally distinct pathways of tyrosine phosphorylation in rat 1a fibroblasts. Shc phosphorylation and receptor endocytosis correlate with activation of Erk kinases. J Biol Chem 272(50):31648–31656
109. Corne SJ, Pickering RW, Warner BT (1963) A method for assessing the effects of drugs on the central actions of 5-hydroxytryptamine. Br J Pharmacol Chemother 20:106–120
110. Corne SJ, Pickering RW (1967) A possible correlation between drug-induced hallucinations in man and a behavioural response in mice. Psychopharmacologia 11(1):65–78
111. Dursun SM, Handley SL (1996) Similarities in the pharmacology of spontaneous and DOI-induced head-shakes suggest 5HT2A receptors are active under physiological conditions. Psychopharmacology 128(2):198–205
112. Gonzalez-Maeso J, Weisstaub NV, Zhou M, Chan P, Ivic L, Ang R, Lira A, Bradley-Moore M, Ge Y, Zhou Q, Sealfon SC, Gingrich JA (2007) Hallucinogens recruit specific cortical 5-HT(2A) receptor-mediated signaling pathways to affect behavior. Neuron 53(3):439–452. https://doi.org/10.1016/j.neuron.2007.01.008
113. Malick JB, Doren E, Barnett A (1977) Quipazine-induced head-twitch in mice. Pharmacol Biochem Behav 6(3):325–329

114. Darmani NA, Martin BR, Glennon RA (1992) Behavioral evidence for differential adaptation of the serotonergic system after acute and chronic treatment with (+/−)-1-(2,5-dimethoxy-4-iodophenyl)-2-aminopropane (DOI) or ketanserin. J Pharmacol Exp Ther 262(2):692–698
115. Fantegrossi WE, Harrington AW, Eckler JR, Arshad S, Rabin RA, Winter JC, Coop A, Rice KC, Woods JH (2005) Hallucinogen-like actions of 2,5-dimethoxy-4-(n)-propylthiophenethylamine (2C-T-7) in mice and rats. Psychopharmacology 181(3):496–503. https://doi.org/10.1007/s00213-005-0009-4
116. Schreiber R, Brocco M, Millan MJ (1994) Blockade of the discriminative stimulus effects of DOI by MDL 100,907 and the 'atypical' antipsychotics, clozapine and risperidone. Eur J Pharmacol 264(1):99–102. [pii]: 0014-2999(94)90643-2
117. Willins DL, Meltzer HY (1997) Direct injection of 5-HT2A receptor agonists into the medial prefrontal cortex produces a head-twitch response in rats. J Pharmacol Exp Ther 282(2):699–706
118. Gonzalez-Maeso J, Yuen T, Ebersole BJ, Wurmbach E, Lira A, Zhou M, Weisstaub N, Hen R, Gingrich JA, Sealfon SC (2003) Transcriptome fingerprints distinguish hallucinogenic and nonhallucinogenic 5-hydroxytryptamine 2A receptor agonist effects in mouse somatosensory cortex. J Neurosci 23(26):8836–8843. [pii]: 23/26/8836
119. Keiser MJ, Setola V, Irwin JJ, Laggner C, Abbas AI, Hufeisen SJ, Jensen NH, Kuijer MB, Matos RC, Tran TB, Whaley R, Glennon RA, Hert J, Thomas KL, Edwards DD, Shoichet BK, Roth BL (2009) Predicting new molecular targets for known drugs. Nature 462(7270):175–181. https://doi.org/10.1038/nature08506. [pii]: nature08506
120. Nakagawasai O, Tadano T, Arai Y, Hozumi S, Oba A, Tan-No K, Yasuhara H, Kisara K, Oreland L (2003) Enhancement of 5-hydroxytryptamine-induced head-twitch response after olfactory bulbectomy. Neuroscience 117(4):1017–1023. [pii]: S0306452202007881
121. Orikasa S, Kisara K (1982) A possible mechanism of the tyramine-induced head-twitch response. Eur J Pharmacol 80(2–3):163–169. [pii]: 0014-2999(82)90050-4
122. Beaulieu JM, Gainetdinov RR (2011) The physiology, signaling, and pharmacology of dopamine receptors. Pharmacol Rev 63(1):182–217. https://doi.org/10.1124/pr.110.002642
123. Garcia EE, Smith RL, Sanders-Bush E (2007) Role of G(q) protein in behavioral effects of the hallucinogenic drug 1-(2,5-dimethoxy-4-iodophenyl)-2-aminopropane. Neuropharmacology 52(8):1671–1677. https://doi.org/10.1016/j.neuropharm.2007.03.013. [pii]: S0028-3908(07)00093-7
124. Deshpande DA, Theriot BS, Penn RB, Walker JK (2008) Beta-arrestins specifically constrain beta2-adrenergic receptor signaling and function in airway smooth muscle. FASEB J 22(7):2134–2141. https://doi.org/10.1096/fj.07-102459. [pii]: fj.07-102459
125. Ferrari SL, Pierroz DD, Glatt V, Goddard DS, Bianchi EN, Lin FT, Manen D, Bouxsein ML (2005) Bone response to intermittent parathyroid hormone is altered in mice null for {beta}-Arrestin2. Endocrinology 146(4):1854–1862. https://doi.org/10.1210/en.2004-1282. [pii]: en.2004-1282
126. Su Y, Raghuwanshi SK, Yu Y, Nanney LB, Richardson RM, Richmond A (2005) Altered CXCR2 signaling in beta-arrestin-2-deficient mouse models. J Immunol 175(8):5396–5402. [pii]: 175/8/5396
127. Wang Y, Tang Y, Teng L, Wu Y, Zhao X, Pei G (2006) Association of beta-arrestin and TRAF6 negatively regulates toll-like receptor-interleukin 1 receptor signaling. Nat Immunol 7(2):139–147. https://doi.org/10.1038/ni1294. [pii]: ni1294
128. Byers MA, Calloway PA, Shannon L, Cunningham HD, Smith S, Li F, Fassold BC, Vines CM (2008) Arrestin 3 mediates endocytosis of CCR7 following ligation of CCL19 but not CCL21. J Immunol 181(7):4723–4732. [pii]: 181/7/4723
129. Kohout TA, Nicholas SL, Perry SJ, Reinhart G, Junger S, Struthers RS (2004) Differential desensitization, receptor phosphorylation, beta-arrestin recruitment, and ERK1/2 activation by the two endogenous ligands for the CC chemokine receptor 7. J Biol Chem 279(22):23214–23222. https://doi.org/10.1074/jbc.M402125200. [pii]: M402125200

130. Yoshida R, Nagira M, Kitaura M, Imagawa N, Imai T, Yoshie O (1998) Secondary lymphoid-tissue chemokine is a functional ligand for the CC chemokine receptor CCR7. J Biol Chem 273(12):7118–7122
131. Zidar DA, Violin JD, Whalen EJ, Lefkowitz RJ (2009) Selective engagement of G protein coupled receptor kinases (GRKs) encodes distinct functions of biased ligands. Proc Natl Acad Sci U S A 106(24):9649–9654. https://doi.org/10.1073/pnas.0904361106. [pii]: 0904361106
132. Bohn LM, Gainetdinov RR, Lin FT, Lefkowitz RJ, Caron MG (2000) Mu-opioid receptor desensitization by beta-arrestin-2 determines morphine tolerance but not dependence. Nature 408(6813):720–723
133. Raehal KM, Walker JK, Bohn LM (2005) Morphine side effects in beta-arrestin 2 knockout mice. J Pharmacol Exp Ther 314(3):1195–1201. https://doi.org/10.1124/jpet.105.087254. [pii]: jpet.105.087254
134. Kim J, Ahn S, Ren XR, Whalen EJ, Reiter E, Wei H, Lefkowitz RJ (2005) Functional antagonism of different G protein-coupled receptor kinases for beta-arrestin-mediated angiotensin II receptor signaling. Proc Natl Acad Sci U S A 102(5):1442–1447. https://doi.org/10.1073/pnas.0409532102. [pii]: 0409532102
135. Ren XR, Reiter E, Ahn S, Kim J, Chen W, Lefkowitz RJ (2005) Different G protein-coupled receptor kinases govern G protein and beta-arrestin-mediated signaling of V2 vasopressin receptor. Proc Natl Acad Sci U S A 102(5):1448–1453. https://doi.org/10.1073/pnas.0409534102. [pii]: 0409534102
136. Meltzer HY, Massey BW (2011) The role of serotonin receptors in the action of atypical antipsychotic drugs. Curr Opin Pharmacol 11(1):59–67. https://doi.org/10.1016/j.coph.2011.02.007
137. Weiner DM, Burstein ES, Nash N, Croston GE, Currier EA, Vanover KE, Harvey SC, Donohue E, Hansen HC, Andersson CM, Spalding TA, Gibson DF, Krebs-Thomson K, Powell SB, Geyer MA, Hacksell U, Brann MR (2001) 5-hydroxytryptamine2A receptor inverse agonists as antipsychotics. J Pharmacol Exp Ther 299(1):268–276
138. Moreno JL, Holloway T, Umali A, Rayannavar V, Sealfon SC, Gonzalez-Maeso J (2013) Persistent effects of chronic clozapine on the cellular and behavioral responses to LSD in mice. Psychopharmacology 225(1):217–226. https://doi.org/10.1007/s00213-012-2809-7
139. Yadav PN, Kroeze WK, Farrell MS, Roth BL (2011) Antagonist functional selectivity: 5-HT2A serotonin receptor antagonists differentially regulate 5-HT2A receptor protein level in vivo. J Pharmacol Exp Ther 339(1):99–105. https://doi.org/10.1124/jpet.111.183780
140. Raote I, Bhattacharyya S, Panicker MM (2013) Functional selectivity in serotonin receptor 2A (5-HT2A) endocytosis, recycling, and phosphorylation. Mol Pharmacol 83(1):42–50. https://doi.org/10.1124/mol.112.078626
141. Willins DL, Berry SA, Alsayegh L, Backstrom JR, Sanders-Bush E, Friedman L, Roth BL (1999) Clozapine and other 5-hydroxytryptamine-2A receptor antagonists alter the subcellular distribution of 5-hydroxytryptamine-2A receptors in vitro and in vivo. Neuroscience 91(2):599–606
142. Dwyer DS, Donohoe D (2003) Induction of hyperglycemia in mice with atypical antipsychotic drugs that inhibit glucose uptake. Pharmacol Biochem Behav 75(2):255–260
143. Lu XH, Bradley RJ, Dwyer DS (2004) Olanzapine produces trophic effects in vitro and stimulates phosphorylation of Akt/PKB, ERK1/2, and the mitogen-activated protein kinase p38. Brain Res 1011(1):58–68. https://doi.org/10.1016/j.brainres.2004.03.018
144. Sutton LP, Rushlow WJ (2011) The effects of neuropsychiatric drugs on glycogen synthase kinase-3 signaling. Neuroscience 199:116–124. https://doi.org/10.1016/j.neuroscience.2011.09.056
145. Jones CA, Watson DJ, Fone KC (2011) Animal models of schizophrenia. Br J Pharmacol 164(4):1162–1194. https://doi.org/10.1111/j.1476-5381.2011.01386.x
146. Fribourg M, Moreno JL, Holloway T, Provasi D, Baki L, Mahajan R, Park G, Adney SK, Hatcher C, Eltit JM, Ruta JD, Albizu L, Li Z, Umali A, Shim J, Fabiato A, MacKerell AD Jr, Brezina V, Sealfon SC, Filizola M, Gonzalez-Maeso J, Logothetis DE (2011) Decoding

the signaling of a GPCR heteromeric complex reveals a unifying mechanism of action of antipsychotic drugs. Cell 147(5):1011–1023. https://doi.org/10.1016/j.cell.2011.09.055

147. McOmish CE, Lira A, Hanks JB, Gingrich JA (2012) Clozapine-induced locomotor suppression is mediated by 5-HT(2A) receptors in the forebrain. Neuropsychopharmacology 37(13):2747–2755. https://doi.org/10.1038/npp.2012.139
148. Williams AA, Ingram WM, Levine S, Resnik J, Kamel CM, Lish JR, Elizalde DI, Janowski SA, Shoker J, Kozlenkov A, Gonzalez-Maeso J, Gallitano AL (2012) Reduced levels of serotonin 2A receptors underlie resistance of Egr3-deficient mice to locomotor suppression by clozapine. Neuropsychopharmacology 37(10):2285–2298. https://doi.org/10.1038/npp.2012.81

5-HT_{2A} Receptor Heterodimerization

Sylwia Łukasiewicz, Ewa Błasiak, and Marta Dziedzicka-Wasylewska

Abstract Interaction of serotonin 5-HT_{2A} receptor with other G protein-coupled receptors (GPCRs) have been shown at the behavioral and/or electrophysiological level. In the present chapter evidence for direct physical interactions of this receptor with various GPCRs have been described. The most interesting in the context of antipsychotic drug action mechanism is the interaction of the serotonin 5-HT_{2A} receptor with dopamine D_2 receptor, which has been shown both in vitro as well as in the native brain tissue. On the other hand, new understanding of hallucinogenic drugs has been proposed by providing data which demonstrate the formation of heterocomplexes by the 5-HT_{2A} receptor with the metabotropic glutamatergic receptor $mGluR_2$. Methodology used in GPCRs heterodimerization studies has evolved, from radioligand binding, receptor crosslinking, receptor complementation, or co-immunoprecipitation approach to biophysical techniques based on resonance energy transfer—each having their pros and cons, however their use still provides new exciting data concerning the complexity of GPCRs physical interactions, which broaden basal knowledge as well as offer new targets for pharmacological intervention.

Keywords Central serotonin 5-HT_{2A} receptor • G protein-coupled receptors (GPCRs) • Heterodimerization • Structural aspects • Resonance energy transfer • Psychotropic drugs

S. Łukasiewicz • E. Błasiak
Faculty of Biochemistry, Biophysics and Biotechnology, Jagiellonian University, Kraków, Poland

M. Dziedzicka-Wasylewska (✉)
Faculty of Biochemistry, Biophysics and Biotechnology, Jagiellonian University, Kraków, Poland

Institute of Pharmacology, Polish Academy of Sciences, Kraków, Poland
e-mail: marta.dziedzicka-wasylewska@uj.edu.pl

B.P. Guiard, G. Di Giovanni (eds.), *5-HT_{2A} Receptors in the Central Nervous System*, The Receptors 32, https://doi.org/10.1007/978-3-319-70474-6_3

Pharmacological Aspects of 5-HT_{2A} Receptor Heteromerization

Serotonin 5-HT_{2A} receptors have been shown to interact with various G protein coupled receptors (GPCRs) at the behavioral and/or electrophysiological level. Moreover, direct physical interactions of this receptor with other GPCRs have been also shown. This is in line with the widely accepted concept of GPCRs homo- and hetero-dimerization. Although many studies have provided data confirming receptor-receptor interactions, there is no consensus as to the exact oligomer size. Recently, Herrick-Davis and co-workers have addressed this issue and, using fluorescence correlation spectroscopy, have shown that GPCRs, including serotonin 5-HT_{2A} (but also adrenergic $alpha_{1b}$ and $beta_2$, muscarinic M_1 and M_2, and dopamine D_1 receptors) naturally exist as homodimers, and this configuration remains stable over a tenfold range of receptor expression level and is not altered by agonist addition [1]. Unfortunately, the researchers have not studied any heterodimers of these GPCRs, but it may be supposed that the potential to dimerize might concern not only an identical partner but also another member of the GPCR family.

The most important operational criterion demonstrating the existence of receptor heteromers is proof of physical association for a given pair of receptors in the native tissue or primary cells within the same cell [2]. However, even the co-expression of a given pair of receptors, for example serotonin 5-HT_{2A} and 5-HT_{1A} receptors in the same neuronal cell does not necessarily mean that these two receptors physically interact with each other, although complex interactions of these two serotonin receptors have been described at the behavioral and/or electrophysiological level [3–5]. In detailed neuroanatomical studies the predominant localization of serotonin 5-HT_{2A} receptors has been ascribed to apical dendrites proximal to the soma of prefrontal glutamatergic pyramidal neuron (where they mediate 5-HT depolarization), and of 5-HT_{1A} receptors—to the axon hillock (which positions them to mediate 5-HT hyperpolarization of the same neuron).

Another example of serotonin 5-HT_{2A} interaction has been described by Marek [6] who has provided behavioral evidence indicating that activation of mu opioid receptors (MORs) may suppress head twitches induced by hallucinogenic drugs, e.g. DOI (2,5-dimethoxy-4-iodoamphetamine). Earlier, the physiological interaction between the 5-HT_{2A} receptors and MOR has been also shown in electrophysiological studies. Since the layer Va of the neocortex possesses high density of both the 5-HT_{2A} receptors and MORs [7, 8] physical interaction between these two receptors may be predicted, however this issue has not been explored any further.

Similarly, interaction between the serotonin 5-HT_{2A} receptor and the dopamine D_1 receptor has been shown at the behavioral level [9] by demonstrating that the 5-HT_{2A} receptor antagonists antagonized d-amphetamine-induced hyperlocomotion and reversed d-amphetamine effect on latent inhibition [10]. However, no further studies have been conducted to look for physical interaction of these two receptors, although it has been shown that the 5-HT_{2A} receptors are present on dopaminergic neurons in the ventral segmental area [11, 12], making such an interaction possible.

On the other hand, direct interaction in the native tissue has been shown for the serotonin 5-HT_{2A} receptors with the dopamine D_2 receptor and the metabotropic glutamate receptor mGlu2.

The serotonin 5-HT_{2A}—dopamine D_2 receptor interaction is especially interesting in the context of antipsychotic drug action mechanism. Recently in a proximity ligation assay Borroto-Escuela have reported 5-HT_{2A}-D_2 heterocomplexes in the ventral and dorsal striatum [13]. In subsequent studies they have shown allosteric facilitating receptor-receptor interaction in the 5-HT_{2A}-D_2 receptor heterocomplex, enhancing the D_2 receptor signaling via Gi/o protein [14]. This finding broadens our understanding of atypical antipsychotic drugs which may counteract the D_2 receptor signaling at low doses in the 5-HT_{2A}-D_2 heterodimers via their combined blockade of both receptors [15]. On the other hand, these data provide a new mechanism of psychotic action of hallucinogenic drugs which may involve enhancement of D_2 receptor signaling. The authors interpret their data by pointing out that hallucinogenics, LSD and DOI, 5-HT_{2A} receptor agonists, induce pathological enhancement of dopamine D_2 receptor signaling, while this receptor forms heterocomplex with 5-HT_{2A} in the nucleus accumbens (core) and the dorsal striatum. Additionally, the significance of 5-HT_{2A}-D_2 receptor heterocomplexes has been recently explored in the context of cannabinoid action mechanisms. Franklin and Carrasco [16] have shown a non-selective cannabinoid receptor agonist, CP55,940, administered for 7 days, enhanced the co-immunoprecipitation of the 5-HT_{2A}-D_2 hetero-complexes in rat prefrontal cortex which indicates further importance of these complexes also in elucidating the cannabinoid action mechanisms.

Different understanding of hallucinogenic drugs was proposed in recent years by providing data which demonstrate the formation of heterocomplexes by the 5-HT_{2A} receptor with the metabotropic glutamatergic receptor $mGlu_2$ [17]. The co-immunoprecipitation of these two receptors has been shown in the lysates from human and mouse brain with no such effect observed in KO mice, both $mGlu_2$–/– and 5-HT_{2A}–/– [18, 19]. Hallucinogenics, 5-HT_{2A} receptor agonists, LSD and DOI induce specific behavior and head twitches which has been shown to be dependent on 5-HT_{2A} receptor expression in cortical neurons [20], and absent in $mGlu_2$ knock-out mice [21]. In further elegant and well-controlled studies Moreno and co-workers have shown that the head-twitches induced by DOI were brought back in $mGlu_2$ KO mice overexpressing the $mGlu_2$ receptor in the prefrontal cortex [17].

The concept of these two receptors acting as heterodimers has been seriously challenged by Delille et al. [22, 23], who have pointed not only shortcomings in the methodology used and lack of direct translation into second messenger systems but also—which is worth considering—lack of unequivocal demonstration of 5-HT_{2A} and $mGlu_2$ receptor co-expression in neurons. Nevertheless, there are strong indications that these receptors physically interact, and even the amino acid residues responsible for this interaction have been identified [17, 19].

The concept that 5-HT_{2A} and $mGlu_2$ receptor heterocomplexes are responsible for a different action of hallucinogenic vs non-hallucinogenic drugs is very interesting, however there are also data pointing to the inherent nature of these

5-HT_{2A} receptor agonists which allows them to activate different signaling pathways upon binding solely to the 5-HT_{2A} receptor [20]. This inherent nature of 5-HT_{2A} agonists has been recently confirmed by molecular dynamics simulations [24].

Structural Aspects of 5-HT_{2A} Receptor Heteromerization

Generally, receptor oligomer has been defined as a macromolecular complex consisting of at least two functional receptor units with biochemical properties differing from those of its individual components [25].

Understanding the heteromers' structure, describing the dimer interface and establishing the mechanisms involved in the receptors association is of fundamental importance. Currently, several models have been proposed. Within the whole GPCRs family, data point to covalent bond formation between the extracellular domains of the receptors [26], interactions between the intracellular domain (i.e. a coiled-coil interaction within the C-tail of the receptor) [27], hydrophobic interactions between the transmembrane domains [28–30] and electrostatic interactions which may occur between characteristic epitopes [31, 32]. However, probably a combination of the mechanisms mentioned above plays a key role in GPCRs oligomerization. Unfortunately, there is still insufficient data on the intermolecular interactions and the precise role of specific regions of receptor molecules involved in the formation of heteromers between 5-HT_{2A} and other receptors. It has been demonstrated that transmembrane (TM) segments (TM4 and TM5) of $mGlu_2$ are necessary for this receptor to be assembled as a heterocomplex with the 5-HT_{2A} receptor [33]. Recent data provide evidence for direct involvement of three residues—for Ala-677, Ala-681, and Ala-685 from the intracellular end of TM4 of the $mGlu_2$ receptors in interaction with the 5-HT_{2A} receptor [17].

Receptor heteromerization via TM 5 and TM 6 has been also described for the 5-HT_{2A} and CB_1 receptors [34].

Based on a bioinformatic approach, Borroto-Escuela and co-workers have suggested the engagement of TM1 and TM3 in the D_2-5-HT_{2A} dimer interface formation [35]. On the other hand, data obtained using site-directed mutagenesis point to the contribution of electrostatic interactions in the formation of D_2-5-HT_{2A} heterocomplexes [32]. These electrostatic interactions occur between characteristic epitopes: one epitope containing mainly two or more adjacent arginine residues is located within the third intracellular loop (ic3) of the D_2 receptor and the second epitope consisting of acidic (two or more adjacent aspartate or glutamate) residues or/and a phosphorylated residue is within the C-terminus of the 5-HT_{2A} receptor. Similar interactions have been also shown for other GPCRs (e.g. D_1-D_2, D_1-NMDA, D_2-CB_1) [31, 36, 37].

Methodology Used in GPCRs Heterodimerization

Several different approaches have been developed and dedicated to the investigation of GPCR oligomerization. Early evidence for oligomer formation comes from radioligand binding, receptor crosslinking, receptor complementation or radiation inactivation experiments. However, a concept concerning oligomeric formation has been ultimately accepted only following direct evidence resulting from biochemical and biophysical experiments.

Availability of cDNA of GPCRs and the development of antibody recognizing receptors or different tags' epitopes led to an identification of many different homo- and heterodimers using immunoprecipitation and co-immunoprecipitation techniques. Although this method has been used to demonstrate the existence of homooligomeric forms of D_3 dopamine receptors in monkey and rat brains [38] or adenosine A_1 receptors in the pig brain cortex [39] and heterooligomeric complexes of AT_1 and B_2 receptors in rat smooth muscle cells [40], human platelets and omental vessels [41], adenosine A_1 and P_2Y_1 in rat brains [42], $mGluR_5$ and calcium-sensing receptor in the bovine brain [43] and others, due to low expression levels, weak GPCR antibody selectivity or, in many cases, lack of receptor specific antibodies, it is used mostly in the cells heterologously expressing tag-labeled receptors.

The experiments using the co-immunoprecipitation approach are conducted in three stages. First, the cells are lysed and membranes are solubilized. Next, GPCRs are precipitated with receptor or epitope-specific antibodies. Immunoprecipitates are then analyzed by SDS gel electrophoresis and Western blotting. The first stage involves the use of detergents to solubilize membrane proteins, which is the biggest drawback of this method. When membrane solubilization is incomplete, small membrane patches containing uninteracting GPCRs may remain in the supernatant. On the other hand, a because of highly hydrophobic nature of GPCRs, artifactual aggregation may occur or by contrast, excessive concentration of detergents may disrupt the existing interactions. Another problem is connected with the level of expression: high overexpression levels, often obtained in heterologous expression systems, can enforce artifactual interactions between GPCRs. The selling point of immunoprecipitation methods is the possibility to detect receptor dimers in *ex vivo* tissue samples. However, this implementation is limited by availability and specificity of antibodies.

Although immunoprecipitation methods are commonly used to examine protein-protein interactions, it is important to note that they can never demonstrate that two proteins are in physical contact rather than being a part of a larger protein complex. A major breakthrough in GPCRs dimerization field has been made following the development of biophysical techniques based on a resonance energy transfer (RET) [44–46]—a non-radiative transfer of energy between electromagnetic dipoles of an energy donor and a suitable acceptor. The RET phenomenon depends on the distance between the donor and acceptor molecules and thus the RET techniques are used to measure interactions of molecules at a distance of less than than 100 Å, which is ideal for monitoring GPCR interactions.

In case of the fluorescence resonance energy transfer (FRET), both the donor and the acceptor of energy are fluorescent molecules. In turn, in the bioluminescence energy transfer (BRET), the energy is transferred to the acceptor from a luciferase enzyme upon oxidation of its substrate. Both techniques allow to study physical interaction between receptor proteins in living cells in real time. They can be measured on cell populations by a microplate reader or scanning microscopy which provides additional benefits of spatial resolution. This is especially advantageous in the case of GPCRs dimerization studies, as it allows investigation in regions limited to plasma membrane and separation from oligomerization which occurs during biosynthesis or internalization in ER, Golgi and endosomes.

In experiments using the FRET, the fluorescent donor and acceptor molecules are fluorescent proteins genetically linked with the receptor protein of interest, or fluorescent dyes combined with the receptor or epitope-specific antibodies. The latter approach allows to study endogenous receptor proteins, but it depends on the accessibility of receptor-specific antibodies. The main problems in data interpretation result from the donor emission leakage to the acceptor's emission channel, excitation of the acceptor by light used to activate the donor molecule, expression levels of the donor and acceptor molecules, photobleaching and autofluorescence. Despite these problems, many GPCRs have been reported to form homo- or heterodimeric complexes [47]. Most FRET techniques cannot distinguish between dimers and higher-order oligomers. Accessibility of various fluorescent proteins led to the development of a three-chromphore FRET, which has been used to show that for α1beta-adrenergic receptors, higher order complexes are formed rather than dimeric complexes [48].

Another modification utilizes fluorescence fading caused by the destruction of the molecule after exposure to the excitation light. This method is called photobleaching FRET (pbFRET) and has beeen successfully used to demonstrate homodimerization of $SSTR_5$ receptors [49] and heterodimerization of D_2 and $SSTR_5$ receptors [50]. In order to reduce the background fluorescence and increase the signal/noise ratio time-resolved FRET (TR-FRET) may be used [51, 52]. Employment of fluorophores with long-lived fluorescence (lanthanide donors such as europium or terbium cryptate complexes with suitable acceptors), allows to perform measurements after short-lived cell autofluorescence has decayed. In this technique, fluorophore labeled antibodies are used which guarantees that only receptors present in the plasma membrane are detected. On the other hand, as described above, the techniques using antibodies are limited by their availability and specificity. The refinement of TR-FRET techniques has led to the introduction of homogenous assays without washing steps, suited for high-throughput screening experiments [53].

Another advantage of the FRET-based techniques is that they can be combined with confocal microscopy imaging. In FRET-FLIM, the energy transfer is calculated from the change in the donor fluorescence lifetime in the presence and in the absence of the acceptor measured in each pixel of the image. Since the fluorescence lifetime is an intrinsic property of a fluorophore and it is sensitive to environmental changes but not to fluorophore concentration, this technique overcomes many problems encountered by the intensity-based methods and benefits from the spatial resolution provided by microscopy. It has been used to show that the 5-HT_{2C} receptors form homooligomers during receptor maturation and processing as energy transfer efficiencies in the plasma membrane were the same as in ER and Golgi [54].

BRET, the technique mentioned earlier, uses bioluminescence resulting from the catalytic degradation of coelenterazine, a substrate for the luciferase enzyme. The energy from this process is then transferred to the green fluorescent proteins [55]. This method allows to avoid problems arising from autofluorescence, photobleaching and excitation causing cell damage, but dim luminescence makes it less sensitive. Despite this fact, this technique has been successfully used in various studies on GPCR dimerization [47, 56].

There are also other, biophysical means to document the GPCRs oligomerization: exploiting ligands conjugated with fluorophores, bimolecular fluorescence of luminescence complmentation (BiFC/BiLC) [57, 58].

Despite continuously developing technologies, a growing number of fluorescent proteins and markers displaying different properties and increasing equipment sensitivity, the understanding of GPCR interactions is hampered by difficulties in detection and manipulation in their native environment. The development of better antibodies will allow to perform studies in tissues/organism models. The *in situ* proximity ligation assay (PLA) can be used to study endogenous GPCR heteromers in the native tissues. This method, in combination with confocal scanning microscopy, is highly selective and sensitive. The most popular format of this technique uses a pair of receptor-specific antibodies from different species, which are recognized by secondary antibodies with attached oligonucleotides. When the probes recognize the target, the attached oligonucleotides are then localized at a sufficiently close distance (less than 40 nm), so proximity dependent ligation forms a circular DNA template, which is thereafter amplified as a result of rolling circle amplification. The product is visualized with a fluorescently labeled probe [13]. This powerful method has been recently used to identify the A_{2A}-D_2 in the mouse striatum [59], or the 5-HT_{1A}-$FGFR_1$ heterocomplexes in the rat hippocampus and the dorsal and median raphe in the midbrain [60].

All the presented examples of the methods used in the GPCRs oligomerization studies illustrate their complexity, each having their pros and cons; no wonder that controversies and ambiguities concerning the obtained results are often encountered in the literature. Nevertheless, the issue of GPCRs forming heterocomplexes is very interesting and still promising, both as far as basal knowledge and novel pharmacological interventions are concerned.

References

1. Herrick-Davis K, Grinde E, Cowan A, Mazurkiewicz JE (2013) Fluorescence correlation spectroscopy analysis of serotonin, adrenergic, muscarinic, and dopamine receptor dimerization: the oligomer number puzzle. Mol Pharmacol 84(4):630–642
2. Pin JP, Neubig R, Bouvier M, Devi L, Filizola M, Javitch JA et al (2007) International Union of Basic and Clinical Pharmacology. LXVII. Recommendations for the recognition and nomenclature of G protein-coupled receptor heteromultimers. Pharmacol Rev 59(1):5–13

3. Amargos-Bosch M, Bortolozzi A, Puig MV, Serrats J, Adell A, Celada P et al (2004) Co-expression and in vivo interaction of serotonin1A and serotonin2A receptors in pyramidal neurons of prefrontal cortex. Cereb Cortex 14(3):281–299
4. Jakab RL, Goldman-Rakic PS (1998) 5-Hydroxytryptamine2A serotonin receptors in the primate cerebral cortex: possible site of action of hallucinogenic and antipsychotic drugs in pyramidal cell apical dendrites. Proc Natl Acad Sci U S A 95(2):735–740
5. de Almeida J, Mengod G (2007) Quantitative analysis of glutamatergic and GABAergic neurons expressing 5-HT(2A) receptors in human and monkey prefrontal cortex. J Neurochem 103(2):475–486
6. Marek GJ (2003) Behavioral evidence for mu-opioid and 5-HT2A receptor interactions. Eur J Pharmacol 474(1):77–83
7. Blue ME, Yagaloff KA, Mamounas LA, Hartig PR, Molliver ME (1988) Correspondence between 5-HT2 receptors and serotonergic axons in rat neocortex. Brain Res 453(1–2):315–328
8. Tempel A, Zukin RS (1987) Neuroanatomical patterns of the mu, delta, and kappa opioid receptors of rat brain as determined by quantitative in vitro autoradiography. Proc Natl Acad Sci U S A 84(12):4308–4312
9. Body S, Cheung TH, Bezzina G, Asgari K, Fone KC, Glennon JC et al (2006) Effects of d-amphetamine and DOI (2,5-dimethoxy-4-iodoamphetamine) on timing behavior: interaction between D1 and 5-HT2A receptors. Psychopharmacology 189(3):331–343
10. Moser PC, Moran PM, Frank RA, Kehne JH (1996) Reversal of amphetamine-induced behaviours by MDL 100,907, a selective 5-HT2A antagonist. Behav Brain Res 73(1–2):163–167
11. Doherty MD, Pickel VM (2000) Ultrastructural localization of the serotonin 2A receptor in dopaminergic neurons in the ventral tegmental area. Brain Res 864(2):176–185
12. Nocjar C, Roth BL, Pehek EA (2002) Localization of 5-HT(2A) receptors on dopamine cells in subnuclei of the midbrain A10 cell group. Neuroscience 111(1):163–176
13. Borroto-Escuela DO, Romero-Fernandez W, Garriga P, Ciruela F, Narvaez M, Tarakanov AO et al (2013) G protein-coupled receptor heterodimerization in the brain. Methods Enzymol 521:281–294
14. Albizu L, Holloway T, Gonzalez-Maeso J, Sealfon SC (2011) Functional crosstalk and heteromerization of serotonin 5-HT2A and dopamine D2 receptors. Neuropharmacology 61(4):770–777
15. Borroto-Escuela DO, Romero-Fernandez W, Narvaez M, Oflijan J, Agnati LF, Fuxe K (2014) Hallucinogenic 5-HT2AR agonists LSD and DOI enhance dopamine D2R protomer recognition and signaling of D2-5-HT2A heteroreceptor complexes. Biochem Biophys Res Commun 443(1):278–284
16. Franklin JM, Carrasco GA (2012) Cannabinoid-induced enhanced interaction and protein levels of serotonin 5-HT(2A) and dopamine D(2) receptors in rat prefrontal cortex. J Psychopharmacol 26(10):1333–1347
17. Moreno JL, Muguruza C, Umali A, Mortillo S, Holloway T, Pilar-Cuellar F et al (2012) Identification of three residues essential for 5-hydroxytryptamine 2A-metabotropic glutamate 2 (5-HT2A.mGlu2) receptor heteromerization and its psychoactive behavioral function. J Biol Chem 287(53):44301–44319
18. Gonzalez-Maeso J, Ang RL, Yuen T, Chan P, Weisstaub NV, Lopez-Gimenez JF et al (2008) Identification of a serotonin/glutamate receptor complex implicated in psychosis. Nature 452(7183):93–97
19. Fribourg M, Moreno JL, Holloway T, Provasi D, Baki L, Mahajan R et al (2011) Decoding the signaling of a GPCR heteromeric complex reveals a unifying mechanism of action of antipsychotic drugs. Cell 147(5):1011–1023
20. Gonzalez-Maeso J, Weisstaub NV, Zhou M, Chan P, Ivic L, Ang R et al (2007) Hallucinogens recruit specific cortical 5-HT(2A) receptor-mediated signaling pathways to affect behavior. Neuron 53(3):439–452
21. Moreno JL, Holloway T, Albizu L, Sealfon SC, Gonzalez-Maeso J (2011) Metabotropic glutamate mGlu2 receptor is necessary for the pharmacological and behavioral effects induced by hallucinogenic 5-HT2A receptor agonists. Neurosci Lett 493(3):76–79

22. Delille HK, Becker JM, Burkhardt S, Bleher B, Terstappen GC, Schmidt M et al (2012) Heterocomplex formation of 5-HT2A-mGlu2 and its relevance for cellular signaling cascades. Neuropharmacology 62(7):2184–2191
23. Delille HK, Mezler M, Marek GJ (2013) The two faces of the pharmacological interaction of mGlu2 and 5-HT(2)A—relevance of receptor heterocomplexes and interaction through functional brain pathways. Neuropharmacology 70:296–305
24. Perez-Aguilar JM, Shan J, LeVine MV, Khelashvili G, Weinstein H (2014) A functional selectivity mechanism at the serotonin-2A GPCR involves ligand-dependent conformations of intracellular loop 2. J Am Chem Soc 136(45):16044–16054
25. Ferre S, Baler R, Bouvier M, Caron MG, Devi LA, Durroux T et al (2009) Building a new conceptual framework for receptor heteromers. Nat Chem Biol 5(3):131–134
26. Kunishima N, Shimada Y, Tsuji Y, Sato T, Yamamoto M, Kumasaka T et al (2000) Structural basis of glutamate recognition by a dimeric metabotropic glutamate receptor. Nature 407(6807): 971–977
27. Pagano A, Rovelli G, Mosbacher J, Lohmann T, Duthey B, Stauffer D et al (2001) C-terminal interaction is essential for surface trafficking but not for heteromeric assembly of GABA(b) receptors. J Neurosci 21(4):1189–1202
28. Gouldson PR, Higgs C, Smith RE, Dean MK, Gkoutos GV, Reynolds CA (2000) Dimerization and domain swapping in G-protein-coupled receptors: a computational study. Neuropsychopharmacology 23(4 Suppl):S60–S77
29. Fotiadis D, Jastrzebska B, Philippsen A, Muller DJ, Palczewski K, Engel A (2006) Structure of the rhodopsin dimer: a working model for G-protein-coupled receptors. Curr Opin Struct Biol 16(2):252–259
30. Overton MC, Chinault SL, Blumer KJ (2003) Oligomerization, biogenesis, and signaling is promoted by a glycophorin A-like dimerization motif in transmembrane domain 1 of a yeast G protein-coupled receptor. J Biol Chem 278(49):49369–49377
31. Lukasiewicz S, Faron-Gorecka A, Dobrucki J, Polit A, Dziedzicka-Wasylewska M (2009) Studies on the role of the receptor protein motifs possibly involved in electrostatic interactions on the dopamine D1 and D2 receptor oligomerization. FEBS J 276(3):760–775
32. Lukasiewicz S, Polit A, Kedracka-Krok S, Wedzony K, Mackowiak M, Dziedzicka-Wasylewska M (2010) Hetero-dimerization of serotonin 5-HT(2A) and dopamine D(2) receptors. Biochim Biophys Acta 1803(12):1347–1358
33. Shan J, Khelashvili G, Mondal S, Mehler EL, Weinstein H (2012) Ligand-dependent conformations and dynamics of the serotonin 5-HT(2A) receptor determine its activation and membrane-driven oligomerization properties. PLoS Comput Biol 8(4):e1002473
34. Vinals X, Moreno E, Lanfumey L, Cordomi A, Pastor A, de La Torre R et al (2015) Cognitive impairment induced by Delta9-tetrahydrocannabinol occurs through heteromers between cannabinoid CB1 and serotonin 5-HT2A receptors. PLoS Biol 13(7):e1002194
35. Borroto-Escuela DO, Romero-Fernandez W, Tarakanov AO, Marcellino D, Ciruela F, Agnati LF et al (2010) Dopamine D2 and 5-hydroxytryptamine 5-HT((2)A) receptors assemble into functionally interacting heteromers. Biochem Biophys Res Commun 401(4):605–610
36. Lee FJ, Xue S, Pei L, Vukusic B, Chery N, Wang Y et al (2002) Dual regulation of NMDA receptor functions by direct protein-protein interactions with the dopamine D1 receptor. Cell 111(2):219–230
37. Jackson SN, Wang HY, Yergey A, Woods AS (2006) Phosphate stabilization of intermolecular interactions. J Proteome Res 5(1):122–126
38. Nimchinsky EA, Hof PR, Janssen WG, Morrison JH, Schmauss C (1997) Expression of dopamine D3 receptor dimers and tetramers in brain and in transfected cells. J Biol Chem 272(46):29229–29237
39. Ciruela F, Casado V, Mallol J, Canela EI, Lluis C, Franco R (1995) Immunological identification of A1 adenosine receptors in brain cortex. J Neurosci Res 42(6):818–828
40. AbdAlla S, Lother H, Quitterer U (2000) AT1-receptor heterodimers show enhanced G-protein activation and altered receptor sequestration. Nature 407(6800):94–98

41. AbdAlla S, Lother H, el Massiery A, Quitterer U (2001) Increased AT(1) receptor heterodimers in preeclampsia mediate enhanced angiotensin II responsiveness. Nat Med 7(9):1003–1009
42. Yoshioka K, Hosoda R, Kuroda Y, Nakata H (2002) Hetero-oligomerization of adenosine A1 receptors with P2Y1 receptors in rat brains. FEBS Lett 531(2):299–303
43. Gama L, Wilt SG, Breitwieser GE (2001) Heterodimerization of calcium sensing receptors with metabotropic glutamate receptors in neurons. J Biol Chem 276(42):39053–39059
44. Jaeger WC, Armstrong SP, Hill SJ, Pfleger KD (2014) Biophysical detection of diversity and bias in GPCR function. Front Endocrinol (Lausanne) 5:26
45. Cottet M, Faklaris O, Maurel D, Scholler P, Doumazane E, Trinquet E et al (2012) BRET and Time-resolved FRET strategy to study GPCR oligomerization: from cell lines toward native tissues. Front Endocrinol (Lausanne) 3:92
46. Zhang R, Xie X (2012) Tools for GPCR drug discovery. Acta Pharmacol Sin 33(3):372–384
47. Pfleger KD, Eidne KA (2005) Monitoring the formation of dynamic G-protein-coupled receptor-protein complexes in living cells. Biochem J 385(Pt 3):625–637
48. Lopez-Gimenez JF, Canals M, Pediani JD, Milligan G (2007) The alpha1b-adrenoceptor exists as a higher-order oligomer: effective oligomerization is required for receptor maturation, surface delivery, and function. Mol Pharmacol 71(4):1015–1029
49. Rocheville M, Lange DC, Kumar U, Sasi R, Patel RC, Patel YC (2000) Subtypes of the somatostatin receptor assemble as functional homo- and heterodimers. J Biol Chem 275(11):7862–7869
50. Rocheville M, Lange DC, Kumar U, Patel SC, Patel RC, Patel YC (2000) Receptors for dopamine and somatostatin: formation of hetero-oligomers with enhanced functional activity. Science 288(5463):154–157
51. Ciruela F, Fernandez-Duenas V, Jacobson KA (2015) Lighting up G protein-coupled purinergic receptors with engineered fluorescent ligands. Neuropharmacology 98:58–67
52. Madiraju C, Welsh K, Cuddy MP, Godoi PH, Pass I, Ngo T et al (2012) TR-FRET-based high-throughput screening assay for identification of UBC13 inhibitors. J Biomol Screen 17(2):163–176
53. Maurel D, Kniazeff J, Mathis G, Trinquet E, Pin JP, Ansanay H (2004) Cell surface detection of membrane protein interaction with homogeneous time-resolved fluorescence resonance energy transfer technology. Anal Biochem 329(2):253–262
54. Herrick-Davis K, Weaver BA, Grinde E, Mazurkiewicz JE (2006) Serotonin 5-HT2C receptor homodimer biogenesis in the endoplasmic reticulum: real-time visualization with confocal fluorescence resonance energy transfer. J Biol Chem 281(37):27109–27116
55. Ayoub MA, Pfleger KD (2010) Recent advances in bioluminescence resonance energy transfer technologies to study GPCR heteromerization. Curr Opin Pharmacol 10(1):44–52
56. Angers S, Salahpour A, Joly E, Hilairet S, Chelsky D, Dennis M et al (2000) Detection of beta 2-adrenergic receptor dimerization in living cells using bioluminescence resonance energy transfer (BRET). Proc Natl Acad Sci U S A 97(7):3684–3689
57. Albizu L, Cottet M, Kralikova M, Stoev S, Seyer R, Brabet I et al (2010) Time-resolved FRET between GPCR ligands reveals oligomers in native tissues. Nat Chem Biol 6(8):587–594
58. Ciruela F, Vilardaga JP, Fernandez-Duenas V (2010) Lighting up multiprotein complexes: lessons from GPCR oligomerization. Trends Biotechnol 28(8):407–415
59. Trifilieff P, Rives ML, Urizar E, Piskorowski RA, Vishwasrao HD, Castrillon J et al (2011) Detection of antigen interactions ex vivo by proximity ligation assay: endogenous dopamine D2-adenosine A2A receptor complexes in the striatum. BioTechniques 51(2):111–118
60. Borroto-Escuela DO, Romero-Fernandez W, Mudo G, Perez-Alea M, Ciruela F, Tarakanov AO et al (2012) Fibroblast growth factor receptor 1- 5-hydroxytryptamine 1A heteroreceptor complexes and their enhancement of hippocampal plasticity. Biol Psychiatry 71(1):84–91

Brain Distribution

Cristiano Bombardi

Abstract The 5-HT_{2A} receptor type (5-HT_{2A}R) is a G-protein-coupled receptor widely distributed in the central nervous system, indicating its participation in numerous neurological effects serotonin-mediated. The 5-HT_{2A}R has attracted interest as a potential drug target for the treatment of several important neurologic and psychiatric disorders, such as epilepsy and depression. The distribution of the 5-HT_{2A}R has been investigated by immunohistochemical stainings, in situ hybridization experiments, and physiologic/pharmacologic procedures. This review summarizes the cellular localization of the 5-HT_{2A}R in the brains, providing the neuronal pathways modulated by serotonin through this specific receptor type.

Keywords 5-HT_{2A} receptor • Serotonin • Thelencephalon • Diencephalon • Brainstem

Introduction

The 5-HT_{2A} receptor type (5-HT_{2A}R) belong to the 5-HT_2 (5-HT_2R) receptor family of metabotropic receptors. The activation of the 5-HT_{2A}R causes phospholipase C-mediated synaptic facilitation by reducing outward potassium current [1–4]. Recently accumulated knowledge demonstrates that the 5-HT_{2A}R has a widespread distribution in the central nervous system, indicating its participation in numerous neurological effects serotonin-mediated [5–18]. Interestingly, the level of expression of the 5-HT_{2A}R varies during early postnatal development [19]. This data suggest that the 5-HT_{2A}r may also modulate neuronal development [19].

The distribution of the 5-HT_2 or 5-HT_{2A} receptors has been investigated by in situ hybridization experiments [20, 21], autoradiography studies [22], [23], physiologic/pharmacologic procedures [10, 24–36] and immunohistochemical experiments [5, 6, 9–11, 18, 37–39]. In particular, using standard immunohistochemical procedures somatodendritic and axonal immunoreactivity for the 5-HT_{2A}R has been

C. Bombardi (✉)
Department of Veterinary Medical Sciences, University of Bologna, Ozzano dell'Emilia, BO, Italy
e-mail: cristiano.bombardi@unibo.it

B.P. Guiard, G. Di Giovanni (eds.), *5-HT_{2A} Receptors in the Central Nervous System*, The Receptors 32, https://doi.org/10.1007/978-3-319-70474-6_4

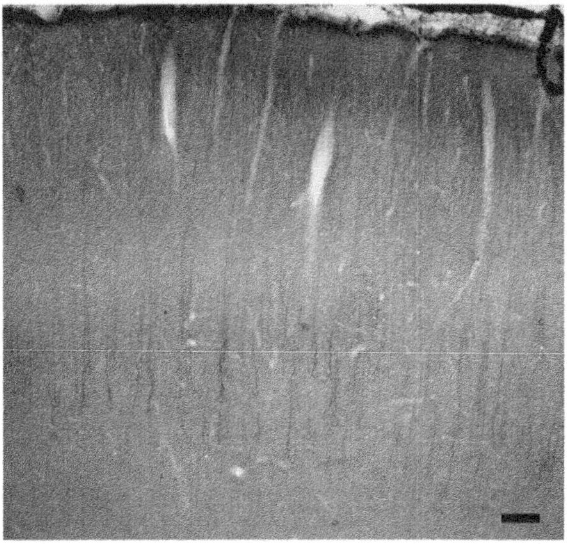

Fig. 1 Distribution of $5\text{-}HT_{2A}$ receptor ($5\text{-}HT_{2A}R$) immunoreactivity in the rat primary motor cortex. Note numerous $5\text{-}HT_{2A}R$-immunoreactive pyramidal neurons in layer V. The apical dendrites of pyramidal cells are clearly visible. Scale bar = 100 μm

located in numerous brain regions [9–11, 18, 37, 38]. It is important to underline that because selective immunohistochemical methods able to differentiate between members of the $5\text{-}HT_2R$ family ($5\text{-}HT_{2A}$, $5\text{-}HT_{2B}$ and $5\text{-}HT_{2C}$ receptors) have not been available until very recently, many of the early immunohistochemical studies did not specifically identify the $5\text{-}HT_{2A}R$ distribution in the brain, but has provided data concerning the immunohistochemical localization of the $5\text{-}HT_2R$ family [37, 38].

Thelencephalon

Neocortex

The rat cerebral cortex presents high levels of $5\text{-}HT_{2A}R$ transcripts, especially in frontal lobe [20]. The hybridization is distributed in layers II, IV, V and VI [20]. In rodents, frontal, parietal, temporal and occipital cortices show a similar distribution of $5\text{-}HT_{2A}r$ immunoreactivity. It has been found that 5HT2Ar-immunoreactive neurons are present throughout the cortex in layers I–VI [18]. Layer V exhibits the highest densities of somatodendritic profiles positive for the $5\text{-}HT_{2A}r$. Also layers II–III contain many $5\text{-}HT_{2A}R$-immunoreactive neurons. In contrast, layer VI includes low or moderate numbers of somatodendritic profiles. Finally, layers I and IV contain only immunopositive dendrites. The strongest $5\text{-}HT_{2A}R$ immunoreactivity is located in the large pyramidal neurons located in the layer V of the frontal and parietal cortex [9, 40]. Strongly labeled somatodendritic profiles are also present in layer V of the cingular cortex [9].

In the neocortex excitatory as well as inhibitory neurons express the $5\text{-HT}_{2A}R$ [9, 18, 40–47]. By immunocytochemical techniques it has been demonstrated that $5\text{-HT}_{2A}r$ is present in pyramidal neurons of the rat (Fig. 1) [40] and monkey (*Macaca mulatta*) cortex (frontal, temporal and parietal lobes) [42, 43]. Accordingly, by means of in situ hybridization histochemistry procedure, it has been shown that $5\text{-HT}_{2A}R$ mRNA is present in pyramidal cells in the rat prefrontal cortex. In addition, using double in situ hybridization it has been shown that many glutamatergic cells of the monkey and human prefrontal cortex express the $5\text{-HT}_{2A}R$. The cells are located especially in layers II–V [48]. In the cerebral cortex serotonin can increase the excitability of pyramidal neurons expressing $5\text{-HT}_{2A}r$ through a focal action in their apical dendritic field [49] where the concentrations of serotonin fibers [50] and $5\text{-HT}_{2A}r$ are both high [42]. Accordingly, the physiological activation of $5\text{-HT}_{2A}R$ excites pyramidal neurons in rat medial prefrontal cortex [51, 52].

In primate prefrontal cortex, the $5\text{-HT}_{2A}R$ is located in non-pyramidal neurons expressing calbindin [42]. Employing immunocytochemical techniques on the frontal, temporal, and parietal lobes of monkeys (*Macaca mulatta*), it has been demonstrated that $5\text{-HT}_{2A}R$ is present in cortical GABA(γ-aminobutyric acid)ergic interneurons [43]. Non-pyramidal neurons are distributed throughout layers II–VI, express calbindin or parvalbumin and correspond to GABAergic interneurons specialize in the perisomatic inhibition of pyramidal cells [43]. In addition, 5-HT_{2A} and 5-HT_3 receptors are expressed by different subpopulations of inhibitory interneurons. In fact, large interneurons and small interneurons express 5-HT_{2A} and 5-HT_3 receptors, respectively [43]. Furthermore, 5-HT_{2A} and 5-HT_{2C} receptors may be co-expressed on pyramidal cells and GABAergic neurons of the rat medial prefrontal cortex [53]. Other immunocytochemical studies have provided evidence that $5\text{-HT}_{2A}R$ is present in parvalbumin-containing interneurons [40] and calbindin-D28k-containing interneurons [42] in the cerebral cortex of rat [40] and monkey (*Macaca mulatta*) [42]. In monkey (*Macaca fascicularis*) and human prefrontal cortex, in situ hybridization studies have also demonstrated that parvalbumin and calbidin-D28k GABAergic interneurons express the $5\text{-HT}_{2A}R$ [48]. It is interesting to note that in the cortex, the $5\text{-HT}_{2A}R$ is preferentially associated with GABA-immunoreactive interneurons (large and medium-size basket cells and chandelier cells), which mediate the perisomatic inhibition of pyramidal cells [43]. By means of an in situ hybridization histochemistry procedure it has been shown that $5\text{-HT}_{2A}r$ mRNA is present in GABAergic cells in rat prefrontal cortex [54]. Consequently, serotonin, can suppress pyramidal neuronal firing by activating the inhibitory interneurons (basket and chandelier cells) which allow a perisomatic inhibition of pyramidal neurons [43]. This aspect is in agreement with previous studies showing that in the rat cortex the activation of $5\text{-HT}_{2A}r$ present in GABAergic interneurons activates the same GABAergic interneurons [55] and inhibits pyramidal neurons [56, 57].

Interestingly, astrocytes of the neocortex are immunoreactive for the $5\text{-HT}_{2A}R$ [18, 41].

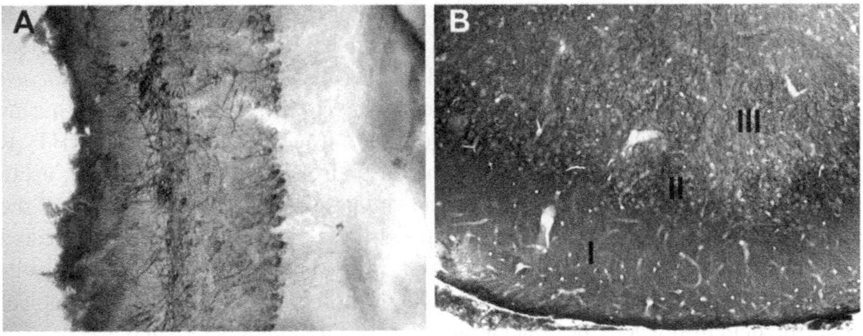

Fig. 2 Distribution of 5-HT_{2A} receptor immunoreactivity in the rat olfactory system. (**a**) In olfactory bulb, mitral cell layer exhibits an evident somatodendritic immunoreactivity. (**b**) Note the strong neuronal immunoreactivity located in layer II of the piriform cortex. Scale bar = 50 μm in **b** (applies to **a**, **b**)

Olfactory System

Different autoradiographic, in situ hybridizationand and immunohistochemical studies have demonstrated the presence of the 5-HT_{2A}r in the olfactory system. Very high levels of 5-HT_{2A}R mRNA are present in the olfactory bulb (especially in the mitral cell layer and external plexiform layer) and in the anterior olfactory nucleus [20]. A lover levels of 5-HT_{2A}r transcript are located in the olfactory tubercle and endopiriform nucleus [20]. In the main olfactory bulb the 5-HT_{2A}r-immunoreactivity is especially located in somata of mitral cell layer and in dendritic profiles of the external plexiform, mitral cell and internal plexiform layers (Fig. 2a) [9, 41, 44]. In the main olfactory bulb a low number of 5-HT_{2A}R positive cells are located in glomerular layer, external plexiform layer, internal plexiform layer and internal granular layer [9, 44]. Very few 5-HT_{2A}R-immunoreactive processes were found within the glomerular layer of the olfactory bulb [44]. In the accessory olfactory bulb the 5-HT_{2A}R-immunoreactivity is more evident in dendrites than in somata. Many somatodendritic profiles are also located in anterior olfactory nucleus, olfactory tubercle, islands of Calleja, piriform cortex (mainly in layers II and III; Fig. 2b), medial and ventral-anterior olfactory nuclei, lateral olfactory tract and endopiriform nucleus [9, 18, 41, 44].

Septum

Ligand binding studies demonstrates the presence of the 5-HT_{2A}R in the medial septal nucleus and in the nuclei of the diagonal band of Broca. However, 5-HT_{2A}R mRNA is present only in the nucleus of the horizontal limb of the diagonal band of Broca [20]. Also immunohistochemical observations demonstrated

5-HT_{2A}r-immunoreactive somata in the in the lateral (dorsal, intermediate, and ventral) and medial septal nuclei and in the nuclei of the diagonal band of Broca [9, 18]. Lateral and triangular septal nuclei exhibit low to moderate densities of 5-HT_{2A}R-immunoreactivity. With the exception of the triangular septal nucleus, a moderate density of dendritic profile immunoreactive for the 5-HT_{2A}r are located in the different area of the septum [9].

Hippocampal Region

The hippocampal region includes the hippocampal formation and the parahippocampal region. The hippocampal formation comprises the dentate gyrus, the hippocampus proper (which is divided into three distinct fields: CA3, CA2 and CA1) and the subiculum. The parahippocampal region includes the presubiculum, the parasubiculum, the entorhinal cortex, the perirhinal cortex and the postrhinal cortex [58–60].

Hippocampal Formation

The presence of the 5-HT_2R or the 5-HT_{2A}R in the hippocampal formation has been demonstrated by in situ hybridization studies, radioligand-binding experiments and recent immunohistochemical studies. Wright et al. [21] demonstrated low, intermediate, and high levels of 5-HT_2R mRNA in the hippocampus proper, subiculum, and dentate gyrus, respectively. 5-HT_{2A}R transcripts have also been observed in the rat hippocampal formation, particularly in the pyramidal cell layer of the CA3 field [20]. In contrast, autoradiographic studies have demonstrated intermediate 5-HT_2R levels of specific binding only in the ventral dentate gyrus [22].

Immunohistochemical experiments have demonstrated that the 5-HT_{2A}R is expressed in the majority of the principal excitatory neurons (granule and pyramidal cells) of the rat hippocampal formation (Fig. 3a, b). In particular, a strong 5-HT_{2A}R immunoreactivity is localized in the apical dendrite of the pyramidal cells where this serotonin receptor may increase excitatory postsynaptic currents [6, 45]. The presence of the 5-HT_{2A}R in pyramidal cells has been demonstrated also electrophysiologically. In fact, in the pyramidal somata of the rat CA1 (ventral field), the outward current induced by serotonin and alpha-methyl-serotonin (a 5-HT_2R agonist) is blocked by ketanserin (a 5-HT_2R antagonist) and spiperone (a 5-HT_{1A} and 5-HT_2 receptors antagonist) in a concentration-dependent manner [61]. Interestingly, the 5-HT_{2A}R is also expressed in the rat mossy fibers [6]. It is known that the mossy fibers arise from the granule cells and leave the dentate gyrus to innervate the pyramidal cells of the CA3 hippocampal field [62]. The 5-HT_{2A}R located at presynaptic level could modulate excitatory neurotransmission in the mossy fibers and consequently act on the hippocampal release of glutamate. This correlates with studies

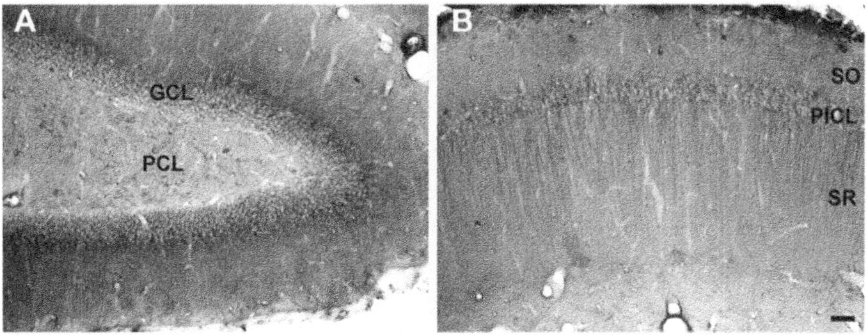

Fig. 3 Distribution of 5-HT$_{2A}$ receptor (5-HT$_{2A}$R) immunoreactivity in the rat dentate gyrus and CA1 field of the hippocampus proper. (**a**) Note the high density of immunopositive neurons in granule cell layer (GCL) and polymorphic cell layer (PCL) of the dentate gyrus. (**b**) In CA1 field of the hippocampus proper, pyramidal cells (located in the pyramidal cell layers, PICL) and some interneurons located in strata oriens (SO) and radiatum (SR) are immunoreactive for the 5-HT$_{2A}$R. Scale bar = 50 μm in **b** (applies to **a**, **b**)

indicating that different subtypes of serotonin receptors can affect presynaptic neurotransmission [63, 64].

Single and double-immunohistochemical experiments have demonstrated that the 5-HT$_{2A}$R is also expressed in a high percentage of GABAergic neurons of the hippocampal formation [6, 8]. Accordingly, stimulation of the 5-HT$_{2A/2C}$ receptors activate GABAergic neurons in the rat dentate gyrus [65] and the rat CA1field of the hippocampus proper [66]. The high density of 5-HT$_{2A}$R–immunoreactive neurons in the deeper portion of the granule cell layer indicate that this serotonin receptor can regulate neurogenesis in the subgranular zone [6, 67]. Since GABA regulates both the progenitor turnover and the integration of newly generated neurons in the dentate gyrus [68], it is reasonable to assume that the GABAergic neurons distributed in the subgranular zone may be involved in 5-HT$_{2A}$R-mediated hippocampal progenitor proliferation [69]. In the dentate gyrus and hippocampus proper, several classifications of GABAergic interneurons have been proposed, based on their morphology, axonal location, neurochemical code and electrophysiological characteristics [60, 62, 70, 71]. In the rat, combining histochemical [45] with morphological/topographical data [6] the 5-HT$_{2A}$R–immunoreactive inhibitory interneurons of the dentate gyrus and hippocampus proper may be classified as following reported. In the dentate gyrus, the interneurons expressing the 5-HT$_{2A}$R may correspond to parvalbumin–immunoreactive pyramidal basket, parvalbumin-immunopositive chandelier (axoaxonic) cells and somatostatin–immunoreactive interneurons with hilar dendrites and ascending axons (HIPP cells). In the hippocampus proper, the GABAergic interneurons containing the 5-HT$_{2A}$R may correspond to parvalbumin–immunopositive pyramidal basket cells, parvalbumin–immunoreactive chandelier cells, calbindin-D28 k–immunopositive neurons (located in strata oriens, radiatum

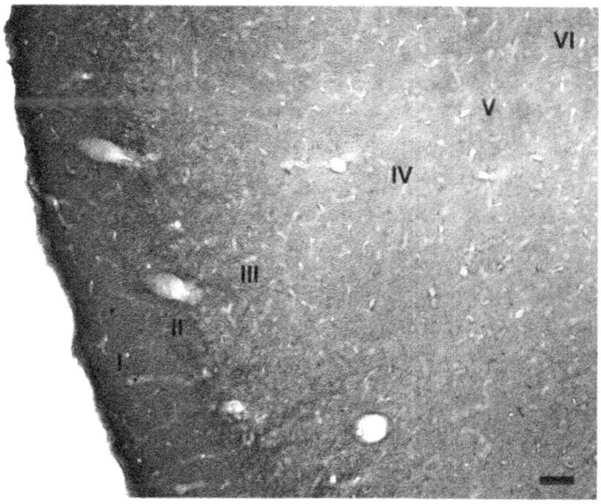

Fig. 4 Distribution of $5\text{-}HT_{2A}$ receptor immunoreactivity in the rat dorsal intermediate entorhinal field (DIE). Immunopositive neurons are distributed in every layer (I–VI). Scale bar = 50 μm

and lacunosum-moleculare), somatostatin/neuropeptide Y–immunoreactive cells terminating in conjunction with entorhinal afferent (O-LM cells), and VIP (vasoactive intestinal peptide)–immunoreactive IS3 interneurons [6, 45].

There are also astrocytes immunolabeled with $5\text{-}HT_{2A}R$ in the hippocampal formation [18].

Parahippocampal Region

The distribution of the $5\text{-}HT_2$ or $5\text{-}HT_{2A}$ receptors in the parahippocampal region has been less studied than in the hippocampal formation. An high density of $5\text{-}HT_2R$ ligand binding sites is present in the rat entorhinal cortex [22]. Accordingly, high levels of $5\text{-}HT_{2A}R$ mRNA are located in the rat entorhinal cortex, particularly in layers V and VI [20]. These data coincide with immunohistochemical experiments showing that a variety of morphological cell types is distributed in the rat entorhinal cortex and elsewhere in the rat parahippocampal region (Fig. 4) [6, 9]. Pyramidal or modified pyramidal cells are the main cell type of the rat parahippocampal region expressing $5\text{-}HT_{2A}R$ [6]. This receptor is strongly expressed on the apical dendrite of pyramidal neurons where it could modulate excitatory glutamate input, as has been demonstrated in the cerebral cortex [52]. In the rat parahippocampal region, $5\text{-}HT_{2A}R$ are also localized in non-pyramidal neurons [6]. In particular, double-immunofluorescence has revealed that a majority of the GABAergic cells in the entorhinal cortex contained $5\text{-}HT_{2A}R$. These non-pyramidal neurons are present in every layer, but are abundant in layers II, III, V, and VI [6]. Interestingly, there is no significant difference in the colocalization pattern of GABA and $5\text{-}HT_{2A}R$ in the different six fields of the entorhinal cortex [6].

Amygdaloid Complex

The amygdaloid complex, or amygdala, is comprised of pallial and subpallial components. The pallial amygdala is composed of deep and cortical pallial nuclei. The deep pallial nuclei are the basolateral amygdala, the anterior amygdaloid area (dorsal region) and the amygdalohippocampal area. The basolateral amygdala includes the lateral, the basolateral (magnocellular, intermediate, and parvicellular subdivisions), and basomedial nuclei. The cortical pallial nuclei consist of the nucleus of the lateral olfactory tract, the bed nucleus of the accessory olfactory tract, the anterior cortical nucleus, the posterolateral cortical nucleus and the posteromedial cortical nucleus. The subpallial amygdala, also called extended amygdala, includes the medial nucleus, the central nucleus, the anterior amygdaloid area (ventral region), the bed nucleus of the *stria terminalis*, and the intercalated nuclei [72].

Deep Pallial Components

The 5-$HT_{2/2A}$ receptors are strongly expressed in the deep nuclei. An in situ hybridization study revealed the presence of the 5-HT_2R mRNA in the lateral, basolateral, and basomedial nuclei [21]. However, in contrast with the results of Wright et al. [21], Pompeiano et al. [20] have not reported the 5-HT_{2A}R mRNA in the basolateral amygdala. Autoradiographic research has demonstrated specific binding sites of the 5-HT_2 receptor in the basolateral amygdala, especially in the lateral nucleus [22]. Immunohistochemical experiments have demonstrated that both pyramidal and non-pyramidal neurons of the basolateral amygdala express the 5-HT_{2A}R [5, 9–11, 18, 37].

Immunohistochemical studies have demonstrated that the pyramidal neurons in the basolateral amygdala represent most cells that are 5-HT_{2A}R-immunoreactive (Fig. 5a) [5, 7, 8, 11]. This receptor appear to be prevalently located in the dendritic processes, especially apical dendrites [5, 11]. Activation of 5-HT_{2A}/5-HT_{2C} receptors, obtained by the local injection of 1-(2,5-dimethoxy-4-iodophenyl)-2-aminopropane (DOI), increases discharge rate [31] and facilitates synaptic plasticity via an NMDA-mediated mechanism [73] in presumptive pyramidal neurons of the rat basolateral amygdala.

In the rat basolateral amygdala the 5-HT_{2A}R is also expressed in a lower percentage of GABAergic non-pyramidal interneurons [5, 7, 8, 11, 37]. In the lateral and basolateral nuclei of the rat amygdala, GABAergic neurons immunopositive for the 5-HT_{2A}R also express parvalbumin and somatostatin [11]. Electrophysiological studies have shown that the 5-HT_{2A}R activates GABAergic non-pyramidal neurons of the basolateral amygdala. In particular, α-methyl-5-hydroxytryptamine (a 5-HT_2R agonist) induces a dose-dependent membrane depolarization in the GABAergic interneurons of the rat basal nucleus [30]. Likewise, activation of the 5-HT_{2A}R enhances frequency and amplitude of spontaneous inhibitory postsynaptic currents (sIPSCs) recorded from pyramidal neurons located in the juvenile rat basolateral

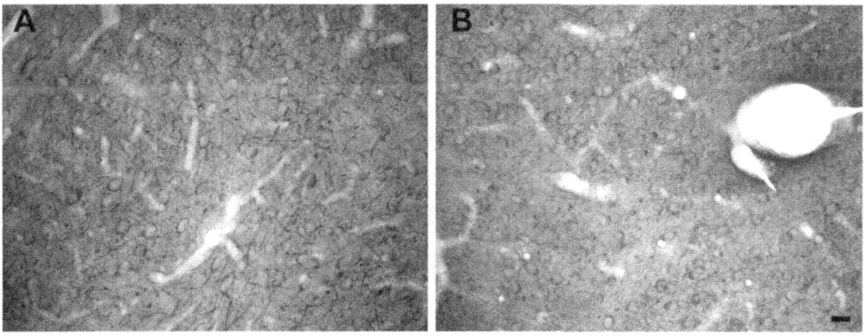

Fig. 5 Distribution of 5-HT_{2A} receptor immunoreactivity in the rat amygdaloid complex. (**a**) Note the high density of immunopositive pyramidal neurons in the basal nucleus (magnocellular division). (**b**) In central nucleus (lateral subdivision), ovoid somata show a strong immunoreactivity (**b**). Scale bar = 20 μm in **b** (applies to **a**, **b**)

amygdala [10]. Accordingly, the inhibition of pyramidal cell firing in the lateral nucleus of the rat amygdala obtained after local application of serotonin is blocked by a simultaneous application of GABA antagonist [32]. Finally, the activation GABAergic non-pyramidal neurons of the rat basolateral amygdala is also induced by DOI [31, 36]. Double-immunofluorescence studies have demonstrated that 5-HT_{2A} and 5-HT_3 receptors virtually do not coexist in the lateral and basolateral nuclei of the rat amygdaloid complex [11, 74]. The possible role of the functional segregation of 5-HT_{2A} and 5-HT_3 receptors in the amygdaloid complex has to be clarified. In the rat basolateral nucleus and along the external and internuclear borders of the rat basolateral amygdala, the 5-HT_{2A}R is also expressed by large GABAergic non-pyramidal neurons that project to the mediodorsal thalamus [11]. Many 5-HT_{2A}R-immunoreactive cells with angular- and ovoid-shaped somata are located in the rat anterior amygdaloid area [5]. Finally, pyramidal and non-pyramidal neurons of the rat amygdalohippocampal area express the 5-HT_{2A}R [5].

5-HT_{2A}R-immunoreactive astrocytes has been observed in the basolateral amygdala [18].

Cortical Pallial Components

In situ hybridization investigations have demonstrated a moderate density of 5-HT_2R mRNA [21] and 5-HT_{2A}R mRNA [20] in the rat cortical nuclei, with the exception of the bed nucleus of the accessory olfactory tract, which has presented high levels of 5-HT_{2A}R mRNA [20]. Several immunohistochemical studies have reported 5-HT_{2A}R-immunoreactive neurons in the rat cortical nuclei [5, 7–9, 37]. In particular, a high density of 5-HT_{2A}R-immunoreactive neurons are located in the nucleus of the lateral olfactory tract and in the bed nucleus of the accessory olfactory tract [5, 9]. In contrast to the pattern of 5-HT_{2A}R immunoreactivity, autoradiographic observations of the binding sites of the 5-HT_2R have demonstrated low receptor levels in the rat cortical amygdaloid nuclei, although the anterior cortical nucleus has shown a high density of binding sites [22].

In the rat cortical nuclei the 5-HT_{2A}R-immunoreactive neurons are heterogeneous in shape and size [5]. Pyramidal neurons are located mainly in the nucleus of the lateral olfactory tract (layer II), the anterior cortical nucleus (layers II and III), the posterolateral cortical nucleus (layers II and III), and the posteromedial cortical nucleus (layers II and III) [5]. In these cells, the 5-HT_{2A}R is abundant in apical dendrites, where it may induce excitatory synaptic currents. Small to large non-pyramidal neurons in many cortical nuclei (nucleus of the lateral olfactory tract, anterior cortical nucleus, posterolateral cortical nucleus and posteromedial cortical nucleus) express 5-HT_{2A}Rs [5]. These interneurons are located in all three layers, but are particularly abundant in layers II and III [5].

Subpallial Components

Immunohistochemical experiments have demonstrated that the rat central nucleus displays ovoid-shaped somata stained for the 5-HT_{2A}R (Fig. 5b) [5, 7–9]. Accordingly, in situ hybridization studies have reported moderate levels of 5-HT_2R mRNA in the rat central nucleus [21].The rat medial nucleus especially contains 5-HT_{2A}R-immunoreactive neurons with ovoid somata [5]. The bed nucleus of the stria terminalis contains a moderate density 5-HT_{2A}R-immunoreactive neurons [5, 9]. These cells are quite similar to those located in the central and medial nuclei [5]. Small and large neurons in the rat intercalated nuclei express the 5-HT_{2A}R [5, 7, 8, 18]. This result is in disagreement with an in situ hybridization studies showing that intercalated nuclei do not present 5-HT_{2A}R mRNA [20].

Basal Ganglia

Radioligand binding [22, 75], in *situ* ibridization [20, 21], functional [76] and immunohistochemical studies [9, 18, 41, 44] indicate a widespread distribution of the 5-HT_{2A}R in the basal ganglia. 5-HT_{2A}R mRNA levels were intermediate in the caudate-putamen, *nucleus accumbens* and in *substantia nigra* (*pars compacta* and *pars lateralis*) [20]. *Globus pallidus* do not show any hybridization [20].With the exception of the compact part of the *substantia nigra*, a relatively abundant number of 5-HT_{2A}R-immunoreactive somata has been observed in the basal ganglia [9, 41]. These immunoreactive neurons are especially numerous in the lateral and dorsal caudate-putamen (Fig. 6) [44]. Somatodendritic profiles immunoreactive for the 5-HT_{2A}R are locate in the *ventral pallidum* [18]. In addition, a widespread distribution of 5-HT_{2A}R-immunopositive dendritic profile are also located in all component of the basal ganglia, including in the *globus pallidus* [44]. In the striatum, in situ ibridization studies indicates that the 5-HT_{2A}R is located striatopallidal and striatonigral neurons containing encephalin and dynorphin, respectively [77, 78].

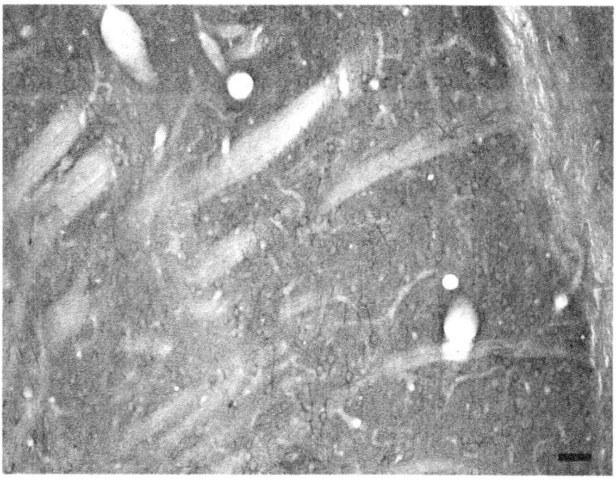

Fig. 6 Distribution of 5-HT_{2A} receptor immunoreactivity in the rat basal ganglia. Note the high number of immunopositive neurons in the caudate-putamen. Scale bar = 50 μm

Diencephalon

Epithalamus

No hybridization are located in the habenular complex [20]. Few somata containing the 5-HT_{2A}R are located only in the lateral habenular nucleus. Accordingly in the whole habenular complex, the density of immunopositive dendrites is low [9].

Thalamus

Using in situ hybridization procedures, the 5-HT_{2A}R seems to be localized only in the reticular nucleus and lateral geniculate nucleus [20]. On the contrary immunohistochemical procedures have found a relatively abundant somatodendritic distribution of the 5-HT_{2A}R in most of thalamic nuclei, with high densities staining in the ventrolateral, gelatinosus, ventral posterolateral (Fig. 7), ventral posteromedial and medial geniculate nuclei. Only midline nuclei show a low number of 5-HT_{2A}R positive neurons [9]. The finding that the 5-HT_{2A}R is expressed by thalamic relay nuclei correlates with an electrophysiological study demonstrating that local administration of DOI (a 5-$HT_{2A/2C}$ receptor agonist) in the ventral posteromedial thalamic nucleus decreases the neocortical high-voltage spindle activity in the rat [79]. Likewise, the systemic administration of ketanserin (a 5-$HT_{2A/2C}$ receptors antagonist) potentiates the inhibitory effects of serotonin in the lateral and medial geniculate nuclei [80].

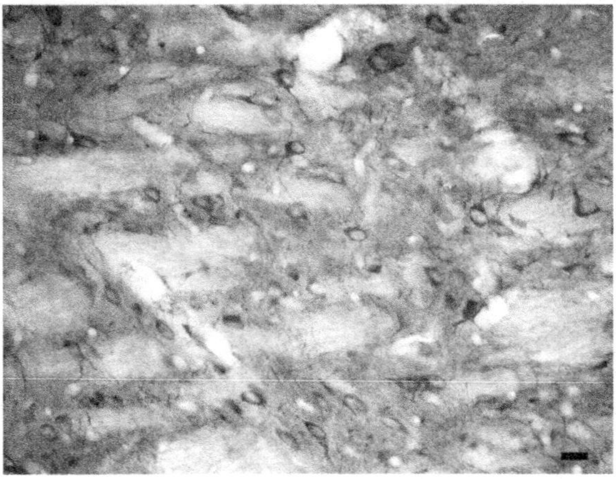

Fig. 7 Distribution of 5-HT_{2A} receptor immunoreactivity in the rat ventral posterolateral nucleus of the thalamus. The neurons appear strongly immunostained. Scale bar = 20 μm

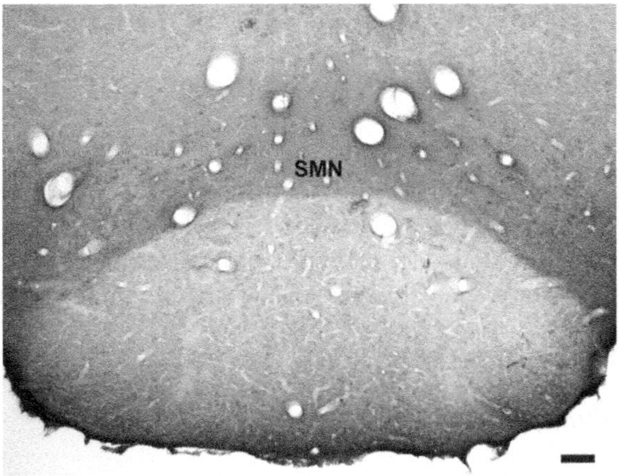

Fig. 8 Distribution of 5-HT_{2A} receptor immunoreactivity in the rat hypothalamus. Note many immunolabeled neurons in supramammillaris nucleus (SMN). Scale bar = 100 μm

Hypothalamus

Autoradiographic, in situ ibridization and immunohistochemical studies have demonstrated that the density of the 5-HT_{2A}R is relatively low in the whole hypothalamus. In this area, 5-HT_{2A}R mRNA are distribute mainly in the medial mammillary nucleus and to some extent in the region of the *tuber cinereum* [20]. Immunoreactive somata are relatively abundant only in the medial mammillary, supramammillaris (Fig. 8) and magnocellular preoptic nuclei [9, 41]. Dendrites expressing the 5-HT_{2A}R are distributed throughout the hypothalamus. These processes are strongly stained and abundant in the parastrial nucleus [9].

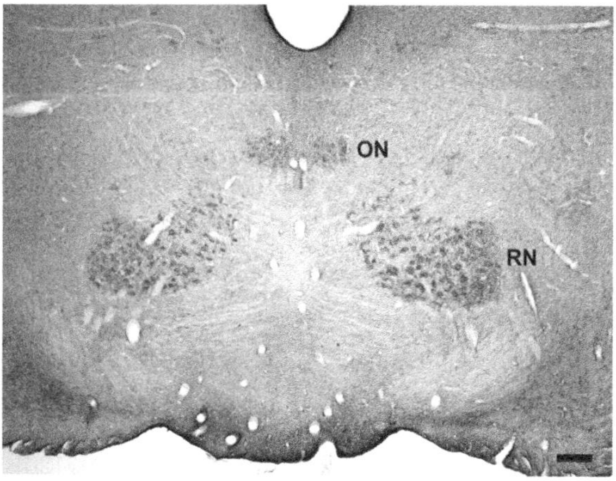

Fig. 9 Distribution of 5-HT_{2A} receptor (5-HT_{2A}R) immunoreactivity in the rat mesencephalon. The motoneurons located in the oculomotor (ON) and red (RN) nuclei are strongly immunoreactive for the 5-HT_{2A}R. Scale bar = 200 μm

Brainstem

Mesencephalon

High level of 5-HT_{2A}R mRNA are present in the oculomotor nucleus, trochlear nucleus, nucleus of Darkschewitsch, interstitial nucleus of Cajal, red nucleus, and peripeduncular nucleus [20]. In the mesencephalon many somatodendritic profiles immunoreative for the 5-HT_{2A}R are numerous especially in the following areas: oculomotor nucleus (Fig. 9), nucleus of Darkschewitsch, anterior pretectal nucleus, mesencephalic trigeminal nucleus and red nucleus (Fig. 9) [9, 41]. A relatively high density of somatodendritic profiles expressing the 5-HT_{2A}R are located in the superior colliculus (deep gray), inferior colliculus, trochlear nucleus, interstitial nucleus of the medial longitudinal fasciculus, magnocellular nucleus of posterior commissure, anterior pretectal nucleus and deep mesencephalic nucleus [9]. Superficial and intermediate gray of the superior colliculus, dorsal and median raphe nuclei, and ventral tegmental area contain a low densities of 5-HT_{2A}R immunoreactivity [9]. Interestingly, double-immunofluorescence experiments have demonstrated that 5-HT_{2A}R colocalize with dopaminergic neurons throughout the A10 cell group [81].

Pons

Pontine nuclei and reticulotegmental nucleus of the pons present high levels of 5-HT_{2A}R mRNA [20]. Intermediate levels of hybridization are located in the dorsal tegmental nucleus, parabrachial nucleus and subcoeruleus nucleus [20]. With the exception of the *locus coeruleus*, many areas located in the pons contains 5-HT_{2A}R-immunoreactive somata. The highest density of immunoreactive cell bodies are located in motor trigeminal nucleus and abducent nucleus [9, 41]. On the contrary, principal sensory trigeminal nucleus and pontine nuclei show rare immunostaining of 5-HT_{2A}R-immunoreactive somata [9]. Throughout the pons there are many dendrites containing the 5-HT_{2A}R. These processes are particularly abundant in dorsal tegmental nucleus, laterodorsal tegmental nucleus, motor trigeminal nucleus and abducens nucleus [9].

Medulla Oblongata

In the medulla oblongata the level of the 5-HT_{2A}R mRNA is low than in the midbrain. In fact, intermediate levels of 5-HT_{2A}R mRNA are located in the vestibular nuclei, prepositus hypoglossal nucleus, inferior olive, cuneate nuclei, gigantocellular reticular nucleus and lateral reticular nucleus [20]. In the medulla oblongata the highest density of 5-HT_{2A}R-immunoreactive somatodendritic profiles are located in some cranial nerves nuclei especially facial, ambiguus and hypoglossal nuclei [9, 41]. Also the vestibular nuclei contained a relatively high density of somatodendritic immunostained profiles [9]. A similar distribution can be observed in some cranial nerve nuclei (spinal trigeminal nucleus, parasympathetic nucleus of the vagus), lateral superior olive, reticular formation (gigantocellular reticular field, intermediate reticular field, parvocellular reticular field and lateral reticular nucleus), gracilis nucleus and cuneate nucleus [9]. A few immunostained cells are located in the ventral and dorsal cochlear nuclei, inferior olive, nucleus of the solitary tract and prepositus hypoglossal nucleus [9].

Cerebellum

Cerebellar nuclei present intermediate levels of 5-HT_{2A}R mRNA [20]. All deep cerebellar nuclei exhibit many somatodendritic profiles expressing the 5-HT_{2A}r [9]. In the cerebellar cortex a relatively high density of somata and dendrites are 5-HT_{2A}R-immunoreactive in the Purkinje cell layer [9]. In the granule cell layer few somatodendritic profiles contain the 5-HT_{2A}R, whereas in the molecular layer this receptor is located only in dendrites [9].

References

1. Aghajanian GK (1995) Electrophysiology of serotonin receptor subtypes and signal transduction pathways. In: Bloom FE, Kupfer DJ (eds) Psychopharmacology: the fourth generation of progress. Raven Press, New York, pp 451–460
2. Barnes NM, Sharp T (1999) A review of central 5-HT receptors and theirs function. Neuropharmacology 38:1083–1152

3. Lambe EK, Aghajanian GK (2001) The role of Kv1.2-containing potassium channels in serotonin-induced glutamate release from thalamocortical terminals in rat frontal cortex. J Neurosci 21:9955–9963
4. Raymond J, Mukhin Y, Gelasco A, Turner J, Collinsworth G, Gettys T, Grewal J, Garnovskaya M (2001) Multiplicity of mechanisms of serotonin receptor signal transduction. Pharmacol Ther 92:179–212
5. Bombardi C (2011) Distribution of 5-HT2A receptor immunoreactivity in the rat amygdaloid complex and colocalization with γ-aminobutyric acid. Brain Res 1370:112–128
6. Bombardi C (2012) Neuronal localization of 5-HT2A receptor immunoreactivity in the rat hippocampal region. Brain Res Bull 87:259–273
7. Bombardi C (2014) Neuronal localization of the 5-HT2 receptor family in the amygdaloid complex. Front Pharmacol 5(68):1–10
8. Bombardi C, Di Giovanni G (2013) Functional anatomy of 5-HT2A receptors in the amygdala and hippocampal complex: relevance to memory functions. Exp Brain Res 230:427–439
9. Cornea-Hébert V, Riad M, Wu C, Singh SK, Descarries L (1999) Cellular end subcellular distribution of the serotonin 5-HT2A receptor in the central nervous system of adult rat. J Comp Neurol 409:187–209
10. Jiang X, Xing G, Yang C, Verma A, Zhang L, Li H (2008) Stress impairs 5-HT2A Receptor-mediated serotonergic facilitation of GABA release in juvenile rat basolateral amygdala. Neuropsychopharmacology 33:1–14
11. McDonald AJ, Mascagni F (2007) Neuronal localization of 5-HT type 2A receptor immunoreactivity in the rat basolateral amygdala. Neuroscience 146:306–320
12. Meneses A (1999) 5-HT system and cognition. Neurosci Biobehav Rev 23:1111–1125
13. Meneses A (2002a) Involvement of 5-HT(2A/2B/2C) receptors on memory formation: simple agonism, antagonism, or inverse agonism? Cell Mol Neurobiol 22:675–688
14. Meneses A (2002b) Tianeptine: 5-HT uptake sites and 5-HT(1–7) receptors modulate memory formation in an autoshaping Pavlovian/instrumental task. Neurosci Biobehav Rev 26:309–319
15. Meneses A (2007) Do serotonin (1–7) receptors modulate short and long-term memory? Neurobiol Learn Mem 87:561–572
16. Meneses A, Hong E (1997) A pharmacological analysis of serotoninergic receptors: effects of their activation of blockade in learning. Prog Neuro-Psychopharmacol Biol Psychiatry 21:243–254
17. Meltzer HY, Huang M (2008) In vivo actions of atypical antispsychotic drug on serononergic and dopaminergic systems. Prog Brain Res 172:177–197
18. Xu T, Pandey SC (2000) Cellular localization of serotonin$_{2A}$ (5-HT$_{2A}$) receptors in the rat brain. Brain Res Bull 51:499–505
19. Li QH, Nakadate K, Tanaka-Nakadate S, Nakatsuka D, Cui Y, Watanabe Y (2004) Unique expression patterns of 5-HT2A and 5-HT2C receptors in the rat brain during post natal development: Western blot and immunohistochemical analyses. J Comp Neurol 469:128–140
20. Pompeiano M, Palacios JM, Mengod G (1994) Distribution of serotonin 5-HT$_2$ receptor family mRNAs: comparison between 5-HT$_{2A}$ and 5-HT$_{2C}$ receptors. Mol Brain Res 23:163–178
21. Wright DE, Seroogy KB, Lundgren KH, Davis BM, Jennes L (1995) Comparative localization of serotonin$_{1A,1C}$, and$_2$ receptor subtype mRNAs in rat brain. J Comp Neurol 351:357–373
22. Pazos A, Cortés R, Palacios JM (1985) Quantitative autoradiographic mapping of serotonin receptors in the rat brain. II. Serotonin-2 receptors. Brain Res 346:231–249
23. Radja F, Laporte A, Daval G, Verge D, Gozlan H, Hamon M (1991) Autoradiography of serotonin receptor subtypes in the central nervous system. Neurochemistry International 18:1–15
24. Aghajanian GK, Marek GJ (1999) Serotonin, via 5-HT2A receptors, increases EPSCs in layer V pyramidal cells of prefrontal cortex by an asynchronous mode of glutamate release. Brain Res 825:161–171
25. Feldman S, Newman ME, Gur E, Weidenfeld J (1998) Role of serotonin in the amygdala in hypothalamo-pituitary-adrenocortical responses. Neuroreport 9:2007–2009

26. Gresch PJ, Strickland LV, Sanders-Bush E (2002) Lysergic acid diethylamide-induced fos expression in rat brain: role of serotonin-2A receptors. Neuroscience 114:707–713
27. Harvey JA (2003) Role of the serotonin 5-HT(2A) receptor in learning. Learn Mem 10:355–362
28. Martín-Ruiz R, Puig MV, Celada P, Shapiro DA, Roth BL, Mengold G, Artigas F (2001) Control of serotoninergic function in medial prefrontal cortex by serotonin-2A receptors through a glutamate-dependent mechanism. J Neurosci 21:9856–9866
29. Nair SG, Gudelsky GA (2004) Activation of 5-HT2A receptors enhances the release of acetylcholine in the prefrontal cortex and hippocampus of the rat. Synapse 53:202–207
30. Rainnie DG (1999) Serotoninergic modulation of neurotransmission in the rat basolateral amygdala. J Neurophysiol 82:69–85
31. Stein C, Davidowa H, Albrecht D (2000) 5-HT_{1A} receptor-mediated inhibition and 5-HT_2 as well as 5-HT_3 receptor-mediated excitation in different subdivisions of the rat amygdala. Synapse 38:328–337
32. Stutzmann GE, LeDoux JE (1999) GABAergic antagonist block the inhibitory effects of serotonin in the lateral amygdala: a mechanism for modulation of sensory inputs related to fear conditioning. J Neurosci 19:1–4
33. Stutzmann GE, McEwen BS, LeDoux JE (1998) Serotonin modulation of sensory inputs to the lateral amygdala: dependency on corticosterone. J Neurosci 18:9529–9538
34. Zangrossi H Jr, Graeff FG (1994) Behavioral effects of intra-amygdala injections of GABA and 5-HT acting drugs in the elevated plus-maze. Braz J Med Biol Res 27:2453–2456
35. Puig MV, Watakabe A, Ushimaru M, Yamamori T, Kawaguchi Y (2010) Serotonin modulates fast-spiking interneuron and synchronous activity in the rat prefrontal cortex through 5-HT1A and 5-HT2A receptors. J Neurosci 30:2211–2222
36. Sokal DM, Giarola AS, Large CH (2005) Effects of $GABA_B$, 5-HT_{1A}, and 5-HT_2 receptor stimulation on activation and inhibition of the rat lateral amygdala following medial geniculate nucleus stimulation in vivo. Brain Res 1031:141–150
37. Morilak DA, Garlow SJ, Ciaranello RD (1993) Immunocytochemical localization and description of neurons expressing $serotonin_2$ receptors in the rat brain. Neuroscience 54:701–717
38. Morilak DA, Somogyi P, Lujan-Miras R, Ciaranello RD (1994) Neurons expressing 5-HT2 receptors in the rat brain: neurochemical identification of cell types by immunocytochemistry. Neuropsychopharmacology 11:157–166
39. Peddie CJ, Davies HA, Colyer FM, Stewart MG, Rodríguez JJ (2008) Colocalisation of serotonin 2A receptors with glutamate receptor subunits NR1 and GluR2 in the dentate gyrus: An ultrastructural study of a modulatory role. Exp Neurol 211:561–573
40. Willins DL, Deutch AY, Roth BL (1997) Serotonin 5-HT_{2A} receptors are expressed on pyramidal cells and interneurons in the rat cortex. Synapse 27:79–82
41. Hamada S, Senzaki K, Hamaguchi-Hamada K, Tabuchi K, Yamamoto H, Yamamoto T, Yoshikawa S, Okano H, Okado N (1998) Localization of 5-HT2A receptor in rat cerebral cortex and olfactory system revealed by immunohistochemistry using two antibodies raised in rabbit and chicken. Mol Brain Res 54:199–211
42. Jakab RL, Goldman-Rakic PS (1998) $5\text{-hydroxytryptamine}_{2A}$ serotonin receptors in the primate cerebral cortex: possible site of action of hallucinogenic and antipsychotic drugs in pyramidal cell apical dendrites. Proc Natl Acad Sci 95:735–740
43. Jakab RL, Goldman-Rakic PS (2000) Segregation of serotonin 5-HT2A and 5-HT3 receptors in inhibitory circuits of the primate cerebral cortex. J Comp Neurol 417:337–348
44. Jansson A, Tinner B, Bancila M, Vergé D, Steinbusch HWM, Agnati LF, Fuxe K (2001) Relationships of 5-hydroxytryptamine immunoreactive terminal-like varicosities to 5-hydroxytryptamine-2A receptor-immunoreactive neuronal processes in the rat forebrain. J Chem Neuroanat 22:185–203
45. Lüttgen M, Ögren SO, Meister B (2004) Chemical identity of 5-HT_{2A} receptor immunoreactive neurons of the rat septal complex and dorsal hippocampus. Brain Res 1010:156–165
46. Miner LAH, Backstrom JR, Sanders-Bush E, Sesack SR (2003) Ultrastructural localization of serotonin 2A receptors in the middle layers of the rat prelimbic prefrontal cortex. Neuroscience 116:107–117

47. Cornea-Hébert V, Watkins KC, Roth BL, Kroeze WK, Gaudreau P, Leclerc N , Descarries L (2002) Similar ultrastructural distribution of the 5-HT2A serotonin receptor and microtubule-associated protein MAP1A in cortical dendrites of adult rat. Neuroscience 113:23–35
48. de Almeida J, Mengod G (2007) Quantitative analysis of glutamatergic and GABAergic neurons expressing 5-HT(2A) receptors in human and monkey prefrontal cortex. J Neurochem 103:475–486
49. Aghajanian GK, Marek GJ (1997) Serotonin induces excitatory postsynaptic potentials in apical dendrites of neocortical pyramidal cells. Neuropharmacology 36:589–599
50. Blue ME, Yagaloff KA, Mamounas LA, Hartig PR, Molliver ME (1988) Correspondence between 5-HT2 receptors and serotoninergic axons in rat neocortex. Brain Res 453:315–328
51. Amargós-Bosch M, Bortolozzi A, Puig MV, Serrats J, Adell A, Celada P, Toth M, Mengod G, Artigias F (2004) Co-expression and in vivo interaction of serotonin1A and serotonin 2A receptors in pyramidal neurons of prefrontal cortex. Cereb Cortex 14:281–299
52. Puig MV, Celada P, Díaz-Mataiax L, Artigias F (2003) *In vivo* modulation of the activity of pyramidal neurons in the rat medial prefrontal cortex by 5-HT2A receptors. Relationship to thalamocortical afferents. Cereb Cortex 13:1870–1882
53. Nocjar C, Alex KD, Sonneborn A, Abbas AI, Roth BL, Pehek EA (2015) Serotonin-2C and 2A receptor co-expression on cells in the rat medial prefrontal cortex. Neuroscience 297:22–37
54. Santana N, Bortolozzi A, Serrats J, Mengod G, Artigias F (2004) Expression of serotonin$_{1A}$ and serotonin$_{2A}$ receptors in pyramidal and GABAergic neurons of the rat prefrontal cortex. Cereb Cortex 14:1100–1109
55. Abi-Saab WM, Bubser M, Roth RH, Deutch AY (1999) 5-HT$_2$ receptor regulation of extracellular GABA levels in the prefrontal cortex. Neuropsychopharmacology 20:92–96
56. Ashby CR, Jiang LH, Kasser RJ, Wang RY (1990) Electrophysiological characterization of 5-hydroxytryptamine-2 receptors in the rat medial prefrontal cortex. J Pharmacol Exp Ther 252:171–178
57. Zhou FM, Hablitz JJ (1999) Activation of serotonin receptors modulates synaptic transmission in rat cerebral cortex. J Neurophysiol 82:2989–2999
58. Furtak SC, Wei S-M, Agster KL, Burwell RD (2007) Functional neuroanatomy of the parahippocampal region in the rat: the perirhinal and postrhinal cortices. Hippocampus 17:709–722
59. Scharfman HE, Witter MP, Schwarcz R (2000) The parahippocampal region. Implications for neurological and psychiatric diseases. Introduction. Ann N Y Acad Sci 911:ix–xiii
60. Witter MP, Amaral DG (2004) Hippocampal formation. In: Paxinos G (ed) The rat nervous system. Elsevier Academic Press, San Diego, pp 635–704
61. Uneyama H, Munakata M, Akaike N (1992) 5-HT response of rat hippocampal pyramidal cell bodies. Neuroreport 3:633–636
62. Amaral DG, Lavenex P (2007) Hippocampal neuroanatomy. In: Andersen P, Morris R, Amaral D, Bliss T, O'Keefe J (eds) The hippocampus book. Oxford University Press, New York, pp 37–114
63. Guo JD, Rainnie DG (2010) Presynaptic 5-HT1B receptor-mediated serotonergic inhibition of glutamate transmission in the bed nucleus of the stria terminalis. Neuroscience 165:1390–1401
64. Hashimoto K, Kita H (2008) Serotonin activates presynaptic and postsynaptic receptors in rat globus pallidus. J Neurophysiol 99:1723–1732
65. Piguet P, Galvan M (1994) Transient and long-lasting actions of 5-HT on rat dentate gyrus neurones in vitro. J Physiol 481:629–639
66. Shen R-Y, Andrade R (1998) 5-hydroxytryptamine 2 receptor facilitates GABAergic neurotransmission in rat hippocampus. J Pharmacol Exp Ther 285:805–812
67. Jha S, Rajendran R, Fernandes KA, Vaidya VA (2008) 5-HT2A/2C receptor blockade regulates progenitor cell proliferation in the adult rat hippocampus. Neurosci Lett 441:210–214
68. Ge S, Goh ELK, Sailor KA, Kitabatake Y, Ming G, Song H (2006) GABA regulates synaptic integration of newly generated neurons in the adult brain. Nature 439:589–593
69. Banasr M, Hery M, Printemps R, Daszuta A (2004) Serotonin induced increases in adult cell proliferation and neurogenesis are mediated through different and common 5-HT receptor subtypes in the dentate gyrus and the subventricular zone. Neuropsychopharmacology 29:450–460

70. Amaral DG, Scharfman HE, Lavenex P (2007) The dentate gyrus: fundamental neuroanatomical organization (dentate gyrus for dummies). Prog Brain Res 163:3–22
71. Freund TF, Buzsáki G (1996) Interneurons of the hippocampus. Hippocampus 6:347–470
72. Olucha-Bordonau FE, Fortes-Marco L, Otero-García M, Lanuza E, Martínez- García F (2015) Amygdala: structure and function. In: Paxinos G (ed) The rat nervous system, 4th edn. Academic Press, San Diego, pp 442–480
73. Chen A, Hough J, Li H (2003) Serotonin type II receptor activation facilitates synaptic plasticity via N-Methyl-D-Aspartate-mediated mechanism in the rat basolateral amygdala. Neuroscience 119:53–63
74. Mascagni F, McDonald AJ (2007) A novel subpopulation of 5-HT type 3A receptor subunit immunoreactive interneurons in the rat basolateral amygdala. Neuroscience 144:1015–1024
75. Radja F, Descarries L, Dewar KM, Reader TA (1993) Serotonin 5-HT1 and 5-HT2 receptors in adult rat brain after neonatal destruction of nigrostriatal dopamine neurons: a quantitative autoradiographic study. Brain Res 606:273–285
76. El Mansari M, Radja F, Ferron A, Reader TA, Molina-Holgado E, Descarries L (1994) Hypersensitivity to serotonin and its agonists in serotonin hyperinnervated neostriatum after neonatal dopamine denervation. Eur J Pharmacol 261:171–178
77. Mijnster MJ, Raimundo AGV, Koskuba K, Klop H, Docter GJ, Groenewegen HJ, Voorn P (1997) Regional and cellular distribution of serotonin 5-hydroxytryptamine 2a receptor mRNA in the nucleus accumbens, olfactory tubercle, and caudate putamen of the rat. J Comp Neurol 389:1–11
78. Ward RP, Dorsa DM (1996) Colocalization of serotonin receptor subtypes 5-HT2A, 5-HT2C, and 5-HT6 with neuropeptides in rat striatum. J Comp Neurol 370:405–414
79. Jäkälä P, Sirviö J, Koivisto E, Björklund, Kaukua MJ, Riekkinen P Jr (1995) Modulation of rat neocortical high-voltage spindle activity by 5-HT1/5-HT2 receptor subtype specific drugs. Eur J Pharmacol 282:39–55
80. Lakoski JM, Aghajanian GK (1985) Effects of ketanserin on neuronal responses to serotonin in the prefrontal cortex, lateral geniculate and dorsal raphe nucleus. Neuropharmacology 24:265–73
81. Nocjar C, Roth BL, Pehek EA (2002) Localization of 5-HT_{2A} receptors on dopamine cells in subnuclei of the midbrain A10 cell group. Neuroscience 111:163–176

PET Imaging of the 5-HT_{2A} Receptor System: A Tool to Study the Receptor's In Vivo Brain Function

Matthias M. Herth and Gitte M. Knudsen

Abstract The serotonergic 5-HT_{2A} receptor system plays a key modulatory role for many brain functions such as regulation of mood, temperature, sex, appetite and emotions. The receptor is also involved in a number of brain disorders, for example, depression, Alzheimer's disease or schizophrenia. This makes it an obvious target for many drugs.

This chapter describes how the in vivo imaging technique positron emission tomography (PET) can be used to investigate 5-HT_{2A} receptors in humans in terms of neurobiology and brain disorders. It also highlights how PET can be used in drug development and in humans in terms of neurobiology and brain disorders. It also highlights how PET can be used in drug development and explains the basic methodology of PET. The chapter discusses currently used 5-HT_{2A} receptor selective PET tracers. This chapter explains with the help of 5-HT_{2A} receptor tracers how PET can be used in vivo to determine a drug's receptor occupancy. Finally, the possibility of 5-HT_{2A} receptor selective PET tracers imaging endogenous serotonin levels in the living brain is discussed.

Keywords PET • 5-HT_{2A} • [^{11}C]MDL 100907 • (R)-[^{18}F]MH.MZ • [^{18}F]altanserin • [^{11}C]Cimbi-36

M.M. Herth (✉)
Center for Integrated Molecular Brain Imaging, Copenhagen University Hospital, Rigshospitalet, Copenhagen, Denmark

Department of Drug Design and Pharmacology, Faculty of Health and Medicine, University of Copenhagen, Copenhagen, Denmark

Department of Clinical Physiology, Nuclear Medicine and PET, Copenhagen University Hospital, Copenhagen, Denmark
e-mail: matthias.herth@sund.ku.dk

G.M. Knudsen
Center for Integrated Molecular Brain Imaging, Copenhagen University Hospital, Rigshospitalet, Copenhagen, Denmark

B.P. Guiard, G. Di Giovanni (eds.), *5-HT_{2A} Receptors in the Central Nervous System*, The Receptors 32, https://doi.org/10.1007/978-3-319-70474-6_5

Abbreviation

$5\text{-}HT_{2A}R$	Serotonin 2A receptor, 5-hydroxytryptamine 2A receptor
$5\text{-}HT_{2B}R$	Serotonin 2B receptor, 5-hydroxytryptamine 2B receptor
$5\text{-}HT_{2C}R$	Serotonin 2C receptor, 5-hydroxytryptamine 2C receptor
AD	Alzheimer's disease
A_s	Specific activity
B_{avail}	Concentration of receptors available for binding
BBB	Blood brain barrier
B_{max}	Maximal concentration of receptors
BP	Binding potential
CNS	Central nervous system
D_2	Dopamine receptor D_2
DAG	Diacylglycerol
DOI	1-[2,5-dimethoxy-4-iodophenyl]-2-aminopropane
f_{ND}	The free fraction in the non-displaceable tissue compartment
HPLC	High-performance liquid chromatography
IP_3	Inositol triphosphate
K_D	Radioligand equilibrium dissociation constant
K_i	Inhibition constant
LSD	Lysergic acid diethylamide
MCI	Mild cognitive impairment
MDMA	3,4-methylenedioxymethamphetamine, "ecstasy"
NIGA	Non-invasive graphical analysis
PET	Positron emission tomography
PLC	Phospholipase C
PKC	Protein kinase
P-pg	P-glycoprotein
SPECT	Single photon emission computed tomography
SRTM	Simplified reference tissue model
SUV	Standard uptake value
TAC	Time-activity curve
TCM	Tissue compartment modeling
$tBR_{target/off\text{-}target}$	Theoretical, observed binding ratio of the target to another off-target

Involvement of the $5\text{-}HT_{2A}$ Receptor in the Neuronal Signal Cascade and Its Role in Pathological Conditions

The G-protein coupled serotonin 2A receptor (5-hydroxytryptamine 2A, or $5\text{-}HT_{2A}$) subtype is the most important excitatory receptor in the serotonergic system and its distribution has been extensively characterized in the central nervous system (CNS) by autoradiography, in situ hybridization, immunocytochemical techniques and with in vivo imaging [1–5]. The 5-HT_{2A} receptor system transduces neuronal signals primarily via the Gαq signal cascade. Upon agonistic receptor stimulation, Gαq and βγ subunits of the G-protein dissociate and initiate downstream effector pathways. For example, the activity of the phospholipase C (PLC) is stimulated, which subsequently promotes a release of diacylglycerol (DAG) and inositol triphosphate (IP_3). Among other things this leads to a stimulation of the protein kinase C (PKC) that ultimately affects the function of other proteins through phosphorylation [6]. A proper balance of 5-HT_{2A} receptor activity at inhibitory and excitatory neurons appears to be required for normal neuronal functioning [7, 8]. The 5-HT_{2A} receptor has been implicated in various physiological and pathological functions (aging, appetite, sexual behavior, pain) and neuropsychiatric disorders, including the Alzheimer's disease (AD), schizophrenia, major depression, anxiety, Asperger's, alcohol addiction, and sleep disorders [5, 8–21]. Evidence for the role of 5-HT_{2A} receptors in these brain disorders and pathological conditions comes both from post-mortem and brain imaging studies [22, 23]. For example, genetically modified mice, which lack 5-HT_{2A} receptors, have changed sleep patterns [19]. Further, short term and long term memory is negatively affected by the 5-HT_{2A} agonist 1-[2,5-dimethoxy-4-iodophenyl]-2-aminopropane (DOI), whereas 5-HT_{2A} antagonists are without any impact on memory [24].

From a pharmacological perspective, 5-HT_{2A} receptors are of interest for many reasons. They are the primary target of psychedelic compounds and contribute to the efficacy of many antipsychotic medications, and are used as treatment for various other psychiatric disorders [25–27]. For example, Nordström et al. showed that relatively low doses (125–200 mg/day) of the atypical antipsychotic drug clozapine lead to 5-HT_{2A} receptor occupancies ranging between 80 and 90% while it only occupied 20–30% of D_2-like dopamine receptors [28]. They suggest that clozapine's "atypical" clinical profile may be explained by its relatively low occupancy of D_2-like dopamine receptors combined with its high affinity for 5-HT_2 receptors [29, 30]. Furthermore, the 5-HT_{2A} receptor is the key receptor involved in hallucinogenic/psychotic effects [31] and many recreational hallucinogens such as lysergic acid diethylamide (LSD), MDMA (3,4-methylenedioxymethamphetamine, "Ecstasy") or DOI elicit their hallucinogenic effects by stimulating the 5-HT_{2A} receptor [32–35].

Positron Emission Tomography (PET)

PET

PET is a non-invasive and quantitative whole body in vivo molecular imaging technique that produces a three-dimensional image of functional processes in the living body. It presents an alternative to studies that otherwise would require the sacrifice of larger number of research animals. Not only do in vivo studies allow for less number of animals needed, research animals can also be used repeatedly for investigations and thus serve as their own controls. Furthermore, PET also permits studies of patients with, e.g. Alzheimer's or Parkinson's disease and may thereby disclose pathophysiological changes, be of diagnostic value or suggest and monitor treatment opportunities. PET data can generate information about pharmacokinetics and pharmacodynamics, metabolism and dose linearity. It can also quantify biological processes at the cellular and subcellular levels within an intact living organism [36, 37]. In addition, PET can be used to monitor pathological processes and environmental factors influencing brain diseases [38]. Moreover, drug effects on receptor up- and down-regulation can be measured [5, 39]. Finally, PET can also be used to determine the occupancy of therapeutic drugs, which can help to estimate the optimal doses in Phase II studies. For example, it has been demonstrated that between 60 and 80% occupancy of D_2 receptors is required for antipsychotic medication efficacy and that beyond this level, side effects are likely to occur [40, 41]. Thus, in theory PET can be used to find the right dose for each individual patient (personalized medicine) by comparing baseline (before treatment) and after intervention conditions (after treatment).

Compared to other in vivo imaging methods, PET has the advantage of high sensitivity (the level of detection approaches 10^{-12} M of tracer) and isotropism (i.e., ability to detect organ accumulation accurately regardless of tissue depth) [42, 43]. In addition, it results in a higher spatial and temporal resolution, sensitivity and better quantification compared to single photon emission computed tomography (SPECT) [5, 42, 43]. For an in-depth review about PET and its applications, see e.g. Herzog et al. or Saha [44, 45].

In short, in vivo quantitative PET studies of 5-HT_{2A} receptor binding may significantly advance the understanding in the living human brain and thus provide a valuable technique for the investigation of 5-HT_{2A} receptors in different subject and patient populations.

Theoretical Background

Basic Physical Principle

PET is based on the unique decay characteristics of positron emitting radionuclides. Thereby, a neutron-deficient isotope converts a proton into a neutron and a positron (β^+-particle). The positron is emitted from the nucleus and travels up to a few millimeters until it encounters an electron. Afterwards, the positron and the electron merge into a positronium. This exotic particle annihilates almost immediately into two γ-photons moving in opposite directions. The coincident detection of numerous of these photon-pairs forms the basis of PET imaging, since computational reconstruction along straight lines between detector pairs allows the determination of the photon's source of origin in a three dimensional space (Fig. 1).

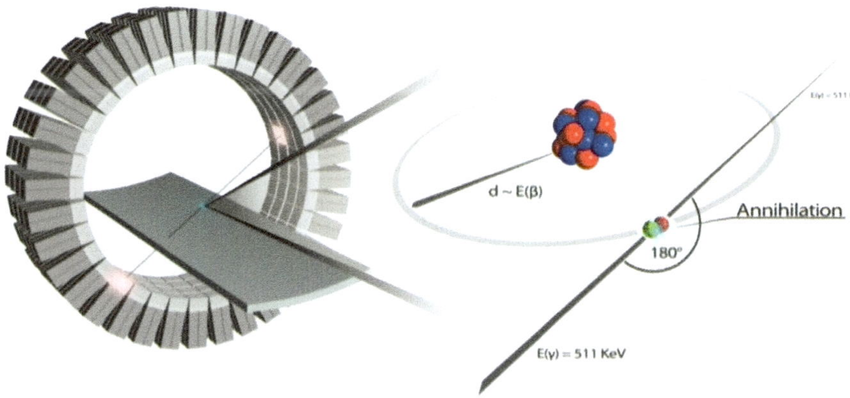

Fig. 1 Basic physical principle of Positron Emission Tomography (PET). After annihilation, coincident detection of numerous photon pairs in a PET scanner provide the necessary data to reconsturct the origin of the photon's source [46]

Methodology and PET Quantifications

Methodology

PET scanning is a rather extensive and costly operation. Usually, such an experiment starts with the production of a radionuclide in a cyclotron, which is afterwards incorporated into a biological relevant molecule. The radiotracer is then analyzed (quality control) and released if the tracer meets the quality criteria such as sufficient high specific activity, radiochemical and chemical purity. Before the tracer can be injected into a living subject (usually into the blood stream), the PET camera has to be calibrated and the subject prepared for injection. During the scan, detectors of the PET camera record the tissue concentration of the radiotracer. For quantification of the specific radiotracer binding in the brain, some models require determination of the concentration of intact radiotracer in arterial blood. After the scan, the data will be reconstructed and then finally analyzed. Usually, 3–8 experimenters are involved in a single PET scan.

Tracer Dose Concept

PET radiotracers aiming to quantify receptor binding have to be applied in trace quantities meaning that the applied concentration of labeled ligand is too low to influence any physiological process. The injected dose of the radioligand is usually ~1.000 fold lower than a pharmacological dose, i.e. the dose needed to evoke a pharmacological response, via the activation of a receptor or inhibition of an enzyme. Usually < 5 μg of tracer (labeled plus unlabeled) are used for human imaging studies. Since PET is sufficiently sensitive to detect trace amounts of a labeled compound, investigations can be done without disturbing the native biological environment. This concept is often referred to as "the tracer principle".

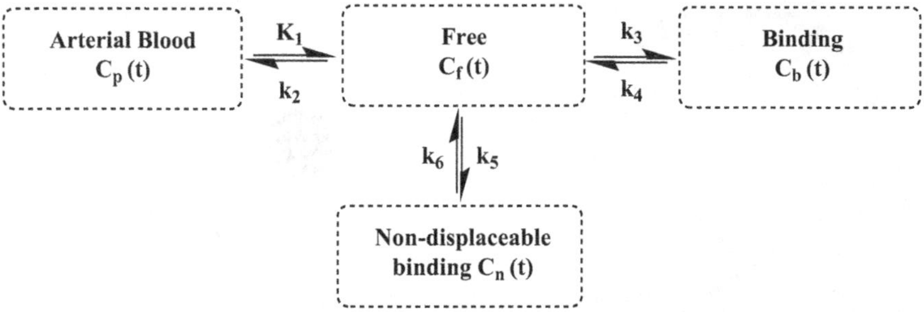

Fig. 2 Three tissue compartmental model consists of four compartments (the plasma, the free ligand in tissue, specific binding and non-displaceable binding compartment) and six transport and binding rates

All brain PET studies seek to measure a target receptor in terms of specific radioligand binding. Specific binding is defined as that associated with the target and distinct from radioligand which is free in solution or nonspecifically associated with other macromolecular components. As mentioned above, the radioligand is administered at tracer doses and thus occupies only a negligible (often defined as <5% to 10%) percentage of target sites. As a consequence, this specific binding will reflect the entire population of target sites, without significantly perturbing the total number of available receptors [47].

PET Kinetic Analysis and the Binding Potential

In order to interpret PET data and thus be able to quantify ligand-neuroreceptor interactions, one has to understand some basic principles behind PET. The data acquired by a PET camera is composed of various signals. In order to isolate the signal component of interest, the data has to be mathematical analyzed by the use of a model. The applied PET kinetics analysis models are based on pharmacokinetics. Assumptions are often made in order to simplify the complex in vivo situation. For some models, one regards the tracer distribution as being assigned to conceptually separate entities, referred to as compartments. It is assumed that once the tracer passes from one to another compartment, the tracer is instantaneously mixed within the compartment. The number of required compartments is determined by the time-activity data from the individual tissue types or brain regions. The most comprehensive compartmental model is the three tissue compartmental model (Fig. 2). The arterial blood constitutes the source from where the tracers pass the blood-brain barrier (BBB) into the first compartment (the free compartment). The second compartment consists of specifically bound radiotracer and the third is a non-displaceable binding compartment that exchanges with the free compartment. It is assumed that the transport and binding rates of the tracer are linearly related to the concentration within the compartments. In this case, the tracer concentration within the free-, the specific binding and non-displaceable binding compartment can be described by the following equations,

Change in concentration within the free compartment

$$\frac{dC_f(t)}{dt} = K_1C_p(t) + k_6C_n(t) + k_4C_b(t) - (k_2 + k_3 + k_5)C_f(t)$$

Change in concentration over time within the specific binding compartment

$$\frac{dC_b(t)}{dt} = k_3C_f(t) - k_4C_b(t)$$

Change in concentration within the non-displaceable binding compartment

$$\frac{dC_n(t)}{dt} = k_5C_f(t) - k_6C_n(t)$$

where $C_p(t)$ [plasma concentration], $C_f(t)$, $C_b(t)$ and $C_n(t)$ are radioactivity concentrations at time (t) for each compartment. The sum of the aforementioned compartments is describing the signal that a PET camera detects, once the blood volume component has been subtracted. The three tissue compartment model consists of six unknown parameters, which are difficult to assess experimentally. Therefore, the three tissue compartment model is often reduced to a two tissue compartment model by assuming that the free and non-displaceable compartments are in instantaneous equilibrium. Rate constants are estimated. This, as well as more simplified two-tissue or one-tissue compartment models enable the estimation of transport and binding rates of the tracer by fitting the measured PET data to the relevant model. In order to avoid arterial cannulation, that is necessary to determine $C_p(t)$, models using a non-displaceable binding region in the brain rather than arterial input functions are often preferred. These include e.g. the simplified reference tissue model (SRTM) or the Logan non-invasive analysis. We here kindly refer the reader to more extensive reviews about PET kinetics analysis and compartment modelling, e.g. Wernick et al. [48, 49].

The most frequently used outcome measure of PET receptor neuroimaging are variants of the binding potential (BP) that is proportional to the receptor concentration. The BP is determined as the ratio of B_{avail} (concentration of receptors available for binding) to K_D (radioligand equilibrium dissociation constant) and can be derived from the Michaelis-Menten equilibrium Eq. (1) [48].

$$BP = \frac{B_{max}}{K_D} \tag{1}$$

In the following, we will refer to the outcome measure BP_{ND}, which is the BP times the fraction of free radioligand in the reference region. BP_{ND} is proportional to the ratio at equilibrium of specifically bound radioligand to that of non-displaceable

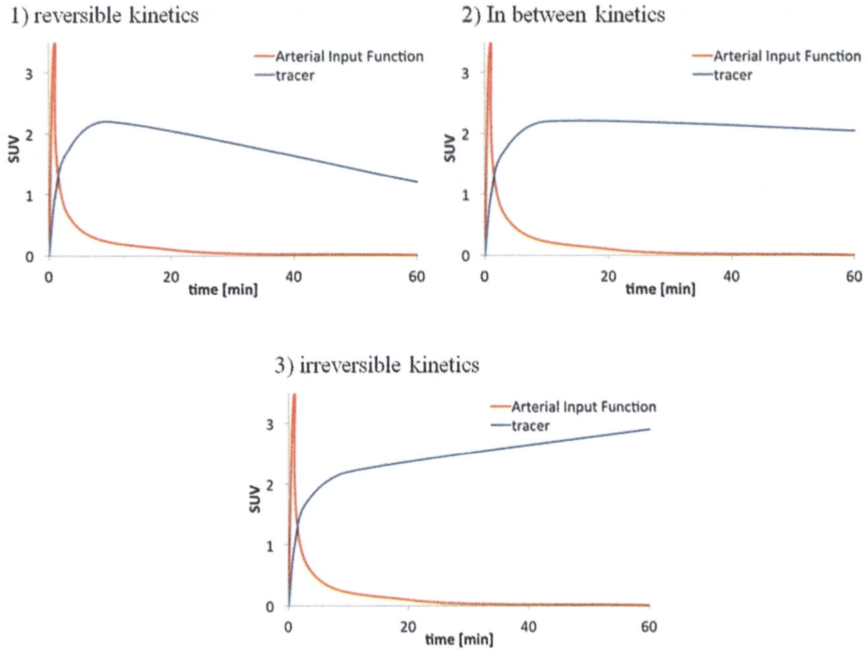

Fig. 3 PET kinetics. In general, kinetics can be divided in three different classes (1) reversible kinetics (2) kinetics in between and (3) irreversible kinetics [51]. Figure 3 displays typical time activity curve (TAC) examples that fall into those three classes [8, 52–55] (To identify which group a given tracer belongs to, one needs to consider both the tissue time-activity curve (TAC) and the arterial input function (AIF))

radioligand in tissue and is the outcome measure from reference tissue methods, as it compares the concentration of radioligand in receptor-rich to receptor-free regions [47]. The specific radioligand binding can be determined at equilibrium relative to either a reference region void of receptors (BP_{ND}), to plasma radioligand (BP_P) or relative to the free (non-protein bound) plasma radioligand concentration (BP_F). PET modeling and the corresponding binding potentials (BP_{ND}, BP_P, BP_F) are reviewed in more detail elsewhere [47, 50].

Tracer Kinetics

In general, the kinetics of a radiotracer falls into three classes: reversible, essentially irreversible and/or in between (Fig. 3). With reversible and completely irreversible radioligands it is easier to identify the model parameters than radioligands with in-between kinetics [51].

In Vivo Selectivity and Occupancy Measurements

In order to address the in vivo selectivity of a radioligand, blocking and/or competition experiments can be carried out. The radioligand is then administered either after or simultaneously with a known selective antagonist (blocking study) or challenged during the PET scan (competition study). If the tracer binds selectively to the same receptor pool as the "cold" drug then a reduction in the binding will be observed. This verifies the specificity of the radioligand binding. By determining the fraction of reduction in binding, one can calculate the occupancy of the target Eq. (2). In this way, the specificity of the radioligand binding but also the drug dose-occupancy relationship can be determined.

$$\text{Occupancy} = \frac{BP_{\text{baseline}} - BP_{\text{challenge}}}{BP_{\text{baseline}}} \quad (2)$$

The occupancy of the target can thus be used to determine a "cold" drug's receptor occupancy at a certain target, which in turns is helpful to determine the maximum and optimal dose of the drug in, e.g., a clinical trial. Only well-established and very selective PET tracers are considered useful for such experiments.

Success Criteria for a 5-HT$_{2A}$ Receptor Ligand

Receptor Availability for 5-HT$_{2A}$ PET Imaging

One of the most relevant criteria for a PET radiotracer to be considered successful is its selectivity for its target receptor or enzyme. Thus, the radioligand must possess high affinity towards the target and low affinity for other receptors and proteins to maximize the target to background ratio. A successful PET tracer has often target affinity in the nanomolar or subnanomolar range. However, as can be seen from Eq. (3) the tissue binding of a radioligand not only depends on the target affinity and selectivity, but also on the number of target receptors available for binding (B_{avail}) compared to off-targets. For example, if B_{avail} for the desired target is much higher than B_{avail} for non-targets, then the observed binding may still reflect the target in question even if the radioligand displays higher affinity towards off-targets. Conversely, if a non-target binding site is abundant compared to the target itself, the binding signal is more likely to reflect the off-target, even if the off-target displays lower affinity than the wanted target [41]. For example, for the 5-HT$_{2A}$ receptor a ~ 35-fold K_d difference over the 5-HT$_{1A}$ receptor is necessary in order to avoid more than 10% PET signal interference from the 5-HT$_{1A}$ receptors in hippocampus regions, whereas in the cortex only a tenfold selectivity is needed. This is due to the B_{avail} value of the 5-HT$_{1A}$ receptor compared to the 5-HT$_{2A}$ target [56]. For most targets, a 10–100 times higher binding of the target compared to other targets is considered acceptable.

In the following, we will describe how in vitro data can be used to estimate a compounds ability to image a certain target. Equation (3) describes the detected PET or autoradiography signal, which is dependent on the selectivity and receptor density (neglecting unspecific binding components). For a good tracer, the first term representing the ability to image the target is much bigger than those of off/targets.

$$\text{Signal}_{\text{PET/Autoradiography}} \underbrace{\frac{B_{\text{max},1}}{K_{d,1}}}_{\text{target}} + \underbrace{\frac{B_{\text{max},2}}{K_{d,2}} + \frac{B_{\text{max},3}}{K_{d,3}} + \ldots\ldots\ldots + \frac{B_{\text{max},n}}{K_{d,n}}}_{\text{off targets}} \quad (3)$$

A calculated measure (determined from in vitro affinity and B_{avail} data), which estimates the theoretical, observed binding ratio of the target to another off-target ($tBR_{target/off\text{-}target}$) is defined by the product of the selectivity (S) and the target to off-target ratio (D).

$$tBR_{\text{target/off target}} = S \times D = \left(\frac{K_{\text{d,off target}}}{K_{\text{d,target}}}\right)\left(\frac{B_{\text{avail,target}}}{B_{\text{avail,off target}}}\right) \quad (4)$$

It allows a selectivity estimation of the theoretical signal of two receptors while correlating their affinity and abundance. The measure neglects any other signal, which does not stem from either the target or the regarding off-target. So to speak, in Eq. (3) only the target and one off-target are taken into account and their corresponding values weighted.

Example: Calculation of the $tBR_{target/off\text{-}target}$ for altanserin in the caudate-putamen region for the target ($5\text{-}HT_{2A}$) and for the off-target ($5\text{-}HT_{1A}$): The affinity of altanserin for the $5\text{-}HT_{2A}$ receptor is 0.13 nM and for the $5\text{-}HT_{1A}$ receptor, it is 1570 nM (Table 1). B_{avail}/B_{max} in caudate putamen is ca. 23 fmol/mg original wet tissue for the $5\text{-}HT_{2A}$ receptor (4) and 4 fmol/mg original wet tissue for the $5\text{-}HT_{1A}$ receptor [56].

$$tBR_{5\ HT_{2A}/5\ HT_{1A}} = S \times D = \left(\frac{1570\ \text{nM}}{0.13\ \text{nM}}\right)\left(\frac{23\frac{\text{fmol}}{\text{mg wet tissue}}\ 5\ HT_{2A}R}{4\frac{\text{fmol}}{\text{mg wet tissue}}\ 5\ HT_{1A}R}\right) \approx 70.000\ \frac{5\ HT_{2A}R}{5\ HT_{1A}R} \quad (5)$$

Thus, the theoretical, observed binding ($tBR_{5\text{-}HT2A/5\text{-}HT1A}$) of altanserin in the caudate-putamen represents ca. 70.000 more the $5\text{-}HT_{2A}$ than the $5\text{-}HT_{1A}$ receptor. Of course, one should critically review the calculated $tBR_{target/off\text{-}target}$, since in vitro binding characteristics may not always predict in vivo binding characteristics. That is, because radioligands which are suitable for in vitro quantification may not necessarily be ideal for in vivo PET imaging. PET tracers have to enter the brain through the BBB. The tracer can be a substrate of efflux pumps. Metabolism or pharmacokinetics could further limit its use [41]. However, when taken these pre-considerations

Table 1 Neuroreceptor densities (B_{max}) of off-targets in relevant brain regions [4, 56, 65, 68, 75–86]

		Frontal cortex	Caudate-putamen	Hippocampus	Thalamus	Cerebellum
5-HT_{2A}	López-Giménez (rat brain)[a]	116[b]	72			
	Kristiansen (rat brain)[c]	525				8
	Hall (human brain)[a]	80	19–26[d]	25		v.d.[e]
	Varnäs (human brain)[a]	56[f]	5–13	24[g]	6–11	
5-HT_{2C}	Marazziti (human brain)[h]	5		15	v.d.[e]	v.d.[e]
5-HT_{1A}	Varnäs (human brain)[a]	54[f]	1–2	85[g]	1–2	
	Hall (human brain)[a]	73	4	82[g]	4	6
D_2	Boyson (rat brain)[h]	0–67	784	95	38	53
α_1	Paermentier (human brain)[i]		33	185	77	

Used ligands: [^{3}H]MDL 100907, [^{3}H]mesulergine, [^{3}H]WAY 100635, [^{3}H]spiroperidol and [^{3}H] prazosin for 5-HT_{2A}, 5-HT_{2C}, 5-HT_{1A}, D_2 and α_1, respectively

[a]B_{max} was obtained from saturation experiments from cryosections, fmol/mg original wet tissue

[b]Average from frontoparietal motor cortex lamina layers I–V

[c]Saturation binding characteristics from homogenates, fmol/mg protein

[d]Only B_{max} values for the nucleus caudatus, putamen showed binding in the range of non-specific binding

[e]Virtually devoid

[f]External layers

[g]CA-12, molecular layer

[h]B_{max} from human brain membranes, fmol/mg protein

[i]B_{max} was obtained from saturation experiments from cryosections, fmol/mg protein

into account, the $tBR_{target/off\text{-}target}$ is a good estimate of the in vivo binding of a given compound and to evaluate its presumable in vivo binding.

Another important issue in PET imaging is that in order to generate robust binding parameters, the density of the target protein/receptor in relevant brain regions needs to be sufficiently high [51, 57]. If the density of the target protein/receptor is very low, it is difficult to obtain a specific image of the target due to low signal-to-noise ratios, for example caused by non-displaceable or unspecific binding components of the tracer itself or its metabolites. Non-displaceable binding refers to the compound's propensity to bind to membranes, proteins, lipids or other cell components without a specific and selective target, whereas unspecific binding refers to interactions with other well defined targets (e.g. receptors or enzymes). Furthermore, PET scanner detection limits as sensitivity and resolution can exclude any reasonable interpretation when analyzing regions with very low target protein/receptors.

It is difficult to determine what the lowest possible B_{avail} number has to be in order to perform PET imaging. This is not surprising since this limit is strongly dependent on the tracer characteristics itself, in particular it depends on its affinity

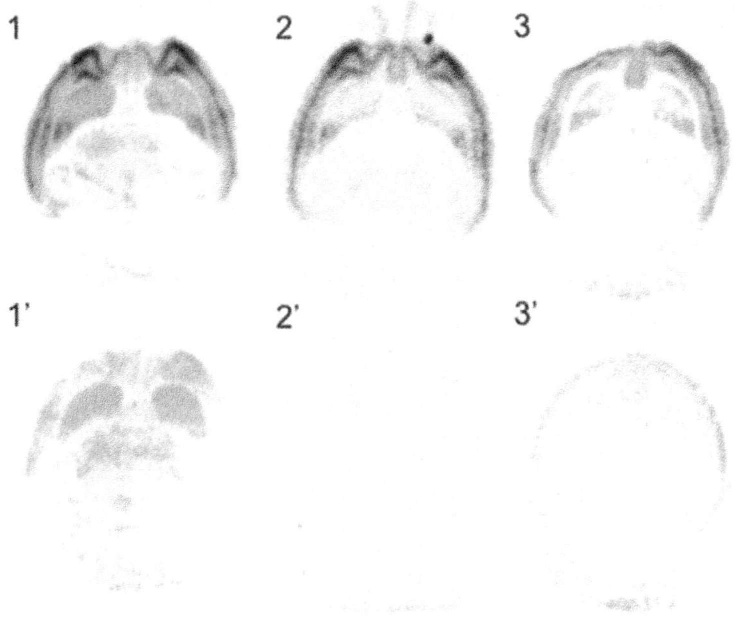

Fig. 4 Autoradiographic images of the total binding and non-specific binding, respectively, of (1/1) [^{18}F]altanserin, (2/2′) [^{18}F]MH.MZ and (3/3′) [^{3}H]MDL 100907 at 14 μm rat brain sections. Images of [^{3}H]MDL 100907 (**b**) and [^{18}F]MH.MZ (**c**) were in complete agreement. The binding of [^{18}F]altanserin (**a**) could not be blocked in striatum with the 5-HT$_{2A}$ antagonist ketanserin, demonstrating the inferior binding characteristics of [^{18}F]altanserin in vitro [62]

to the target as well as the degree of non-displaceable binding. However, PET studies of the serotonin transporter or extrastriatal dopamine D_2/D_3 receptors suggest that a target density (B_{max}) of 30–100 fmol/mg original wet tissue in human brain suffices for detection by PET [58–60].

The In Vitro 5-HT$_{2A}$ Receptor Distribution

The 5-HT$_{2A}$ receptor distribution has been determined by autoradiography and 5-HT$_{2A}$ receptor mRNA distribution by in situ hybridization studies. In rats, the two methods generate similar outcomes in the neocortex (strong labeling in lamina V), the caudate-putamen (stronger signal in caudal parts of the nucleus), the olfactory tubercle, and in several brainstem nuclei (pontine nuclei, motor trigeminal nucleus, facial nucleus) [61]. The highest 5-HT$_{2A}$ receptor density was found in the frontal cortex, medium in the caudate-putamen, less in the olfactory system, hippocampus, thalamus and the mesencephalon and lowest in the cerebellum [2, 62–66] (Fig. 4).

In the human brain, 5-HT_{2A} receptors distribute slightly differently with the highest binding in the frontal cortex (56–80 fmol/mg original wet tissue), medium binding in the caudate, the hypothalamus and the hippocampal formation (10–30 fmol/mg original wet tissue) and very low binding in cerebellum [4, 56]. Interestingly, Danish Landrace pigs showed a distinctly different distribution pattern in that cerebellar 5-HT_{2A} receptor binding constituted up to 50% of that of neocortex [4, 67]. In comparison, it constituted below 2% compared to that in frontal cortex in rats [68].

In the absence of a suitable brain region void of 5-HT_{2A} receptors, reference tissue modeling could be viable with rat PET data while for pigs, correct quantification necessitates some kind of arterial input function to be measured [67]. For use in humans, results are not quite consistent in that PET studies with [^{18}F]altanserin have validated the cerebellum as a suitable reference region [69]. In contrast, the 5-HT_{2A} receptor antagonist [^{11}C]MDL 100907 have in some [70–72], but not in all PET studies [73] displayed signs of specific binding in cerebellum. In rats, the cerebellum has been verified as a suitable reference region. Nevertheless, Maeshima et al. suggest omitting the outermost parts of the cerebellum when defining the reference region because of specific 5-HT_{2A} receptor binding at the outermost Purkinje cell and molecular layer of the rat cerebellum [74].

In Table 1, we summarize published 5-HT_{2A} receptor densities in human and rat brain tissue [4, 62, 65, 68, 75] as measured with either autoradiography or tissue homogenate. In addition, B_{max} values of off-target receptors in the same brain area are displayed. It is important to keep in mind that because of laboratories' different experimental set-ups and conditions (e.g. different radioligands, temperatures or buffer systems), the B_{max} values in Table 1 may be difficult to compare directly. Furthermore, in contrast to tissue homogenate studies, autoradiographic experiments determine the receptor density from one slice of the target region from where a region is drawn. Thus, depending on exactly where the slice is taken from, B_{max} values may not accurately reflect the receptor distribution in the entire targeted brain volume, as tissue homogenate binding measures do [65]. In addition, the B_{max} values determined in fmol/mg protein and in fmol/mg wet tissue cannot be directly compared or converted. The B_{max} value in fmol/mg protein is equivalent to ~ 5–10 times of the value determined in fmol/mg wet tissue. The conversion factor is depended on the protein content of the tissue, but the exact conversion factor is usually unknown.

In summary, B_{max} and K_D values can come out very differently and the absolute values should be taken with a grain of salt. However, these values may still reasonably well reflect target to off-target ratios (D). Table 1 is not an exhaustive summary of available studies but display the neuroreceptor densities that are comparable because of the radioligand used. The relevant neuroreceptor densities will be discussed further.

Table 2 Published $LogD_{7.4}$ data of 5-HT_{2A} receptor tracers

	Altanserin	MDL 100907	(R)-MH.MZ	Cimbi-36
$LogD_{7.4}$	2.1–3.1 [68, 93, 94]	1.9–3.8 [68, 70, 93, 94]	2.8 [93]	3.43 [95][a], 3.2[b]

[a]Data measured in our lab
[b]clogD7.4 with Pallas

Lipophilicity and Non-Displaceable Binding

Among various physical properties, the lipophilicity plays an important role for the success of a radiotracer. Lipophilicity can for example be determined experimentally as the ratio between octanol/water and measured by the shake-flask [87] or by the HPLC method [88]. According to Lipinski's "rule of five," a $logD_{7.4}$ value >5 is optimal for drug absorption and permeation into the CNS. If lipophilicity gets too high, the risk of prohibitively high non-displaceable tissue binding may result. Low lipophilicity prevents cell membrane permeability [51].

Rowley et al. suggested the ideal interval for small molecules to penetrate the BBB to be 2–3 [89]. However, published lipophilicity values often differ (see e.g. Table 2). Thus, a direct comparison of published values should be taken with a grain of salt if they are not tested within the same laboratory. Moreover, it is impossible from $logD_{7.4}$ values to predict in vivo non-displaceable binding and there are also successful PET tracers described with lower or higher $logD_{7.4}$ values [90, 91]. Consequently, lipophilicity determinations just allow a course assessment whether a compound possesses reasonably good BBB passage and low non-displaceable binding [41]. Rigid adherence to the "lipophilicity rule" may result in a self-fulfilling prophesy in the sense that the inclination to attempt developing radiotracers outside the "safe" range of lipophilicities.

In conclusion, estimated or measured lipophilicity data can give a rough estimate of BBB permeability and non-displaceable binding, and the data may be particularly useful to optimize a good tracer candidate, when alternative analogue candidates are available [92]. Table 2 displays $logD_{7.4}$ for some 5-HT_{2A} receptor PET tracers that will be discussed in this review.

Recently, immobilized artificial membrane (IAM) chromatography was proposed to be a good measure to determine BBB penetration. Preliminary results showed a good correlation between the penetration and the determined permeability by IAM [96, 97]. Future studies are needed to verify this observation.

Metabolism and Clearance Rate

Radiotracer metabolism and clearance rates are both crucial factors for the fate of a PET tracer. Rapid clearance results in difficulties with accurate determination of the input curve and reduces the accuracy of the subsequent mathematical modeling with an arterial input function [41]. Radiometabolites can interfere with the PET signal of the intact radiotracer. Either they can increase the non-specific signal or even worse, bind to the same target as the parent compound and falsify the specific binding reading. Polar metabolites are less likely to cross the BBB and are therefore generally preferable [41, 51, 91, 98, 99]. Metabolites that cross the BBB should not have affinity for any target within the brain, but if they do, a correction for non-specific binding interference caused by lipophilic radiometabolites can be done. Use of, for example, a bolus–infusion design can be helpful. In this experimental setup, the radioligand is injected as a bolus followed by constant infusion to attain steady-state in plasma and brain. Alternatively, the parent compound and its radiometabolites can be entered in a dual-input function analysis [41, 69]. Inactive hydrophilic metabolites must also be measured to correct the plasma input function. Typically, a successful radiotracer has a limited number of polar metabolites that do not constitute to the majority of the brain radioactivity in the early frames of the scan and show 50–90% metabolism towards the end of the imaging process [51].

Furthermore, it is important to remember that PET radiometabolites should be determined at tracer levels. Reaction and enzyme kinetics are concentration dependent. Finally, it should be kept in mind that radioligand metabolism often shows some interspecies differences and therefore do not readily translate from one species to another.

Finally, PET tracers that are substrates for efflux transporters such as the permeability glycoprotein (P-gp) could potentially confound the usefulness of the radioligand, particularly if efflux transporters are expressed differently within various brain regions.

Current 5-HT_{2A} Receptor PET Tracers

Today, several PET radioligands for the 5-HT_{2A} receptor system have been evaluated [4, 41, 61, 62, 64–66, 68, 75, 93, 100–102], e.g. [^{11}C]ketanserin [103], [^{18}F]fluoroethylketanserin [104], [^{11}C]NMSP [105] [^{11}C]MBL [106], [^{18}F]setoperone [107], [^{18}F]fananserin [108], [^{18}F]MH.MZ [64, 109, 110], (R)-[^{18}F]MH.MZ [64], [^{18}F]altanserin [111], [^{18}F]deuteroaltanserin [112], [^{11}C]MDL 100907 [73], and [^{11}C]Cimbi-36 [95]. Especially, the last five structures appear to be very promising due to their selectivity profile, functional imaging possibilities and metabolism (Table 3). This chapter will focus on these structures in respect to their in vitro profile, their in vivo tracer behaviour as well as their application in 5-HT_{2A}R PET imaging. A more thorough review about synthesis and labeling strategies of these ligands was recently published [113].

Table 3 Chemical structures and selectivity profile of MDL 100907, (R)-MH.MZ, altanserin, and Cimbi-36

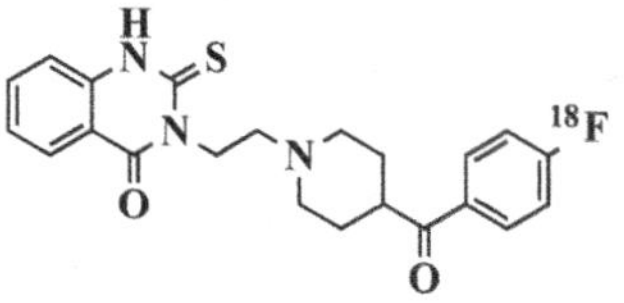

[18F]altanserin

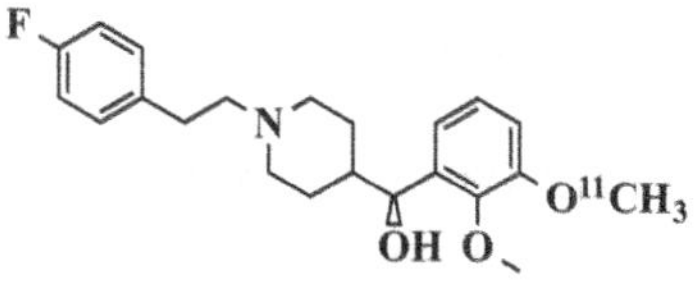

[11C]MDL 100907

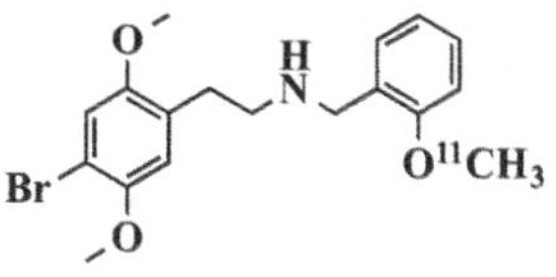

[11C]Cimbi-36

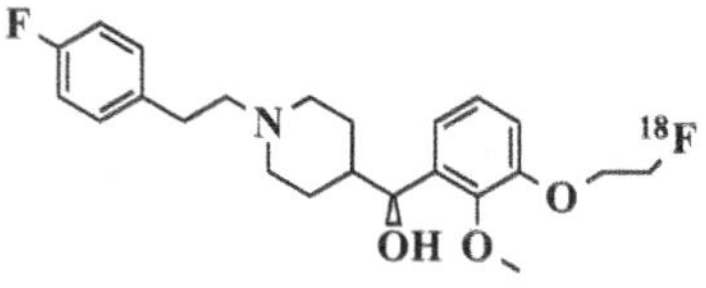

(R)-[18F]MH.MZ

	5-HT$_{2A}$	5-HT$_{2C}$	5-HT$_{1A}$	D$_2$	α_1
MDL 100907	0.36 nM	107 nM	> 10.000 nM	2250 nM	128 nM
(R)-MH.MZ	0.72 nM	53 nM	> 10.000 nM	2686 nM	335 nM
altanserin	0.13 nM	6 nM	1570 nM	62 nM	4.55 nM
Cimbi-36	1.01 nM	1.7 nM	1255 nM	> 10.000 nM	1256 nM

More selectivity data are reported in following references [4, 64, 75, 93, 95, 114, 115]

Antagonistic PET Tracers

[18F]Altanserin and [18F]deuteroaltanserin

In Vitro Profile

Altanserin (K_i = 0.13 nM) is a fluorobenzoylpiperidine derivative structurally related to ketanserin and historically, it is by far the most applied 5-HT$_{2A}$ PET tracer [41]. However, previous studies have questioned the selectivity of altanserin in PET studies [62, 115], since a moderate affinity of altanserin for both the α_1-adrenergic receptor (K_i = 4.55 nM) and the 5-HT$_{2C}$ receptor (K_i = 6.0 nM) have been found (Table 2). In addition, the relatively high abundance of D$_2$ receptors combined with altanserin's affinity for this receptor (K_i = 62 nM) rise concerns in striatal regions (nucleus caudatus and putamen).

However, a closer look at the $tBR_{target/off\text{-}target}$ for altanserin debilitates these speculations to some extent. Even in a worst case scenario, a $tBR_{5\text{-}HT2A/5\text{-}HT2C}$ of >10 is calculated in frontal cortex and thalamus regions. In contrast, the $tBR_{5\text{-}HT2A/\alpha1}$ is calculated to be <10 in all other regions. Because of this concern, Kristiansen et al. carried out an in vitro study in rat brain homogenate (frontal cortex and cerebellum) to examine to which extent non-specific binding influences the signal of altanserin. The binding pattern of [18F]altanserin was not altered by a 5-HT$_{2B/2C}$ selective antagonist (SB 206553) or by prazosin, a α_1 selective ligand [68, 116]. However, in an autoradiography study with [^{18}F]altanserin strong binding of the tracer was seen in rat striatum. The 5-HT$_{2A}$ selective antagonist ketanserin could not displace the binding, but prazosin and raclopride (a selective D$_2$ antagonist) to some extent (Fig. 4) [62]. As mentioned above, autoradiography studies may not always be representative of the images obtained at tracer doses and consistently, [^{18}F]altanserin PET images seems to largely represent 5-HT$_{2A}$ receptor binding [69].

Metabolism

In humans, four radiolabeled metabolites of [^{18}F]altanserin have been described [69, 115, 117], whereas in pigs only three were detected [67]. Unfortunately, all radiometabolites are able to cross the BBB in pigs, non-human primates and humans [69, 118]. However, only two have been demonstrated to significantly contribute to the total amount of measured radioactivity within the brain [69, 118–120]). They were identified as [^{18}F]altanserinol and [^{18}F]4-(4-fluorobenzoyl)piperidine ([^{18}F]FBP) [115] (Scheme 1).

Scheme 1 Major metabolic pathways of [^{18}F]altanserin

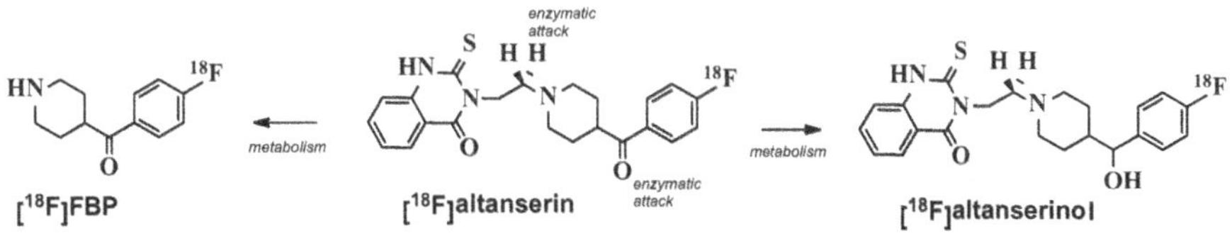

Both radiometabolites have negligible specific binding to 5-HT$_{2A}$ receptors in vitro [115], but they cross the BBB and bind non-specifically and fairly uniformly across brain regions in humans [119]. Especially in regions with low levels of 5-HT$_{2A}$ receptors like the cerebellum [2, 121], the presence of radiometabolites can complicate modeling [122]. In addition, these lipophilic metabolites contribute to the non-specific binding and thus lead to a lower non-displaceable binding potential. Another drawback of altanserin is that its metabolism is highly variable between individuals (Fig. 5a) [67], and thus, population based input functions do not work. This is of particular interest for tracers where reference-tissue models do not apply. However, in contrast to pigs, cerebellum seems to be a fairly valid reference region in the rodent, monkey and human brain [62, 67].

Finally, it should be mentioned that [^{18}F]altanserin shows huge interspecies differences in radioligand metabolism. Compared to humans, [^{18}F]altanserin is slowly metabolized in rodents and the lipophilic metabolite altanserinol or other lipophilic species are negligible. In rodents; only polar metabolites were detected over 3 h both in plasma and brain [123–125].

In 1999, a first attempt was carried out to reduce the metabolic rate of [^{18}F]altanserin by introducing a carbon-deuterium (C-D) bond trying to suppress N-dealkylation. This isotopic effect relies on the greater C-D bond strength. As a result, [^{18}F]deuteroaltanserin showed 29% higher plasma parent-to-metabolite ratios than [^{18}F]altanserin [115].

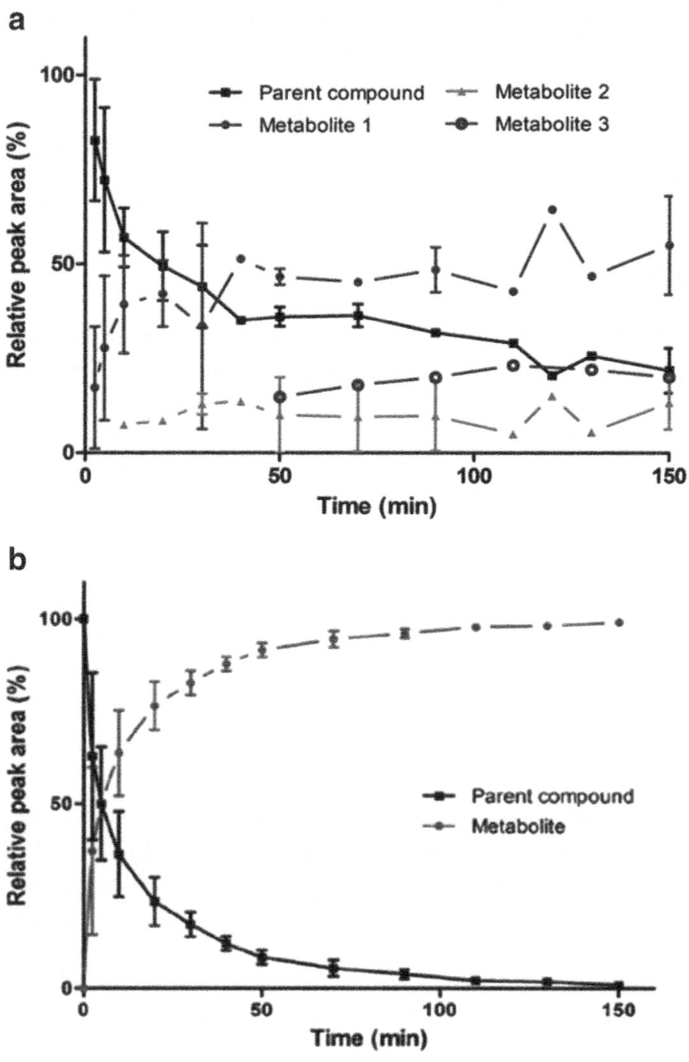

Fig. 5 (**a**) Metabolism profile of [^{18}F]altanserin in pigs: A high inter-individual variation is observed [67]. The metabolism profile in other species is comparable (**b**) Representative metabolism profile of MDL 100907 derivatives; displayed is the metabolism of (R)-[^{18}F]MH.MZ [67]

In Vivo Profile

As mentioned earlier, [^{18}F]altanserin has so far been the most frequently applied 5-HT$_{2A}$ receptor PET radiotracer [5, 41, 99]. In the rat brain, it showed in vivo selectivity by blocking and challenge experiments with the 5-HT$_{2A}$ receptor antagonists ritanserin, setoperone and ketanserin [111, 116, 126, 127]. Furthermore, both Riss and Kroll et al. showed that [^{18}F]altanserin is suitable for quantification of 5-HT$_{2A}$ receptors in Lister hooded rats as well as in Sprague-Dawley rats and that no radioactive metabolites were observed within the rat brain. SRTM was possible [124, 126]. However, the detected cerebral uptake of [^{18}F]altanserin was very low (Standard Uptake Value (SUV) ~ 0.7, compared to that observed in humans (SUV = 1.26) [128]). This is probably due to species differences in P-glycoprotein (P-gp) activity. Syvänen et al. demonstrated that the SUV of [^{18}F]altanserin increased by a factor of 2.6 in cerebellum in rats and 1.8 in mini-pigs after P-gp inhibition [129]. These results could be repeated by Kroll et al. in 2014 [130]. However, the observed low uptake in rats might cause restrictions in experimental usage and reduce test-retest reliability. In general, the observed strong P-gp depency of [^{18}F]altanserin in rats questions its use in this particular animal model. Nevertheless, a BP_{ND} of 1.9–2.0 was reported for rat frontal cortex [124, 126]. In 2013, Martín et al. reported the first biological evaluation study of [^{18}F]altanserin after focal cerebral ischemia in rats [131]. A dramatic decrease in [^{18}F]altanserin binding in ischemic cortex and striatum was observed. These changes are in accordance with neurological and behavioral recovery over time.

In 1994, the first human [^{18}F]altanserin study was conducted by Biver et al. [122]. Further studies established– without taking the radiolabeled metabolites into account—that the BP_{ND} was 2.3–2.9 in human cortex [20, 132, 133]. The binding of [^{18}F]altanserin was successfully blocked by ketanserin [69, 128]. As discussed before, [^{18}F]altanserin produces radiometabolites, which cross the BBB in humans and contribute to the non-specific binding. Thus, complex kinetic modeling is required for quantification. For example, a dual-input function approach was used by Price et al. in baboons and humans [119, 120]. However, the statistical quality of PET and HPLC data complicates the extraction of binding parameters making dual input functions difficult by increasing the experimental complexity (scan times >90 min, many arterial blood samples for HPLC analysis) and patient discomfort due to the scan duration. Thus, new and simpler methods to quantify the binding of [^{18}F]altanserin were needed. As a result, a bolus–infusion protocol was developed accounting for radiometabolite binding within the brain. This is possible because within the resulting steady-state situation the non-specific binding of radiolabeled altanserin and its metabolite(s) can directly be subtracted from each other [112, 134]. Furthermore, Pinborg et al. improved this steady-state approach by reducing the infusion time to just 2 h. This resulted in a higher feasibility to conduct [^{18}F]altanserin human PET studies [69]. In general, the bolus–infusion paradigm was shown to have excellent test–retest reliability in large brain regions with high binding [134]. Figure 6 displays a typical distribution profile of [^{18}F]altanserin in humans.

[^{18}F]deuteroaltanserin was developed to prevent or substantially decrease the production of a lipophilic metabolite. The resulted product also led to higher brain uptake in baboon and humans, as compared with [^{18}F]altanserin. In this way, the cortical-to-cerebellar ratio in humans was increased by 26%, suggesting it might be a superior PET radioligand [112]. Test-retest reliability and 5-HT_{2A} receptor specificity was essentially equivalent to that of [^{18}F]altanserin in baboons [135, 136]. Since then, only two further studies have been published using [^{18}F]deuteroaltanserin in humans [137, 138]. It remains to be seen whether this ligand will become as successful as its predecessor [41].

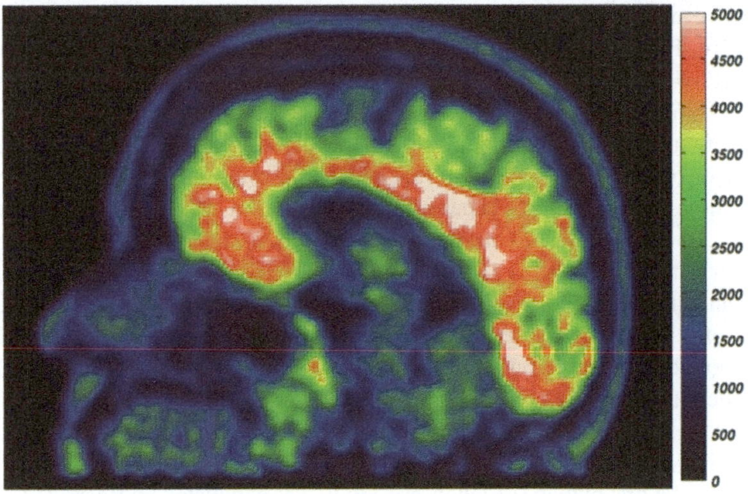

Fig. 6 Specific 5-HT_{2A} receptor binding by [^{18}F]altanserin. Highest densities of this receptor are detected in cortical areas, very low ones in hippocampus and the cerebellum. The color table indicates receptor binding potential

[^{11}C]MDL 100907 and [^{18}F](R)-MH.MZ

[^{11}C]MDL 100907

In Vitro Profiles

MDL 100907 ($K_d = 0.14$–0.19 nM) belongs as altanserin to the 4-piperidine derivative class. It is a reversible, highly selective 5-HT_{2A} ligand with subnanomolar affinity. But compared to altanserin, MDL 100907 shows a superior $tBR_{target/off\text{-}target}$ in all relevant brain regions. Radioligand binding and autoradiography studies confirmed its in vitro selectivity and high specific binding in rat, pig, nonhuman primate, and human brain, making it the first truly selective 5-HT_{2A} receptor ligand [4, 41, 61, 65, 67, 68, 75]. Thus, MDL 100907 should allow direct PET visualization and characterization of 5-HT_{2A} receptors without the need of blocking additional sites.

Remark: The difference in MDL 100907's selectivity profile compared to altanserin's could be caused by a different binding mode of both tracers. Whereas altanserin binds to the 5-HT_{2A} receptor with the p-fluorobenzoyl moiety into the hydrophobic binding pocket of the receptor [139], it appears that MDL 100907 related 4-piperidinemethanol derivatives binds with the p-fluorophenylethyl residue in this hydrophobic binding pocket, so to speak vice versa [110].

Metabolism

Scott et al. studied the metabolism of MDL 100907 in rats and dogs extensively. These pharmacokinetic studies revealed that the drug is crossing the BBB and binds in the brain. It undergoes extensive first-pass metabolism to an active metabolite

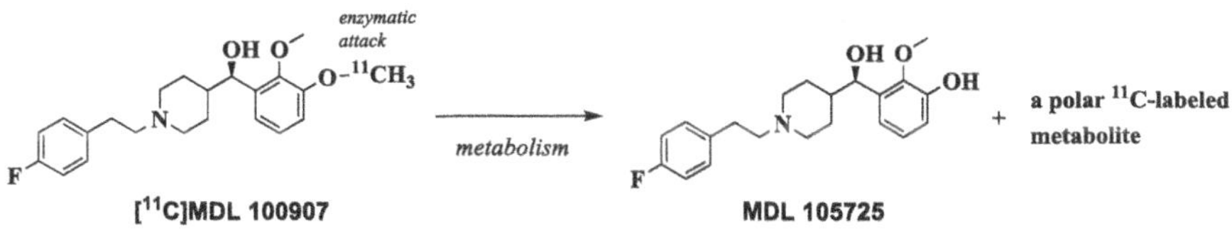

Fig. 7 Metabolic pathway of MDL 100907 derivatives

(MDL 105725) (Fig. 7), but the permeability of MDL 100907 is more than four times higher than that of MDL 105725. Furthermore, no significant metabolism to MDL 100907 could be detected in the brain. Thus, their data suggest that MDL 100907 is the predominant active species present within the brain, even at high doses of MDL 100907 [140]. In humans, [^{11}C]MDL 100907 displays a rapid metabolism and has thus a similar metabolism profile [70, 72, 141]. Figure 5 compares the metabolism of a [^{11}C]MDL 100907 (Fig. 5b), with that of [^{18}F]altanserin (Fig. 5b).

In Vivo Profile

[^{11}C]MDL 100907 showed selective 5-HT_{2A} binding in various species, e.g. in rat, non-human primate and human brain tissue [4, 27, 61, 75]. Cortex-to-cerebellum ratio of 3.5–4.5 in both rats and non-human primates were reported. The binding could be blocked with stable MDL 100907 and ketanserin [73, 142]. The first human [^{11}C]MDL 100907 study was conducted in 1998 by Ito et al. and analyzed with SRTM [70]. Further studies have validated the methodology for modeling [^{11}C]MDL 100907 binding in PET studies, and identified that two-tissue compartment modeling (2-TCM) using arterial input is superior to reference tissue models [8, 9, 41, 72, 143–147]. Recently, Meyer et al. demonstrated the feasibility of a non-invasive graphical analysis (NIGA) of the bindings kinetics of [^{11}C]MDL 100907 [8]. In general, 2-TCM or NIGA resulted in a detected BP_{ND} of 2.2–3.0 in cortex regions for [^{11}C]MDL 100907 [8, 9, 71, 72, 144, 146–149] and test–retest variability for the BP_{ND} in cortex was very good [146]. However, [^{11}C]MDL 100907 showed slow kinetics and thus quantification may become problematic. This is maybe the reason why so far only a limited number of clinical [^{11}C]MDL 100907 have been reported compared to those of [^{18}F]altanserin. However, in recent years the number of human PET scan with [^{11}C]MDL 100907 increased (PubMed search). Figure 8 displays a typical [^{11}C]MDL 100907 baseline and block human PET scan. So far, no dependency of [^{11}C]MDL 100907 on P-gp has been reported. However, it is very likely that [^{11}C]MDL 100907 displays the same behavior as (R)-[^{18}F]MH.MZ (see next section).

(R)-[^{18}F]MH.MZ

In Vitro Profile

(R)-MH.MZ is structurally related to MDL 100907 and was developed to combine the superior selectivity of MDL 100907 with the superior isotopic characters of fluorine-18 for clinical studies.

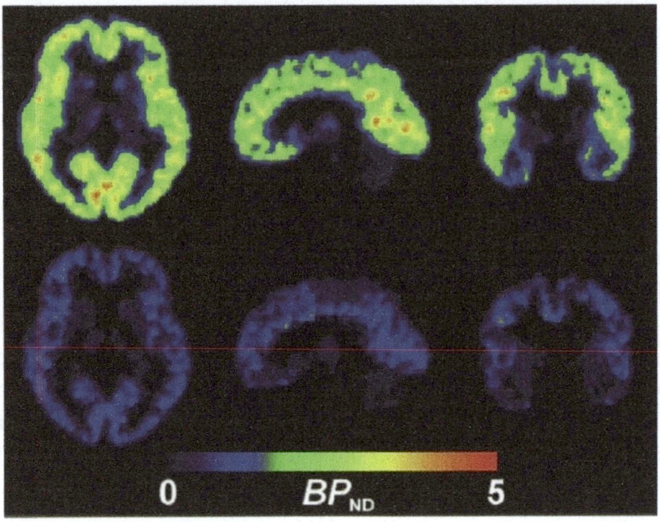

Fig. 8 Parametric binding potential maps generated by voxel-wise non-invasive graphical analyses. Upper row, baseline study; lower row, blocked study after pretreatment with 30 mg mirtazapine. Identical color scaling was used for both images [8].

Regarding the $tBR_{target/off-target}$, (R)-MH.MZ shows very high values in all relevant human brain regions. Thus, the $tBR_{target/off-target}$ of (R)-[^{18}F]-MH.MZ is comparable to that of MDL 100907 in humans.

Metabolism

(R)-[^{18}F]MH.MZ displays a similar metabolism profile compared to [^{11}C]MDL 100907 in mice, rats and pigs [62, 64, 67, 109]. Only one polar radioactive metabolite was detected, which was not able to cross the BBB, at least in rats and pigs [64, 67]. Furthermore, (R)-[^{18}F]MH.MZ displays, in contrast to [^{18}F]altanserin, an inter-species and inter-individual stable and reproducible metabolism possibly allowing for population based metabolite correction of the input function [67]. At the moment, it is unknown if humans metabolize (R)-[^{18}F]MH.MZ in the same way. Finally, [^{18}F]MDL 100907 has recently been synthesized and evaluated [150–152]. [^{18}F]MDL 100907 is most likely metabolized to its 3-OH-analogue [^{18}F]MDL 105725 (see Fig. 7), which potentially enters the brain and interferes with the interpretation of [^{18}F]MDL 100907 uptake. Thus, [^{18}F]MDL 100907 does not seem to be a useful alternative to [^{11}C]MDL 100907 or (*R*)-[^{18}F]MH.MZ.

In Vivo Profile

(R)-[^{18}F]MH.MZ reveals comparable in vivo binding characteristics compared to [^{11}C]MDL 100907. μPET studies in rats and mice showed a BP_{ND} of 2.6 in frontal cortical regions for (R)-[^{18}F]MH.MZ using SRTM [64, 109]. In pigs, the racemic version of (R)-[^{18}F]MH.MZ revealed a BP_{ND} of 3.3 in cortex using one-tissue

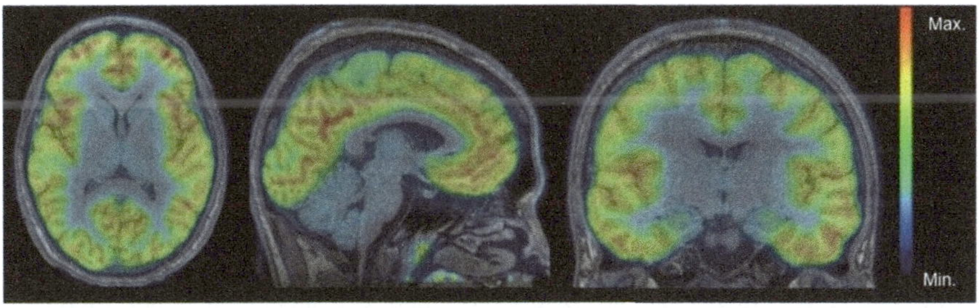

Fig. 9 The PET/MRI fusion image (averaged 90–105 min. p.i.) shows (*R*)-[^{18}F]MH.MZ binding to cortical 5-HT_{2A} receptors in the human brain. *Yellow* and *orange* areas represent regions with high tracer retention and receptor density (predominantly cortex) whereas colder areas like cerebellum show very low uptake. From left to right are displayed transversal, sagittal and coronal views (Courtesy of Dr. Vasko Kramer)

compartment modeling (1-TCM) [67]. In addition, the non-specific binding was low. Unfortunately, the time–activity curves showed a very slow washout from rat and pig brain complicating modeling [64, 67, 109]. As it is the case for [^{18}F]altanserin, (R)-[^{18}F]MH.MZ is also a P-gp substrate and in P-gp knockout mice, brain concentrations of (R)-MH.MZ were about fivefold higher than in wild-type animals [153]. Recently, the first human PET study with (R)-[^{18}F]MH.MZ was published by Kramer et al. (Fig. 9) [154]. Preliminary results indicate that (R)-[^{18}F]MH.MZ is indeed able to image the 5-HT_{2A} receptor system accurately. It remains open whether (R)-[^{18}F]MH.MZ has better characteristics than [^{18}F]altanserin in human PET experiments.

Comparison of Labeled MDL 100907 Derivatives and [^{18}F]altanserin for PET Imaging

MDL 100907 vs. (R)-[^{18}F]MH.MZ: Carbon-11 vs. Flourine-18

The selectivity and in vivo binding profile of (R)-[^{18}F]MH.MZ and [^{11}C]MDL 100907 is very comparable. The advantage of [^{11}C]MDL100907 over (R)-[^{18}F]MH.MZ is that it allows for conduction of test-retest experiments in the same subject on the same day. This is possible because of the 20.4 min half-live of carbon-11. Test-retest experiments, often performed on the same day and with an intervention in between, is an elegant way to minimize examination time for the patient (1 day instead of 2) and can also be used to evaluate the robustness of quantification methods. Caution, however, needs to be taken to avoid spill-over effects, e.g., if non-tracer doses accumulate from scan 1 to 2. The potential confound of circadian rhythm should also be considered. Test-retest reliability thus constitutes an elementary part of every tracer evaluation. For [^{11}C]MDL 100907, test-retest variability was tested in nine healthy volunteers on the same day, with a 60 min interval between the two scans. Test–retest variability was very good (7–11%) in most neocortical regions.

The advantage of (R)-[^{18}F]MH.MZ over [^{11}C]MDL100907 is that the longer half life of flourine-18 ($t_{½}$ = 110 min) allows the user to transport the tracer to other PET-scanner facilities within a range of several hundred kilometers. Furthermore, more individuals can be scanned since more activity can be produced with a single production and finally, higher specific activities (A_s) compared to carbon-11 ($t_{½}$ = 20.4 min) can be reached [62, 64, 93, 109]. In principle, fluorine-18 also allows for a better resolution due to its lower β^{+-}energy of the emitted positron [44]. It also allows performing longer PET scans, which is particularly helpful for tracers like (R)-[^{18}F]MH.MZ, with slow, but reversible kinetics. Longer scan times should in principle lead to better statistics, which allow more precise modeling. Obviously, the half-life of fluorine-18 also restricts the possibility of test-retest studies at the same day. However, ^{18}F–tracers allow performing intervention studies within one scan, thereby reducing time, radiosynthesis efforts and possibly radiation exposure. Unfortunately, tracers with slow kinetics i.e. tracers that have a slow dissociation constant (in Fig. 2, k_4) and thus bind relatively long to the receptor, are often not very susceptible to such a study design. This is for example the case for [^{18}F]MH.MZ [155]. Test-retest studies at different days are inconvenient for patients (especially for patients with neurodegenerative or mental disorders). Even test-retest experiments in animals at different days are tedious. For example, the facility needs a suitable infrastructure to house research animals. Furthermore, care should be taken to avoid stressing the animals, which may bias the second scan.

[^{18}F]altanserin vs. (R)-[^{18}F]MH.MZ

Recently, we conducted a direct comparison of [^{18}F]MH.MZ and [^{18}F]altanserin in pigs to validate their tracer characteristics [67]. As expected, slow kinetics of [^{18}F]MH.MZ complicated the PET modeling. But its higher affinity allows for quantification of brain areas with low 5-HT_{2A} receptor density whereas [^{18}F]altanserin is better suited for high-binding regions. The relatively high non-specific binding component (caused by lipophilic radiometabolites and altanserin's lipophilicity) and the less selective 5-HT_{2A} receptor profile interfere with 5-HT_{2A} receptor quantification and modeling in low binding regions [67]. However, in high binding regions [^{18}F]altanserin's fast and reversible kinetics simplifies kinetic modeling and the non-specific binding signal can be neglected compared to the selective 5-HT_{2A} binding [51, 67]. Figure 10 compares the time-activity curves (TACs) of [^{18}F]MH.MZ and [^{18}F]altanserin in Danish landrace pigs.

[^{11}C]MDL 100907, (R)-[^{18}F]MH.MZ and [^{18}F]altanserin in μPET Studies

Quantification of receptor binding in small animals is of special interest as it can be directly related to the animals' behavior. Several well established behavioral rat models exist, e.g. for addiction [156] and thus PET imaging in such models allows a direct in vivo insight into the molecular differences.

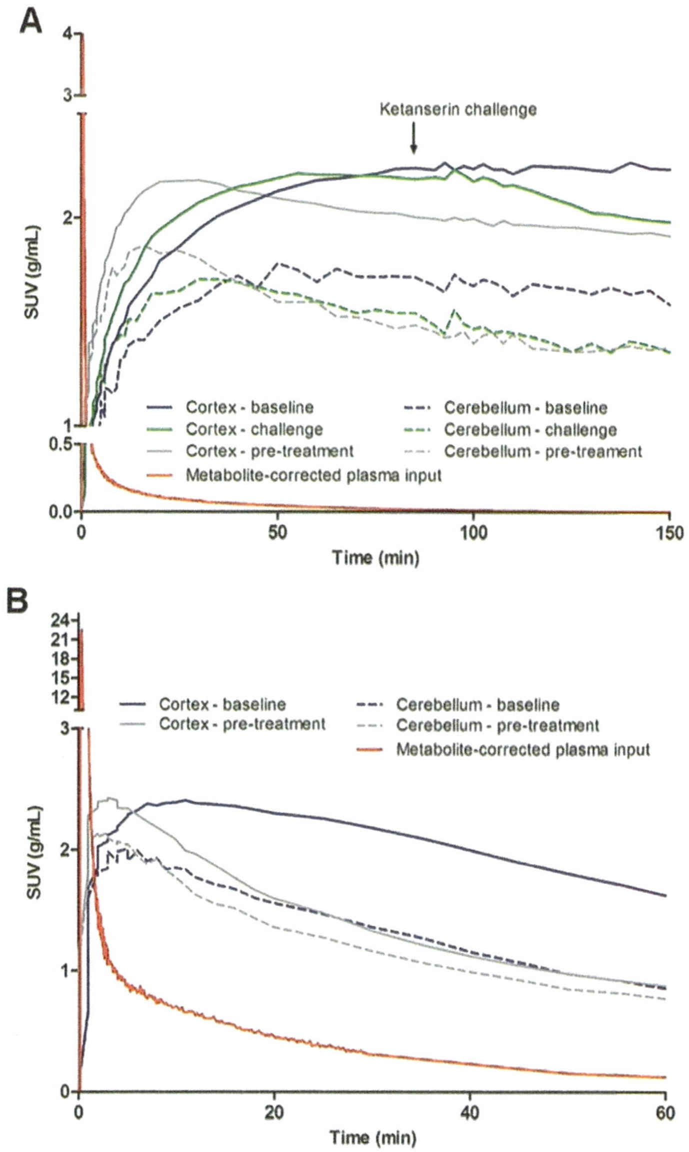

Fig. 10 TACs and metabolism-corrected input curves for [^{18}F]MH.MZ (**a**) and [^{18}F]altanserin (**b**). Data are presented as SUV. [^{18}F]MH.MZ: Baseline (n = 1), challenge (n = 2), pre-treatment (n = 1). [^{18}F]Altanserin: Baseline (n = 1), pre-treatment (n = 1) [67]

[^{11}C]MDL 100907, (R)-[^{18}F]MH.MZ and [^{18}F]altanserin can be provided in reasonable specific activities allowing small animal experiments. They were successfully applied to rodents using SRTM [64, 126, 157]. However, test-retest experiments in the same small animal are usually not practical. Thus, a ^{18}F–tracer appears to be more practical since one single production provides enough activity for several μPET experiments throughout the day. In a direct comparison of [^{18}F]altanserin and (R)-[^{18}F]MH.MZ, the latter appears to be superior for small animal PET molecular imaging, mainly due to its higher in vivo selectivity profile in frontal cortex (BP_{ND} = 2.6 ((R)-[^{18}F]MH.MZ) against BP_{ND} = 1.9 ([^{18}F]altanserin)) and its higher cerebral uptake (max SUV = 1.8 ((R)-[^{18}F]MH.MZ) against SUV 0.69 ([^{18}F]altanserin)) [64, 126, 127]. A higher BP_{ND} enables one to detect smaller changes or manipulations of the receptor status. Higher uptake gives better count statistics. In addition, [^{18}F]altanserin also appears to be more sensitive to P-gp in rodents favoring (R)-[^{18}F]MH.MZ as the PET tracer of choice in this animal model. However, (R)-[^{18}F]MH.MZ is also a P-gp substrate. A highly selective PgP-insensitive 5-HT_{2A} receptor PET ^{18}F–tracer is still not available for rodent PET studies. However, we believe that (R)-[^{18}F]MH.MZ shows a reasonable tracer profile to be used in rodents.

Agonist PET Tracers

Today, the vast majority of PET tracers available are antagonists. This is not too surprising given that it is easier to identify selective antagonists than agonists/inverse agonists and accordingly the number of antagonists by far outnumbers available agonists/inverse agonists discovered in pharmaceutical drug screening programs [41]. However, agonistic radiotracers could have some advantages in respect to functional imaging. According to the extended ternary model (Fig. 11), agonists only bind to the receptors` high-affinity state, whereas antagonists label the whole population of receptors. Thus, antagonists provide a good indication of the total available receptor number. In contrast, radiolabeled agonists reflect the number of receptors, which are able to induce neuroreceptor signaling [41, 158–160]. However, it is questionable if PET agonists can directly quantify the number of these active states since agonists should also be more sensitive to the endogenous neurotransmitter. This is because agonists and the endogenous neurotransmitter compete according to the extended ternary model towards the same binding side. Obviously, these

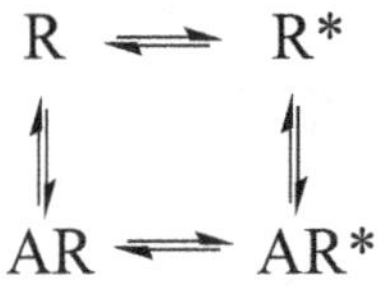

Fig. 11 The extended ternary complex model explains the existence of different affinity states of the same receptor. The model postulates that a partially activated receptor conformation (R*) exists in equilibrium with a ground state (R) and an activated G protein-coupled (AR*) conformation. Agonists bind with higher affinity to the partially activated (G protein coupled) R* state, also known as the high affinity state (R_{HIGH}) and with lower affinity to the uncoupled ground state (R), also known as the low-affinity state (R_{LOW}). Antagonists bind with equal affinity to all states. This explains why agonists label a smaller fraction of the same receptor compared to antagonists [159, 166]

considerations only play a role if a relatively small proportion of the total available receptor states are in the receptors` high-affinity state. For the 5-HT_{2A} receptor, some in vitro data suggest that 40–60% of the receptors are in the high affinity state [161, 162] and thus, 5-HT_{2A} agonists should indeed be more susceptible to competition with endogenous serotonin [41, 159, 163, 164]. The theory has been verified for the D_2 receptor system, which is assumed to be comparable to the 5-HT_{2A} receptor system (50% of D_2 receptors are in the high affinity state and 50% are in the low affinity state). For example, Narendran et al. and Cumming et al. reported that D_2 receptor agonist radiotracers are superior to antagonist radiotracers in measuring dopamine release in vivo in monkeys and mice [164, 165].

[^{11}C]Cimbi-36

In Vitro Profile

Cimbi-36 is a 5-HT_{2A} phenethylamine agonist structurally related to 1-(2,5-dimeth oxy-4-iodophenyl)-2-aminopropane (DOI) with an affinity of 1.01 nM. The intrinsic activity showed nearly full agonistic activity, with 87% activation of the 5-HT_{2A} receptor compared to 5-HT itself. The EC_{50} was determined to be 0.51 nM [95]. The selectivity profile of Cimbi-36 showed highest affinities for the 5-HT_{2A}, 5-HT_{2B} and 5-HT_{2C} receptor. Against targets other than 5-HT_2 receptors, Cimbi-36 showed at least a 30-fold lower affinity (Table 1) [95]. $tBR_{target/off-target}$ revealed selectivity of Cimbi-36 against most relevant neuroreceptors other than the 5-HT_{2B} and 5-HT_{2C} receptor. Especially in the striatum, but also in the hippocampus the measured PET signal may reflect more 5-HT_{2C} binding than 5-HT_{2A} binding. However, in cortical regions where the highest abundance of 5-HT_{2A} is localized, Cimbi-36's $tBR_{target/off-target}$ should allow selective quantitative PET 5-HT_{2A} imaging. As mentioned above, Cimbi-36 also shows a high affinity towards 5-HT_{2B} receptors (K_i = 0.5 nM) [95]. Since the brain 5-HT_{2B} receptor density is very low [167–170], binding to this receptor should not interfere. Consequently, the cortical Cimbi-36 binding signal stems most likely from 5-HT_{2A} receptor binding, whereas binding in some other brain regions may stem from a combination of receptors.

Metabolism

[^{11}C]Cimbi-36 is primarily metabolized via 5′-demethylation, followed by conjugation to glucuronic acid (Fig. 12). Both metabolic steps (demethylation and glucuronidation) are very fast, with only minute levels of intermediate phenol M1 present at any time point. The glucuronide M2 is eliminated much slower from plasma [171]. This radiolabeled metabolite reached a maximum in plasma at around 20–40 min after injection and then dropped off slightly (Fig. 12). HPLC analysis of homogenized pig brain tissue taken 20 min after [^{11}C]Cimbi-36 injection only showed negligible amounts of this metabolite in frontal cortex tissue compared to plasma [95, 171].

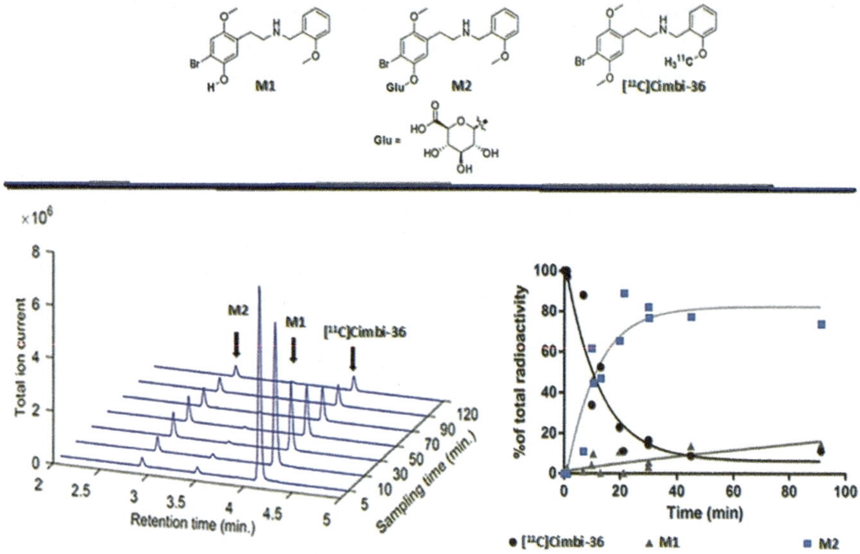

Fig. 12 Metabolic profile of [^{11}C]Cimbi-36. HPLC analysis of plasma (*left*), metabolism rate in pig plasma (*right*) [95, 171]

In Vivo Profile

Several attempts to develop a 5-HT$_{2A}$ selective agonist tracer have been done in the past [95, 172]. [^{11}C]Cimbi-36 is the most promising tracer today. It was evaluated first in pigs and showed an accumulation pattern in accordance with the known 5-HT$_{2A}$ receptor distribution. Highest uptake was observed in cortical areas, lowest in cerebellum. The cortical binding could be blocked by pretreatment with ketanserin and observed kinetics appeared to be reversible. Further, the cortical BP$_{ND}$ was determined to be 0.82 using SRTM. One lipophilic metabolite was formed. Unfortunately, brain tissue analysis 60 min after i.v. injection of [^{11}C]Cimbi-36 contained insufficient radioactivity to determine whether this metabolite could enter the brain to any appreciable amount [95]. Future studies have to be carried out to determine the influence of this metabolite on PET quantification. In 2013, a preclinical safety assessment of [^{11}C]Cimbi-36 was reported [173]. Administration of the tracer seemed to be associated with an average radiation burden and no adverse effects were seen in the animals. The same year, the first study with [^{11}C]Cimbi-36 in non-human primates was reported [174]. Strong binding in cortex as well as in the choroid plexus was observed. Since the choroid plexus is known to be an area with a high 5-HT$_{2C}$ receptor concentration and Cimbi-36 displays low nanomolar affinity towards these receptors, the observed binding pattern suggests that in regions with high 5-HT$_{2C}$ receptor density, this receptor may be imaged with [^{11}C]Cimbi-36 [77]. Subsequently, the selectivity profile of [^{11}C]Cimbi-36 was further investigated in the

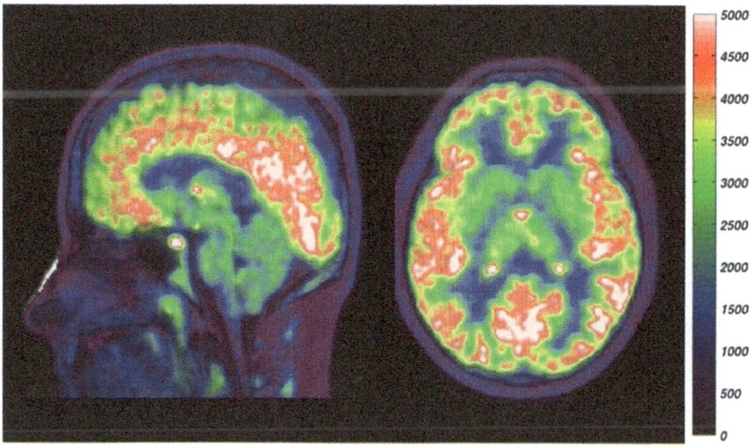

Fig. 13 Summed PET images of [^{11}C]Cimbi-36 (20–80 min. p.i.). Strong binding to cortical 5-HT_{2A} receptors in the human brain is observed. From left to right are displayed sagittal and transversal views

non-human primate brain. No in vivo 5-HT_{2C} binding was detected in cortex. However, the radiotracer accumulation in the choroid plexus was not displaceable by the 5-HT_{2A} selective antagonist, MDL 100907, but by the 5-HT_{2C} receptor selective antagonist, SB 242084. This suggests that [^{11}C]Cimbi-36 offers the possibility to image selectively two receptors in distinct brain areas, namely the 5-HT_{2A} receptor in the cortex and the 5-HT_{2C} in the choroid plexus. In other brain regions (e.g. in the hippocampus or in the striatum), the BP_{ND} was partly altered by SB 242084 as expected from the calculated $tBR_{target/off\text{-}target}$ and selective 5-HT_{2A} imaging may be restricted [174]. Recently, we published the first human PET study of [^{11}C]Cimbi-36 showing high cortical brain binding, displaceable with ketanserin (Fig. 13). Reference tissue modeling resulted in a small predictable bias in PET outcome measures compared to two-tissue compartment modeling using arterial input [175].

^{18}F–Labeled Derivatives of Cimbi-36

As mentioned earlier, fluorine-18 displays several advantages from a clinical perspective. Therefore, several groups have tried to develop a ^{18}F–version of [^{11}C]Cimbi-36 [176, 177]. Unfortunately, none of the presented derivatives showed suitable tracer characteristics for in vivo PET neuroimaging [177]. Low brain uptake and extensive bone uptake limited the use of these compounds. Very recently, Prabhakaran et al. reported an extended in vitro evaluation of one of these compounds [178].

PET Studies of the 5-HT_{2A} Receptor System

Imaging in Brain Disorders

In Europe 23% of healthy years are lost due to brain diseases, while the amount is 50% for "years lost with disability", meaning that approximately one-third of the burden of illness is caused by brain ailments [179]. Although important achievements have been made, brain disorders constitute an area where new and effective treatments are most needed. As the average age of the population in industrial countries continues to rise, an increasing number of neurodegenerative and psychiatric illnesses will place a significant burden on society due the associated health care costs [179].

The 5-HT_{2A} receptor displays a potential target for researchers and pharmaceutical companies, because of its involvement in many disorders, including depression, schizophrenia, obesity and AD. We will here summarize the results gained by in vivo PET studies that have been carried out to study this involvement.

Depression

Today depression affects an estimated 121 million people worldwide. Many studies have shown that the 5-HT_{2A} receptor system is implicated in this disease. Postmortem studies suggest that there is increased 5-HT_{2A} receptors in prefrontal cortex in patients with depression and in suicide victims [149, 180–183]. In contrast to the postmortem work, in vivo imaging studies of depressed patients have been more conflicting, decreased [133, 183–185], or increased [149, 186] 5-HT_{2A} receptor binding in the cerebral cortex has been observed. However, this discrepancy could be explained either due to methodological problems related to inappropriate radioligands used or to psychotropic medications prior to scanning [115, 136, 149, 183]. In this regard, Meyer et al. carried out in 2003 a study with depressed patients who were drug free for more than 3 months. In their work, increased 5-HT_{2A} receptor binding in the frontal cortex could be detected with PET for the first time [186]. Bhagwagar et al. reported the same observation 3 years later [149].

Schizophrenia

The 5-HT_{2A} receptor has been shown to be involved in symptoms of schizophrenia. In particular, it is thought to critically contributing to the pharmacological action of atypical antipsychotics such as olanzapine, risperidone or clozapine [25, 187–189]. Furthermore, hallucinogenic drugs, including mescaline, psilocybin and LSD, exert their effect by stimulating the 5-HT_{2A} receptor. Intake of hallucinogens results in symptoms that partly are overlapping with those of a schizophrenic psychosis [190–194]. In murine models, 5-HT_{2A} receptor-regulated pathways on cortical pyramidal neurons mediate the signaling pattern and behavioral responses to hallucinogenic drugs [195, 196]. Post mortem studies to determine the 5-HT_{2A} receptor protein content or mRNA expression in the frontal cortex of schizophrenic subjects [196–201] have reported conflicting results. Whereas some studies suggest an up-regulation of 5-HT_{2A} receptor binding sites, others point toward the absence of an alteration or even towards 5-HT_{2A} receptor down-regulation [196]. The discrepancies were attempted to be explained by factors such as too small sample sizes, treatment effects or neglecting age effects [25, 195, 199, 202]. Recently, Muguruza et al. suggested that functional selectivity could be the reason for the discrepancies gained with different radiolabeled compounds such as ketanserin (antagonist) or LSD (partial agonist) [196].

Since 1998, several in vivo PET studies with [^{11}C]NMSP and [^{18}F]setoperone were published [203–206]. In three studies, no difference between schizophrenic patients and healthy controls were found in cerebral 5-HT_{2A} receptor binding [203–205] whereas one study reported decreased binding in the left lateral frontal cortex in six patients [206]. However, the studies include relatively small sample sizes and they used PET tracers that were not 5-HT_{2A} receptor selective.

The first in vivo PET study with a 5-HT_{2A} receptor selective tracer ([^{18}F]altanserin) was published in 2005 [207]. Decreased prefrontal 5-HT_{2A} receptor binding in drug-naïve patients with schizophrenia could be observed. Five years later, Rasmussen et al. verified these results in an extended study, also using [^{18}F]altanserin. Instead of 13 subjects, 30 first-episode, antipsychotic-naïve schizophrenic patients, and 30 matched healthy controls were included. Samples sizes of this order of magnitude are essential to obtain sufficient power to detect differences in the order of 10% [134]. Significant lower 5-HT_{2A} receptor binding in the frontal cortex of schizophrenic patients was also observed in this study [208].

Body Weight Changes and Obesity

Previous data have shown that the 5-HT_{2A} receptor is involved in weight gain and obesity. For example, stimulation of the 5-HT_{2A} receptor induces satiety in rodents [209]. Furthermore, obese mice showed increased 5-HT_{2A} receptor density concentration in comparison to obese-resistant mice [210, 211]. A positive association between body-mass index and neocortical 5-HT_{2A} receptor binding has been reported [212] and replicated twice in subsequent studies [213, 214]. Second-generation antipsychotics, blocking the 5-HT_{2A} receptor are also known to be more liable to induce weight gain than first-generation antipsychotics [215]. Therefore, Rasmussen et al. investigated the 5-HT_{2A} receptor with [^{18}F]altanserin PET and related the findings to weight gain compared before and after six months of antipsychotic monotherapy [216] in 15 antipsychotic-naive first-episode schizophrenia patients. A significant positive correlation both between neocortical 5-HT_{2A} receptor binding prior to treatment and subsequent weight gain was found. Conversely, the predictive value of neocortical 5-HT_{2A} receptor binding was also seen in a cohort of obese individuals undergoing by-pass surgery; higher presurgical neocortical

5-HT_{2A} receptor binding predicted greater weight loss after by-pass surgery and the change in 5-HT_{2A} receptor binding correlated with weight loss after surgery [214].

Alzheimer's Disease (AD)

AD is the most common neurodegenerative disease and is ranked as the third most costly disorder. The prevalence of AD in Western societies appears to double every 10 years after the age of 65, with estimates indicating that 16–29% over the age of 85 are sufferers [217]. For AD, there is currently no curative treatment. However, there is strong evidence from postmortem studies that the 5-HT_{2A} receptor is profoundly reduced in AD patients. For example, Lai et al. reported that the loss of 5-HT_{2A} receptors in temporal cortex correlated with the rate of cognitive decline [218]. These finding have now consistently been supported by in vivo functional imaging studies, showing large reductions in 5-HT_{2A} receptor binding in mild to moderately demented AD patients [22, 138, 219–222]. Importantly, the studies found profound and widespread reductions of 5-HT_{2A} binding in neocortical regions in both patients with mild cognitive impairment (MCI) and AD (MCI: 22–29% and AD: 28–39%) [223]. Furthermore, the reduction in binding could already be observed early in the time course of the disease [22]. Interestingly, this postsynaptic 5-HT_{2A} receptor reduction is not accompanied by a similar reduction of the presynaptic serotonin transporter [224], suggesting that the reduction cannot be explained by the loss of serotonergic neurons projecting to neocortical regions. Further studies are needed to reveal why the neocortical 5-HT_{2A} receptor is so prominently affected already early on in AD. One plausible explanation could be a beta-amyloid/5-HT_{2A} receptor association [222, 223, 225–228] since the observed 5-HT_{2A} reduction inversely mirrors the small increase in beta-amyloid accumulation in most early AD studies and follows a similar spatial distribution [225, 226]. Tau accumulation, on the other hand, follows a different pattern [223, 225]. Unfortunately, no correlation was seen between the clinical severity of MCI/AD and the reduction in 5-HT_{2A} receptor binding [22, 220], but this may simply reflect lack of relevance and accuracy of the employed cognitive tests. However, these studies included rather small sample sizes, and also, the limited range of cognitive dysfunction may have made it difficult to determine correlations between serotonergic and cognitive dysfunction [223].

It is long recognized that alterations in the serotonergic system can affect behavioral symptoms that occur in AD, particularly depressive and psychotic symptoms [229]. Since the 5-HT_{2A} receptor is strongly involved in both behavioral effects, several groups tried to correlate the reduced receptor binding to these symptoms. No correlation has been found in five small sample size PET studies with [^{18}F]altanserin [22, 138, 220, 222, 224]. There was not found any difference between AD patients with and without depressive or psychotic symptoms, but this observation warrants further investigation in larger patient samples.

5-HT_{2A} PET Imaging in Drug Development and Discovery

PET is increasingly used in drug development programs and discovery process of novel drug candidates since PET is, for example, able to examine the drug's biological characteristics (BBB passage, target engagement, non-specific binding or metabolism) or able to determine the optimal drug dosage. In general, there are two ways to utilize PET molecular imaging in the drug discovery and development process. Either the drug candidate can be directly labeled and used to determine its pharmacokinetic profile (such as BBB passage, metabolism, biodistribution and reversibility of target binding) or a well-described and specific PET tracer can be used to find a suitable drug dose for subsequent clinical trials.

As an example for both approaches, which have been applied for drugs targeting the 5-HT_{2A} receptor, the drug pimavanserin will be discussed. Pimavanserin is known as a 5-HT_{2A} selective inverse agonist [230]. It has been developed to treat Parkinson's disease psychosis and to improve the treatment of schizophrenia [231]. To study the in vivo behavior of this compound, the relationship between oral dose, plasma levels, and uptake of pimavanserin was studied in humans with the PET tracer [^{11}C]NMSP at baseline and after drug administration [232]. [^{11}C]NMSP is a dual D_2/5-HT_2 receptor ligand and thus it is imaging both receptors at the same time. However, the $tBR_{5\text{-}HT2A/D2}$ in cortical regions and the $tBR_{D2}/_{5\text{-}HT2A}$ in the striatum are high enough to distinguish between both subtypes. Cortical [^{11}C]NMSP binding was blocked in a dose-dependent manner. For example, an oral dose of 10 mg of pimavanserin resulted in an almost complete displacement.

In 2015, Andersen et al. published the synthesis and in vivo evaluation of [^{11}C]pimavanserin [233]. In Danish Landrace pigs the radioligand readily entered the brain and displayed binding in the cortex in accordance with the distribution of 5-HT_{2A} receptors. This binding could not be blocked by either ketanserin or pimavanserin itself, indicating high non-specific binding or irreversible binding. Surprisingly and in addition to the cortical binding, high accumulation of [^{11}C]pimavanserin was observed in thalamus. This binding suggests additional and unknown binding sites of [^{11}C]pimavanserin within the pig brain.

As mentioned earlier, it is important to be able to determine target occupancy levels of the drug molecules at different doses and correlate these data with in vivo potency. Incorrect dosage is one of the major reasons why potential drug molecules fail in clinical trials [46]. The ideal dose of a drug is high enough to result in the desired effect, but not so high to cause side-effects. [^{18}F]altanserin was used to investigate the relationship between 5-HT_{2A} receptor occupancy and treatment effect after treatment with quetiapine, a specific atypical antipsychotic [234]. The expected nonlinear relationship was found between 5-HT_{2A} receptor occupancy and quetiapine dose and a receptor occupancy level between 60 and 70% appeared to be the optimal. No further treatment effect was obtained above a receptor occupancy of 70%.

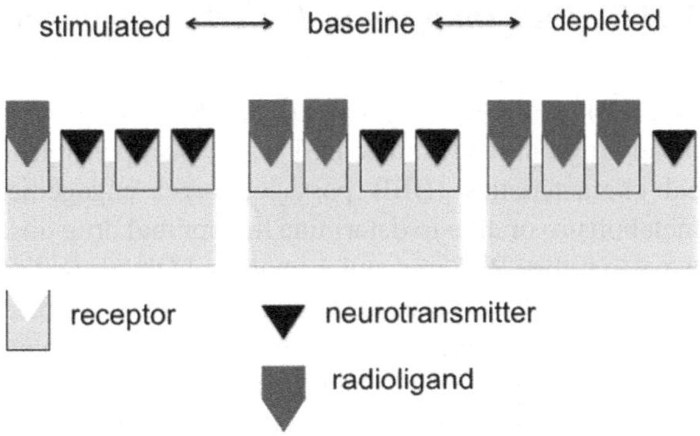

Fig. 14 The occupancy model describes BP changes in response to a challenge: A competition for receptor occupancy between the neurotransmitter and the radioligand results in either reduced BP after neurotransmitter release or increased BP after neurotransmitter depletion [159, 166]

Measurement of Endogenous Serotonin Levels

Being able to determine the in vivo 5-HT neurotransmitter release or to measure 5-HT fluctuations in a non-invasive way would be extremely useful to improve our understanding of brain functions potentially leading to new treatment options. PET has been shown to be able to measure dynamic fluctuation in the context of dopamine release [235–237] and led to a novel understanding of dopaminergic mechanism of action, for example, in schizophrenia [238] and the Parkinson's disease [239].

The basic principle to measure the endogenous neurotransmitter concentration in the synaptic cleft with PET relies on the differential occupation of target receptors by a neurotransmitter, after concentration fluctuation after a challenge (Fig. 14) [159, 166]. Altered neurotransmitter concentrations will change the receptor availability, which can be detected with an exogenous radiotracer by comparing the changes in BP under baseline and challenge conditions. A thorough review on the topic was published by Paterson et al. [159, 166] and Tyacke et al. [240].

Susceptibility of [^{18}F]altanserin/[^{18}F]deuteroaltanserin Towards Endogenous Serotonin

The influence of synaptic serotonin levels on [^{18}F]altanserin binding to 5-HT$_{2A}$ receptors in man is controversial. In 2003, Larisch et al. reported that clomipramine challenge decreased the BP$_{ND}$ in cortex. Clomipramine is a 5-HT re-uptake inhibitor and thus, increases synaptic serotonin levels, which could possibly compete with altanserin and lead to a lower BP$_{ND}$ [241]. However, clomipramine also displays nanomolar 5-HT$_{2A}$ receptor affinity and could thus compete directly with [^{18}F]altanserin [242–244]. One year later, Pinborg et al. reported that cortical [^{18}F]altanserin binding was insensitive to the selective serotonin reuptake inhibitor, citalopram [245]. Rodent studies suggest that citalopram causes only a two- to three-fold increase of cortical 5-HT levels [246], which might be insufficient for a displacement of [^{18}F]altanserin [247]. This is in line with the data of Kristiansen et al. who showed that [^{18}F]altanserin can only be displaced at relatively high 5-HT levels [68]. In 2007, Matusch et al. reported the insensitivity of [^{18}F]altanserin to acute neurotransmitter fluctuations under ketamine after a radioligand bolus-infusion paradigm [248] and in 2001, Staley et al. reported that cortical [^{18}F]deuteroaltanserin binding was likewise unaltered in baboons when the more potent 5-HT releaser dexfenfluramine (10- to 25-fold higher serotonin concentrations [249–251] was administered after a bolus–infusion application [136]. By contrast, a similar study of Quednow et al. showed, that dexfenfluramine-induced 5-HT release, decreased cortical [^{18}F]altanserin receptor binding in humans after a single bolus injection [247]. The authors speculate that the different pharmacological challenge approach (bolus-infusion vs. single bolus injection) explained the different outcomes. They believe that their challenge approach resulted in a higher and thus sufficient endogenous 5-HT level dose to displace [^{18}F]altanserin [136, 247]. This is indeed a likely explanation since it has also been shown that the 5-HT releaser fenfluramine much more potently increases 5-HT levels in the pig brain with a single bolus injection [252].

Susceptibility of [^{11}C]MDL 100907 and (R)-[^{18}F]MH.MZ Towards Endogenous Serotonin

The susceptibility of [^{11}C]MDL 100907 to changes in endogenous serotonin was determined in two studies. The 5-HT releaser, fenfluramin, did not reveal any effect on the BP_{ND} of [^{11}C]MDL 100907 in rat cortical regions [142]. Furthermore, acute reduction of endogenous 5-HT by rapid tryptophan depletion did also not alter the specific binding of [^{11}C]MDL 100907 in humans [146]. These experiments suggest that [^{11}C]MDL 100907 is not susceptible to competition with at least small changes in endogenous 5-HT. The susceptibility of (R)-[^{18}F]MH.MZ towards endogenous serotonin was so far not tested. Most likely, it will behave as [^{11}C]MDL 100907 and not be sensitive to serotonin.

[^{11}C]Cimbi-36´s Susceptibility Towards Endogenous Serotonin

As mentioned earlier, agonist 5-HT$_{2A}$ PET tracers should be more sensitive to endogenous neurotransmitter fluctuation than antagonists. Finnema et al. [174] published a study in non-human primates and found that fenfluramin induced 5-HT release was associated with a decrease in the BP_{ND} of [^{11}C]Cimbi-36. In a recent study by Jorgensen et al. [252], it was shown that in pigs, [^{11}C]Cimbi-36 significantly decreases in response to various pharmacological challenges that increase 5-HT. Thus, it appears that [^{11}C]Cimi-36 binding is sensitive towards endogenously released serotonin, particularly if the increase is substantial.

Summary

To date, several promising 5-HT_{2A} receptor PET radiopharmaceutical have been developed and successfully applied in vivo. Especially the antagonists [^{11}C]MDL 100907 and [^{18}F]altanserin have been used to study the cerebral 5-HT_{2A} receptor in humans. However, both tracers are disadvantaged in some aspects and thus, attempts have been made to develop tracers, which can circumvent these shortcomings. For example, (R)-[^{18}F]MH.MZ and [^{11}C]Cimbi-36 were developed as a next generation of 5-HT_{2A} tracers. Both tracers are currently evaluated in greater detail and further experiments are still needed to validate their beneficial tracer characteristics. Table 4 summarizes some important properties of tracers discussed in here.

Outlook

The search for an ideal antagonist or agonist 5-HT_{2A} receptor PET tracer continues. The selective antagonists, [^{11}C]MDL 100907 and (R)-[^{18}F]MH.MZ, are disadvantaged by their slow kinetics. [^{18}F]altanserin displays low specific-to-unspecific uptake and generates BBB permeating radiolabeled metabolites, which complicate receptor quantification. The ideal combination of an antagonistic PET radiopharmaceutical with high selectivity, non-BBB penetrating radiometabolites, high specific-to-non-specific binding and fast kinetics remains to be identified.

[^{11}C]Cimbi-36 is the most promising agonistic 5-HT_{2A} receptor tracer so far. However, the selectivity profile is still not perfect and this may limit its use in high 5-HT_{2C} receptor binding regions. In addition, having a ^{18}F–labeled version would be a good addition. Finally, no data on an inverse agonist 5-HT_{2A} receptor PET tracer has been reported. This tracer could help to quantify the number of receptors in the high affinity state by applying first a PET scan with an antagonistic tracer followed up by a second scan with an inverse agonist. The subtraction of the second from the first scan should, in principle, allow a quantification of receptors in their high affinity state. Of course, several problems coming along with such a study. For example, the free fraction in the non-displaceable tissue compartment (f_{ND}), the non-specific binding and the in vivo affinity of both ligands have to be determined. Otherwise, a direct comparison of the tracers binding profiles is not possible.

Table 4 Overview of 5-HT$_{2A}$ PET tracer characteristics

	[^{18}F]altanserin	[^{18}F] deutero-altanserin	[^{11}C]MDL 100907	(R)-[^{18}F] MH.MZ	[^{11}C]Cimbi-36
Reference cmpd	+ + +	– – –	+ + +	+	–
	Commercially available	Stable isotope synthesis	Commercially available	Single-step synthesis	Multi-step Synthesis
Precursor	+ + +	– – –	+ + +	+ + +	–
	Commercially available	Stable isotope synthesis	Commercially available	Commercially available	Multi-step Synthesis
Radiolabeling	–	–	+ + +	+	+
	Demanding radiosynthesis	Demanding radiosynthesis	Single step radiosynthesis	Two-step radiosynthesis	Two-step radiosynthesis
Selectivity	+	+	+ + +	+ + +	+
	Affinity towards α_1, D_2	Affinity towards α_1, D_2	Selective	Selective	Affinity towards 5-HT$_{2C}$
Metabolism	– – –	+	+ + +	+ + +	+
	4 BBB crossing metabolites	Reduced metabolism	No metabolite in the brain	No metabolite in the brain	Negligible amounts in the brain
Non-specific binding (NB)	– – –	–	+ + +	+ + +	+ +
	Metabolites cause high NB	Further studies to be conducted	Low NB	Low NB	Low NB, but higher than MDL 100907
Cortical binding potential[a]	2.3–2.9 (human)	~ 0.6 (human)	2.2–3.0	so far n.r.	1.8 (rhesus monkey) 0.82 (pig)
	0.7 (pig)	– –	(human) ~ 4 (rhesus monkey) n.r.[b]	so far n.r.	
	1.9–2.0 (rat)			2.5 (rat)	–
Kinetics	+ + +	+ + +	–	–	+ + +
	Fast, reversible	Fast, reversible	Slow, reversible	Slow, reversible	Fast, reversible
Max SUV	1.26 (human)	n.r.[b]	5.6 (rhesus monkey)	so far n.r.	3.2 (rhesus monkey)
	2 (pig)	–		so far n.r.	
	0.7 (rat)	n.r.[b]	- n.r.[b]	1.8 (rat)	2.2 (pig)
					–
Functionality	Antagonist	Antagonist	Antagonist	Antagonist	Agonist
Susceptibility towards 5-HT	Different outcomes	Different outcomes	None	Probably none	Probably yes

[a]Different kinetic modeling was applied in some cases
[b]n.r. not reported

References

1. Rosel P, Arranz B, San L, Vallejo J, Crespo JM, Urretavizcaya M et al (2000) Altered 5-HT$_{2A}$ binding sites and second messenger inositol trisphosphate (IP3) levels in hippocampus but not in frontal cortex from depressed suicide victims. Psychiat Res Neuroim 99(3):173–181
2. Pazos A, Probst A, Palacios JM (1987) Serotonin receptors in the human-brain .4. Autoradiographic mapping of serotonin-2 receptors. Neuroscience 21(1):123–139
3. Cornea-Hebert V, Riad M, Wu C, Singh SK, Descarries L (1999) Cellular and subcellular distribution of the serotonin 5-HT$_{2A}$ receptor in the central nervous system of adult rat. J Comp Neurol 409(2):187–209
4. Hall H, Farde L, Halldin C, Lundkvist C, Sedvall G (2000) Autoradiographic localization of 5-HT$_{2A}$ receptors in the human brain using [^{3}H]M100907 and [^{11}C]M100907. Synapse 38(4):421–431
5. Saulin A, Savli M, Lanzenberger R (2012) Serotonin and molecular neuroimaging in humans using PET. Amino Acids 42(6):2039–2057
6. Urban JD, Clarke WP, von Zastrow M, Nichols DE, Kobilka B, Weinstein H et al (2007) Functional selectivity and classical concepts of quantitative pharmacology. J Pharmacol Exp Ther 320(1):1–13
7. Leysen I, Van der Gucht E, Eysel UT, Huybrechts R, Vandesande F, Arckens L (2004) Time-dependent changes in the expression of the MEF2 transcription factor family during topographic map reorganization in mammalian visual cortex. Eur J Neurosci 20(3):769–780
8. Meyer PT, Bhagwagar Z, Cowen PJ, Cunningham VJ, Grasby PM, Hinz R (2010) Simplified quantification of 5-HT$_{2A}$ receptors in the human brain with [^{11}C]MDL 100907 PET and non-invasive kinetic analyses. NeuroImage 50(3):984–993
9. Girgis RR, Slifstein M, Xu XY, Frankle WG, Anagnostou E, Wasserman S et al (2011) The 5-HT$_{2A}$ receptor and serotonin transporter in Asperger's disorder: a PET study with [^{11}C] MDL 100907 and [^{11}C]DASB. Psychiat Res Neuroim 194(3):230–234
10. Akash K, Balarama K, Paulose C (2008) Enhanced 5-HT$_{2A}$ receptor status in the hypothalamus and corpus striatum of ethanol-treated rats. Cell Mol Neurobiol 28(7):1017–1025
11. Jones BJ, Blackburn TP (2002) The medical benefit of 5-HT research. Pharmacol Biochem Behav 71(4):555–568
12. Landolt HP, Wehrle R (2009) Antagonism of serotonergic 5-HT$_{2A/2C}$ receptors: mutual improvement of sleep, cognition and mood? Eur J Neurosci 29(9):1795–1809
13. Burnet PWJ, Eastwood SL, Harrison PJ (1996) 5-HT$_{1A}$ and 5-HT$_{2A}$ receptor mRNAs and binding site densities are differentially altered in schizophrenia. Neuropsychopharmacology 15(5):442–455
14. Burnet PWJ, Chen CPLH, McGowan S, Franklin M, Harrison PJ (1996) The effects of clozapine and haloperidol on serotonin-1A, -2A and -2C receptor gene expression and serotonin metabolism in the rat forebrain. Neuroscience 73(2):531–540
15. Erritzoe D, Rasmussen H, Kristiansen KT, Frokjaer VG, Haugbol S, Pinborg L et al (2008) Cortical and subcortical 5-HT$_{2A}$ receptor binding in neuroleptic-naive first-episode schizophrenic patients. Neuropsychopharmacology 33(10):2435–2441
16. Erritzoe DF, Frokjaer VB, Christoffersen MV, Baare W, Ramsoy T, Svarer C et al (2008) Decreased serotonin-2A binding in MDMA and hallucinogen users: an [^{18}F]altanserin PET study. NeuroImage 41:46
17. Haahr MT, Erritzoe D, Lindqvist D, Baare W, Hojgaard L, Almdal T et al (2008) Obesity is associated with increased cortical 5-HT$_{2A}$ receptor binding and decreased serotonin transporter binding in humans: a pilot study. NeuroImage 41:152
18. Naughton M, Mulrooney JB, Leonard BE (2000) A review of the role of serotonin receptors in psychiatric disorders. Hum Psychopharm Clin 15(6):397–415
19. Popa D, Lena C, Fabre V, Prenat C, Gingrich J, Escourrou P et al (2005) Contribution of 5-HT$_{2}$ receptor subtypes to sleep-wakefulness and respiratory control, and functional adaptations in knock-out mice lacking 5-HT$_{2A}$ receptors. J Neurosci 25(49):11231–11238

20. Sheline YI, Mintun MA, Moerlein SM, Snyder AZ (2002) Greater loss of 5-HT_{2A} receptors in midlife than in late life. Am J Psychiat 159(3):430–435
21. Abbas A, Roth BL (2008) Pimavanserin tartrate: a 5-HT_{2A} inverse agonist with potential for treating various neuropsychiatric disorders. Expert Opin Pharmaco 9(18):3251–3259
22. Hasselbalch SG, Madsen K, Svarer C, Pinborg LH, Holm S, Paulson OB et al (2008) Reduced 5-HT_{2A} receptor binding in patients with mild cognitive impairment. Neurobiol Aging 29(12):1830–1838
23. McKeith IG, Marshall EF, Ferrier IN, Armstrong MM, Kennedy WN, Perry RH et al (1987) 5-HT receptor-binding in postmortem brain from patients with affective-disorder. J Affect Disorders 13(1):67–74
24. Meneses A (2007) Stimulation of 5-HT_{1A}, 5-HT_{1B}, 5-$HT_{2A/2C}$, 5-HT_3 and 5-HT_4 receptors or 5-HT uptake inhibition: short- and long-term memory. Behav Brain Res 184(1):81–90
25. Gonzalez-Maeso J, Sealfon SC (2009) Psychedelics and schizophrenia. Trends Neurosci 32(4):225–232
26. Kaye WH, Frank GK, Bailer UF, Henry SE, Meltzer CC, Price JC et al (2005) Serotonin alterations, in anorexia and bulimia nervosa: new insights from imaging studies. Physiol Behav 85(1):73–81
27. Leysen JE (2004) 5-HT_2 receptors. Curr Drug Targets CNS Neurol Disord 3:11–26
28. Nordstrom AL, Farde L, Halldin C (1993) High 5-HT_2 receptor occupancy in clozapine treated patients demonstrated by PET. Psychopharmacology 110(3):365–367
29. Grunder G, Yokoi F, Offord SJ, Ravert HT, Dannals RF, Salzmann JK et al (1997) Time course of 5-HT_{2A} receptor occupancy in the human brain after a single oral dose of the putative antipsychotic drug MDL 100907 measured by positron emission tomography. Neuropsychopharmacology 17(3):175–185
30. Meltzer HY, Bastani B, Ramirez L, Matsubara S (1989) Clozapine—new research on efficacy and mechanism of action. Eur Arch Psychiatry Neurol Sci 238(5–6):332–339
31. Halberstadt AL (2015) Recent advances in the neuropsychopharmacology of serotonergic hallucinogens. Behav Brain Res 277:99–120
32. Liechti ME, Geyer MA, Hell D, Vollenwieder FX (2000) MDMA ("ecstasy") effects on sensorimotor gating after selective pretreatments in humans. Biol Psychiatry 47(8):148
33. Liechti ME, Vollenweider FX (2000) Acute psychological and physiological effects of MDMA ("Ecstasy") after haloperidol pretreatment in healthy humans. Eur Neuropsychopharmacol 10(4):289–295
34. Geyer MA, Vollenweider FX (2008) Serotonin research: contributions to understanding psychoses. Trends Pharmacol Sci 29(9):445–453
35. Gonzalez-Maeso J, Weisstaub NV, Zhou MM, Chan P, Ivic L, Ang R et al (2007) Hallucinogens recruit specific cortical 5-HT_{2A} receptor-mediated signaling pathways to affect behavior. Neuron 53(3):439–452
36. Massoud TF, Gambhir SS (2003) Molecular imaging in living subjects: seeing fundamental biological processes in a new light. Genes Dev 17(5):545–580
37. Antoni GKT, Langström B (2003) 11C-Labeling chemistry and labeled compounds. In: Vértes A, Nagy S, Klencsár Z, Lovas RG, Rösch F (eds) Handbook of nuclear chemistry, vol 4. Kluwer Academic Publishers, Dordrecht, pp 119–165
38. Lanzenberger R, Wadsak W, Spindelegger C, Mitterhauser M, Akimova E, Mien LK et al (2010) Cortisol plasma levels in social anxiety disorder patients correlate with serotonin-1A receptor binding in limbic brain regions. Int J Neuropsychopharmacol 13(9):1129–1143
39. Spindelegger C, Lanzenberger R, Wadsak W, Mien LK, Stein P, Mitterhauser M et al (2009) Influence of escitalopram treatment on 5-HT_{1A} receptor binding in limbic regions in patients with anxiety disorders. Mol Psychiatry 14(11):1040–1050
40. Farde L, Nordstrom AL (1992) PET analysis indicates atypical central dopamine receptor occupancy in clozapine-treated patients. Br J Psychiatry 160:30–33
41. Paterson LM, Kornum BR, Nutt DJ, Pike VW, Knudsen GM (2013) 5-HT radioligands for human brain imaging with PET and SPECT. Med Res Rev 33(1):54–111

42. Herth MM, Andersen VL, Lehel S, Madsen J, Knudsen GM, Kristensen JL (2013) Development of a ^{11}C-labeled tetrazine for rapid tetrazine-trans-cyclooctene ligation. Chem Commun 49(36):3805–3807
43. Herth MM, Barz M, Moderegger D, Allmeroth M, Jahn M, Thews O et al (2009) Radioactive Labeling of Defined HPMA-Based Polymeric Structures Using [^{18}F]FETos for In Vivo Imaging by Positron Emission Tomography. Biomacromolecules 10(7):1697–1703
44. Herzog HRF (2005) PET- und SPECT-Technik. Pharm Unserer Zeit 34(6):468–473
45. Saha G (2010) The basics of PET imaging. Springer, New York. 978-1-4419-0804-9
46. Piel M, Vernaleken I, Rosch F (2014) Positron emission tomography in CNS drug discovery and drug monitoring. J Med Chem 57(22):9232–9258
47. Innis RB, Cunningham VJ, Delforge J, Fujita M, Giedde A, Gunn RN et al (2007) Consensus nomenclature for in vivo imaging of reversibly binding radioligands. J Cerebr Blood Flow Metab 27(9):1533–1539
48. Wernick MN, Aarsvold JN (eds) (2004) Emission tomography: the fundamentals of PET and SPECT. Academic Press, San Diego. ISBN: 978-0127444826
49. Salinas C, Weinzimmer D, Searle G, Labaree D, Ropchan J, Huang Y et al (2013) Kinetic analysis of drug-target interactions with PET for characterization of pharmacological hysteresis. J Cereb Blood Flow Metab 33(5):700–707
50. Watabe H, Ikoma Y, Kimura Y, Naganawa M, Shidahara M (2006) PET kinetic analysis—compartmental model. Ann Nucl Med 20(9):583–588
51. Wong DF (2003) Predicting the success of a radiopharmaceutical for in vivo imaging of central nervous system neuroreceptor systems. Mol Imaging Biol 5(6):350–362
52. Logan J (2000) Graphical analysis of PET data applied to reversible and irreversible tracers. Nucl Med Biol 27(7):661–670
53. Marshall RC, Powers-Risius P, Huesman RH, Reutter BW, Taylor SE, Maurer HE et al (1998) Estimating glucose metabolism using glucose analogs and two tracer kinetic models in isolated rabbit heart. Am J Physiol 275(2):668–679
54. Patlak CS, Blasberg RG (1985) Graphical evaluation of blood-to-brain transfer constants from multiple-time uptake data—generalizations. J Cerebr Blood Flow Metab 5(4):584–590
55. Patlak CS, Blasberg RG, Fenstermacher JD (1983) Graphical Evaluation of Blood-to-Brain Transfer Constants from Multiple-Time Uptake Data. J Cerebr Blood Flow Metab 3(1):1–7
56. Varnas K, Halldin C, Hall H (2004) Autoradiographic distribution of serotonin transporters and receptor subtypes in human brain. Hum Brain Mapp 22(3):246–260
57. Sullivan GM, Parsey RV, Kumar JSD, Arango V, Kassir SA, Huang YY et al (2007) PET Imaging of CRF1 with [^{11}C]R121920 and [^{11}C]DMP696: is the target of sufficient density? Nucl Med Biol 34(4):353–361
58. Zeng ZZ, Chen TB, Miller PJ, Dean D, Tang YS, Sur C et al (2006) The serotonin transporter in rhesus monkey brain: comparison of DASB and citalopram binding sites. Nucl Med Biol 33(4):555–563
59. Christian BT, Narayanan T, Shi B, Morris ED, Mantil J, Mukherjee J (2004) Measuring the in vivo binding parameters of [^{18}F]fallypride in monkeys using a PET multiple-injection protocol. J Cereb Blood Flow Metab 24(3):309–322
60. Farde L, Hall H, Ehrin E, Sedvall G (1986) Quantitative-Analysis of D_2 Dopamine Receptor-Binding in the Living Human-Brain by Pet. Science 231(4735):258–261
61. Lopez-Gimenez JF, Vilaro MT, Palacios JM, Mengod G (1998) [^{3}H]MDL 100907 labels 5-HT_{2A} serotonin receptors selectively in primate brain. Neuropharmacology 37(9):1147–1158
62. Herth MM, Debus F, Piel M, Palner M, Knudsen GM, Luddens H et al (2008) Total synthesis and evaluation of [^{18}F]MH.MZ. Bioorg Med Chem Lett 18(4):1515–1519
63. Pazos A, Probst A, Palacios JM (1987) Serotonin Receptors in the Human-Brain 3. Autoradiographic Mapping of Serotonin-1 Receptors. Neuroscience 21(1):97–122
64. Debus F, Herth MM, Piel M, Buchholz HG, Bausbacher N, Kramer V et al (2010) ^{18}F-Labeling and evaluation of novel MDL 100907 derivatives as potential 5-HT_{2A} antagonists for molecular imaging. Nucl Med Biol 37(4):487–495

65. Lopez-Gimenez JF, Mengod G, Palacios JM, Vilaro MT (1997) Selective visualization of rat brain 5-HT_{2A} receptors by autoradiography with [^{3}H]MDL 100907. Naunyn Schmiedebergs Arch Pharmacol 356(4):446–454
66. Malgouris C, Flamand F, Doble A (1993) Autoradiographic studies of RP 62203, a potent 5-HT_2 receptor antagonist—pharmacological characterization of [^{3}H]RP 62203 binding in the rat-brain. Eur J Pharmacol 233(1):37–45
67. Hansen HD, Ettrup A, Herth MM, Dyssegaard A, Ratner C, Gillings N et al (2013) Direct comparison of [^{18}F]MH.MZ and [^{18}F]altanserin for 5-HT_{2A} receptor imaging with PET. Synapse 67(6):328–337
68. Kristiansen H, Elfving B, Plenge P, Pinborg LH, Gillings N, Knudsen GM (2005) Binding characteristics of the 5-HT_{2A} receptor antagonists altanserin and MDL 100907. Synapse 58(4):249–257
69. Pinborg LH, Adams KH, Svarer C, Holm S, Hasselbalch SG, Haugbol S et al (2003) Quantification of 5-HT_{2A} receptors in the human brain using [^{18}F]altanserin. PET and the bolus/infusion approach. J Cereb Blood Flow Metab 23(8):985–996
70. Ito H, Nyberg S, Halldin C, Lundkvist C, Farde L (1998) PET imaging of central 5-HT_{2A} receptors with [^{11}C]MDL 100,907. J Nucl Med 39(1):208–214
71. Talvik-Lotfi M, Nyberg S, Nordstrom AL, Ito H, Halldin C, Brunner F et al (2000) High 5-HT_{2A} receptor occupancy in MDL 100907 treated schizophrenic patients. Psychopharmacology 148(4):400–403
72. Watabe H, Channing MA, Der MG, Adams HR, Jagoda E, Herscovitch P et al (2000) Kinetic analysis of the 5-HT_{2A} ligand [^{11}C]MDL 100907. J Cereb Blood Flow Metab 20(6):899–909
73. Lundkvist C, Halldin C, Ginovart N, Nyberg S, Swahn CG, Carr AA et al (1996) [^{11}C]MDL 100907, a radioligand for selective imaging of 5-HT_{2A} receptors with positron emission tomography. Life Sci 58(10):Pl87–Pl92
74. Maeshima T, Shutoh F, Hamada S, Senzaki K, Hamaguchi-Hamada K, Ito R et al (1998) Serotonin2A receptor-like immunoreactivity in rat cerebellar Purkinje cells. Neurosci Lett 252(1):72–74
75. Johnson MP, Siegel BW, Carr AA (1996) [^{3}H]MDL 100907: a novel selective 5-HT_{2A} receptor ligand. Naunyn Schmiedebergs Arch Pharmacol 354(2):205–209
76. Herth MM, Volk B, Pallagi K, Bech LK, Antoni FA, Knudsen GM et al (2012) Synthesis and in vitro evaluation of oxindole derivatives as potential radioligands for 5-HT_7 receptor imaging with PET. ACS Chem Neurosci 3(12):1002–1007
77. Marazziti D, Rossi A, Giannaccini G, Zavaglia KM, Dell'Osso L, Lucacchini A et al (1999) Distribution and characterization of [^{3}H]mesulergine binding in human brain postmortem. Eur Neuropsychopharmacol 10(1):21–26
78. Varnas K, Thomas DR, Tupala E, Tiihonen J, Hall H (2004) Distribution of 5-HT_7 receptors in the human brain: a preliminary autoradiographic study using [^{3}H]SB-269970. Neurosci Lett 367(3):313–316
79. Grossisseroff R, Dillon KA, Fieldust SJ, Biegon A (1990) Autoradiographic analysis of alpha-1-noradrenergic receptors in the human brain postmortem—effect of suicide. Arch Gen Psychiatry 47(11):1049–1053
80. DePaermentier F, Mauger JM, Lowther S, Crompton MR, Katona CLE, Horton RW (1997) Brain alpha-adrenoceptors in depressed suicides. Brain Res 757(1):60–68
81. Boyson SJ, Mcgonigle P, Molinoff PB (1986) Quantitative autoradiographic localization of the D_1 and D_2 subtypes of dopamine-receptors in rat-brain. J Neurosci 6(11):3177–3188
82. Boyson SJ, Adams C (1991) A detailed quantitative autoradiographic study of D_2 receptors in parkinsons-disease and progressive supranuclear palsy. Ann Neurol 30(2):256
83. Boyson SJ, Adams CE (1997) D_1 and D_2 dopamine receptors in perinatal and adult basal ganglia. Pediatr Res 41(6):822–831
84. Boyson SJ, Adams C (1990) Receptors for dopaminergic and serotonergic drugs in human choroid-plexus. Ann Neurol 28(2):250
85. Hall H, Lundkvist C, Halldin C, Farde L, Pike VW, McCarron JA et al (1997) Autoradiographic localization of 5-HT_{1A} receptors in the post-mortem human brain using [^{3}H]WAY 100635 and [^{11}C]WAY 100635. Brain Res 745(1–2):96–108

86. Hall H, Halldin C, Farde L, Sedvall G (1998) Whole hemisphere autoradiography of the postmortem human brain. Nucl Med Biol 25(8):715–719
87. Mo HP, Balko KM, Colby DA (2010) A practical deuterium-free NMR method for the rapid determination of 1-octanol/water partition coefficients of pharmaceutical agents. Bioorg Med Chem Lett 20(22):6712–6715
88. OECD (1989) Guideline for testing of chemicals. OECD, Paris
89. Rowley M, Kulagowski JJ, Watt AP, Rathbone D, Stevenson GI, Carling RW et al (1997) Effect of plasma protein binding on in vivo activity and brain penetration of glycine/NMDA receptor antagonists. J Med Chem 40(25):4053–4068
90. Bouchard G, Pagliara A, Carrupt PA, Testa B, Gobry V, Girault HH (2002) Theoretical and experimental exploration of the lipophilicity of zwitterionic drugs in the 1,2-dichloroethane/water system. Pharm Res 19(8):1150–1159
91. Pike VW (2009) PET radiotracers: crossing the blood-brain barrier and surviving metabolism. Trends Pharmacol Sci 30(8):431–440
92. Waring MJ (2010) Lipophilicity in drug discovery. Expert Opin Drug Dis 5(3):235–248
93. Herth MM, Kramer V, Piel M, Palner M, Riss PJ, Knudsen GM et al (2009) Synthesis and in vitro affinities of various MDL 100907 derivatives as potential ^{18}F-radioligands for 5-HT$_{2A}$ receptor imaging with PET. Bioorgan Med Chem 17(8):2989–3002
94. Muehlhausen UEJ, Coenen HH (2008) Synthesis, labeling and first evaluation of [^{18}F]R91150 as a serotonin 5-HT$_{2A}$ receptor antagonist for PET. J Label Compd Radiopharm 52(1):13–22
95. Ettrup A, Hansen M, Santini MA, Paine J, Gillings N, Palner M et al (2011) Radiosynthesis and in vivo evaluation of a series of substituted ^{11}C-phenethylamines as 5-HT$_{2A}$ agonist PET tracers. Eur J Nucl Med Mol Imaging 38(4):681–693
96. Reichel A, Begley DJ (1998) Potential of immobilized artificial membranes for predicting drug penetration across the blood-brain barrier. Pharm Res 15(8):1270–1274
97. Vraka CNL, Hendl M, Zeilinger M, Savli M, Lanzenberger R, Mitterhauser M, Wadsak W (2013) Immobilized artificial membrane (IAM) chromatographie—ine geeignete, präklinische Methode zur Vorhersage der Bluthirnschranken Penetration? AGRR Abstract Book
98. Ishiwata K, Yanai K, Iwata R, Takahashi T, Hatazawa J, Ithoh M, Watabe T, Ido T (1996) Analysis of plasma metabolites during human PET-studies with three receptor ligands, [^{11}C]YM-09151-2, [^{11}C]doxepin and [^{11}C]pyrilamine. Tohoku J Exp Med 178(2):129–136
99. Wong DF, Gjedde A, Wagner HN (1986) Quantification of neuroreceptors in the living human-brain 1. Irreversible binding of ligands. J Cerebr Blood Flow Metab 6(2):137–146
100. Doble A, Girdlestone D, Piot O, Allam D, Betschart J, Boireau A et al (1992) Pharmacological characterization of RP-62203, a novel 5-hydroxytryptamine 5-HT$_2$ receptor antagonist. Br J Pharmacol 105(1):27–36
101. Ashworth S, Hume SP, Lammertsma AA, OpackaJuffry J, Shah F, Pike VW (1996) Development of central 5-HT$_{2A}$ receptor radioligands for PET: comparison of [^{3}H]RP 62203 and [^{3}H]SR 46349B kinetics in rat brain. Nucl Med Biol 23(3):245–250
102. Fajolles C, Boireau A, Ponchant M, Laduron PM (1992) [^{3}H]RP-62203, a ligand of choice to label invivo brain 5-HT$_2$ receptors. Eur J Pharmacol 216(1):53–57
103. Baron JC, Samson Y, Comar D, Crouzel C, Deniker P, Agid Y (1985) An invivo study of central serotonin receptors in humans using ^{11}C-labeled ketanserin and positron tomography. Rev Neurol 141(8–9):537–545
104. Moerlein SM, Perlmutter JS (1991) Central serotonergic S2 binding in Papio-Anubis measured invivo with N-omega-[^{18}F]Fluoroethylketanserin and PET. Neurosci Lett 123(1):23–26
105. Lyon RA, Titeler M, Frost JJ, Whitehouse PJ, Wong DF, Wagner HN et al (1986) [^{3}H]3-N-methylspiperone labels D$_2$ dopamine-receptors in basal ganglia and S$_2$ serotonin receptors in cerebral-cortex. J Neurosci 6(10):2941–2949
106. Lever JR, Dannals RF, Wilson AA, Ravert HT, Scheffel U, Hoffman BJ et al (1989) Synthesis and invivo characterization of D-(+)-(N1-[^{11}C]Methyl)-2-Br-Lsd—a radioligand for positron emission tomographic studies of Serotonin 5-HT$_2$ receptors. Nucl Med Biol 16(7):697–704

107. Blin J, Pappata S, Kiyosawa M, Crouzel C, Baron JC (1988) [^{18}F]Setoperone—a new high-affinity ligand for positron emission Tomography study of the serotonin-2 receptors in baboon brain invivo. Eur J Pharmacol 147(1):73–82
108. Besret L, Dauphin F, Huard C, Lasne MC, Vivet R, Mickala P et al (1996) Specific in vivo binding in the rat brain of [^{18}F]RP 62203: a selective 5-HT_{2A} receptor radioligand for positron emission tomography. Nucl Med Biol 23(2):169–171
109. Herth MM, Piel M, Debus F, Schmitt U, Luddens H, Rosch F (2009) Preliminary in vivo and ex vivo evaluation of the 5-HT_{2A} imaging probe [^{18}F]MH.MZ. Nucl Med Biol 36(4):447–454
110. Kramer V, Herth MM, Santini MA, Palner M, Knudsen GM, Rosch F (2010) Structural combination of established 5-HT_{2A} receptor ligands: new aspects of the binding mode. Chem Biol Drug Des 76(4):361–366
111. Lemaire C, Cantineau R, Guillaume M, Plenevaux A, Christiaens L (1991) [^{18}F]Altanserin—a radioligand for the study of serotonin receptors with PET—radiolabeling and invivo biologic behavior in rats. J Nucl Med 32(12):2266–2272
112. van Dyck CH, Malison RT, Seibyl JP, Laruelle M, Klumpp H, Zoghbi SS et al (2000) Age-related decline in central serotonin transporter availability with [^{123}I]beta-CIT SPECT. Neurobiol Aging 21(4):497–501
113. Herth MM, Knudsen GM (2015) Current radiosynthesis strategies for 5-HT_{2A} receptor PET tracers. J Labelled Comp Radiopharm 58(7):265–273
114. Ettrup A, Hansen M, Santini MA, Paine J, Gillings N, Palner M et al (2010) In vivo evaluation of a series of substituted ^{11}C-phenetylamines as 5-HT_{2A} agonist PET tracers. NeuroImage 52:48
115. Tan PZ, Baldwin RM, Van Dyck CH, Al-Tikriti M, Roth B, Khan N et al (1999) Characterization of radioactive metabolites of 5-HT_{2A} receptor PET ligand [^{18}F]altanserin in human and rodent. Nucl Med Biol 26(6):601–608
116. Biver F, Lotstra F, Monclus M, Dethy S, Damhaut P, Wikler D et al (1997) In vivo binding of [^{18}F]altanserin to rat brain 5-HT_2 receptors: a film and electronic autoradiographic study. Nucl Med Biol 24(4):357–360
117. Hasler F, OF K, Krasikova RN, Cservenyak T, Quednow BB, Vollenweider FX et al (2009) GMP-compliant radiosynthesis of [^{18}F]altanserin and human plasma metabolite studies. Appl Radiat Isot 67(4):598–601
118. Smith GS, Price JC, Lopresti BJ, Huang YY, Simpson N, Holt D et al (1998) Test-retest variability of serotonin 5-HT_{2A} receptor binding measured with positron emission tomography and [^{18}F]altanserin in the human brain. Synapse 30(4):380–392
119. Price JC, Lopresti BJ, Mason NS, Holt DP, Huang Y, Mathis CA (2001) Analyses of [^{18}F] altanserin bolus injection PET data. I: consideration of radiolabeled metabolites in baboons. Synapse 41(1):1–10
120. Price JC, Lopresti BJ, Meltzer CC, Smith GS, Mason NS, Huang Y et al (2001) Analyses of [^{18}F]altanserin bolus injection PET data. II: consideration of radiolabeled metabolites in humans. Synapse 41(1):11–21
121. Eastwood SL, Burnet PWJ, Gittins R, Baker K, Harrison PJ (2001) Expression of serotonin 5-HT_{2A} receptors in the human cerebellum and alterations in schizophrenia. Synapse 42(2):104–114
122. Biver F, Goldman S, Luxen A, Monclus M, Forestini M, Mendlewicz J et al (1994) Multicompartmental study of [^{18}F]Altanserin binding to brain 5-HT_2 receptors in humans using positron emission tomography. Eur J Nucl Med 21(9):937–946
123. Riss PJ, Hong YT, Williamson D, Caprioli D, Sitnikov S, Ferrari V et al (2011) Validation and quantification of [^{18}F]altanserin binding in the rat brain using blood input and reference tissue modeling. J Cereb Blood Flow Metabol 31(12):2334–2342
124. Kroll T, Elmenhorst D, Matusch A, Wedekind F, Weisshaupt A, Beer S et al (2013) Suitability of [^{18}F]altanserin and PET to determine 5-HT_{2A} receptor availability in the rat brain: in vivo and in vitro validation of invasive and non-invasive kinetic models. Mol Imaging Biol 15(4):456–467

125. Kroll T, Elmenhorst D, Matusch A, Celik AA, Wedekind F, Weisshaupt A et al (2014) [^{18}F] Altanserin and small animal PET: impact of multidrug efflux transporters on ligand brain uptake and subsequent quantification of 5-HT$_{2A}$ receptor densities in the rat brain. Nucl Med Biol 41(1):1–9
126. Riss PJ, Hong YT, Williamson D, Caprioli D, Sitnikov S, Ferrari V et al (2011) Validation and quantification of [F-18]altanserin binding in the rat brain using blood input and reference tissue modeling. J Cerebr Blood Flow Metab 31(12):2334–2342
127. Shrestha SS, Liow J, Lu S, Jenko K, Gladding RL, Svenningsson P, Morse CL, Zoghbi SS, Pike VW, Innis RB (2014) [^{11}C]CUMI-101, a PET Radioligand, behaves as a serotonin 1A recepor antagonist and also binds to alpha-1 adrenoceptors in brain. J Nucl Med 55:141–146
128. Sadzot B, Lemaire C, Maquet P, Salmon E, Plenevaux A, Degueldre C et al (1995) Serotonin 5-HT$_{2A}$ receptor imaging in the human brain using positron emission tomography and a new radioligand, [^{18}F]Altanserin—results in young normal controls. J Cereb Blood Flow Metab 15(5):787–797
129. Syvanen S, Lindhe O, Palner M, Kornum BR, Rahman O, Langstrom B et al (2009) Species differences in blood-brain barrier Transport of three positron emission tomography radioligands with emphasis on P-Glycoprotein transport. Drug Metab Dispos 37(3):635–643
130. Froklage FE, Syvänen S, Hendrikse NH, Huisman MC, Molthoff CMF, Tagawa Y, Reijneveld JC, Heimans JJ, Lammertsma AA, Eriksson J, de Lange ECM, Voskuyl RA (2012) [^{11}C] Flumazenil brain uptake is influenced by the blood-brain barrier efflux transporter P-glycoprotein. EJNMMI Res 2:12
131. Martin A, Szczupak B, Gomez-Vallejo V, Plaza S, Padro D, Cano A et al (2013) PET imaging of serotoninergic neurotransmission with [^{11}C]DASB and [^{18}F]altanserin after focal cerebral ischemia in rats. J Cereb Blood Flow Metab 33(12):1967–1975
132. Moses EL, Drevets WC, Smith G, Mathis CA, Kalro BN, Butters MA et al (2000) Effects of estradiol and progesterone administration on human serotonin 2A receptor binding: a PET study. Biol Psychiatry 48(8):854–860
133. Mintun MA, Sheline YI, Moerlein SM, Vlassenko AG, Huang YY, Snyder AZ (2004) Decreased hippocampal 5-HT$_{2A}$ receptor binding in major depressive disorder: in vivo measurement with [^{18}F]Altanserin positron emission tomography. Biol Psychiatry 55(3):217–224
134. Haugbol S, Pinborg LH, Arfan HM, Frokjaer VM, Madsen J, Dyrby TB et al (2007) Reproducibility of 5-HT$_{2A}$ receptor measurements and sample size estimations with [^{18}F] altanserin PET using a bolus/infusion approach. Eur J Nucl Med Mol Imaging 34(6):910–915
135. Soares JC, van Dyck CH, Tan PZ, Zoghbi SS, Garg P, Soufer R et al (2001) Reproducibility of in vivo brain measures of 5-HT$_{2A}$ receptors with PET and [^{18}F]deuteroaltanserin. Psychiatry Res Neuroim 106(2):81–93
136. Staley JK, Van Dyck CH, Tan PZ, Al Tikriti M, Ramsby Q, Klump H et al (2001) Comparison of [^{18}F]altanserin and [^{18}F]deuteroaltanserin for PET imaging of serotonin(2A) receptors in baboon brain: pharmacological studies. Nucl Med Biol 28(3):271–279
137. Kugaya A, Epperson CN, Zoghbi S, van Dyck CH, Hou Y, Fujita M et al (2003) Increase in prefrontal cortex serotonin(2A) receptors following estrogen treatment in postmenopausal women. Am J Psychiat 160(8):1522–1524
138. Santhosh L, Estok KM, Vogel RS, Tamagnan GD, Baldwin RM, Mitsis EM et al (2009) Regional distribution and behavioral correlates of 5-HT$_{2A}$ receptors in Alzheimer's disease with [^{18}F]deuteroaltanserin and PET. Psychiatry Res 173(3):212–217
139. Dezi C, Brea J, Alvarado M, Ravina E, Masaguer CF, Loza MI et al (2007) Multistructure 3D-QSAR studies on a series of conformationally constrained butyrophenones docked into a new homology model of the 5-HT$_{2A}$ receptor. J Med Chem 50(14):3242–3255
140. Scott DO, Heath TG (1998) Investigation of the CNS penetration of a potent 5-HT$_{2A}$ receptor antagonist (MDL 100907) and an active metabolite (MDL 105725) using in vivo microdialysis sampling in the rat. J Pharm Biomed Anal 17(1):17–25
141. Ito H, Nyberg S, Halldin C, Lundkvist C, Farde L (1997) 5-HT$_{2A}$ receptor imaging in the human brain using [^{11}C]MDL 100,907 and PET. J Nucl Med 38(5):297

142. Hirani E, Sharp T, Sprakes M, Grasby P, Hume S (2003) Fenfluramine evokes 5-HT_{2A} receptor-mediated responses but does not displace [^{11}C]MDL 100907: small animal PET and gene expression studies. Synapse 50(3):251–260
143. Hinz R, Bhagwagar Z, Cowen PJ, Cunningham VJ, Grasby PM (2007) Validation of a tracer kinetic model for the quantification of 5-HT_{2A} receptors in human brain with [^{11}C]MDL 100907. J Cereb Blood Flow Metab 27(1):161–172
144. Bhagwagar Z, Hinz R, Cunningham VJ, Cowen PJ, Grasby PM (2003) Conference: Summer Meeting of the British-Association-for-Psychopharmacology. J Psychopharmacology 17
145. Bhagwayar Z, Hinz R, Taylor M, Fancy S, Cowen PJ, Grasby PM (2006) Increased 5-HT_{2A} receptor binding in euthymic, medciation free recovered depressed patients: a positron emission study with [^{11}C]MDL 100907. Am J Psychiatry 163(9):1580–1587
146. Talbot PS, Slifstein M, Hwang DR, Huang YY, Scher E, Abi-Dargham A et al (2012) Extended characterisation of the serotonin 2A (5-HT_{2A}) receptor-selective PET radiotracer [^{11}C]MDL 100907 in humans: Quantitative analysis, test-retest reproducibility, and vulnerability to endogenous 5-HT tone. NeuroImage 59(1):271–285
147. Urban NBL, Girgis RR, Talbot PS, Kegeles LS, Xu X, Frankle WG et al (2012) Sustained recreational use of ecstasy is associated with altered pre and postsynaptic markers of serotonin transmission in neocortical areas: a PET study with [^{11}C]DASB and [^{11}C]MDL 100907. Neuropsychopharmacology 37(6):1465–1473
148. Bhagwagar Z, Hinz R, Taylor M, Fancy S, Cowen PJ, Grasby PM (2005) Increased 5-HT_{2A} receptor binding in euthymic, medication free recovered depressed patients: a positron emission study with [^{11}C]MDL 100,907. J Psychopharmacol 19(5):A52–A5A
149. Bhagwagar Z, Hinz R, Taylor M, Fancy S, Cowen P, Grasby P (2006) Increased 5-HT_{2A} receptor binding in euthymic, medication-free patients recovered from depression: a positron emission study with [^{11}C]MDL 100907. Am J Psychiatry 163(9):1580–1587
150. Muhlhausen U, Ermert J, Herth MM, Coenen HH (2009) Synthesis, radiofluorination and first evaluation of (+/−)-[^{18}F]MDL 100907 as serotonin 5-HT_{2A} receptor antagonist for PET. J Labelled Comp Radiat 52(1–2):6–12
151. Ren H, Wey HY, Strebl M, Neelamegam R, Ritter T, Hooker JM (2014) Synthesis and imaging validation of [^{18}F]MDL 100907 enabled by Ni-mediated fluorination. ACS Chem Neurosci 5(7):611–615
152. Ren H, Strebl M, Neelamegam R, Hooker J, Ritter T (2014) Advancing PET imaging of central 5-HT_{2A} receptors with [^{18}F]MDL100907: passing the stumbling blocks in late-stage Ni-mediated ^{18}F-fluorination. J Nucl Med 55:554
153. Schmitt U, Leed DE, Herth MM, Piel M, Buchholz HG, Roesch F et al (2011) P-Glycoprotein influence on the brain uptake of a 5-HT_{2A} ligand: [^{18}F]MH.MZ. Neuropsychobiology 63(3):183–190
154. Kramer VHM, Hernandez E, Juri C, Pruzzo R, Rösch F, Amaral H (2015) Quantification of 5-HT_{2A}-receptors in the human brainwith [^{18}F]MH.MZ. EANM. Dent Abstr 427:545
155. Hansen H, Herth MM, Ettrup A, Dyssegaard A, Knudsen GM (2013) Comparison of the 5-HT_{2A} PET tracers, [^{18}F]MH.MZ and [^{18}F]altanserin. J Labelled Comp Radiat 56:S338
156. Dalley JW, Fryer TD, Brichard L, Robinson ESJ, Theobald DEH, Laane K et al (2007) Nucleus accumbens $D_{2/3}$ receptors predict trait impulsivity and cocaine reinforcement. Science 315(5816):1267–1270
157. Visser AK, De Vries EF, Ramakrishnan NK, Willemsen AT, Bosker FJ, den Boer JA, Dierckx RA, van Waarde A (2013) Analysis of 5-HT_{2A} receptor binding with [^{11}C]MDL 100907 in rats: optimization of kinetic modeling. Mol Imaging Biol 15(s):730–738
158. Fitzgerald LW, Conklin DS, Krause CM, Marshall AP, Patterson JP, Tran DP et al (1999) High-affinity agonist binding correlates with efficacy (intrinsic activity) at the human serotonin 5-HT_{2A} and 5-HT_{2C} receptors: Evidence favoring the ternary complex and two-state models of agonist action. J Neurochem 72(5):2127–2134
159. Paterson LM, Tyacke RJ, Nutt DJ, Knudsen GM (2010) Measuring endogenous 5-HT release by emission tomography: promises and pitfalls. J Cereb Blood Flow Metab 30(10):1682–1706

160. Verdurand M, Berod A, Le Bars D, Zimmer L (2011) Effects of amyloid-beta peptides on the serotoninergic 5-HT_{1A} receptors in the rat hippocampus. Neurobiol Aging 32(1):103–114
161. Lopez-Gimenez JF, Villazon M, Brea J, Loza MI, Palacios JM, Mengod G et al (2001) Multiple conformations of native and recombinant human 5-hydroxytryptamine(2A) receptors are labeled by agonists and discriminated by antagonists. Mol Pharmacol 60(4):690–699
162. Branchek T, Adham N, Macchi M, Kao HT, Hartig PR (1990) [^{3}H]DOB (4-Bromo-2,5-dimethoxyphenylisopropylamine) and [^{3}H]Ketanserin label 2 affinity states of the cloned human 5-hydroxytryptamine2 receptor. Mol Pharmacol 38(5):604–609
163. Narendran R, Hwang DR, Slifstein M, Hwang YC, Huang Y, Ekelund J et al (2004) Measurement of in vivo affinity of [^{11}C]NPA and the proportion of D_2 receptors configured in agonist high affinity state (%Rhigh) in baboons using PET. NeuroImage 22:T19–T20
164. Narendran R, Mason NS, Laymon CM, Lopresti BJ, Velasquez ND, May MA et al (2010) A comparative evaluation of the dopamine $D_{2/3}$ agonist radiotracer [^{11}C](−)-N-propyl-norapomorphine and antagonist [^{11}C]raclopride to measure amphetamine-induced dopamine release in the human striatum. J Pharmacol Exp Ther 333(2):533–539
165. Cumming P, Wong DF, Gillings N, Hilton J, Scheffel U, Gjedde A (2002) Specific binding of [^{11}C]raclopride and N-[^{3}H]propyl-norapomorphine to dopamine receptors in living mouse striatum: occupancy by endogenous dopamine and guanosine triphosphate-free G protein. J Cereb Blood Flow Metab 22(5):596–604
166. Paterson LM, Nutt DJ, Knudsen GM (2010) A critical review of studies measuring endogenous 5-HT release by emission tomography. NeuroImage 52:107
167. Diaz SL, Doly S, Narboux-Neme N, Fernandez S, Mazot P, Banas SM et al (2012) 5-HT_{2B} receptors are required for serotonin-selective antidepressant actions. Mol Psychiatry 17(2):154–163
168. Bonhaus DW, Bach C, Desouza A, Salazar FHR, Matsuoka BD, Zuppan P et al (1995) The pharmacology and distribution of human 5-hydroxytryptamine(2b) (5-HT_{2B}) receptor gene-products—comparison with 5-HT_{2A} and 5-HT_{2C} receptors. Br J Pharmacol 115(4):622–628
169. Choi DS, Maroteaux L (1996) Immunohistochemical localisation of the serotonin 5-HT_{2B} receptor in mouse gut, cardiovascular system, and brain. FEBS Lett 391(1–2):45–51
170. Duxon MS, Flanigan TP, Reavley AC, Baxter GS, Blackburn TP, Fone KCF (1997) Evidence for expression of the 5-hydroxytryptamine-2B receptor protein in the rat central nervous system. Neuroscience 76(2):323–329
171. Leth-Petersen S, Gabel-Jensen C, Gillings N, Lehel S, Hansen HD, Knudsen GM et al (2016) Metabolic Fate of Hallucinogenic NBOMes. Chem Res Toxicol 29(1):96–100
172. Ettrup A, Palner M, Gillings N, Santini MA, Hansen M, Kornum BR et al (2010) Radiosynthesis and evaluation of [^{11}C]Cimbi-5 as a 5-HT_{2A} receptor agonist radioligand for PET. J Nucl Med 51(11):1763–1770
173. Ettrup AHS, Hansen M, Wasim M, Santini MA, Palner M, Madsen J, Svarer C, Kristensen JL, Knudsen GM (2013) Preclinical safety assessment of the 5-HT_{2A} receptor agonist PET radioligand [^{11}C]Cimbi-36. Mol Imaging Biol 15(4):376–383
174. Finnema SJ, Stepanov V, Ettrup A, Nakao R, Amini N, Svedberg M et al (2014) Characterization of [^{11}C]Cimbi-36 as an agonist PET radioligand for the 5-HT_{2A} and 5-HT_{2C} receptors in the nonhuman primate brain. NeuroImage 84:342–353
175. Ettrup A, da Cunha-Bang S, McMahon B, Lehel S, Dyssegaard A, Skibsted AW et al (2014) Serotonin 2A receptor agonist binding in the human brain with [^{11}C]Cimbi-36. J Cereb Blood Flow Metab 34(7):1188–1196
176. Ettrup A, Hansen M, Gillings N, Dyssegaard A, Lehel S, Hansen HD et al (2012) Development of an ^{18}F-labeled 5-HT_{2A} receptor agonist PET radioligand. J Cereb Blood Flow Metab 32:109
177. Herth MM, Petersen IN, Hansen HD, Hansen M, Ettrup A, Jensen AA, Lehel S, Dyssegaard A, Gillings N, Knudsen GM, Kristensena JL (2016) Synthesis and evaluation of 18F-labeled 5-HT2A receptor agonists as PET ligands. Nucl Med Biol 43(8):455–462
178. Prabhakaran J, Underwood MD, Kumar JSD, Simpson NR, Kassir SA, Bakalian MJ et al (2015) Synthesis and in vitro evaluation of [^{18}F]FECIMBI-36: a potential agonist PET ligand for 5-$HT_{2A/2C}$ receptors. Bioorg Med Chem Lett 25(18):3933–3936

179. Olesen J, Leonardi M (2003) The burden of brain diseases in Europe. Eur J Neurol 10(5):471–477
180. Turecki G, Briere R, Dewar K, Antonetti T, Lesage A, Seguin M et al (1999) 5-HTR_{2A} genetic variation and level of 5-HT_{2A} receptor binding in completed suicide cases. Biol Psychiatry 45(8):123
181. Arango V, Ernsberger P, Marzuk PM, Chen JS, Tierney H, Stanley M et al (1990) Autoradiographic demonstration of increased serotonin 5-HT_2 and beta-adrenergic-receptor binding-sites in the brain of suicide victims. Arch Gen Psychiatry 47(11):1038–1047
182. Hrdina PD, Demeter E, Vu TB, Sotonyi P, Palkovits M (1993) 5-HT uptake sites and 5-HT_2 receptors in brain of antidepressant-free suicide victims depressives—increase in 5-HT_2 sites in cortex and amygdala. Brain Res 614(1–2):37–44
183. Stockmeier CA (2003) Involvement of serotonin in depression: evidence from postmortem and imaging studies of serotonin receptors and the serotonin transporter. J Psychiatr Res 37(5):357–373
184. Biver F, Wikler D, Lotstra F, Damhaut P, Goldman S, Mendlewicz J (1997) Serotonin 5-HT_2 receptor imaging in major depression: focal changes in orbito-insular cortex. Br J Psychiatry 171:444–448
185. Sheline YI, Mintun MA, Barch DM, Wilkins C, Snyder AZ, Moerlein SM (2004) Decreased hippocampal 5-HT_{2A} receptor binding in older depressed patients using [^{18}F]altanserin positron emission tomography. Neuropsychopharmacology 29(12):2235–2241
186. Meyer JH, McMain S, Kennedy SH, Korman L, Brown GM, DaSilva JN et al (2003) Dysfunctional attitudes and 5-HT_2 receptors during depression and self-harm. Am J Psychiatry 160(1):90–99
187. Kurita M, Holloway T, Garcia-Bea A, Kozlenkov A, Friedman AK, Moreno JL et al (2012) HDAC2 regulates atypical antipsychotic responses through the modulation of mGlu2 promoter activity. Nat Neurosci 15(9):1245–1254
188. Miyamoto S, Duncan GE, Marx CE, Lieberman JA (2005) Treatments for schizophrenia: a critical review of pharmacology and mechanisms of action of antipsychotic drugs. Mol Psychiatry 10(1):79–104
189. Fribourg M, Moreno JL, Holloway T, Provasi D, Baki L, Mahajan R et al (2011) Decoding the signaling of a GPCR heteromeric complex reveals a unifying mechanism of action of antipsychotic drugs. Cell 147(5):1011–1023
190. Young BG (1974) A phenomenological comparison of LSD and schizophrenic states. Br J Psychiatry 124(578):64–74
191. Hermle L, Funfgeld M, Oepen G, Botsch H, Borchardt D, Gouzoulis E et al (1992) Mescaline-induced psychopathological, neuropsychological, and neurometabolic effects in normal subjects: experimental psychosis as a tool for psychiatric research. Biol Psychiatry 32(11):976–991
192. Vollenweider FX, Vollenweider-Scherpenhuyzen MF, Babler A, Vogel H, Hell D (1998) Psilocybin induces schizophrenia-like psychosis in humans via a serotonin-2 agonist action. Neuroreport 9(17):3897–3902
193. Gouzoulis-Mayfrank E, Arnold S, Heekeren K (2006) Deficient inhibition of return in schizophrenia-further evidence from an independent sample. Prog Neuro-Psychopharmacol Biol Psychiatry 30(1):42–49
194. Quednow BB, Ettinger U, Mossner R, Rujescu D, Giegling I, Collier DA et al (2011) The schizophrenia risk allele C of the TCF4 rs9960767 polymorphism disrupts sensorimotor gating in schizophrenia spectrum and healthy volunteers. J Neurosci 31(18):6684–6691
195. Gonzalez-Maeso J, Sealfon SC (2009) Agonist-trafficking and hallucinogens. Curr Med Chem 16(8):1017–1027
196. Muguruza C, Moreno JL, Umali A, Callado LF, Meana JJ, Gonzalez-Maeso J (2013) Dysregulated 5-HT_{2A} receptor binding in postmortem frontal cortex of schizophrenic subjects. Eur Neuropsychopharmacol 23(8):852–864
197. Bennett JP Jr, Enna SJ, Bylund DB, Gillin JC, Wyatt RJ, Snyder SH (1979) Neurotransmitter receptors in frontal cortex of schizophrenics. Arch Gen Psychiatry 36(9):927–934

198. Gurevich EV, Joyce JN (1997) Alterations in the cortical serotonergic system in schizophrenia: a postmortem study. Biol Psychiatry 42(7):529–545
199. Dean B, Crossland N, Boer S, Scarr E (2008) Evidence for altered post-receptor modulation of the serotonin 2a receptor in schizophrenia. Schizophr Res 104(1–3):185–197
200. Kang K, Huang XF, Wang Q, Deng C (2009) Decreased density of serotonin 2A receptors in the superior temporal gyrus in schizophrenia—a postmortem study. Prog Neuro-Psychopharmacol Biol Psychiatry 33(5):867–871
201. Gonzalez-Maeso J, Ang RL, Yuen T, Chan P, Weisstaub NV, Lopez-Gimenez JF et al (2008) Identification of a serotonin/glutamate receptor complex implicated in psychosis. Nature 452(7183):93–97
202. Dean B (2003) The cortical serotonin2A receptor and the pathology of schizophrenia: a likely accomplice. J Neurochem 85(1):1–13
203. Lewis R, Kapur S, Jones C, DaSilva J, Brown GM, Wilson AA et al (1999) Serotonin 5-HT_2 receptors in schizophrenia: a PET study using [^{18}F]setoperone in neuroleptic-naive patients and normal subjects. Am J Psychiatry 156(1):72–78
204. Okubo Y, Suhara T, Suzuki K, Kobayashi K, Inoue O, Terasaki O et al (2000) Serotonin 5-HT_2 receptors in schizophrenic patients studied by positron emission tomography. Life Sci 66(25):2455–2464
205. Trichard C, Paillere-Martinot ML, Attar-Levy D, Blin J, Feline A, Martinot JL (1998) No serotonin 5-HT_{2A} receptor density abnormality in the cortex of schizophrenic patients studied with PET. Schizophr Res 31(1):13–17
206. Ngan ET, Yatham LN, Ruth TJ, Liddle PF (2000) Decreased serotonin 2A receptor densities in neuroleptic-naive patients with schizophrenia: a PET study using [^{18}F]setoperone. Am J Psychiatry 157(6):1016–1018
207. Hurlemann R, Boy C, Meyer PT, Scherk H, Wagner M, Herzog H et al (2005) Decreased prefrontal 5-HT_{2A} receptor binding in subjects at enhanced risk for schizophrenia. Anat Embryol 210(5–6):519–523
208. Rasmussen H, Erritzoe D, Andersen R, Ebdrup BH, Aggernaes B, Oranje B et al (2010) Decreased frontal serotonin(2A) receptor binding in antipsychotic-naive patients with first-episode schizophrenia. Arch Gen Psychiatry 67(1):9–16
209. Roerig JL, Steffen KJ, Mitchell JE (2011) Atypical antipsychotic-induced weight gain: insights into mechanisms of action. CNS Drugs 25(12):1035–1059
210. Huang XF, Han M, Storlien LH (2004) Differential expression of 5-HT_{2A} and 5-HT_{2C} receptor mRNAs in mice prone, or resistant, to chronic high-fat diet-induced obesity. Brain Res Mol Brain Res 127(1–2):39–47
211. Huang XF, Huang X, Han M, Chen F, Storlien L, Lawrence AJ (2004) 5-$HT_{2A/2C}$ receptor and 5-HT transporter densities in mice prone or resistant to chronic high-fat diet-induced obesity: a quantitative autoradiography study. Brain Res 1018(2):227–235
212. Pinborg LH, Adams KH, Svarer C, Holm S, Hasselbalch SG, Haugbol S et al (2003) Quantification of 5-HT_{2A} receptors in the human brain using [^{18}F]altanserin-PET and the bolus/infusion approach. J Cereb Blood Flow Metab 23(8):985–996
213. Erritzoe D, Frokjaer VG, Haugbol S, Marner L, Svarer C, Holst K et al (2009) Brain serotonin 2A receptor binding: relations to body mass index, tobacco and alcohol use. NeuroImage 46(1):23–30
214. Haahr ME, Hansen DL, Fisher PM, Svarer C, Stenbaek DS, Madsen K et al (2015) Central 5-HT neurotransmission modulates weight loss following gastric bypass surgery in obese individuals. J Neurosci 35(14):5884–5889
215. Newcomer JW, Ratner RE, Eriksson JW, Emsley R, Meulien D, Miller F et al (2009) A 24-week, multicenter, open-label, randomized study to compare changes in glucose metabolism in patients with schizophrenia receiving treatment with olanzapine, quetiapine, or risperidone. J Clin Psychiatry 70(4):487–499
216. Rasmussen H, Ebdrup BH, Oranje B, Pinborg LH, Knudsen GM, Glenthoj B (2014) Neocortical serotonin2A receptor binding predicts quetiapine associated weight gain in

antipsychotic-naive first-episode schizophrenia patients. Int J Neuropsychopharmacol 17(11):1729–1736
217. WHO (2003) http://www.who.int/mediacentre/news/releases/2003/pr27/en/
218. Lai MK, Tsang SW, Alder JT, Keene J, Hope T, Esiri MM et al (2005) Loss of serotonin 5-HT_{2A} receptors in the postmortem temporal cortex correlates with rate of cognitive decline in Alzheimer's disease. Psychopharmacology 179(3):673–677
219. Blin J, Baron JC, Dubois B, Crouzel C, Fiorelli M, Attarlevy D et al (1993) Loss of brain 5-HT_2 receptors in Alzheimers-disease—in-vivo assessment with positron emission tomography and [^{18}F]setoperone. Brain 116:497–510
220. Meltzer CC, Price JC, Mathis CA, Greer PJ, Cantwell MN, Houck PR et al (1999) PET imaging of serotonin type 2A receptors in late-life neuropsychiatric disorders. Am J Psychiatry 156(12):1871–1878
221. Versijpt J, Van Laere KJ, Dumont F, Decoo D, Vandecapelle M, Santens P et al (2003) Imaging of the 5-HT_{2A} system: age-, gender-, and Alzheimer's disease-related findings. Neurobiol Aging 24(4):553–561
222. Marner L, Knudsen GM, Madsen K, Holm S, Baare W, Hasselbalch SG (2011) The reduction of baseline serotonin 2A receptors in mild cognitive impairment is stable at two-year follow-up. J Alzheimers Dis 23(3):453–459
223. Dierckx RAJO, Otte A, de Vries EFJ, van Waarde A, Luiten PGM (2014) S.G. PET and SPECT of neurobiological systems. Springer, New York. Chapter 23 (ISBN 978-3-642-42014-6)
224. Marner L, Frokjaer VG, Kalbitzer J, Lehel S, Madsen K, Baare WFC et al (2012) Loss of serotonin 2A receptors exceeds loss of serotonergic projections in early Alzheimer's disease: a combined [^{11}C]DASB and [^{18}F]altanserin-PET study. Neurobiol Aging 33(3):479–487
225. Braak H, Braak E (1991) Demonstration of amyloid deposits and neurofibrillary changes in whole brain sections. Brain Pathol 1(3):213–216
226. Jack CR, Knopman DS, Jagust WJ, Shaw LM, Aisen PS, Weiner MW et al (2010) Hypothetical model of dynamic biomarkers of the Alzheimer's pathological cascade. Lancet Neurol 9(1):119–128
227. Holm P, Ettrup A, Klein AB, Santini MA, El-Sayed M, Elvang AB et al (2010) Plaque deposition dependent decrease in 5-HT_{2A} serotonin receptor in a beta PPswe/PS1dE9 amyloid overexpressing mice. J Alzheimers Dis 20(4):1201–1213
228. Christensen R, Marcussen AB, Wortwein G, Knudsen GM, Aznar S (2008) A beta((1-42)) injection causes memory impairment, lowered cortical and serum BDNF levels, and decreased hippocampal 5-HT_{2A} levels. Exp Neurol 210(1):164–171
229. Meltzer CC, Smith G, DeKosky ST, Pollock BG, Mathis CA, Moore RY et al (1998) Serotonin in aging, late-life depression, and Alzheimer's disease: the emerging role of functional imaging. Neuropsychopharmacology 18(6):407–430
230. Friedman JH (2013) Pimavanserin for the treatment of Parkinson's disease psychosis. Expert Opin Pharmacol 14(14):1969–1975
231. Vanover KE, Weiner DM, Makhay M, Veinbergs I, Gardell LR, Lameh J et al (2006) Pharmacological and behavioral profile of N-(4-fluorophenylmethyl)-N-(1-methylpiperidin-4-yl)-N '-(4-(2-methylpropyloxy)phenylmethyl) carbamide (2R, 3R)-dihydroxybutanedioate (2, 1) (ACP-103), a novel 5-hydroxytryptamine(2A) receptor inverse agonist. J Pharmacol Exp Ther 317(2):910–918
232. Nordstrom AL, Mansson M, Jovanovic H, Karlsson P, Halldin C, Farde L et al (2008) PET analysis of the 5-HT_{2A} receptor inverse agonist ACP-103 in human brain. Int J Neuropsychopharmacol 11(2):163–171
233. Andersen VL, Hansen HD, Herth MM, Dyssegaard A, Knudsen GM, Kristensen JL (2015) ^{11}C-labeling and preliminary evaluation of pimavanserin as a 5-HT_{2A} receptor PET-radioligand. Bioorg Med Chem Lett 25(5):1053–1056
234. Rasmussen H, Ebdrup BH, Erritzoe D, Aggernaes B, Oranje B, Kalbitzer J et al (2011) Serotonin 2A receptor blockade and clinical effect in first-episode schizophrenia patients treated with quetiapine. Psychopharmacology 213(2–3):583–592

235. Laruelle M, Abi-Dargham A, van Dyck C, Gil R, D'Souza DC, Krystal J et al (2000) Dopamine and serotonin transporters in patients with schizophrenia: an imaging study with [^{123}I]beta-CIT. Biol Psychiatry 47(5):371–379
236. Egerton A, Demjaha A, Grasby PM, McGuire PK, Murray RM, Howes OD (2009) Longitudinal assessment of presynaptic dopamine function in the striatum: a [^{18}F]DOPA PET test-retest study. Eur Neuropsychopharmacol 19:S313–S3S4
237. Egerton A, McGuire PK, Howes OD, Egerton A, Howes OD (2009) Functionally defined regions may aid interpretation of striatal dopamine elevation in schizophrenia. Schizophr Res 109(1–3):200
238. Abi-Dargham A (2004) Antipsychotics across the spectrum: an overview of their mechanisms of action. Int J Neuropsychopharmacol 7:100
239. Brooks DJ, Piccini P (2006) Imaging in Parkinson's disease: the role of monoamines in behavior. Biol Psychiatry 59(10):908–918
240. Tyacke RJ, Nutt DJ (2015) Optimising PET approaches to measuring 5-HT release in human brain. Synapse 69(10):505–511
241. Larisch R, Klimke A, Hamacher K, Henning U, Estaji S, Hohlfeld T et al (2003) Influence of synaptic serotonin level on [^{18}F]altanserin binding to 5-HT_{2A} receptors in man. Behav Brain Res 139(1–2):21–29
242. Toll LB-GI, Polgar WE, Brandt SR, Adapa ID, Rodriguez L, Schwartz RW, Haggart D, O'Brien A, White A, Kennedy JM, Craymer K, Farrington L, Auh JS (1998) Standard binding and functional assays related to medications development division testing for potential cocaine and opiate narcotic treatment medications. NIDA Res Monogr 178:440–466
243. Millan MJGA, Lejeune F, Newman-Tancredi A, Rivet JM, Auclair A, Peglion JL (2001) S33005, a novel ligand at both serotonin and norepinephrine transporters: I. Receptor binding, electrophysiological, and neurochemical profile in comparison with venlafaxine, reboxetine, citalopram, and clomipramine. J Pharmacol Exp Ther 298(2):565–580
244. Palvimaki EP, Roth BL, Majasuo H, Laakso A, Kuoppamaki M, Syvalahti E et al (1996) Interactions of selective serotonin reuptake inhibitors with the serotonin 5-HT_{2C} receptor. Psychopharmacology 126(3):234–240
245. Pinborg LH, Adams KH, Yndgaard T, Hasselbalch SG, Holm S, Kristiansen H et al (2004) [^{18}F]altanserin binding to human 5-HT_{2A} receptors is unaltered after citalopram and pindolol challenge. J Cereb Blood Flow Metab 24(9):1037–1045
246. Mork A, Kreilgaard M, Sanchez C (2003) The R-enantiomer of citalopram increase in extracellular 5-HT counteracts escitalopram-induced in the frontal cortex of freely moving rats. Neuropharmacology 45(2):167–173
247. Quednow BB, Treyer V, Hasler F, Dorig N, Wyss MT, Burger C et al (2012) Assessment of serotonin release capacity in the human brain using dexfenfluramine challenge and [^{18}F] altanserin positron emission tomography. NeuroImage 59(4):3922–3932
248. Matusch A, Hurlemann R, Kops ER, Winz OH, Elmenhorst D, Herzog H et al (2007) Acute S-ketamine application does not alter cerebral [^{18}F]altanserin binding: a pilot PET study in humans. J Neural Transm 114(11):1433–1442
249. Rothman RB, Jayanthi S, Wang XY, Dersch CM, Cadet JL, Prisinzano T et al (2003) High-dose fenfluramine administration decreases serotonin transporter binding, but not serotonin transporter protein levels, in rat forebrain. Synapse 50(3):233–239
250. Rothman RB, Baumann MH (2003) Monoamine transporters and psychostimulant drugs. Eur J Pharmacol 479(1–3):23–40
251. Rothman RB, Clark RD, Partilla JS, Baumann MH (2003) (+)-Fenfluramine and its major metabolite, (+)-norfenfluramine, are potent substrates for norepinephrine transporters. J Pharmacol Exp Ther 305(3):1191–1199
252. Jorgensen LM, Weikop P, Villadsen J, Visnapuu T, Ettrup A, Hansen HD et al (2016) Cerebral 5-HT release correlates with [^{11}C]Cimbi36 PET measures of 5-HT_{2A} receptor occupancy in the pig brain. J Cereb Blood Flow Metab 37(2):425–434

Part II
5-HT_{2A}R Related Brain Functions

Serotonergic Receptor 5-HT$_{2A}$ in the Cardiosympathovagal System

M. Kermorgant, A. Pavy-Le Traon, J.M. Senard, and D.N. Arvanitis

Abstract The serotonin receptor 5-HT$_{2A}$ is widely expressed throughout the central nervous system. While abundant evidence exits implicating 5-HT$_{2A}$ receptors in regulating central nervous system, in particular stress responses and that their expression levels or signaling can contribute to stress-related disorders such as anxiety, depression and aggression; the 5-HT$_{2A}$ receptors is also gaining importance in regulating the activity of the autonomic nervous system. Elucidating the functional specificity and significance of the 5HT$_{2A}$ receptor in autonomic function is a challenge given the existence and often co-localization of other 5HT$_2$ receptor subtypes, the central and peripheral expression pattern of the 5HT$_{2A}$ receptor, and the relative poor selectivity of the pharmacological agents used to identify their function. Data has long been accumulated indicating that the 5-HT$_{2A}$ receptor-induced regulation of the autonomic nervous system function appears to be mediated, at least in part, through the regulation of the serotoninergic afferents and efferents to the *nucleus tractus solitarius*. In this article, we review the role of 5-HT$_{2A}$ receptor function in the modulation of cardiac sympathovagal balance with special emphasis on the networks by which 5-HT$_{2A}$ receptors modulate the function of the *nucleus tractus solitarius* in regulating the baroreflex and autonomic function.

Keywords Baroreflex • Autonomic function • Cardiac sympathovagal balance • Nucleus tractus solitarius • Serotoninergic afferents

---

M. Kermorgant • D.N. Arvanitis (✉)
Institut des Maladies Métaboliques et Cardiovasculaires, INSERM UMR1048, Université Paul Sabatier, Toulouse, France
e-mail: constandina.arvanitis@univ-tlse3.fr

A. Pavy-Le Traon
Institut des Maladies Métaboliques et Cardiovasculaires, INSERM UMR1048, Université Paul Sabatier, Toulouse, France

Department of Neurology, Institute for Neurosciences, University Hospital of Toulouse, Toulouse, France

J.M. Senard
Institut des Maladies Métaboliques et Cardiovasculaires, INSERM UMR1048, Université Paul Sabatier, Toulouse, France

Department of Clinical Pharmacology, University Hospital of Toulouse, Toulouse, France

B.P. Guiard, G. Di Giovanni (eds.), *5-HT$_{2A}$ Receptors in the Central Nervous System*, The Receptors 32, https://doi.org/10.1007/978-3-319-70474-6_6

Introduction

Ample evidence shows that serotonin (5-hydroxytryptamine; 5-HT) is a unique hormone and neurotransmitter that can act as an, auto- and/or paracrine factor, or intracellular signaling molecule to provoke a plethora of functions throughout the body [1]. 5-HT exerts its functions through its interaction with a minimum of 15 receptors that are further divided into 7 groups (5-HT_1 to 5-HT_7). Among these receptor subtypes the 5HT_2 receptors are further subdivided into A, B and C classes, which are differentially localized and orchestrate the complex signaling effects of 5-HT. In the periphery, the 5-HT_{2A} receptor is expressed in platelets [2, 3], where it has been largely defined in facilitating platelet aggregation [4, 5]. Peripheral 5-HT_{2A} receptors are also found in blood vessels [6], monocytes [7], in the heart [8], and, along the vagus nerve [9]. While the 5-HT_{2A} receptors are best characterized for their roles in the central nervous system, where their function in a wide spectrum of behaviors implicating a number of mental illnesses with complex etiologies [10], there is growing evidence for an important role in the regulation of the cardiovascular and autonomic systems. These effects include bradycardia or tachycardia, hypotension or hypertension, and vasodilation or vasoconstriction. Here we review the involvement of 5-HT_{2A} in the brain-to-heart axis with particular attention to the reported findings of its role in control of the baroreflex and cardiac sympathovagal balance.

Central 5-HT_{2A} Signal Integration in Baroreflex and Autonomic Control

Multiple visceral afferents terminate in the brainstem and in particular in the *nucleus tractus solitarius* (NTS). While glutamate is the principal neurotransmitter at the afferent-NTS junction, other neuromodulators can affect NTS function. For example, 5-HT is as a neuromodulator in the NTS and adjacent dorsal motor nucleus of the vagus (DMNV) that alters presynaptic glutamate release or postsynaptic activity [11–14]. 5-HT positive terminals and fibers are found in significant numbers throughout the NTS [15–17], with the rodent, specifically the rat, displaying marked concentrations of 5-HT-positive terminals in the medial NTS [18, 19]. Retrograde tracing has demonstrated 5-HT neuronal networks between the NTS and *raphe nuclei* [20]. The neuronal connections were further shown to project from the *raphe magnus* and dorsal raphe to the NTS [21]. Additional works showed that vagal afferents and the nodose ganglia converged to form the peripheral 5-HT inputs into the NTS [22, 23], and transmit 5–HT containing afferent projections to the *nucleus ambiguus*, which was shown as the main site of vagal parasympathetic preganglionic cholinergic neurons for the heart [24].

Moreover, differential application of specific agonists and antagonists to serotoninergic receptors induces a wide range of effects that included either the

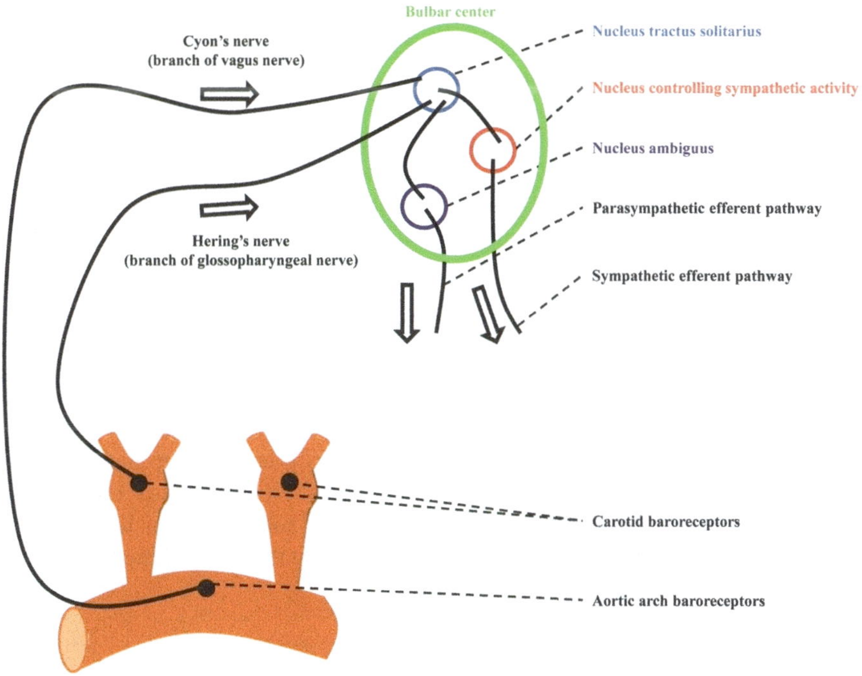

Fig. 1 Simplified schematic of baroreflex loop showing afferent nerves from arterial baroreceptors making their primary synapse in the NTS and the efferent connections from the NTS to the efferent vagal fibers

excitation or inhibition, or both [25]. Early studies demonstrated that 5-HT$_{2A}$ receptor activation increased synaptic transmission through postsynaptic mechanisms [26]. However, in other systems the activation of the 5-HT$_{2A}$ receptor led to synaptic transmission and downstream neuronal activity inhibition [27, 28]. Nonetheless, in both cases the 5-HT$_{2A}$ receptors are shown to modulate synaptic transmission in the NTS and these modifications can regulate the cardiorespiratory reflexes under physiological and pathophysiological circumstances as described below.

5-HT$_{2A}$ in Baroreflex Modulation and Autonomic Function

The baroreceptor reflex represents the major mechanism for the rapid adjustment of heart activity to blood pressure changes. A simplified schematic of the baroreflex is represented in Fig. 1; and a comprehensive, historical perspective on our understanding of baroreceptor physiology and its therapeutic relevance in resistant hypertension is recently described [29]. Variations encountered in carotid and aortic areas are accompanied by an activation of arterial baroreceptors. Baroreceptors are stretch-activated receptors located in the aortic arch and carotid sinus. They monitor blood pressure in order to modulate sympathetic and parasympathetic outputs and maintain physiological blood pressure ranges for correct tissue perfusion.

Baroreceptor afferences transmit messages to the central nervous system enabling the maintenance of arterial pressure in a narrow physiological range [30]. The afferents from the baroreceptors of the carotid sinus and aortic arch terminate in the NTS, *nucleus ambiguus* and *dorsal vagal nucleus*. The baroreflex strongly correlates with the functioning of the regulation of both branches of the autonomic nervous system, even if there is a bias toward the sympathetic branch. As such, this regulation of both sympathetic and parasympathetic afferents and efferents has proven difficult to dissociate one from the other. For example, neuronal pathways connect the afferents to parasympathetic efferents in the DMNV and the *nucleus ambiguus* and sympathetic efferents, in the *rostral ventrolateral medulla* (RVLM), so that stimulation of the carotid or aortic stretch receptors by a blood pressure increase is followed by a rise in vagal tone and a decrease in sympathetic tone, and vice versa [31–35]. Many neurotransmitters play a key role in this mechanism, notably 5-HT which is present within nerve terminals in NTS [17]. Moreover, in a single neuron there may be several kinds of serotoninergic receptors, which potentially mediate opposite cellular responses further cofounding the study of the 5HT receptors in this system.

Baroreflex Control

It is well known that 5-HT administered by the central route provokes an elevation in arterial pressure due to a release of vasopressin and a reflex bradycardia [36]. In anesthetized rats, the microinjection of 5-HT in NTS induces hypotension and bradycardia. These effects are blocked by ketanserin and ritanserin (the preferential 5-HT_{2A} receptor antagonist) [37]. Further studies using the rat a a model showed that during shock-evoked passive behavior the 5-HT_{2A} receptors in the NTS provoked the cardiosympathovagal component of the baroreflex [38]. The specificity for the activation and function of the 5-HT_{2A} receptor in inhibiting the baroreflex and dissociating this effect from sympathetic activity in rats was shown by Comet and colleagues [39]. Herein, using intra-NTS microinjections of the 5-HT_{2A} receptor agonist 1-(2,5-dimethoxy-4-iodophenyl)-2-aminopropane (DOI) the authors showed that *in vivo*, the activation of 5-HT_{2A} receptors in the NTS produced hypotension and bradycardia [39]. More precisely, the investigators found that direct activation of 5-HT_{2B} and 5-HT_{2C} receptors in the NTS did not result in changes in heart rate or mean blood pressure. Instead, only local administration of the preferential $5HT_{2A}$ receptor antagonist MDL-100907 prevented the cardiovascular responses to DOI. Together, these data were the first to show an explicit role for 5-HT_{2A} receptors in mediating baroreflex and not the sympathetic responses in anesthetized rats. Shen et al. [40] demonstrated that central 5-HT_{2A} receptors mediated the baroreflex enhancing effects of ketanserin; an antihypertensive drug best characterized for its sympathoinhibitory action. What's more is the authors demonstrated that ketanserin treatment led to the persistence of the hypertensive effect even with the destruction of the central 5-HT neurons. These results strongly suggest that the ketanserin induced sympathoinhibition is either 5-HT_{2A} independent or regulated by way of peripheral 5-HT_{2A} receptors.

Autonomic Function

The 5-HT_{2A} receptors are well established to cause sympathoexcitation [41]. Ultrastructural analyses have confirmed the localization of 5-HT_{2A} receptors in separate populations of axons, axon terminals and glia DMNV [9]. Centrally, 5-HT_{2A} receptor activation causes sympathoexcitation and vasopressin release [42]. The activation of 5-HT_{2A} receptor produces a sympathoexcitation leading to an increase of arterial pressure and a tachycardia following the sympathetic discharge. These hypertensive effects are attenuated after intra-cerebroventricular injections of spiperone (preferential 5-HT_{2A} receptor antagonist). The major site for this action is the RVLM [36]. The authors further note colocalization of 5-HT_{2A} receptors on catecholaminergic neurons, suggesting that 5-HT_{2A} receptor mediated control of autonomic functions involves catecholaminergic neurons [43]. In particular, the 5-HT_{2A} receptor is expressed on GABAergic interneurons in the *locus coeruleus* and the ventral tegmental area [44]. 5-HT_{2A} activation leads to a decrease in noradrenergic and dopaminergic neuron activities. More recent reports show that infusion of volinanserin (5-HT_{2A} receptor antagonist) led to a decrease in blood pressure [45]. To elucidate the function of the 5-HT_{2A} receptors mediated by the neurons in the NTS, the receptor agonist DOI was applied to neurons receiving vagal input and this produced mixed excitation and inhibition [46]. A later study showed that via extracellular recordings of NTS neurons, 5-HT_{2A} receptors could facilitate the excitatory response [47]. While most data suggests that 5-HT_{2A} receptors action is excitatory in the NTS, more investigation is needed to determine the effects of 5-HT_{2A} receptors on autonomic function. Indeed recent studies have begun to elucidate the function of other 5-HT receptors in the regulation of the sympathetic nervous system. For example, by use of adeno-associated viral vector encoding a small hairpin RNA (shRNA) to selectively reduce 5-HT_{1A}-receptor message in neurons of the NTS, Vantrease et al. [48] showed that 5-HT_{1A} receptor in the caudal NTS contributed to sympathoexcitation following hemorrhage. While such studies have helped decipher how 5-HT and 5-HT receptors can regulate the autonomic nervous system, more studies are required to identify the explicit role of the 5HT_{2A} receptor in autonomic functioning.

Cardiac 5-HT_{2A}

Even though a large portion of information regarding 5-HT_{2A} activity in the baroreflex or autonomic function concerns its expression in the NTS, one cannot exclude its function in the heart. It has long been recognized that 5-HT plays an important role in cardiac development and function. The 5-HT_{2A} receptor shows widespread expression in the human cardiovascular system. It is present on arterial smooth muscle [49], endothelial cells and on cardiomyocytes [50]; and best characterized for mediating vasoconstriction [51]. Functional analyses of 5-HT_{2A} in the heart are limited. Using sheep aortic valve interstitial cells, it was demonstrated that the 5-HT_{2A} is responsible for 5-HT-mediated increased the transforming growth factor beta-1 activity that may contribute to progression of 5-HT-related heart valve disease [52]; even though 5-HT_{2B} receptors have been widely implicated in heart valve diseases [53]. What's interesting is that 5-HT_{2A} receptor mRNA is significantly expressed in rat arteries, and to a lesser extent, in veins [6], yet their distinctive function herein remains unknown. In the rodent, accumulating evidence points toward 5-HT_2 receptors as potential candidates in cardiac diseases. For instance, in a rat model for congestive heart failure, two independent reports demonstrated an increased expression of the 5-HT_{2A} receptor mRNA. More recently, in a mouse model for cardiac hypertrophy, as induced by transverse aortic constriction, 5-HT_{2A} receptor expression was transiently increased [54]. Herein the authors showed that selective blockade of 5-HT_{2A} receptors prevented the development of cardiac hypertrophy through inhibition of the CamKII/HDAC4 pathway [54]. Together, these studies show an important cardiac-autonomic role of the 5-HT_{2A} receptor, yet it is not known if changes in 5-HT_{2A} receptor expression and/or function are causal or compensatory in cardiac disease. Further studies are therefore required to decipher and dissociate the role of 5-HT_{2A} in the heart, in the vagus, and in the NTS in maintaining proper cardiac sympathovagal balance.

Synopsis

The many implications of 5-HT_{2A} receptor demonstrate its importance in several physiological functions and in particularly in cardiovascular homeostasis. The current knowledge of 5-HT_{2A} receptor-specific actions within the autonomic nervous system is in its infancy but, several findings suggest that it is extremely important in modulating cardiovascular function in health and disease. Evidence to date suggests that 5-HT_{2A} receptor subtypes have complex interactions with the baroreflex and autonomic nervous system at the pre- and post-junctional level. While most studies have focused on the role of central 5-HT_{2A} receptor, the peripheral role for this receptor is also gaining importance. Several important questions remain unanswered. For example, one unresolved issue is to dissect the role of 5-HT_{2A} receptor in the vessel and or along the nerve in order to fully appreciate the neuronal or vascular component in modulating the baroreflex and autonomic function. Another important question surrounds the use of the preferred but not selective agonists and antagonists used to evaluate 5-HT_{2A} function as opposed to other 5-HT_{2A}-receptor subtypes. Importantly, elucidating the yet unknown functional consequences of the 5-HT_{2A} receptor hetero- (5-$HT_{2A/2C}$) and homo-dimerization (5-$HT_{2A/2A}$) should help clarify many physiological and pathophysiological responses. Further studies are needed in order to determine the neural mechanisms involved in the cardiovascular response induced by 5-HT_{2A} receptor. In reviewing the literature, we could find no evaluation of the baroreflex or autonomic function in animal studies that focused on selective ablation of 5-HT_{2A} receptors such as knockout models or optogenetic approaches. Future works on decoding the role of 5-HT_{2A} receptors will need to encompass studies on cardiovascular variables and hemodynamic aspects; as such these studies will provide compelling evidence for 5-HT_{2A} receptors as potential therapeutic targets in cardiovascular and baroreceptor-related diseases.

References

1. El-Merahbi R, Löffler M, Mayer A, Sumara G (2015) The roles of peripheral serotonin in metabolic homeostasis. FEBS Lett 589(15):1728–1734
2. De Clerck F, Xhonneux B, Leysen J, Janssen PA (1984) Evidence for functional 5-HT2 receptor sites on human blood platelets. Biochem Pharmacol 33(17):2807–2811
3. De Chaffoy de Courcelles D, Leysen JE, De Clerck F, Van Belle H, Janssen PA (1985) Evidence that phospholipid turnover is the signal transducing system coupled to serotonin-S2 receptor sites. J Biol Chem 260(12):7603–7608
4. Duerschmied D, Bode C (2009) The role of serotonin in haemostasis. Hamostaseologie 29(4):356–359
5. Henninger DD, Panés J, Eppihimer M, Russell J, Gerritsen M, Anderson DC, Granger DN (1997) Cytokine-induced VCAM-1 and ICAM-1 expression in different organs of the mouse. J Immunol 158(4):1825–1832
6. Kato S, Kumamoto H, Hirano M, Akiyama H, Kaneko N (1998) Expression of 5-HT2A and 5-HT1B receptor mRNA in blood vessels. Mol Cell Biochem 199(1–2):57–61
7. Dürk T, Panther E, Müller T, Sorichter S, Ferrari D, Pizzirani C, Di Virgilio F, Myrtek D, Norgauer J, Idzko M (2005) 5-Hydroxytryptamine modulates cytokine and chemokine production in LPS-primed human monocytes via stimulation of different 5-HTR subtypes. Int Immunol 17(5):599–606
8. Kaumann AJ, Levy FO (2006) 5-hydroxytryptamine receptors in the human cardiovascular system. Pharmacol Ther 111(3):674–706
9. Huang J, Pickel VM (2003) Ultrastructural localization of serotonin 2A and N-methyl-D-aspartate receptors in somata and dendrites of single neurons within rat dorsal motor nucleus of the vagus. J Comp Neurol 455(2):270–280
10. Zhang G, Stackman RW Jr (2015) The role of serotonin 5-HT_{2A} receptors in memory and cognition. Front Pharmacol 6:225
11. Browning KN, Travagli RA (1999) Characterization of the in vitro effects of 5-hydroxytryptamine (5-HT) on identified neurones of the rat dorsal motor nucleus of the vagus (DMV). Br J Pharmacol 128(6):1307–1315
12. Cui RJ, Roberts BL, Zhao H, Zhu M, Appleyard SM (2012) Serotonin activates catecholamine neurons in the solitary tract nucleus by increasing spontaneous glutamate inputs. J Neurosci 32(46):16530–16538

13. Feldman PD (1994) Electrophysiological effects of serotonin in the solitary tract nucleus of the rat. Naunyn Schmiedeberg's Arch Pharmacol 349(5):447–454
14. Feldman PD (1995) Effects of serotonin-1 and serotonin-2 receptor agonists on neuronal activity in the nucleus tractus solitarius. J Auton Nerv Syst 56(1–2):119–124
15. Maley B, Elde R (1982) Immunohistochemical localization of putative neurotransmitters within the feline nucleus tractus solitarii. Neuroscience 7(10):2469–2490
16. Pickel VM, Joh TH, Chan J, Beaudet A (1984) Serotoninergic terminals: ultrastructure and synaptic interaction with catecholamine-containing neurons in the medial nuclei of the solitary tracts. J Comp Neurol 225(2):291–301
17. Steinbusch HW (1981) Distribution of serotonin immunoreactivity in the central nervous system of the rat-cell bodies and terminals. Neuroscience 6(4):557–618
18. Calzà L, Giardino L, Grimaldi R, Rigoli M, Steinbusch HW, Tiengo M (1985) Presence of 5-HT-positive neurons in the medial nuclei of the solitary tract. Brain Res 347(1):135–139
19. Curtis JT, Anderson MB, Curtis KS (2013) Regional differences in serotonin content in the nucleus of the solitary tract of male rats after hypovolemia produced by polyethylene glycol. J Physiol Sci 63(1):39–46
20. Thor KB, Helke CJ (1987) Serotonin- and substance P-containing to the nucleus tractus solitarii of the rat. J Comp Neurol 265(2):275–293
21. Schaffar N, Kessler JP, Bosler O, Jean A (1988) Central serotonergic projections to the nucleus tractus solitarii: evidence from a double labeling study in the rat. Neuroscience 26(3):951–958
22. Nosjean A, Compoint C, Buisseret-Delmas C, Orer HS, Merahi N, Puizillout JJ, Laguzzi R (1990) Serotonergic projections from the nodose ganglia to the nucleus tractus solitarius: an immunohistochemical and double labeling study in the rat. Neurosci Lett 114(1):22–26
23. Sykes RM, Spyer KM, Izzo PN (1994) Central distribution of substance P, calcitonin gene-related peptide and 5-hydroxytryptamine in vagal sensory afferents in the rat dorsal medulla. Neuroscience 59:195–210
24. Shapiro RE, Miselis RR (1985) The central neural connections of the area postrema of the rat. J Comp Neurol 234(3):344–364
25. Jordan D (2005) Vagal control of the heart: central serotonergic (5-HT) mechanisms. Exp Physiol 90(2):175–181
26. Austgen JR, Kline DD (2013) Endocannabinoids blunt the augmentation of synaptic transmission by serotonin 2A receptors in the nucleus tractus solitarii (nTS). Brain Res 1537:27–36
27. Brozoski DT, Dean C, Hopp FA, Hillard CJ, Seagard JL (2009) Differential endocannabinoid regulation of baroreflex-evoked sympathoinhibition in normotensive versus hypertensive rats. Auton Neurosci 150(1–2):82–93
28. Kano M, Ohno-Shosaku T, Hashimotodani Y, Uchigashima M, Watanabe M (2009) Endocannabinoid-mediated control of synaptic transmission. Physiol Rev 89(1):309–380
29. Victor RG (2015) Carotid baroreflex activation therapy for resistant hypertension. Nat Rev Cardiol 12(8):451–463
30. Sapru HN (1991) Baroreceptor reflex components and their alteration in hypertension. In: Zucker IH, Gilmore JP (eds) Reflex control of circulation. CRC, Boca Raton, pp 195–214
31. Aicher SA, Milner TA, Pickel VM, Reis DJ (2000) Anatomical substrates for baroreflex sympathoinhibition in the rat. Brain Res Bull 51(2):107–110
32. Dampney RAL (1994) Functional organization of central pathways regulating the cardiovascular system. Physiol Rev 74(2):323–364
33. Dampney RAL, Coleman MJ, Fontes MAP, Hirooka Y, Horiuchi J, Li YW, Polson JW, Potts PD, Tagawa T (2002) Central mechanisms underlying short- and long-term regulation of the cardiovascular system. Clin Exp Pharmacol Physiol 29(4):261–268
34. Singewald N, Philippu A (1996) Involvement of biogenic amines and amino acids in the central regulation of cardiovascular homeostasis. Trends Pharmacol Sci 17(10):356–363
35. Stauss HM (2002) Baroreceptor reflex function. Am J Physiol Regul Integr Comp Physiol 283(2):R284–R286

36. Knowles ID, Ramage AG (1999) Evidence for a role for central 5-HT2B as well as 5-HT2A receptors in cardiovascular regulation in anaesthetized rats. Br J Pharmacol 128(3):530–542
37. Shvaloff A, Laguzzi R (1986) Serotonin receptors in the rat nucleus tractus solitarii and cardiovascular regulation. Eur J Pharmacol 132(2–3):283–288
38. Sevoz-Couche C (2006) Cardiac baroreflex facilitation evoked by hypothalamus and prefrontal cortex stimulation: role of the nucleus tractus solitarius 5-HT2A receptors. AJP: Regul Integr Comp Physiol 291(4):R1007-R1015
39. Comet MA, Vernard JF, Hamon M, Laguzzi R, Sévoz-Couche C (2007) Activation of nucleus tractus solitarius 5-HT_{2A} but not other 5-HT_2 receptor subtypes inhibits the sympathetic activity in rats. Eur J Neurosci 26(2):345–354
40. Shen FM, Wang J, Ni CR, JG Y, Wang WZ, DF S (2007) Ketanserin-induced baroreflex enhancement in spontaneously hypertensive rats depends on central 5-HT(2A) receptors. Clin Exp Pharmacol Physiol 34(8):702–707
41. Mc Call RB, Clement ME (1994) Role of serotonin1A and serotonin2 receptors in the central regulation of the cardiovascular system. Pharmacol Rev 46(3):231–243
42. Ramage AG (2001) Central cardiovascular regulation and 5-hydroxytryptamine receptors. Brain Res Bull 56(5):425–439
43. Nosjean A, Hamon M, Darmon M (2002) 5-HT2A receptors are expressed by catecholaminergic neurons in the rat nucleus tractus solitarii. Neuroreport 13(17):2365–2369
44. Guiard BP, El Mansari M, Blier P (2009) Prospect of a dopamine contribution in the next generation of antidepressant drugs: the triple reuptake inhibitors. Curr Drug Targets 10(11): 1069–1084
45. Mbaki Y, Gardiner J, Mc Murray G, Ramage AG (2012) 5-HT 2A receptor activation of the external urethral sphincter and 5-HT 2C receptor inhibition of micturition: a study based on pharmacokinetics in the anaesthetized female rat. Eur J Pharmacol 682(1–3):142–152
46. Wang Y, Ramage AG, Jordan D (1997) In vivo effects of 5-hydroxytryptamine receptor activation on rat nucleus tractus solitarius neurones excited by vagal C-fibre afferents. Neuropharmacology 36(4–5):489–498
47. Sévoz-Couche C, Spyer KM, Jordan D (2000) In vivo modulation of vagal-identified dorsal medullary neurones by activation of different 5-Hydroxytryptamine(2) receptors in rats. Br J Pharmacol 131(7):1445–1453
48. Vantrease JE, Dudek N, Don Carlos LL, Scrogin KE (2015) 5-HT1A receptors of the nucleus tractus solitarii facilitate sympathetic recovery following hypotensive hemorrhage in rats. Am J Physiol Heart Circ Physiol 309(2):H335–H344
49. Ullmer C, Schmuck K, Kalkman HO, Lübbert H (1995) Expression of serotonin receptor mRNAs in blood vessels. FEBS Lett 370(3):215–221
50. Bush E, Fielitz J, Melvin L, Martinez-Arnold M, Mc Kinsey TA, Plichta R, Olson EN (2004) A small molecular activator of cardiac hypertrophy uncovered in a chemical screen for modifiers of the calcineurin signaling pathway. Proc Natl Acad Sci U S A 101(9):2870–2875
51. Kaumann AJ, Parsons AA, Brown AM (1993) Human arterial constrictor serotonin receptors. Cardiovasc Res 27(12):2094–2103
52. Xu J, Jian B, Chu R, Lu Z, Li Q, Dunlop J, Rosenzweig-Lipson S, Mc Gonigle P, Levy RJ, Liang B (2002) Serotonin mechanisms in heart valve disease II—the 5-HT_2 receptor and its signaling pathway in aortic valve interstitial cells. Am J Pathol 161(6):2209–2218
53. Hutcheson JD, Setola V, Roth BL, Merryman WD (2011) Serotonin receptors and heart valve disease—it was meant 2B. Pharmacol Ther 132(2):146–157
54. Lairez O, Cognet T, Schaak S, Calise D, Guilbeau-Frugier C, Parini A, Mialet-Perez J (2013) Role of serotonin 5-HT2A receptors in the development of cardiac hypertrophy in response to aortic constriction in mice. J Neural Transm (Vienna) 120(6):927–935

Crosstalk Between 5-HT$_{2A}$ and mGlu2 Receptors: Implications in Schizophrenia and Its Treatment

José L. Moreno and Javier González-Maeso

Abstract Schizophrenia is a psychiatric disorder that affects 1% of the population worldwide. The serotonin and glutamate receptor systems have been implicated in schizophrenia and its treatment. Serotonin 5-HT$_{2A}$ receptor is target of hallucinogens such as lysergic acid diethylamide (LSD) and psilocin, as well as involved in the mechanism of action of atypical antipsychotic drugs such as clozapine and risperidone. The metabotropic glutamate 2 (mGlu2) receptor modulates the physiological responses induced by the 5-HT$_{2A}$ receptor, and preclinical and clinical work suggests that this glutamate receptor may represent a new approach to treat schizophrenia. Here we review recent advances in our understanding of the crosstalk between these two receptors, as well as their implication in schizophrenia and antipsychotic drug action.

Keywords G protein-coupled receptor (GPCR) • Serotonin • 5-HT$_{2A}$ receptor • Glutamate • Metabotropic glutamate 2 receptor • mGlu2 • Schizophrenia • Psychosis • Antipsychotic • Clozapine • Lysergic acid diethylamide (LSD) • GPCR dimer • GPCR heterocomplex

Introduction

Schizophrenia is a devastating mental illness that affects 1% of the population in all cultures [1–5]. Its impairment of mental and social functioning often leads to the development of comorbid diseases. These changes disrupt the lives of patients as well as their families and friends. Although serendipitously discovered in the first half of the twentieth century [6–8], the two main families of antipsychotic medications, which include typical or first generation antipsychotics such as chlorpromazine and haloperidol, and atypical or second generation antipsychotics such as

J.L. Moreno • J. González-Maeso (✉)
Department of Physiology and Biophysics, Virginia Commonwealth University School of Medicine, Richmond, VA, USA
e-mail: jgmaeso@vcu.edu

B.P. Guiard, G. Di Giovanni (eds.), *5-HT$_{2A}$ Receptors in the Central Nervous System*, The Receptors 32, https://doi.org/10.1007/978-3-319-70474-6_7

clozapine, olanzapine and risperidone, remain the standard treatment for schizophrenia [9–12]. Nevertheless, recent clinical studies sponsored by the NIH highlighted that three-fourths of schizophrenia patients stop using antipsychotic medications within 18 months of starting therapy [13]. The reasons given for discontinuing prescribed drugs included reduced efficacy, poor tolerability, and severe side effects. This serious issue for a major mental illness requires understanding the cause of non-compliance as well as developing more effective therapeutic treatments for schizophrenia.

The principal brain target that all antipsychotic drugs bind to is the dopamine receptor system. Additionally, atypical antipsychotic drugs all have in common a high affinity for the serotonin receptor system, as well as a modest affinity for the dopamine receptor system. In recent years, novel compounds that regulate signaling by glutamate (the major excitatory neurotransmitter in the brain) are emerging as a promising new approach for the treatment of this disorder. In this book chapter we focus on two neurotransmitter receptors that have been implicated in the pathophysiology of schizophrenia and other psychotic disorders, as well as in the molecular mechanism of action of antipsychotic drugs: serotonin 5-HT_{2A} and metabotropic glutamate 2 (mGlu2) receptors.

Serotonin Receptors

Serotonin (5-hydroxytryptamine, 5-HT) appeared very early in evolution [14]. The indoleamines serotonin and the closely related melatonin are present in some protozoans and in almost all metazoans, where they play important roles in development and plasticity. The isolation of serotonin was achieved after decades of investigation to characterize the chemical properties of a substance that was suspected to be contained in the platelets [15–17]. In 1937, Vittorio Erspamer working at the University of Pavia in Italy extracted a chemical compound from enterochromaffin cells in the gastrointestinal tract that was capable for inducing smooth muscle contraction [18]. This compound was named enteramine. In the decade of the 1940s, Maurice Rapport reported a similar effect of a substance responsible for the vasoconstrictor activity of serum, and named the substance serotonin after the Latin word *serum* and the Greek work *tonic* [19–22]. A few years later, in 1952, it was demonstrated that enteramine and serotonin corresponded to the same compound [23]. Experiments using serotonin (enteramine) and newly discovered indolealkylamine derivatives showed their effects on contraction of a variety of smooth muscle-containing tissues, such as carotid artery, jejunum, uterus, and nictitating membranes in animal models that included sheep, ox, rabbits, cats and dogs [24].

However, it was not until 1943 when the serendipitous discovery of the psychedelic properties of lysergic acid diethylamide (LSD) by the Swiss chemist Albert Hoffman (1906–2008) [25, 26] and the fascinating structural similarities between LSD and the monoamine serotonin captivated the interest of basic neuroscientists (Fig. 1). Thus in 1968, George Aghajanian at Yale University showed that LSD

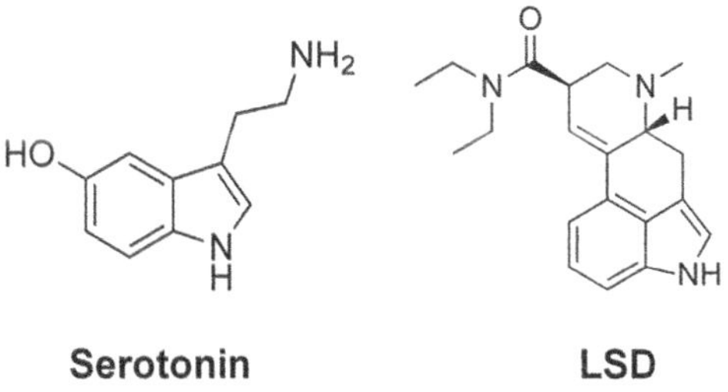

Fig. 1 Chemical structure of the neurotransmitter serotonin (5-hydroxytryptamine, 5-HT) and the psychedelic 5-HT_{2A} receptor agonist LSD (lysergic acid diethylamide)

affects the activity of midbrain neurons containing serotonin [27]. Later in 1976, Solomon Snyder at Johns Hopkins University suggested the presence of 5-HT binding sites in rat cortical membrane preparations [28]. Using the radioligands [^{3}H]5-HT, [^{3}H]spiperone and [^{3}H]LSD, the same research group demonstrated in 1979 the presence of two distinct 5-HT binding sites [29, 30]. The use of classical pharmacological studies in the 1980s and 1990s, along with cloning approaches (the first serotonin receptor, 5-HT_{1A}, was cloned and characterized in 1988 [31]), led to the discovery of 13 distinct genes encoding for serotonin receptors of the G protein-coupled receptor (GPCR) family [32–35]. In addition, there is one ligand-gated ion channel, the 5-HT_3 receptor.

Currently the G protein-coupled 5-HT receptors are divided into 6 subfamilies based on their pharmacological properties, amino acid sequences, gene organization and second messenger coupling pathways: 5-HT_1 family (5-HT_{1A}, 5-HT_{1B} and 5-HT_{1D} receptors, which are principally $G_{i/o}$ protein-coupled), 5-HT_2 family (5-HT_{2A}, 5-HT_{2B} and 5-HT_{2C} receptors, which are principally $G_{q/11}$ protein-coupled), 5-HT_4 family (5-HT_4 receptor, which is principally G_s protein-coupled), 5-HT_5 family (5-HT_5 receptor, which is principally $G_{i/o}$ protein-coupled), 5-HT_6 family (5-HT_6 receptor, which is principally G_s protein-coupled), and the 5-HT_7 family (5-HT_7 receptor, which is principally G_s protein-coupled).

The 5-HT_{2C} receptor was originally classified as a "5-HT_{1C}" receptor in 1985 by Angel Pazos, Daniel Hoyer and Jose Palacios at Sandoz Ltd. in Basle, Switzerland [36]. This study used radioligand binding approaches with [^{3}H]5-HT, [^{3}H]8-OH-DPAT, [^{3}H]LSD, [^{3}H]ketanserin and [^{3}H]mesulergine in plasma membrane preparations of porcine choroid plexus. However, after the demonstration by Jeffrey Conn and Elaine Sanders-Bush that the "5-HT_{1C}" site in choroid plexus affects the phosphatidylinositol pathway [37, 38], and based on the comparison of its primary sequence with those of 5-HT_{2A} and 5-HT_{2B} receptors, the "5-HT_{1C}" site was consequently named 5-HT_{2C} receptor. Additionally, this serotonin receptor is thus far the only GPCR that undergoes post-transcriptional pre-mRNA editing, which represents a post-transcriptional regulatory mechanism by which RNA transcripts are covalently modified on specific nucleosides in a way that the encoded product may be subject to possible functional alterations [39]. RNA editing of the *5-HT_{2C}* receptor has been shown to affect the pattern of 5-HT_{2C} receptor-dependent G protein coupling and signaling [40–42].

The 5-HT$_5$ receptor family has two known members: 5-HT$_{5A}$ and 5-HT$_{5B}$. The 5-HT$_{5A}$ and 5-HT$_{5B}$ receptors were first cloned in mouse and rat [43]. The human 5-HT$_{5A}$ receptor homolog was subsequently cloned [44], whereas the human *5-HT$_{5B}$* receptor gene results in an expression of a short and not functional protein due to the presence of stop codons in the coding sequence.

Alternative splicing, a process by which exons can be either excluded or included in or from a pre-mRNA resulting in multiple mRNA isoforms, affects 60–70% of human genes and is a key factor underlying protein diversity. Functional splice variants have been reported for 5-HT receptors including 5-HT$_{2A}$, 5-HT$_4$ and 5-HT$_7$ receptors [45].

The 5-HT$_{2A}$ receptor, which was cloned in 1994 [46], is principally coupled to G$_{q/11}$ proteins and is expressed in brain regions involved in cognition, perception, sensorimotor gating and mood, such as prefrontal cortex, striatum and thalamus [47–50]. It is the main target of psychoactive drugs such as the psychedelics LSD, mescaline and psilocin [51, 52]. Additionally, drugs used to treat psychiatric disorders, such as schizophrenia (clozapine, olanzapine and risperidone) [9], Parkinson's disease psychosis (pimavanserin) [53], and depression (mirtazapine and mianserin) [54], and neurological disorders, such as migraine (ergotamine) [55] and Parkinson's (lisuride) [56], also bind with high affinity to the 5-HT$_{2A}$ receptor. It has also been involved in some of the psychoactive effects of efavirenz, a drug used to treat human immunodeficiency virus (HIV) [57]. Findings in the last decade have identified a functional crosstalk between the 5-HT$_{2A}$ receptor and the mGlu2 receptor that might open new avenues for the design of therapeutic compounds to improve the treatment of a number of psychiatric and neurological disorders.

Functional Interaction Between 5-HT$_{2A}$ and mGlu2 Receptors

Glutamate is the main excitatory neurotransmitter in the CNS, and it binds to two main structural groups of neurotransmitter receptors: ion channels and G protein-coupled receptors (GPCRs). Ionotropic glutamate receptors include N-methyl-D-aspartate (NMDA) receptors [58–60], α-amino-3-hydroxy-5-methyl-4-isoxazolepropionic acid (AMPA) receptors [61, 62], and kainic acid (KA) receptors [63, 64]. Until the mid-1980s, the actions of glutamate in the mammalian brain were thought to be mediated exclusively via glutamate-gated cation channels. It was in 1985 when Sladeczek et al. provided the first evidence that glutamate stimulated inositol phosphate formation in striatal neuronal cultures [65], suggesting that glutamate might activate metabotropic receptors along with the classical ligand-gated ion channel receptors. Similar findings were observed by Ferdinando Nicoletti and his team in brain slices from young rats [66]. In 1991, two groups independently reported the primary sequence of the *mGlu1* gene [67, 68]. At present, eight different mGlu subtypes (mGlu1-8) and a number of splice variants have been reported from rodents and human. Classification of mGlu receptors is further divided into three groups according to sequence similarity, pharmacology and preferred G protein signaling mechanism [69–72]. Group I mGlu receptors (mGlu1 and mGlu5) are preferentially coupled to $G_{q/11}$ proteins and hence positively affect the function of phospholipase C, whereas group II (mGlu2 and mGlu3) and group III (mGlu4, mGlu6, mGlu7 and mGlu8) receptors are typically coupled to $G_{i/o}$ proteins, which leads to inhibition of adenylate cyclase activity.

Using an ex-vivo model to record electrophysiological responses in rat brain slices, it was reported that bath application of serotonin (5-HT) produced an increase in the frequency and amplitude of spontaneous excitatory post-synaptic potential/currents (EPSPs/EPSCs) in layer V pyramidal cells of neocortical neurons, including frontal, cingulate and frontoparietal cortex [73, 74]. This 5-HT-mediated electrophysiological response was blocked by the 5-HT_{2A} receptor antagonists MDL100,907 and SR463B, suggesting for the first time that activation of the 5-HT_{2A} receptor induced excitatory postsynaptic potentials in dendrites of cortical pyramidal neurons. Follow-up work by the same research team provided evidence that the selective group II mGlu (mGlu2 and mGlu3) receptor agonist LY354740 suppressed the 5-HT_{2A} receptor-dependent induction of postsynaptic EPSPs/EPSCs in layer V cortical pyramidal neurons [75]. Additionally, autoradiography assays with the mGlu2/3 receptor ligand [^{3}H]LY354740 and the 5-HT_{2A} receptor ligand [^{125}I]DOI showed an evident overlap of the laminar distribution of mGlu2/3 and 5-HT_{2A} receptors in frontal cortex that was not observed in other cortical regions [75]. These findings, along with the implication of the 5-HT_{2A} receptor in the physiological effects of psychedelics and antipsychotic drugs (see above), suggested that the signaling crosstalk between mGlu2/3 and 5-HT_{2A} receptors might be relevant for the treatment of schizophrenia and other psychotic disorders, opening a new line of basic and translational research focused on signaling and neural circuit mechanisms involved in this crosstalk.

Although rodent models of psychiatric alterations have limitations [76–81], head-twitch behavior, which is a rapid side-to-side movement of the head, is a behavior model observed in rats and mice after administration of serotonergic psychedelics [82–84]. Thus, this behavior is observed in rodents injected with drugs such as LSD, psilocin, mescaline, DOI (1-(2,5-dimethoxy-4-iodophenyl)-2-aminopropane), DOM (1-(2,5-dimethoxy-4-bromophenyl)-2-aminopropane), and DOB (1-(2,5-dimethoxy-4-bromophenyl)-2-aminopropane), and is absent in 5-HT_{2A} knockout mice [85, 86]. Based on previous findings using transgenic Cre mice in which 5-HT_{2A} was expressed only in forebrain pyramidal neurons [85, 87], it was suggested that cortical pyramidal 5-HT_{2A} was necessary and sufficient to induce head-twitch behavior by psychedelics in mice [85] (Fig. 2). This hypothesis was further supported by the use of independent experimental approaches such as electrophysiological recordings in mouse cortical slices [88] and electrolytic lesions in the thalamic nuclei [89, 90]. Although these data support that the 5-HT_{2A} receptor expressed in cortical pyramidal neurons is necessary for at least some of the phenotypes induced by psychedelics, they do not exclude the involvement of the 5-HT_{2A} receptor in sub-cortical regions in other processes [91, 92]—the role of presynaptic 5-HT_{2A} receptors in the modulation of thalamocortical plasticity and associative learning has been supported by recent studies based on virally mediated-overexpression approaches [93].

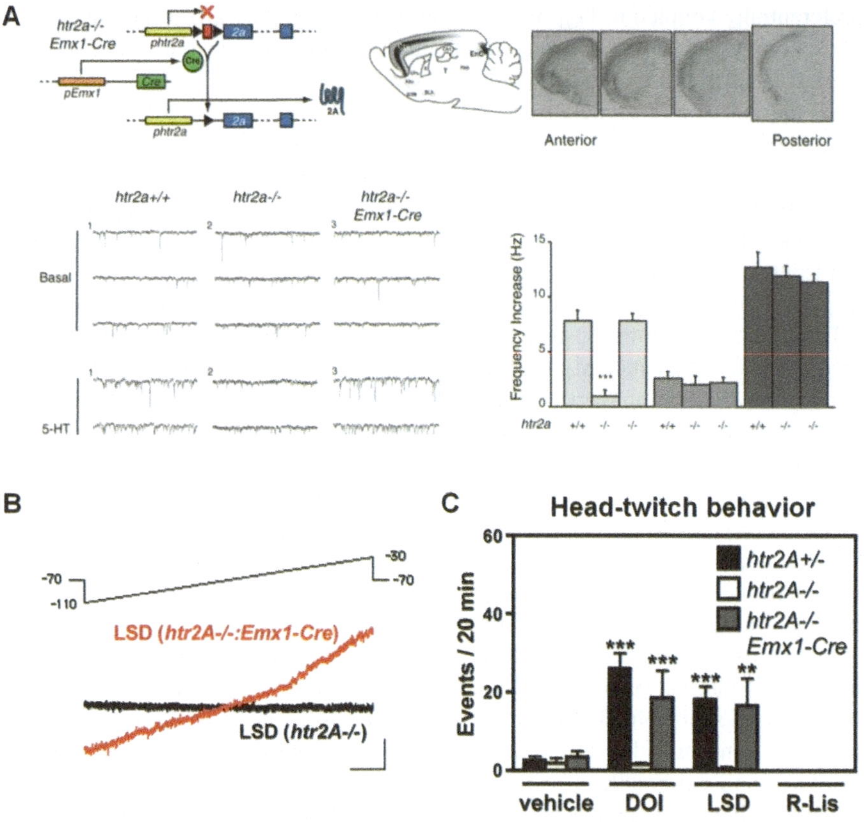

Fig. 2 Psychedelics induce head-twitch behavior via activation of the 5-HT$_{2A}$ receptor in cortical pyramidal neurons. (**a**) Cortical restoration of 5-HT$_{2A}$ receptor expression in 5-HT$_{2A}$ receptor knockout mice. By insertion of a "stop" flanked by *lox-p* sites (*black triangles*), it was interrupted the transcription and translation of the *Htr2a* gene. *Emx1-Cre* excises the stop cassette and restores expression of *Htr2a* to areas where their promoter activities overlap. Density of 5-HT$_{2A}$ receptor was tested by [^{125}I]DOI autoradiography (see [85, 87]). (**b**) Effects of LSD on voltage ramp-activated ionic currents in 5-HT$_{2A}$ knockout mice (*htr2a-/-*) and in mice that express the 5-HT$_{2A}$ receptor only in cortical pyramidal neurons (*htr2a$^{-/-}$:Emx1-Cre*) (see [85]). (**c**) Control mice (*htr2a+/-*), 5-HT$_{2A}$ knockout mice (*htr2a-/-*), and cortical rescue mice (*htr2a$^{-/-}$:Emx1-Cre*) were injected with DOI, LSD, R-lisuride (R-Lis) or vehicle, and the head-twitch behavioral response was scored (see [85])

Nevertheless, using head-twitch behavior as a model of psychedelic-like behavioral responses induced by the 5-HT$_{2A}$ receptor agonist DOI, it was demonstrated that administration of the mGlu2/3 agonist LY354740 prevented DOI-induced head-twitch behavior in rats [94]. One possible explanation for the effects of mGlu2/3 receptor agonists on head-twitch behavior and electrophysiological responses induced by DOI is related to the act on a presynaptic site to decrease glutamate release induced by activation of the 5-HT$_{2A}$ receptor. Thus, changes in the frequency of synaptic currents are generally attributed to a presynaptic site, whereas changes in the amplitude of synaptic currents can be attributed to either a presynaptic or a postsynaptic locus [95]. Considering that activation of either mGlu2 or mGlu3 receptors negatively regulates glutamate release at cortical synapses [96], the effect of mGlu2/3 receptor agonist suppressing 5-HT_{2A} receptor-dependent glutamate release has been proposed to be mediated via presynaptic mGlu2/3 receptors [75]. An alternative, although not mutually exclusive, pathway by which activation of mGlu2/3 receptors modulates physiological and behavioral phenotypes induced by activation of the 5-HT_{2A} receptor has been proposed recently, suggesting that a close molecular interaction between 5-HT_{2A} and mGlu2 receptors in cortical pyramidal neurons also affect their basic mechanism of functional crosstalk.

Heteromerization Between 5-HT_{2A} and mGlu2 Receptors

Although most of the plasma membrane receptor proteins, including ion channels and enzyme-linked receptors, behave and function as dimeric or oligomeric complexes, it was assumed for decades that GPCRs were physiologically active as monomers. This concept of monomeric GPCRs was challenged in the 2000s by two key observations:

Using chimeric receptor constructs of the muscarinic M_3 receptor and the α_{2C} adrenergic receptor, it was reported that the chimeric receptors α_{2C}/M_3 and M_3/α_{2C}, in which the portion containing transmembrane domains VI and VII along with the C-terminal were exchanged between the muscarinic M_3 and the α_{2C} adrenergic receptors, did not show detectable binding activity when expressed alone in COS-7 cells [97]. However, binding activity of the muscarinic ligand [^{3}]methyl-scopolamine or the adrenergic ligand [^{3}H]rauwolscine was rescued in cells co-transfected with α_{2C}/M_3 and M_3/α_{2C} [97]. These results led to the first speculation that the formation of GPCR dimers may underlie a mechanism of inter-molecular crosstalk.

A second fundamental finding supporting close molecular proximity between GPCRs was that related to expression and function of the $GABA_B$ receptor. Thus the $GABA_B$ receptor (now named as $GABA_B$-R1) was cloned in 1997 [98], but recombinant expression of the $GABA_B$-R1 in cell lines such as COS, HEK293 and BHK showed intracellular retention of the construct [99]. This was followed by four publications in 1998 showing that $GABA_B$-R1 is able to form a heterodimeric receptor with $GABA_B$-R2 [100–103]. Neither of these two constructs is functionally active when expressed individually, but $GABA_B$ receptor function is rescued in cells co-transfected with the $GABA_B$-R1 and $GABA_B$-R2 constructs. Further work elegantly demonstrated that $GABA_B$-R1 is retained intracellularly at the endoplasmic reticulum, and that $GABA_B$-R2 is able to traffic $GABA_B$-R1 toward the cell surface via a coiled-coil molecular interaction at the C-terminus [104–106].

These two findings were followed by numerous publications supporting the existence of GPCR homo- and hetero-dimerization as well as higher order oligomerization [107–111]. Indeed, experiments in living cells suggested that the minimal signaling unit required for dopamine D_2 function is composed of two receptor molecules and one G protein [112–115], and that homomerization of the α_{1B}-adrenergic receptor is required receptor maturation, surface delivery and receptor function [116, 117]. Today it is widely accepted that family C GPCRs, particularly the mGlu receptors, behave and function as strict homodimers [118–120]. Thus, monomeric mGlu receptors are unable to couple to and activate G proteins [119]. However, the functional significance of family A homo-dimerization/homo-oligomerization remains controversial. For example, Roger Sunahara and his team demonstrated that purification and reconstitution of a monomeric family A GPCR, including rhodopsin [121], β_2-adrenergic [122] and μ-opioid [123] receptors, into a phospholipid bilayer in the form of high-density lipoprotein particles results in functional G protein coupling. These findings suggested that oligomerization of family A GPCRs is not required for agonist and antagonist binding, and that a monomeric family A GPCR is the minimal functional unit in regard to G protein activation. This hypothesis has been supported by the relatively recent crystal structure of the active ternary complex composed of agonist-occupied monomeric β_2-adrenergic receptor and the G_s protein heterotrimer [124]. Thus it seems clear that further work will be necessary to establish the molecular basis responsible for the functional differences observed between family A GPCRs expressed in unnatural soluble nanoscale phospholipid bilayers (nanodisc) and in heterologous expression systems such as living cells.

A different, although related, question is that focused on the potential molecular interaction between individual GPCR molecules that belong to distinct GPCR subtypes. The first example was reported by Lakshmi Devi in 1999 with the demonstration that two fully functional opioid receptors, κ and δ, are able to form heteromeric complexes, and that heteromerization modulated opioid receptor function [125]. Since then, several other groups have shown that GPCRs form heteromeric receptor complexes that affect pharmacology and function. One of these examples is the heteromer between 5-HT_{2A} and mGlu2 receptors.

In 2008, it was shown that immunoprecipitation using anti-c-Myc antibodies in cells transfected to co-express c-Myc-tagged 5-HT_{2A} and HA-tagged mGlu2 constructs resulted in co-immunoprecipitation of anti-HA immunoreactivity [126]. Co-immunoprecipitation of anti-HA immunoreactivity was not observed when either c-Myc-5HT_{2A} or HA-mGlu2 were expressed individually, or when plasma membrane preparations from cells transfected with either the c-Myc- or the HA-tagged forms of the receptors were combined before immunoprecipitation [126] (Fig. 3). Additionally, the specificity of this protein complex formation was supported by the findings showing absence of co-immunoprecipitation signal in cells co-expressing either c-Myc-5-HT_{2A} and HA-mGlu3 or 5-HT_{2C}-c-Myc and HA-mGlu2 [126]. These findings, which have been validated recently in a different experimental system (see below), suggest that 5-HT_{2A} and mGlu2, but not 5-HT_{2C} and mGlu2 or 5-HT_{2A} and mGlu3, form part of the same protein complex in tissue cultures.

The close molecular proximity between 5-HT_{2A} and mGlu2 receptors at the plasma membrane of living HEK293 cells was supported independently by biophysical approaches that included co-immunoprecipiation [127, 128], biolumininescence resonance energy transfer (BRET) [126], fluorescence resonance

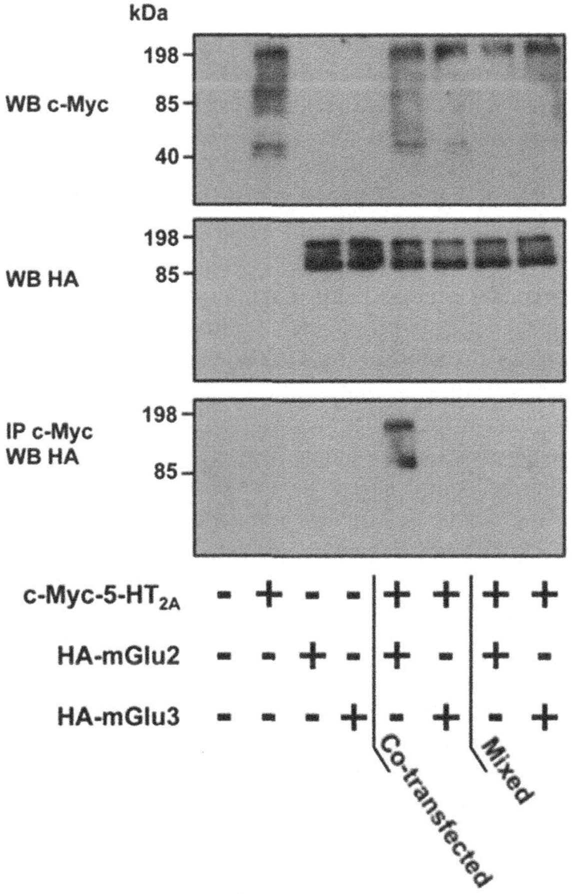

Fig. 3 Co-immunoprecipiation (IP) experiments of c-Myc-5-HT$_{2A}$, HA-mGlu2 or HA-mGlu3 in co-transfected HEK293 cells. The mGlu2, but not the mGlu3, co-immunoprecipitates with the 5-HT$_{2A}$ receptor. For a control, cells separately expressing the c-Myc- or HA-tagged forms were mixed. *WB* Western blot (see [126, 127])

energy transfer (FRET) [126], flow cytometric analysis of FRET signal (FCM-based FRET) [127], antibody-based time-resolved FRET (TR-FRET) [128], and combination of TR-FRET and SNAP-tag approach [129]. These data convincingly support at least a fraction of the population of 5-HT$_{2A}$ and mGlu2 receptors are expressed in close molecular proximity in heterologous expression systems such as living HEK293 cells.

One of the limitations of heterologous systems, however, is that related to the translational significance of the results observed, and, particularly, whether the components of the heteromeric receptor complex are co-expressed in the same cells in native tissue. Independent findings have demonstrated that the 5-HT$_{2A}$ receptor is expressed at high density in layer II/III and layer V cortical pyramidal neurons, as well as in certain groups of cortical GABAergic interneurons and in sub-cortical regions such as striatum and thalamus [47–50, 85]. These findings were based on neuroanatomical approaches that included autoradiography assays with 5-HT$_{2A}$ receptor ligands, *in situ* hybridization and immunohistological tools. Using autoradiographic assays with the mGlu2/3 receptor ligand [^{3}H]LY349415, it was reported that mGlu2/3 receptors show a similar laminar distribution as that observed for the 5-HT_{2A} receptor in the frontal cortex of rats [75]. The use of mGlu2 knockout and mGlu3 knockout mice suggested that it is the mGlu2 receptor, and not the mGlu3 receptor, the one that overlaps in terms of laminar distribution with the 5-HT_{2A} receptor in rodent frontal cortex [130]. Although interesting, however, these data do not provide evidence as to whether 5-HT_{2A} and mGlu2 are co-expressed in the same cells.

This was first suggested with the use of a multiple oligonucleotide based-fluorescent in situ hybridization (FISH) neuroanatomical approach [131, 132]. Thus, it was demonstrated that *5-HT_{2A}* mRNA and *mGlu2* mRNA co-localized in layer V cortical pyramidal neurons [126]. Expression of *mGlu3* mRNA was undetected in mouse frontal cortex, whereas *mGlu3* mRNA signal was observed in sub-cortical regions such as thalamus [126]. One of the limitations of neuroanatomical studies with GPCRs is the specificity and selectivity of anti-GPCR antibodies. Using anti-5-HT_{2A} and anti-mGlu2 antibodies whose specificity and selectivity had been validated in frontal cortex tissue samples of 5-HT_{2A} knockout and mGlu2 knockout mice, immunohistochemical assays showed co-localization of 5-HT_{2A} immunoreactivity and mGlu2 immunoreactivity in mouse cortical neurons [133]. These data suggest that 5-HT_{2A} and mGlu2 are co-expressed in the same population of cortical pyramidal neurons in mouse. They do not, however, provide evidence as to whether these two particular receptor subtypes interact at the sub-cellular level. This was approached with the use of electron microscopy to explore the ultrastructural localization of 5-HT_{2A} and mGlu2 receptors in the frontal cortex [127]. Labeling for 5-HT_{2A} receptor was observed most commonly in dendrites, near synapsis and extrasynaptically [127]. Labeling for mGlu2 receptor was observed in presynaptic terminals and in dendrites at or near postsynaptic sites [127]. Interestingly, these electron microscopy assays showed that labeling for both 5-HT_{2A} and mGlu2 receptors was observed in close sub-cellular proximity, particularly at or near synaptic junctions [127]. Using a sub-cellular fractionation approach to purify fractions enriched in presynaptic active zone (PAZ) and postsynaptic density (PSD) proteins, it has recently been confirmed that the 5-HT_{2A} receptor is detected only in the PSD, whereas the mGlu2 receptor is detected in both the PSD and the PAZ fractions [134]. In addition, 5-HT_{2A} and mGlu2 receptors can be co-immunoprecipitated from mouse and human frontal cortex plasma membrane preparations [133] (Fig. 4). Controls to validate the specificity of the co-immunoprecipitation approach in mouse frontal cortex tissue samples included 5-HT_{2A} knockout and mGlu2 knockout mice, as well as frontal cortex tissue samples from 5-HT_{2A} knockout and mGlu2 knockout mice which were homogenized together (mixed) and processed identically for immunoprecipitation and immunoblot [133]. Together, these data suggest that 5-HT_{2A} and mGlu2 may form part of a protein complex at the PSD in mouse frontal cortex.

Structure of the 5-HT_{2A}-mGlu2 Receptor Complex

It is well accepted that hetero-dimerization of $GABA_B$-R1 and $GABA_B$-R2, which are assembled to form a functionally active $GABA_B$ receptor, is mediated via a parallel coiled-coil interaction at their C-termini [110]. It has also been shown that the

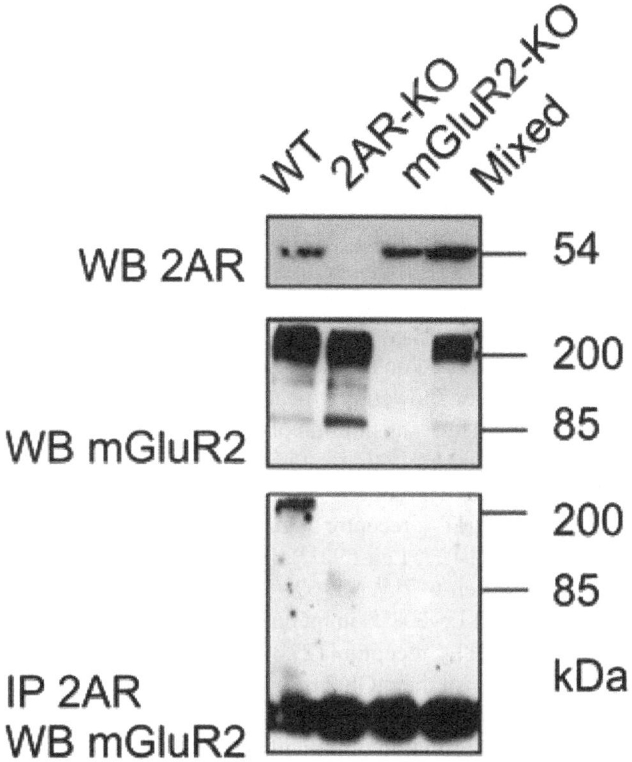

Fig. 4 Mouse frontal cortex membrane preparations were immunoprecipiated (IP) with anti-5-HT$_{2A}$ antibody. Immunoprecipitates were analyzed by Western blot (WB) with anti-mGlu2 antibody (*lower blot*). Mouse frontal cortex membrane preparation were also directly analyzed by WB with anti-5HT$_{2A}$ antibody (*upper blot*) or with the anti-mGlu2 antibody (*middle blot*). 5-HT$_{2A}$ knockout and mGlu2 knockout mouse frontal cortex tissue samples were processed identically and used as negative controls. Frontal cortex tissue samples from 5-HT$_{2A}$ knockout and mGlu2 mice were also homogenized together (mixed) and processed identically for immunoprecipitation and WB (see [133])

mGlu receptors form strict homodimers whose protomers are covalently bound via a disulfide bridge located at the extracellular Venus flytrap domains [110]. The residues and domains that are involved in homomeric formation of family A GPCRs remains however a topic of much controversy, and the conclusions seem to depend on the experimental approach used for the study of the structure of GPCR homodimers and oligomers. These approaches included electron microscopy [135], co-immunoprecipitation and site-directed mutagenesis [117, 136], disulfide cross-linking [112–115], fusion of a peptide derived from the HIV trans-activator of transcription (TAT) to transmembrane amino acid sequences [137], and X-ray crystallography [138–140] (for review, see [110]).

With regards to the transmembrane domains (TM) and residues located at the heteromeric interface between 5-HT$_{2A}$ and mGlu2 receptors, the differences in the capacity of mGlu2 and mGlu3 receptors to interact with the 5-HT$_{2A}$ receptor, as well as the capacity of 5-HT$_{2A}$ and 5-HT$_{2C}$ receptors to interact with the mGlu2 receptor, were used as an strategy to identify the structural components of the 5-HT_{2A}-mGlu2 receptor heteromer. Using a series of molecular chimeras, it has been demonstrated that the TM4 of mGlu2 is necessary and sufficient to form a 5-HT_{2A}-mGlu2 receptor heteromer [126, 127]. Thus, the mGlu2 receptor chimera containing the sequence corresponding to TM4 of the mGlu3 receptor (mGlu2ΔTM4) is not able to form a GPCR heteromer with the 5-HT_{2A} receptor [127]. On the contrary, it was demonstrated that the mGlu3 receptor chimera containing the sequence corresponding to TM4 of the mGlu2 receptor (mGlu3ΔTM4) is located in close molecular proximity with the 5-HT_{2A} receptor [127]. Together, these data suggest that TM4 of the mGlu2 receptor is necessary to form a GPCR heteromeric complex with the 5-HT_{2A} receptor.

Additional approaches based on the combination of single point mutations at the TM4 of the mGlu2 receptor showed a critical role of the residues located near the intracellular end of TM4. Thus, substitution of restudies Ala-$677^{4.40}$, Ala-$681^{4.44}$, and Ala-$685^{4.48}$ in mGlu2 for Ser-$686^{4.40}$, Phe-$690^{4.44}$ and Gly-$694^{4.48}$ in mGlu3 (mGlu2ΔTM4N) significantly reduced co-immunoprecipitation and FCM-based FRET signal with the 5-HT_{2A} receptor [127]. Additionally, substitution of Ser-$688^{4.51}$, Gly-$689^{4.52}$, Leu-$691^{4.54}$, Leu-$692^{4.55}$, Ile-$693^{4.56}$, Val-$695^{4.58}$, Ala-$696^{4.59}$ and Val-$699^{4.62}$ in mGlu2 for Leu-$697^{4.51}$, Val-$698^{4.52}$, Ile-$700^{4.54}$, Val-$701^{4.55}$, Met-$702^{4.56}$, Ser-$704^{4.58}$, Val-$705^{4.59}$ and Ile-$708^{4.62}$ in mGlu3 (mGlu2ΔTM4C) did not affect heteromerization with the 5-HT_{2A} receptor [127]. These data suggested that Ala-$677^{4.40}$, Ala-$681^{4.44}$, and Ala-$685^{4.48}$ of the mGlu2 receptor are responsible for GPCR heteromeric formation with the 5-HT_{2A} receptor.

When exploring the contribution of individual residues at the cytoplasmic end of TM4 of the mGlu2 receptor in GPCR heteromeric formation with the with the 5-HT_{2A} receptor, it was reported that the single point mutations A$677^{4.40}$S, A$681^{4.44}$F or A$685^{4.48}$G showed FRET signal when they were co-expressed with 5-HT_{2A} [127]. These FCM-based FRET data suggested that two or more of these residues are necessary at the heteromeric interface between 5-HT_{2A} and mGlu2 receptors. Remarkably, when each of the double mutations were introduced into the mGlu2-eYFP construct (A$677^{4.40}$S and A$681^{4.44}$F, A$681^{4.44}$F and A$685^{4.48}$G, or A$677^{4.40}$S and A$685^{4.48}$G), it was observed that any two of the three residues located at the intracellular end of TM4 are responsible for the difference in GPCR heteromeric formation between the 5-HT_{2A} and mGlu2 or mGlu3 receptor in living mammalian cells [127] (Fig. 5). Using an equivalent experimental approach with 5-HT_{2A} and 5-HT_{2C} chimeric constructs, more recent data suggest that TM4 of 5-HT_{2A} is also necessary to form the 5-HT_{2A}-mGlu2 heteromeric complex in HEK293 cells [134]. Together, these data suggest that a TM4-TM4 interface is at least in part responsible for stabilizing the quaternary structure of the 5-HT_{2A}-mGlu2 heteromeric complex.

As discussed above, previous findings suggested that family A GPCRs form higher order oligomers in live cells. Additionally, mGlu receptors are expressed as strict homodimers. Using a sequential three-color FRET imaging approach

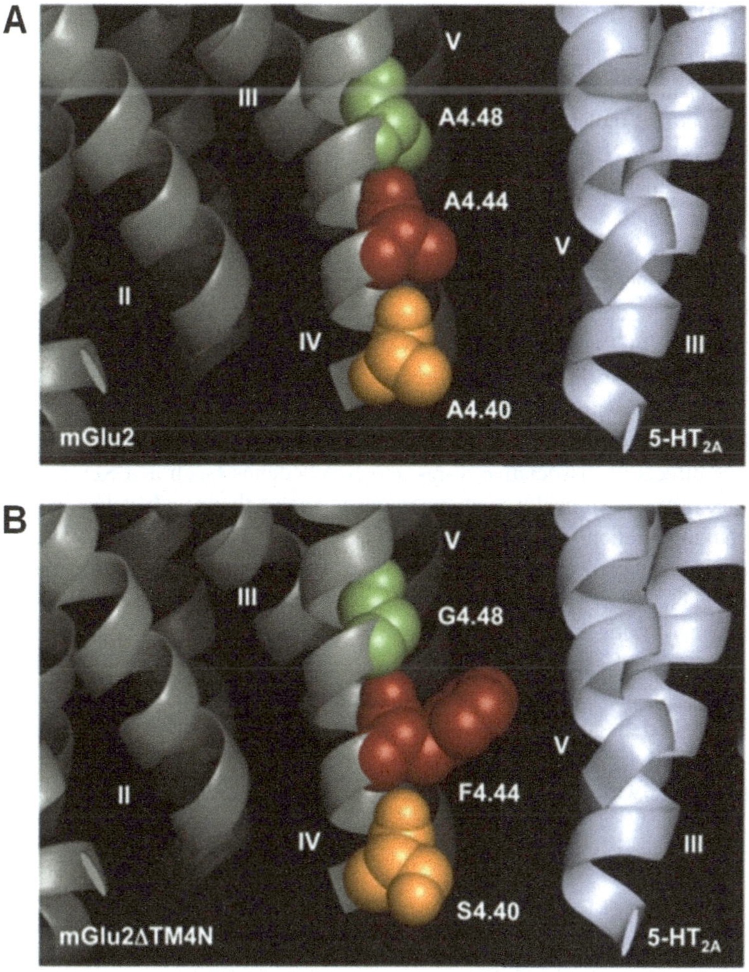

Fig. 5 Ala-677$^{4.40}$, Ala-681$^{4.44}$, and Ala-685$^{4.48}$ are critical for the mGlu2 receptor to form a GPCR heteromer with the 5-HT$_{2A}$ receptor. (**a**) Ribbon backbone representation of the transmembrene helices of the 5-HT$_{2A}$-mGlu2 heteromer model. Residues Ala-677$^{4.40}$, Ala-681$^{4.44}$, and Ala-685$^{4.48}$ are shown as spheres. (**b**) Ribbon backbone representation of the transmembrane helices of the 5-HT$_{2A}$-mGlu2ΔTM4N model. Residues Ser-677$^{4.40}$, Phe-681$^{4.44}$ and Gly-685$^{4.48}$ are shown as spheres (see [127])

(3-FRET) [116], it has been demonstrated that at least a portion of 5-HT$_{2A}$ and mGlu2 receptors are within an oligomeric complex in which the components were at distances that enable FRET [134]. Although further work will be necessary to elucidate the exact location of the promoters of the 5-HT$_{2A}$-mGlu2 heteromer, these results are consistent with mounting evidence that in living cells 5-HT$_{2A}$ and mGlu2 receptors are able to form a higher-order oligomeric complex.

Psychedelics and Antipsychotic Drugs

From a translational point of view, the 5-HT_{2A} receptor has been involved in psychiatric disorders that include schizophrenia [141], depression [54], anxiety [87] and suicidal behavior [142]. More recent studies have also suggested that the 5-HT_{2A} receptor plays a fundamental role in cognition and memory [93, 143–145]. With regards to schizophrenia and its treatment, all atypical antipsychotic drugs present a high affinity for the 5-HT_{2A} receptor as well as a modest affinity for the dopamine D2 receptor and other monoaminergic receptors [10]. Modeling in rodents the psychotic symptoms observed in schizophrenia patients remains controversial. Dissociative drugs such as phencyclidine (PCP), ketamine and MK801, which induce their effects via non-competitive antagonism on the NMDA receptor [146, 147], induce in control subjects psychosis and cognitive deficits similar to those observed in schizophrenic patients [148–155]. Although there are noted dissimilarities between the hallucinogenic and cognitive effects of dissociative drugs and psychedelics (such as LSD, psilocin and mescaline), studies in healthy volunteers also suggest that psychedelics induced certain behavioral deficits that model schizophrenia symptoms [152, 156–163].

Using dissociative drugs and psychedelics in rodent models of schizophrenia, the use of 5-HT_{2A} receptor ligands and antipsychotic drugs suggested that at least part of the antipsychotic-like behavioral phenotypes induced by atypical antipsychotic drugs is mediated via 5-HT_{2A} receptor signaling [164, 165]. This hypothesis has been further suggested with the use of 5-HT_{2A} knockout mice [133]. Thus, it was first established the lowest dose of clozapine that prevented MK801-induced hyperlocomotor activity [133]. Notably, this antipsychotic-like behavior observed in wild-type mice was absent in 5-HT_{2A} knockout mice [133].

Psychedelics

Activation of the 5-HT_{2A} receptor has also been involved in the unique effects of psychedelics on perception, cognition and sensorimotor gating. Thus, the effects of psychedelics such as LSD, mescaline, psilocin and TCB-2 are both diminished in 5-HT_{2A} knockout mice [85, 86, 166] and prevented by 5-HT_{2A} receptor antagonists [166–172]. Similar findings implicating the 5-HT_{2A} receptor in the behavioral effects or psychedelics have been observed in healthy volunteers [156]. Regarding rodent models, it has been demonstrated that head-twitch behavior represents mouse behavioral proxy of human psychedelic potential. Thus, only psychedelic 5-HT_{2A} agonists induce head-twitch behavior, whereas this behavior is not induced by closely related non-psychedelic 5-HT_{2A} agonists such as lisuride and ergotamine [85]. Similarly, the 5-HT_{2A} receptor has been involved in the effects of psychedelics affecting paradigms of sensorimotor gating such as prepulse inhibition of startle (PPI) [173, 174]. These data, however, do not exclude the possibility that other monoaminergic receptors, such as 5-HT_{2C} and dopamine D2, are also involved in some of the behavioral effects of psychedelics. Thus, the psychedelic 5-HT_{2A} agonist DOI, which also binding with high affinity to the 5-HT_{2C} receptor, has been shown to induce an inverted U-shaped dose-response effect on locomotor behavior [166]. These data indicated that high doses of DOI (10 mg/kg) decrease locomotor activity via 5-HT_{2C}, whereas low doses of DOI (0.625–5.0 mg/kg) increase locomotor activity via 5-HT_{2A} [166]. Similarly, using an animal behavior model in which rats were trained to discriminate LSD from saline at two different pre-injection times (30-min and 90-min) using a two-lever, food-reinforced operant conditioning task, it was demonstrated that the discriminative stimulus of LSD occurs in two phases, and that the second temporal phase involves principally dopamine D2 receptor-dependent pathways [175, 176].

The close molecular interaction between mGlu2 and 5-HT_{2A} has also been supported by findings in rodent models of signaling mechanism and behavior. Using microarray studies followed by high-throughput quantitative real-time PCR (qRT-PCR) showed that specific 5-HT_{2A} receptor-dependent signaling pathways lead to a unique pattern of gene expression (transcriptome) in the somatosensory cortex that predicted psychedelic potential [86]. Thus, psychedelics, such as LSD, psilocin, mescaline, DOI, DOM and DOB induced expression of *c-fos*, *egr-1* and *egr-2* in mouse somatosensory cortex [85, 86]. This effect was not observed in the somatosensory cortex of 5-HT_{2A} knockout mice [85, 86]. Similar findings were observed with LSD in rat frontal cortex [177, 178]. Additionally, non-psychedelic ligands, such as ergotamine and two isomers of lisuride (R-lisuride and S-lisuride) induced a 5-HT_{2A} receptor dependent expression of *c-fos*, but not *egr-1* or *egr-2* [85, 86]. This pattern of gene expression is currently used as a tool to test for psychedelic-like cellular events in rodent models [179–183]. Induction of expression of *egr-1* and *egr-2* also served as an experimental tool to show that psychedelic 5-HT_{2A} receptor agonists induced activation of both $G_{q/11}$- and $G_{i/o}$-dependent signaling pathways, whereas 5-HT_{2A} receptor agonists that lack psychedelic properties induced activation of $G_{q/11}$-dependent, but not and $G_{i/o}$-dependent, signaling pathways [85]. These findings have recently been supported using a phosphoproteomics analysis in tissue cultures [184]. Additionally, experiments using G_q knockout mice showed that mice in which G_q-dependent signaling was eliminated show a partial decrease in DOI-induced head-twitches [185]. It has also been suggested that the effects of the serotonin precursor L-5-hydroxytryptophan (5-HTP), but not those of the psychedelic 5-HT_{2A} receptor agonist DOI, on head-twitch behavior require expression of β-arrestin in mouse [186].

Administration of the mGlu2/3 receptor agonist LY379268 prevented the induction of *egr-1* or *egr-2* by DOI, whereas the induction of *c-fos* by DOI was unaffected in mouse somatosensory cortex [126]. Similar findings showing absence of effect of LY379268 on DOI-dependent *c-fos* expression have been reported in mouse somatosensory and frontoparietal cortex [187], whereas LY379268 prevented the effect of DOI on the transcriptional activity of the *c-fos* gene in other regions such as the frontal cortex [187]. Additionally, DOI-dependent induction of expression of *egr-1* and *egr-2* was absent in the frontal cortex of mGlu2 knockout mice [183]. This effect was not observed when induction of expression of *c-fos* by DOI was tested

[183]. Thus, intraperitoneal injection of DOI induced expression of *c-fos*, but not *egr-1* and *egr-2* in the frontal cortex of mGlu2 knockout mice [183].

As discussed above, activation of mGlu2/3 receptors by the mGlu2/3 agonist LY354740 prevented the effect of DOI on head-twitch behavior in rats [94]. Similar findings have been observed with the mGlu2/3 agonist LY379268 in mouse [126, 183] and with a selective positive allosteric modulator of the mGlu2 receptor (BINA) in rats [188]. Remarkably, the 5-HT_{2A} receptor-dependent effect of DOI and LSD on head-twitch behavior was significantly reduced in mGlu2 knockout mice as compared to wild-type littermates [183] (Fig. 6a). These data suggest that activation of mGlu2 receptor by either orthosteric agonists or positive allosteric modulators prevent head-twitch behavior induced by psychedelics in rodents. They also demonstrate the intriguing finding that head-twitch behavior induced by psychedelics requires mGlu2 receptor function. Additionally, using a virally mediated over-expression approach with herpes simplex virus (HSV), it was demonstrated that HSV-mediated over-expression of mGlu2 in the frontal cortex of mGlu2 knockout mice rescued the head-twitch behavior induced by DOI. This did not occur after HSV-mediated over-expression of mGlu2ΔTM4N (a chimeric construct that does form the 5-HT_{2A}-mGlu2 heteromeric receptor complex) [127] (Fig6b–d).

Antipsychotics

This molecular and behavioral crosstalk between 5-HT_{2A} and mGlu2 receptors has been shown in models of antipsychotic drug action. Thus, using *Xenopus* oocytes as a heterologous expression system, it was suggested that changes in $G_{q/11}$ and $G_{i/o}$ activity predict psychoactive behavioral effects [133]. Antipsychotic drugs such as clozapine, risperidone or LY379268 increased $G_{i/o}$ activity and decreased $G_{q/11}$ activity, whereas psychedelic drugs such as DOI and LY341495 increased $G_{q/11}$ activity and decreased $G_{i/o}$ activity [133]. This did not occur with ritanserin, methysergide and the mGlu2/3 receptor agonist eGlu, which do not behave as antipsychotic drugs in rodent models [133]. This work provided a metric (BI) that allows quantification and prediction of antipsychotic and pro-psychotic effects of serotonin or glutamate drugs acting through the 5-HT_{2A}-mGlu2 heteromer (Fig. 7).

This crosstalk was challenged by a follow up work in a recombinant HEK293 cell line stably expressing mGlu2 and 5-HT_{2A} receptors [129] (for a review article, see [189]). Thus, although the authors validated that mGlu2 and 5-HT_{2A} receptors are expressed in close molecular proximity in HEK293 cells [129], they showed that co-expression of 5-HT_{2A} does not affect the pharmacological properties of mGlu2/3 receptor agonists inhibiting cAMP activity [129], and that co-expression of mGlu2 does not affect the pharmacological properties of 5-HT_{2A} receptor agonists activating Ca^{2+} release [129]. Nevertheless, the potential effects of co-expression of 5-HT_{2A} and mGlu2 receptors on signaling outcomes tested by Delille et al. [129] were different when compared to those tested by Fribourg et al. [133].

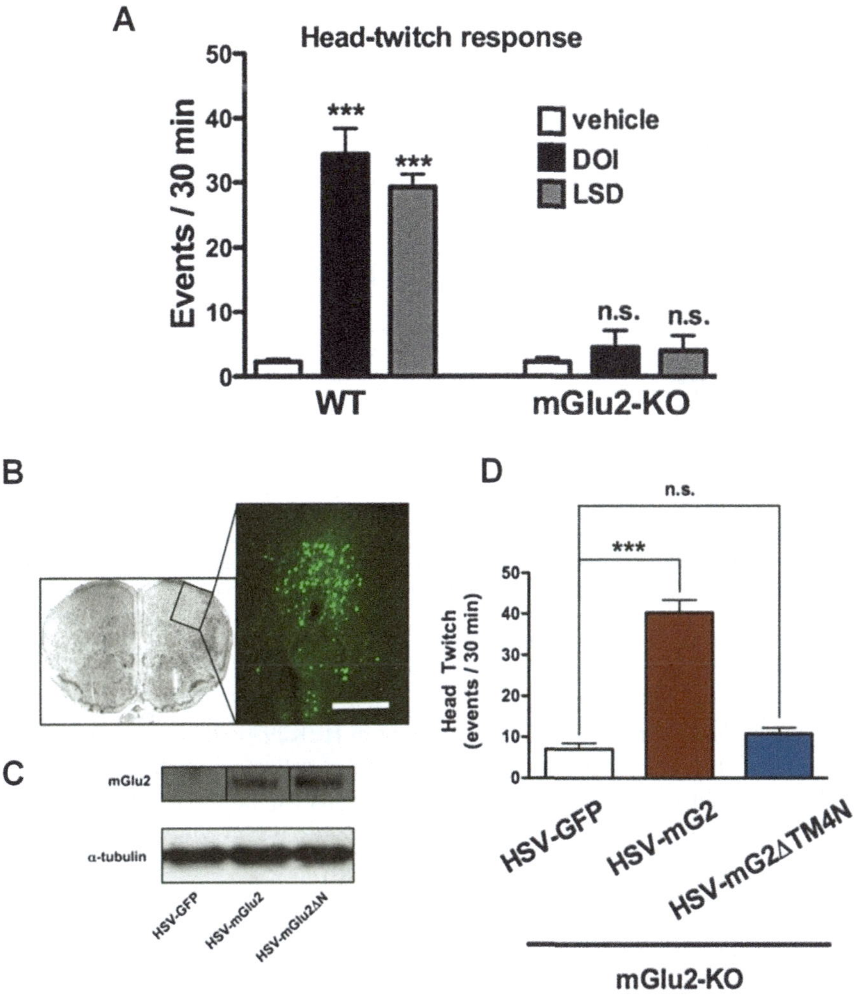

Fig. 6 (**a**) Wild-type and mGlu2 knockout mice were injected with DOI, LSD or vehicle, and the head-twitch response was scored 15 min after injection for 30 min (see [183]). (**b,c**) Expression of mGlu2 as a receptor heterocomplex with 5-HT$_{2A}$ is necessary for head-twitch behavior induced by psychedelics. Representative HSV-mediated transgene expression in frontal cortex (**b**). HSV-GFP, HSV-mGlu2 and HSV-mGlu2ΔTM4N were injected into the frontal cortex of mGlu2 knockout mice, and anti-mGlu2 expression was measured by Western blot (**c**). HSV-mediated over-expression of mGlu2, but not mGlu2ΔTM4N, in the frontal cortex of mGlu2 knockout mice rescues the head-twitch behavior induced by the psychedelic 5-HT$_{2A}$ receptor agonist DOI (**d**) (see [127])

Fribourg et al. tested the effects of activation of the G$_{i/o}$-coupled mGlu2 receptor on G$_{q/11}$ signaling in *Xenopus* oocytes expressing mGlu2 alone or mGlu2 and 5-HT$_{2A}$ and together, and those of activation of the G$_{q/11}$-coupled 5-HT$_{2A}$ receptor on G$_{i/o}$ signaling in *Xenopus* oocytes expressing 5-HT$_{2A}$ alone or 5-HT$_{2A}$ and mGlu2 together [133]. Delille et al., on the contrary, tested the effects of activation of the

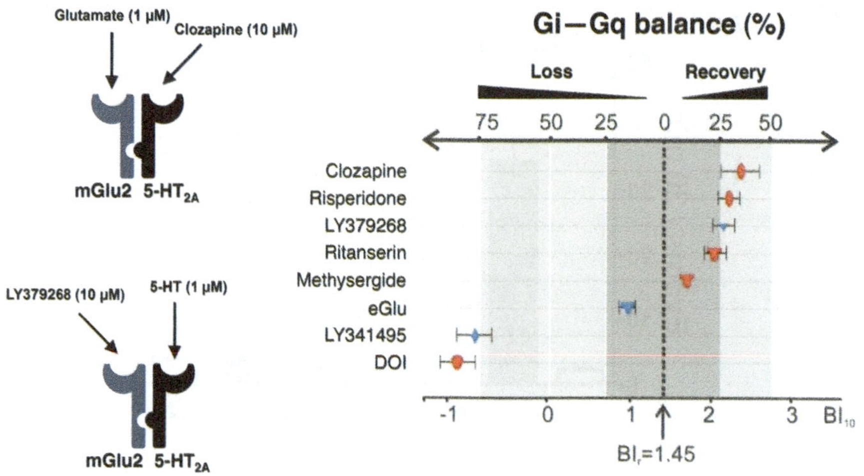

Fig. 7 Use of BI to classify the anti-/pro-psychotic activity of drugs targeting the 5-HT_{2A}-mGlu2 heteromeric complex. Correlation maps between the BI and percentage of G_i-G_q balance loss of recovery for different grants. BIs were calculated for 10 μM concentrations of the drug together with 1 μM glutamate or 1 μM serotonin. Effects of the different between G_i and G_q signaling are shown for drugs with known antipsychotic effects, such as clozapine, risperidone and LY379268, for ritanserin, an antidepressant, for the neutral antagonists methysergide and eGlu, for the psychedelic DOI and for the pro-psychotic LY341495 (see [133])

$G_{i/o}$-coupled mGlu2 receptor on $G_{i/o}$ signaling in HEK293 cells expressing mGlu2 alone of mGlu2 and 5-HT_{2A} together, and those of activation of the $G_{q/11}$-coupled 5-HT_{2A} receptor on $G_{q/11}$ signaling in HEK293 cells expressing 5-HT_{2A} alone and 5-HT_{2A} and mGlu2 together [129]. Nevertheless, in a recent study Baki, Fribourg et al. stably co-expressed in HEK293 cells 5-HT_{2A} and mGlu2 receptors, and found that crosstalk-positive phenotype in these clones correlated with biochemical and pharmacological factors such as co-localization of the two receptors at the cell surface and absolute and relative levels of expression [190]. These findings [190], along with previous publications by the same group [126, 133, 134], support the hypothesis that the presence or absence crosstalk in cells co-expressing 5-HT_{2A} and mGlu2 receptors depends profoundly on the absolute and relative levels of expression of these two constructs in both HEK293 cells and *Xenopus* oocytes. Although further investigation will be needed to better understand the underlying mechanisms by which absolute and relative expression of 5-HT_{2A} and mGlu2 receptors defines the existence of their heteromeric crosstalk, these data (see [126, 127, 133, 134, 190]) provide an explanation of why the Delille et al. study produced clones in which crosstalk between 5-HT_{2A} and mGlu2 receptors was absent. This is further supported by more recent findings providing a mechanistic explanation of the heteromeric crosstalk between 5-HT_{2A} and mGlu2 receptors in HEK293 cells.

Stimulation of $G_{q/11}$ is known to elicit a transient increase of intracellular calcium ($[Ca^{2+}]_i$) via an IP_3-mediated Ca^{2+} release from the endoplasmic reticulum that can be recorded using fluorescent calcium-sensitive dyes, such as Fura-2 [191].

Importantly, recent studies using Fura-2 to test the effect of the mGlu2/3 receptor agonist LY379268 on Ca^{2+} release in HEK293 cells co-expressing mGlu2 and 5-HT_{2A} suggest that the addition of LY379268 resulted in elevation of intracellular Ca^{2+} [134]. It was also explored a possible functional complementation between two non-functional mGlu2 protomers and the 5-HT_{2A} receptor. It was combined the use of mGlu2 constructs carrying mutations in lobe II of the Venus flytrap domain (Y216A, D295A; YADA) [118], which abolishes ligand binding and receptor function without disrupting plasma membrane expression, or in intracellular loop 3 (F756S) [192], which abolishes G protein activation. Notably, the effect of LY379268 on Ca^{2+} release was rescued in cells co-expressing YADA-mGlu2 and mGlu2-F756S together with 5-HT_{2A} [134]. These findings suggest that a fully functional mGlu2 homodimeric receptor complex and its intrinsic G protein coupling properties are fundamental for this component of the 5-HT_{2A}-mGlu2 heteromeric receptor complex to crosstalk with the 5-HT_{2A} component.

These results, together with the previous demonstration that the intracellular end of TM4 of mGlu2 is necessary to form a protein complex with the 5-HT_{2A} receptor [127], provided the rationale to explore the relative location of the mGlu2 protomer within the mGlu2 receptor homodimer that needs to contact the $G_{i/o}$ heterotrimer in order to cross-talk with 5-HT_{2A} receptor and consequently initiate $G_{q/11}$-dependent signaling. This question was addressed by combining mGlu2/mGlu3 chimeric constructs that disrupt 5-HT_{2A}-mGlu2 heteromeric formation [127] with mutations in either the Venus flytrap or the intracellular loop 3 of mGlu2 that specifically affect ligand binding or G protein coupling [134], respectively (Fig. 8). Thus, if the mGlu2 protomer within the two components of the mGlu2 homodimer that contacts directly with the 5-HT_{2A} receptor is fundamental for interacting with $G_{i/o}$ proteins, then the effect of the mGlu2/3 agonist LY379268 on Ca^{2+} release through the 5-HT_{2A}-mGlu2 heteromeric receptor complex would be disrupted in cells co-expressing YADA-mGlu2ΔTM4N (mGlu2 mutant that does not bind orthosteric agonists and that does not form the 5-HT_{2A}-mGlu2 heteromeric receptor complex) and mGlu2-F756S (mGlu2 mutant that does not activate $G_{i/o}$ proteins) together with 5-HT_{2A}-mCherry [134]. Alternatively, if the mGlu2 protomer within the mGlu2 receptor homodimer located distantly from the 5-HT_{2A} receptor is necessary for $G_{i/o}$ protein coupling, then the effect of the mGlu2/3 agonist LY379268 on Ca^{2+} release through the 5-HT_{2A}-mGlu2 heteromeric receptor complex would be disrupted in cells co-expressing YADA-mGlu2 (mGlu2 mutant that does not bind orthosteric agonists) and mGlu2ΔTM4N-F756S (mGlu2 mutant that does not activate $G_{i/o}$ proteins and that does not form the 5-HT_{2A}-mGlu2 heteromeric receptor complex) together with 5-HT_{2A}-mCherry [134]. The results for the mutant constructs showed that LY379268 induces Ca^{2+} release in cells co-expressing YADA-mGlu2ΔTM4N and mGlu2-F756S together with 5-HT_{2A}, but not in cells co-expressing YADA-mGlu2 and mGlu2ΔTM4N-F756S together with 5-HT_{2A}. These data suggest that the mGlu2 promoter that binds LY379268 needs to have an intact TM4 to communicate with the 5-HT_{2A} component [134]. They also suggest that $G_{i/o}$ protein coupling to the mGlu2 protomer located distantly from the 5-HT_{2A} receptor within the 5-HT_{2A}-mGlu2 protomer receptor complex is necessary to induce Ca^{2+} release via $G_{q/11}$ proteins [134].

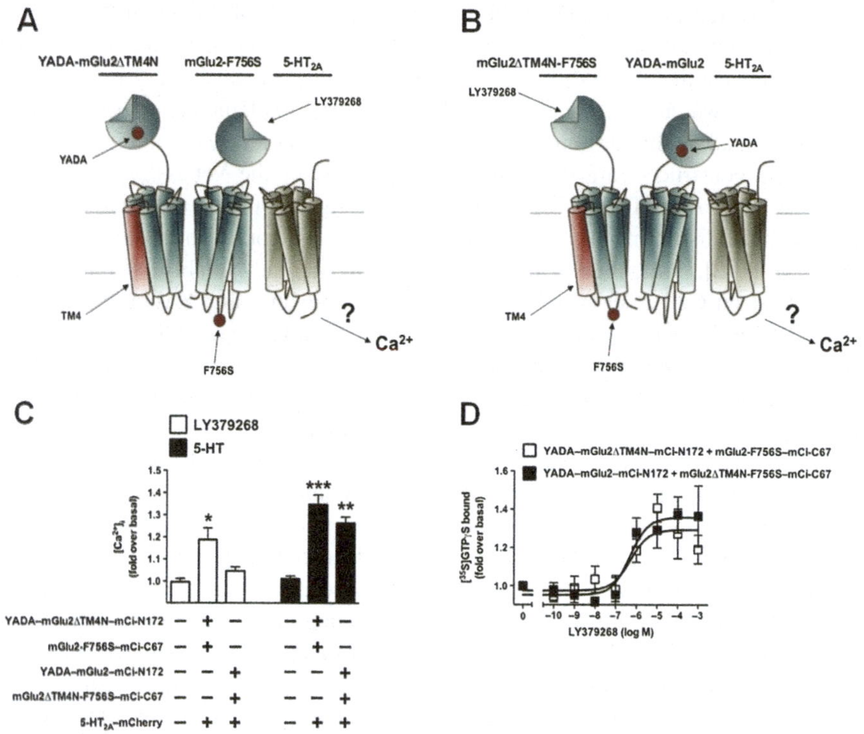

Fig. 8 (**a**,**b**) Cartoon representation of relative location of the components of the 5-HT$_{2A}$-mGlu2 heteromeric receptor complex that are necessary to induce Ca^{2+} release in the presence of the mGlu2/3 agonist LY379268 based upon the mGlu2/mGlu3 chimeric and single point mutations constructs co-transfected in HEK293 cells. In order to manipulate orthosteric ligand binding, G protein activation and/or heteromeric formation with the 5-HT$_{2A}$ receptor of each protomer within the mGlu2 homodimeric unit, the following panel of mGlu2 mutants were constructed: substitution of residues Ala-677$^{4.40}$, Ala-681$^{4.44}$ and Ala-685$^{4.48}$ in mGlu2 for Ser-686$^{4.40}$, Phe-690$^{4.44}$ and Gly-694$^{4.48}$ in mGlu3 (mGlu2ΔTM4N), which significantly reduces the formation of a complex with the 5-HT$_{2A}$ receptor; the YADA-mGlu2 construct, which carries mutations in lobe II of the Venus flytrap domain that abolish orthosteric ligand binding (Y216A, D295A); and a single point mutation into the intracellular loop 3 of mGlu3 (F756S), which abolishes G protein activation. These constructs allowed the control of the relative location of the mGlu2 protomer within the two components of the mGlu2 homodimer that needs to couple to and activate $G_{i/o}$ proteins in order to *trans*-activate the 5-HT$_{2A}$ component of the 5-HT$_{2A}$-mGlu2 heteromeric receptor complex, leading to $G_{q/11}$ coupling and consequently Ca^{2+} release from the endoplasmic reticulum. (**c**) Ca^{2+} release after stimulation with LY379268 and subsequently with 5-HT in cells mock-transfected or co-transfected with mCi-N172- or mCi-C67- and mCherry-tagged constructs (**d**), LY379268-stimulated [^{35}S]GTPγS binding in plasma membrane preparations of cells co-transfected with mCi-N172- and mCi-C67-tagged receptors (see [134])

Genetic Crosstalk

Single nucleotide polymorphism (SNP) is a single nucleotide variation that occurs when a single nucleotide, for example adenine (A), replaces one of the other three nucleotides: thymine (T), cytosine (C), and guanine (G) [193]. This is an important variation for the diversity among individuals, as well as for genetic variations that lead to phenotypes, traits and diseases. Most SNPs (93%) reported by genome-wide association (GWA) studies discovered appear to contribute to human disease risk, and they are not located in protein-coding regions [194]. This suggests that SNP regulates gene transcription levels indirectly through promoter-distal regulatory elements and the regulation of spatial genome architectures.

A recent pharmacogenetic study examined the influence of genetic variants on the response to the mGlu2/3 agonist LY2140023 (pro-drug of LY404039) in schizophrenia patients [195]. Interestingly, results from this study suggested an association between the 5-HT_{2A} receptor SNP rs7330461 and response to LY2140023. Thus T/T homozygous patients showed a significantly greater improvement in positive and negative symptoms in response to treatment with LY2140023 as compared to A/A homozygous patients [195]. Additionally, T/T homozygous patients under LY2140023 showed greater improvement in positive and negative symptoms as compared to placebo [195]. These findings, which have been validated recently in a different pharmacogenetic study [196], suggest a genetic association between SNPs at the *5-HT_{2A}* gene and the therapeutic response to LY2140023 treatment in schizophrenia treatment. Additional experimentation will be needed to unravel the basic mechanism underlying how this SNP at the *5-HT_{2A}* gene affects the dynamic regulation of chromosomal conformations critical for mGlu2-dependnent antipsychotic-related phenotypes.

Epigenetic Crosstalk

As discussed above, LY2140023 showed significant improvement in the Positive and Negative Syndrome Scale (PANSS) total score in schizophrenia patients [197]. This promising outcome encouraged the introduction of a new class of potential antipsychotic drugs acting as agonists of mGlu2/3 receptors. Unfortunately, follow-up studies with LY2140023 produced either inconclusive results [198], or clinical outcomes that were not different from placebo [199, 200], whereas improvement in PANSS total score was significantly greater in the standard of care (SOC: olanzapine, risperidone or aripiprazole) group [199, 200]. Interestingly, results in mouse models and postmortem human brain samples suggested that chronic antipsychotic treatment with atypical antipsychotic drugs might prevent the therapeutic efficacy of LY2140023 [201]. Thus, chronic treatment with atypical antipsychotic drugs, such as clozapine and risperidone, but not with the typical antipsychotic drug haloperidol, down-regulated expression of *mGlu2* mRNA in mouse frontal cortex [201]. This effect was associated with repressive histone modifications at the *mGlu2* promoter in mouse and human frontal cortex. Thus, acetylation of histone H3 (marker of transcriptional activation) was decreased in the frontal cortex of mice treated chronically with clozapine [201]. This epigenetic change did not occur in the frontal cortex of 5-HT_{2A} knockout mice [201]. Considering that this repressive histone modification at the *mGlu2* promoter was observed in the frontal cortex of schizophrenia patients previously treated with atypical antipsychotic drugs, but not in untreated schizophrenia patients [201], these results suggest that down-regulation of mGlu2 expression represents a consequence of atypical antipsychotic drug medication, and not a biochemical marker of schizophrenia in postmortem human brain samples.

Remarkably, this hypothesis raised by preclinical and postmortem human brain studies has been validated by clinical results from a recent post-hoc analysis where the antipsychotic effects of LY2140023 were comparable to those induced by risperidone in schizophrenia patients previously treated with typical antipsychotics (i.e., haloperidol), whereas previous exposure to atypical antipsychotics (e.g., clozapine and olanzapine) led to an effect of LY2140023 that did not separate from placebo [202]. Previous data also suggest that 5-HT$_{2A}$ receptor-dependent up-regulation of HDAC2 and increased binding of HDAC2 to the *mGlu2* promoter might be involved in the signaling mechanism responsible for epigenetic repression of the *mGlu2* gene after chronic atypical antipsychotic drug treatment [201]. This study may lead to the identification of epigenetic drugs that prevent the effect of chronic atypical antipsychotic treatment on mGlu2 expression, hence improving the currently limited antipsychotic efficacy of mGlu2/3 receptor agonists in certain cohorts of schizophrenia patients previously treated with atypical antipsychotic medication.

Postmortem Schizophrenia Brain

Although it is possible and useful to model some aspects of schizophrenia and other psychiatric disorders in animals (see above), most of the psychiatric disorders are uniquely human and therefore it is difficult to judge the similitude of findings obtained in rodent models that attempt to recapitulate molecular and/or neurochemical alterations without validation in human studies. In vivo imaging studies, such as positron emission tomography (PET), allow the visualization of the brain in living subjects. Although interesting, these studies do not permit either the level of resolution or the variety of studies afforded by postmortem examination of the human brain. Consequently, investigation of psychiatric and neurological illnesses using postmortem human brain tissue remains the gold standard for identifying molecular, neurochemical, genetic and epigenetic alterations that are not addressable by in vivo studies [203, 204].

Density of 5-HT$_{2A}$ receptor in the frontal cortex of schizophrenic subjects has been studied by a number of laboratories. Most of these studies have been carried out based on the use of radioligands that included [^{3}H]ketanserin [126, 205–217] and [^{3}H]LSD [218–221], as well as in untreated first-episode schizophrenic patients by positron emission tomography (PET) with [^{18}F]altanserin [222, 223]. Remarkably, there are striking differences in the results obtained: some studies suggested up-regulation of 5-HT$_{2A}$ receptor binding sites, whereas others pointed toward absence of alterations or down-regulation in the number of binding sites (Table 1). Recent findings suggest that these differences are governed in part by experimental factors such as antemortem treatment with antipsychotic drugs, age of the subject at the time of death, and cause of death.

In this regard, previous studies using the 5-HT_{2A} receptor [^{3}H]ketanserin suggested that density of the 5-HT_{2A} receptor is increased in frontal cortex of subjects that were tested negative for antipsychotic treatment at the time of death (antipsychotic-free) [126, 217]. Density of the 5-HT_{2A} receptor was comparable to control subjects in frontal cortex samples of schizophrenic subjects treated with antipsychotic drugs at the time of death [126, 217]. The hypothesis of a selective up-regulation of [^{3}H]ketanserin binding in frontal cortex of antipsychotic-free schizophrenic subjects is further supported by some [205, 207–214, 216] but not all [206, 207] of the studies suggesting that [^{3}H]ketanserin binding is decreased in the frontal cortex of treated schizophrenic subjects. Overall, these findings suggest that up-regulation of [^{3}H]ketanserin binding may predispose to psychosis, an hypothesis that has also been supported by findings in postmortem temporal cortex of parkinsonian subjects with visual hallucinations treated with medications that were not antipsychotics [226].

Similar findings have been observed with other 5-HT_{2A} receptor ligands, such as the psychedelic 5-HT_{2A} agonist [^{3}H]LSD. Thus, [^{3}H]LSD was up-regulated in in postmortem frontal cortex of untreated, but not treated, schizophrenic subjects [219–221]. Interestingly, PET imaging studies with the 5-HT_{2A} receptor ligand [^{18}F] altanserin convincingly demonstrate that [^{18}F]altanserin binding is decreased in frontal cortex of drug-naïve first-episode schizophrenic patients [223]. A potential explanation for these apparently discrepant findings is the different functional outcomes of LSD-like drugs and altanserin, as well as their affinity for all the structural conformations of the 5-HT_{2A} receptor. Thus, pharmacological and signaling findings in postmortem frontal cortex suggest that, opposite the 5-HT_{2A} receptor agonist LSD, the ligand altanserin behaves as 5-HT_{2A} receptor inverse agonist. Thus, LSD presents a higher affinity for the active G protein-coupled conformation of the 5-HT_{2A} receptor, whereas altanserin presents a higher affinity for the inactive G protein-uncoupled conformation of the 5-HT_{2A} receptor. In postmortem frontal cortex of schizophrenia patients, it has also been suggested that functional uncoupling of heterotrimeric G proteins led to increased fraction of high-affinity sites of altanserin displacing [^{3}H]ketanserin binding to the 5-HT_{2A} receptor in schizophrenic subjects, but not in controls. Together, these results suggest that up-regulation in the fraction of active G protein-coupled 5-HT_{2A} receptor will lead to both increased binding of LSD, and, possibly, decreased binding of altanserin in schizophrenic subjects. Further work studying 5-HT_{2A} receptor-G protein coupling is definitely needed to validate this hypothesis. Similarly, further work will be needed with the newly developed 5-HT_{2A} receptor agonist [227, 228].

Previous studies suggest a potential role of the serotonergic system in suicidal behavior. Therefore suicide as a cause of death might be involved in the alterations in frontal cortex 5-HT_{2A} receptor density is schizophrenic subjects that committed suicide [229]. Although this hypothesis cannot be excluded, previous studies suggested that there were no differences in [^{3}H]ketanserin binding to the 5-HT_{2A} receptor in a group suicide victims with psychiatric disorders such as dysthymic disorder, alcoholism, and anorexia nervosa, among others. Similarly, previous work has shown that [^{3}H]ketanserin binding is unaffected in postmortem frontal cortex of suicide victims with major depression [230–237]. Few studies using [^{3}H]ketanserin

Table 1 Radioligand binding studies of 5-HT_{2A} receptor in human cortex obtained from schizophrenic and control subjects

Authors	Ref.	Radioligand	Tissue preparation	Cohort (Sz/C)	Treat.	BA	Results
Bennet et al. (1979)	[218]	[^{3}H]LSD vs LSD	Membrane	12/12	Yes	Various	Downregulated
Whittaker et al. (1981)	[219]	[^{3}H]LSD vs LSD	Membrane	5/8	No	4,10,11	Upregulated
Whittaker et al. (1981)	[219]	[^{3}H]LSD vs LSD	Membrane	8/8	Yes	4,10,11	Decreased K_D
Reynolds et al. (1983)	[205]	[^{3}H] Ketanserin vs LSD	Membrane	11/10	Yes	10	No change
Mita et al. (1986)	[206]	[^{3}H]LSD vs pipamperone	Membrane	11/16	Yes	9	Downregulated
Arora and Metlzer (1991)	[224]	[^{3}H] spiperone vs cinanserin	Membrane	11/11	Yes	8,9	Downregulated
Joyce et al. (1993)	[220]	[^{125}I]LSD vs ketanserin	Tissue section	10/8	Yes	Various	Upregulated
Laruelle et al. (1993)	[207]	[^{3}H]LSD vs pipamperone	Membrane	6/13	Yes	10,17,18	Downregulated
Dan et al. (1996)	[209]	[^{3}H]LSD vs spiperone	Membrane	20/20	Yes	9	No change
Burnet et al. (1996)	[208]	[^{3}H] Ketanserin vs methysergide	Tissue section	13/15	Yes	46	Downregulated
Dean and Hayes (1996)	[209]	[^{3}H] Ketanserin vs spiperone	Tissue section	20/20	Yes	8,9,10	Downregulated
Gurevich and Joyce et al. (1997)	[221]	[^{125}I]LSD vs ketanserin	Tissue section	5/12	No	Various	Downregulated
Gurevich and Joyce et al. (1997)	[221]	[^{125}I]LSD vs ketanserin	Tissue section	5/12	Yes	Various	Downregulated
Dean et al. (1998)	[210]	[^{3}H] Ketanserin vs spiperone	Tissue section	55/55	Yes	9	Downregulated
Dean et al. (1999)	[211]	[^{3}H] Ketanserin vs spiperone	Tissue section	19/19	Yes	9	Downregulated

(continued)

Table 1 (continued)

Authors	Ref.	Radioligand	Tissue preparation	Cohort (Sz/C)	Treat.	BA	Results
Pralong et al. (2000)	[225]	[^{3}H] Ketanserin vs spiperone	Membrane	10/10	Yes	22	Downregulated
Pralong et al. (2000)	[225]	[^{3}H] Ketanserin vs spiperone	Tissue section	20/20	Yes	22	Downregulated
Marazziti et al. (2003)	[216]	[^{3}H] Ketanserin saturation	Membrane	15/15	Yes	Frontal cortex	Upregulated
Matsumoto et al. (2005)	[212]	[^{3}H] Ketanserin	Tissue section	6/6	Yes	9	Downregulated
González-Maeso et al. (2008)	[126]	[^{3}H] Ketanserin saturation	Membrane	13/13	No	9	Upregulated
González-Maeso et al. (2008)	[126]	[^{3}H] Ketanserin saturation	Membrane	12/12	Yes	9	No change
Dean et al. (2008)	[213]	[^{3}H] Ketanserin saturation	Tissue section	14/14	Yes	9	Downregulated
Dean et al. (2008)	[213]	[^{3}H] Ketanserin saturation	Membrane	14/14	Yes	9	Downregulated
Erritzoe et al. (2008)	[222]	[^{18}F] Altanserin	PET scan	15/15	No	N/A	No change
Kang et al. (2009)	[214]	[^{3}H] Ketanserin	Tissue section	8/8	Yes	22	Downregulated
Rasmussen et al. (2010)	[223]	[^{18}F] Altanserin	PET scan	30/30	No	N/A	Downregulated
Muguruza et al. (2012)	[217]	[^{3}H] Ketanserin saturation	Membrane	29/29	No	9	Upregulated
Muguruza et al. (2012)	[217]	[^{3}H] Ketanserin vs DOI	Membrane	29/29	No	9	Increased affinity
Muguruza et al. (2012)	[217]	[^{3}H] Ketanserin vs altanserin	Membrane	29/29	No	9	Increased affinity

(continued)

Table 1 (continued)

Authors	Ref.	Radioligand	Tissue preparation	Cohort (Sz/C)	Treat.	BA	Results
Muguruza et al. (2012)	[217]	[^{3}H] Ketanserin saturation	Membrane	16/16	Yes	9	No change
Muguruza et al. (2012)	[217]	[^{3}H] Ketanserin vs DOI	Membrane	16/16	Yes	9	No change
Muguruza et al. (2012)	[217]	[^{3}H] Ketanserin vs altanserin	Membrane	16/16	Yes	9	Increased affinity
Moreno et al. (2012)	[127]	[^{3}H] LY341495 vs LY379268/ DOI	Membrane	27/27	Yes	9	Increased crosstalk
Moreno et al. (2015)	[134]	[^{35}S]GTPγS binding	Membrane	27/27	Yes	9	Dysregulated

Membrane (Membrane plasma preparation)
Treat. (treatment)
BA (Brodmann area)
Sz (schizophrenia)
C (control)
PET (positron emission tomography)

reported that the 5-HT$_{2A}$ receptor is increased in suicide victims with depressive disorders [238], and in suicide victims without psychiatric diagnosis [239, 240]. [^{3}H]Ketanserin binding has also been shown to correlate with lifetime aggression in suicide [142]. Although further investigation will be needed to determine the impact, if any, of suicidal behavior on 5-HT$_{2A}$ receptor binding, these data suggest that suicide does not affect density of 5-HT$_{2A}$ receptor binding in postmortem human frontal cortex tissue samples.

Possible alterations of 5-HT$_{2A}$ and mGlu2 as a GPCR heteromeric complex have also been reported in postmortem frontal cortex of schizophrenic subjects. Competition binding assays provide a sensitive measure of functional interactions within a receptor heterocomplex [241]. In classical pharmacology, displacement curves of [^{3}H]antagonist binding in the presence of increasing concentrations of an agonist show pattern, with high-affinity ($K_{i\text{-high}}$) and low-affinity ($K_{i\text{-low}}$) values [242, 243]. Importantly, experiments in tissue culture (HEK293 cells) and in mouse frontal cortex membrane preparations show that the biphasic displacement curve of the mGlu2/3 receptor antagonist [^{3}H]LY341495 by the mGlu2/3 receptor agonist LY379268 becomes monophasic in the presence of the 5-HT$_{2A}$ receptor agonist DOI [126]. Experiments with mGlu2/mGlu3 chimeric constructs that form or do not form the 5-HT$_{2A}$-mGlu2 heteromer suggest that this allosteric crosstalk between 5-HT$_{2A}$ and mGlu2 receptor needs their close molecular proximity at the plasma membrane [126, 127, 190]. Importantly, it has been reported that the difference between the high-affinities of LY379268 displacing [^{3}H]LY341495 binding in the presence and in the absence of DOI was significantly increased in frontal cortex of schizophrenic subjects [127]. Together with the results obtained with chimeric constructs in tissue culture, these findings suggest that the allosteric crosstalk between the components of the 5-HT_{2A}-mGlu2 receptor heterocomplex is dysregulated in frontal cortex of schizophrenic subjects.

This schizophrenia-related alteration in the crosstalk between mGlu2 and 5-HT_{2A} receptor has been supported by functional assays testing receptor-G protein coupling. In agreement with the studies in heterologous expression systems, such as HEK293 cells [134, 190] and *Xenopus* oocytes [133], the mGlu2/3 receptor agonist was able to activate both $G_{i/o}$ and $G_{q/11}$ proteins in mouse frontal cortex [134]. It was also found that the effect of LY379268 on activation of $G_{i1,2,3}$ and $G\alpha_{q/11}$ was absent in frontal cortex of mGlu2 knockout mice [134]. Importantly, LY379268 was able to activate $G_{i1,2,3}$, but not $G_{q/11}$, in the frontal cortex of 5-HT_{2A} knockout mice [134]. These data demonstrate that $G_{i1,2,3}$ coupling as the canonical signal transduction mediated by activation of the mGlu2 receptor remains uninfluenced in the absence of 5-HT_{2A} receptor-dependent signaling, whereas mGlu2-dependent activation of $G_{q/11}$ proteins requires expression of the 5-HT_{2A} receptor in mouse frontal cortex. Similar effects have been observed in postmortem human frontal cortex, with activation of both $G_{i1,2,3}$ and $G_{q/11}$ by LY379268 [134]. Importantly, the impact of LY379268 on activation of $G_{q/11}$ was significantly reduced in schizophrenic subjects as compared to controls, whereas LY379268-dependent coupling of mGlu2 to $G\alpha_{i1,2,3}$ was unaffected [134]. These data may be potentially useful for a better understanding of the biochemical alterations responsible for schizophrenia and other psychotic disorders.

Prenatal Insults and Schizophrenia

Genetics plays an important role in the etiology of schizophrenia. GWA studies conducted in the mid-2000s showed genetic alterations associated with schizophrenia risk that included large recurrent microdeletions [244], variations in copy number [245], and rare duplications and microdeletions [246], and structural variants of genes involved in neurodevelopmental pathways [247]. It was also suggested that schizophrenia risk is associated with polygenic pathways that involve a great number of common alleles each of which with a very small effect [248]. More recent genomic studies have narrowed down the number of genetic loci potentially associated with schizophrenia. These genes include dopamine D_2 (*DRD2*) and serotonin 5-HT_{2A} (*Htr2a*) receptors, as well as genes involved in glutamatergic neurotransmission [249], voltage-gated ion channel, the signaling complex formed by activity-regulated cytoskeleton-associated scaffold protein (ARC) at the postsynaptic density [250], and complement component 4 [251].

These findings provide convincing evidence that genetics plays a fundamental role in the basic molecular mechanisms responsible for schizophrenia. These genetic factors, however, are not the only cause responsible for this psychiatric disorder. As an example, monozygotic twins, whose genetic material is ~100% identical, have a concordance for schizophrenia of nearly 50% [252–254]. These results support an important contribution of genetics to the development of schizophrenia. At the same time, however, they also indicate that environmental factor may play a fundamental role in schizophrenia risk. Related to this topic, epidemiological studies have indicated that maternal infection during pregnancy with virus, such as influenza [255–257] and rubella [258], bacteria, such as bronchopneumonia [259], and protozoa, such as *Toxoplasma gondii* [260], affect profoundly the risk of developing schizophrenia in the adult offspring. Another factor associated to schizophrenia risk is related to maternal severe stress during pregnancy. Thus, severe adverse life events during pregnancy, such as war [261, 262], famine [263] and death or illness in a first-degree relative [264] have been shown to increase schizophrenia risk in the adult offspring.

Interestingly, using a mouse-adapted influenza virus, it was shown that maternal infection during pregnancy induced schizophrenia-related phenotypes in the adult offspring [265]. Among these phenotypes, it was reported that maternal influenza A/WSN/33 (H1N1) viral infection during pregnancy was able to up-regulate 5-HT_{2A} receptor and down-regulate mGlu2 receptor in the frontal cortex of the adult offspring [265]. Of note, a similar pattern of dysregulation was observed in postmortem frontal cortex of untreated schizophrenic subjects [126, 217], suggesting that a mouse model of prenatal viral infection may facilitate targeting therapies for treatment of this psychiatric disorder. Importantly, this study has been followed by other reports showing a similar pattern of dysregulation in expression and function of 5-HT_{2A} and mGlu2 receptors in rodent models of environmental insults. These included maternal stress during pregnancy [266–268], maternal administration of poly-(I:C) during pregnancy as a model of viral infection [182, 268], maternal administration of lipopolysaccharide (LPS) during pregnancy as a model of Gram-negative bacterial infection [181], administration of kynurenic acid during pregnancy as a model of negative modulation of alpha 7 nicotinic acetylcholine receptors [269], chronic restraint stress in adult mice [270], sleep deprivation as a model of stress in adult mice [271], comparison between Roman Low- (RLA) and High- (RHA) avoidance rat strains [272], transgenic mice with a knock-in of a tryptophan hydroxylase 2 (Tph2) R439H [273], repeated administration of methamphetamine [274], chronic exposure to valproate in young rats [275], and perinatal exposure to bisphenol A (an estrogen that mimics endocrine disruption) [276]. These findings, together with results in postmortem human brain samples (see above), suggest that dysregulation of 5-HT_{2A} and mGlu2 expression and their function in the frontal cortex might be responsible for some of the schizophrenia-related phenotypes observed in rodent models of prenatal and postnatal insults. Although there are a great many more unexplored genes affected by environmental insults, these experimental tools in animal models might help understand the role, if any, of the crosstalk between 5-HT_{2A} and mGlu2 receptor in frontal cortex as principal component of a signaling pathway that is responsible for schizophrenia-related phenotypes. These findings may ultimately lead to the identification of new therapeutic approaches not only for treatment but also prevention of cognitive deficits in schizophrenia and other psychiatric conditions.

References

1. Adam D (2013) On the spectrum. Nature 496:416–418
2. Lewis DA, Gonzalez-Burgos G (2006) Pathophysiologically based treatment interventions in schizophrenia. Nat Med 12:1016–1022
3. van Os J, Kapur S (2009) Schizophrenia. Lancet 374:635–645
4. Sawa A, Snyder SH (2002) Schizophrenia: diverse approaches to a complex disease. Science 296:692–695
5. Freedman R (2003) Schizophrenia. N Engl J Med 349:1738–1749
6. Lehmann HE, Hanrahan GE (1954) Chlorpromazine; new inhibiting agent for psychomotor excitement and manic states. AMA Arch Neurol Psychiatry 71:227–237
7. Granger B, Albu S (2005) The haloperidol story. Ann Clin Psychiatry 17:137–140
8. Hippius H (1999) A historical perspective of clozapine. J Clin Psychiatry 60(Suppl 12):22–23
9. Miyamoto S, Miyake N, Jarskog LF, Fleischhacker WW, Lieberman JA (2012) Pharmacological treatment of schizophrenia: a critical review of the pharmacology and clinical effects of current and future therapeutic agents. Mol Psychiatry 17:1206–1227
10. Meltzer HY (2013) Update on typical and atypical antipsychotic drugs. Annu Rev Med 64:393–406
11. Lieberman JA, Bymaster FP, Meltzer HY, Deutch AY, Duncan GE, Marx CE, Aprille JR, Dwyer DS, Li XM, Mahadik SP, Duman RS, Porter JH, Modica-Napolitano JS, Newton SS, Csernansky JG (2008) Antipsychotic drugs: comparison in animal models of efficacy, neurotransmitter regulation, and neuroprotection. Pharmacol Rev 60:358–403
12. Ibrahim HM, Tamminga CA (2011) Schizophrenia: treatment targets beyond monoamine systems. Annu Rev Pharmacol Toxicol 51:189–209
13. Lieberman JA, Stroup TS, McEvoy JP, Swartz MS, Rosenheck RA, Perkins DO, Keefe RS, Davis SM, Davis CE, Lebowitz BD, Severe J, Hsiao JK (2005) Effectiveness of antipsychotic drugs in patients with chronic schizophrenia. N Engl J Med 353:1209–1223
14. Turlejski K (1996) Evolutionary ancient roles of serotonin: long-lasting regulation of activity and development. Acta Neurobiol Exp (Wars) 56:619–636
15. Janeway T, Richarson H, Park E (1918) Experiments on the vasoconstrictor action of blood serum. Arch Intern Med 21:565–571
16. Reid G, Bick M (1942) Pharmacologically active substances in serum. Aust J Exp Biol Med Sci 20:33–46
17. Zucker M (1944) A study of the substances in blood serum and platelets which stimulate smooth muscle. Am J Physiol 142:12–26
18. Erspamer V, Asero B (1952) Identification of enteramine, the specific hormone of the enterochromaffin cell system, as 5-hydroxytryptamine. Nature 169:800–801
19. Rapport MM, Green AA, Page IH (1947) Purification of the substance which is responsible for the vasoconstrictor activity of serum. Fed Proc 6:184
20. Rapport MM, Green AA, Page IH (1948) Serum vasoconstrictor, serotonin; isolation and characterization. J Biol Chem 176:1243–1251
21. Rapport MM, Green AA, Page IH (1948) Partial purification of the vasoconstrictor in beef serum. J Biol Chem 174:735–741
22. Rapport MM, Green AA, Page IH (1948) Crystalline serotonin. Science 108:329–330
23. Reid G, Rand M (1952) Pharmacological actions of synthetic 5-hydroxytryptamine (serotonin, thrombocytin). Nature 169:801–802
24. Mohammad-Zadeh LF, Moses L, Gwaltney-Brant SM (2008) Serotonin: a review. J Vet Pharmacol Ther 31:187–199
25. Hofmann A (1979) How LSD originated. J Psychedelic Drugs 11:53–60
26. Hofmann A (1980) LSD: my problem child. McGraw-Hill, New York
27. Aghajanian GK, Foote WE, Sheard MH (1968) Lysergic acid diethylamide: sensitive neuronal units in the midbrain raphe. Science 161:706–708

28. Bennett JP Jr, Snyder SH (1976) Serotonin and lysergic acid diethylamide binding in rat brain membranes: relationship to postsynaptic serotonin receptors. Mol Pharmacol 12:373–389
29. Peroutka SJ, Lebovitz RM, Snyder SH (1979) Serotonin receptor binding sites affected differentially by guanine nucleotides. Mol Pharmacol 16:700–708
30. Peroutka SJ, Snyder SH (1979) Multiple serotonin receptors: differential binding of [3H]5-hydroxytryptamine, [3H]lysergic acid diethylamide and [3H]spiroperidol. Mol Pharmacol 16:687–699
31. Kobilka BK, Frielle T, Collins S, Yang-Feng T, Kobilka TS, Francke U, Lefkowitz RJ, Caron MG (1987) An intronless gene encoding a potential member of the family of receptors coupled to guanine nucleotide regulatory proteins. Nature 329:75–79
32. Barnes NM, Sharp T (1999) A review of central 5-HT receptors and their function. Neuropharmacology 38:1083–1152
33. Nichols DE, Nichols CD (2008) Serotonin receptors. Chem Rev 108:1614–1641
34. Hannon J, Hoyer D (2008) Molecular biology of 5-HT receptors. Behav Brain Res 195:198–213
35. McCorvy JD, Roth BL (2015) Structure and function of serotonin G protein-coupled receptors. Pharmacol Ther 150:129–142
36. Pazos A, Hoyer D, Palacios JM (1984) The binding of serotonergic ligands to the porcine choroid plexus: characterization of a new type of serotonin recognition site. Eur J Pharmacol 106:539–546
37. Conn PJ, Sanders-Bush E (1985) Serotonin-stimulated phosphoinositide turnover: mediation by the S2 binding site in rat cerebral cortex but not in subcortical regions. J Pharmacol Exp Ther 234:195–203
38. Conn PJ, Sanders-Bush E, Hoffman BJ, Hartig PR (1986) A unique serotonin receptor in choroid plexus is linked to phosphatidylinositol turnover. Proc Natl Acad Sci U S A 83:4086–4088
39. Burns CM, Chu H, Rueter SM, Hutchinson LK, Canton H, Sanders-Bush E, Emeson RB (1997) Regulation of serotonin-2C receptor G-protein coupling by RNA editing. Nature 387:303–308
40. Berg KA, Cropper JD, Niswender CM, Sanders-Bush E, Emeson RB, Clarke WP (2001) RNA-editing of the 5-HT(2C) receptor alters agonist-receptor-effector coupling specificity. Br J Pharmacol 134:386–392
41. McGrew L, Price RD, Hackler E, Chang MS, Sanders-Bush E (2004) RNA editing of the human serotonin 5-HT2C receptor disrupts transactivation of the small G-protein RhoA. Mol Pharmacol 65:252–256
42. Price RD, Weiner DM, Chang MS, Sanders-Bush E (2001) RNA editing of the human serotonin 5-HT2C receptor alters receptor-mediated activation of G13 protein. J Biol Chem 276:44663–44668
43. Nelson DL (2004) 5-HT5 receptors. Curr Drug Targets CNS Neurol Disord 3:53–58
44. Rees S, den Daas I, Foord S, Goodson S, Bull D, Kilpatrick G, Lee M (1994) Cloning and characterisation of the human 5-HT5A serotonin receptor. FEBS Lett 355:242–246
45. Martin GR, Eglen RM, Hamblin MW, Hoyer D, Yocca F (1998) The structure and signalling properties of 5-HT receptors: an endless diversity? Trends Pharmacol Sci 19:2–4
46. Cook EH Jr, Fletcher KE, Wainwright M, Marks N, Yan SY, Leventhal BL (1994) Primary structure of the human platelet serotonin 5-HT2A receptor: identify with frontal cortex serotonin 5-HT2A receptor. J Neurochem 63:465–469
47. Lopez-Gimenez JF, Mengod G, Palacios JM, Vilaro MT (1997) Selective visualization of rat brain 5-HT2A receptors by autoradiography with [3H]MDL 100,907. Naunyn Schmiedebergs Arch Pharmacol 356:446–454
48. Lopez-Gimenez JF, Vilaro MT, Palacios JM, Mengod G (2001) Mapping of 5-HT2A receptors and their mRNA in monkey brain: [3H]MDL100,907 autoradiography and in situ hybridization studies. J Comp Neurol 429:571–589
49. Jakab RL, Goldman-Rakic PS (1998) 5-Hydroxytryptamine2A serotonin receptors in the primate cerebral cortex: possible site of action of hallucinogenic and antipsychotic drugs in pyramidal cell apical dendrites. Proc Natl Acad Sci U S A 95:735–740

50. Weber ET, Andrade R (2010) Htr2a Gene and 5-HT(2A) receptor expression in the cerebral cortex studied using genetically modified mice. Front Neurosci 4:36
51. Nichols DE (2004) Hallucinogens. Pharmacol Ther 101:131–181
52. Nichols DE (2016) Psychedelics. Pharmacol Rev 68:264–355
53. Cummings J, Isaacson S, Mills R, Williams H, Chi-Burris K, Corbett A, Dhall R, Ballard C (2014) Pimavanserin for patients with Parkinson's disease psychosis: a randomised, placebo-controlled phase 3 trial. Lancet 383:533–540
54. Celada P, Puig M, Amargos-Bosch M, Adell A, Artigas F (2004) The therapeutic role of 5-HT1A and 5-HT2A receptors in depression. J Psychiatry Neurosci 29:252–265
55. Reddy DS (2013) The pathophysiological and pharmacological basis of current drug treatment of migraine headache. Expert Rev Clin Pharmacol 6:271–288
56. Agundez JA, Garcia-Martin E, Alonso-Navarro H, Jimenez-Jimenez FJ (2013) Anti-Parkinson's disease drugs and pharmacogenetic considerations. Expert Opin Drug Metab Toxicol 9:859–874
57. Gatch MB, Kozlenkov A, Huang RQ, Yang W, Nguyen JD, Gonzalez-Maeso J, Rice KC, France CP, Dillon GH, Forster MJ, Schetz JA (2013) The HIV antiretroviral drug Efavirenz has LSD-Like properties. Neuropsychopharmacology 38(12):2373–2384
58. Traynelis SF, Wollmuth LP, McBain CJ, Menniti FS, Vance KM, Ogden KK, Hansen KB, Yuan H, Myers SJ, Dingledine R (2010) Glutamate receptor ion channels: structure, regulation, and function. Pharmacol Rev 62:405–496
59. Paoletti P (2011) Molecular basis of NMDA receptor functional diversity. Eur J Neurosci 33:1351–1365
60. Paoletti P, Bellone C, Zhou Q (2013) NMDA receptor subunit diversity: impact on receptor properties, synaptic plasticity and disease. Nat Rev Neurosci 14:383–400
61. Wang G, Gilbert J, Man HY (2012) AMPA receptor trafficking in homeostatic synaptic plasticity: functional molecules and signaling cascades. Neural Plast 2012:825364. https://doi.org/10.1155/2012/825364
62. Hunt DL, Castillo PE (2012) Synaptic plasticity of NMDA receptors: mechanisms and functional implications. Curr Opin Neurobiol 22:496–508
63. Contractor A, Mulle C, Swanson GT (2011) Kainate receptors coming of age: milestones of two decades of research. Trends Neurosci 34:154–163
64. Matute C (2011) Therapeutic potential of kainate receptors. CNS Neurosci Ther 17:661–669
65. Sladeczek F, Pin JP, Recasens M, Bockaert J, Weiss S (1985) Glutamate stimulates inositol phosphate formation in striatal neurones. Nature 317:717–719
66. Nicoletti F, Iadarola MJ, Wroblewski JT, Costa E (1986) Excitatory amino acid recognition sites coupled with inositol phospholipid metabolism: developmental changes and interaction with alpha 1-adrenoceptors. Proc Natl Acad Sci U S A 83:1931–1935
67. Masu M, Tanabe Y, Tsuchida K, Shigemoto R, Nakanishi S (1991) Sequence and expression of a metabotropic glutamate receptor. Nature 349:760–765
68. Houamed KM, Kuijper JL, Gilbert TL, Haldeman BA, O'Hara PJ, Mulvihill ER, Almers W, Hagen FS (1991) Cloning, expression, and gene structure of a G protein-coupled glutamate receptor from rat brain. Science 252:1318–1321
69. Nicoletti F, Bockaert J, Collingridge GL, Conn PJ, Ferraguti F, Schoepp DD, Wroblewski JT, Pin JP (2011) Metabotropic glutamate receptors: from the workbench to the bedside. Neuropharmacology 60:1017–1041
70. Pin JP, Galvez T, Prezeau L (2003) Evolution, structure, and activation mechanism of family 3/C G-protein-coupled receptors. Pharmacol Ther 98:325–354
71. Pin JP, Kniazeff J, Goudet C, Bessis AS, Liu J, Galvez T, Acher F, Rondard P, Prezeau L (2004) The activation mechanism of class-C G-protein coupled receptors. Biol Cell 96:335–342
72. Conn PJ, Pin JP (1997) Pharmacology and functions of metabotropic glutamate receptors. Annu Rev Pharmacol Toxicol 37:205–237
73. Aghajanian GK, Marek GJ (1997) Serotonin induces excitatory postsynaptic potentials in apical dendrites of neocortical pyramidal cells. Neuropharmacology 36:589–599

74. Aghajanian GK, Marek GJ (1999) Serotonin, via 5-HT2A receptors, increases EPSCs in layer V pyramidal cells of prefrontal cortex by an asynchronous mode of glutamate release. Brain Res 825:161–171
75. Marek GJ, Wright RA, Schoepp DD, Monn JA, Aghajanian GK (2000) Physiological antagonism between 5-hydroxytryptamine(2A) and group II metabotropic glutamate receptors in prefrontal cortex. J Pharmacol Exp Ther 292:76–87
76. Fernando AB, Robbins TW (2011) Animal models of neuropsychiatric disorders. Annu Rev Clin Psychol 7:39–61
77. Arguello PA, Gogos JA (2006) Modeling madness in mice: one piece at a time. Neuron 52:179–196
78. Nestler EJ, Hyman SE (2010) Animal models of neuropsychiatric disorders. Nat Neurosci 13:1161–1169
79. Kellendonk C, Simpson EH, Kandel ER (2009) Modeling cognitive endophenotypes of schizophrenia in mice. Trends Neurosci 32:347–358
80. Jones CA, Watson DJ, Fone KC (2011) Animal models of schizophrenia. Br J Pharmacol 164:1162–1194
81. Young JW, Zhou X, Geyer MA (2010) Animal models of schizophrenia. Curr Top Behav Neurosci 4:391–433
82. Canal CE, Morgan D (2012) Head-twitch response in rodents induced by the hallucinogen 2,5-dimethoxy-4-iodoamphetamine: a comprehensive history, a re-evaluation of mechanisms, and its utility as a model. Drug Test Anal 4(7–8):556–576
83. Hanks JB, Gonzalez-Maeso J (2013) Animal models of serotonergic psychedelics. ACS Chem Neurosci 4:33–42
84. Moreno JL, Gonzalez-Maeso J (2013) Preclinical models of antipsychotic drug action. Int J Neuropsychopharmacol 16(10):2131–2144
85. Gonzalez-Maeso J, Weisstaub NV, Zhou M, Chan P, Ivic L, Ang R, Lira A, Bradley-Moore M, Ge Y, Zhou Q, Sealfon SC, Gingrich JA (2007) Hallucinogens recruit specific cortical 5-HT(2A) receptor-mediated signaling pathways to affect behavior. Neuron 53:439–452
86. Gonzalez-Maeso J, Yuen T, Ebersole BJ, Wurmbach E, Lira A, Zhou M, Weisstaub N, Hen R, Gingrich JA, Sealfon SC (2003) Transcriptome fingerprints distinguish hallucinogenic and nonhallucinogenic 5-hydroxytryptamine 2A receptor agonist effects in mouse somatosensory cortex. J Neurosci 23:8836–8843
87. Weisstaub NV, Zhou M, Lira A, Lambe E, Gonzalez-Maeso J, Hornung JP, Sibille E, Underwood M, Itohara S, Dauer WT, Ansorge MS, Morelli E, Mann JJ, Toth M, Aghajanian G, Sealfon SC, Hen R, Gingrich JA (2006) Cortical 5-HT2A receptor signaling modulates anxiety-like behaviors in mice. Science 313:536–540
88. Béïque J-C, Imad M, Mladenovic L, Gingrich JA, Andrade R (2007) Mechanism of the 5-hydroxytryptamine 2A receptor-mediated facilitation of synaptic activity in prefrontal cortex. Proc Natl Acad Sci U S A 104:9870–9875
89. Puig MV, Celada P, Diaz-Mataix L, Artigas F (2003) In vivo modulation of the activity of pyramidal neurons in the rat medial prefrontal cortex by 5-HT2A receptors: relationship to thalamocortical afferents. Cereb Cortex 13:870–882
90. Celada P, Puig MV, Diaz-Mataix L, Artigas F (2008) The hallucinogen DOI reduces low-frequency oscillations in rat prefrontal cortex: reversal by antipsychotic drugs. Biol Psychiatry 64:392–400
91. Marek GJ, Wright RA, Gewirtz JC, Schoepp DD (2001) A major role for thalamocortical afferents in serotonergic hallucinogen receptor function in the rat neocortex. Neuroscience 105:379–392
92. Scruggs JL, Patel S, Bubser M, Deutch AY (2000) DOI-induced activation of the cortex: dependence on 5-HT2A heteroceptors on thalamocortical glutamatergic neurons. J Neurosci 20:8846–8852
93. Barre A, Berthoux C, De Bundel D, Valjent E, Bockaert J, Marin P, Becamel C (2016) Presynaptic serotonin 2A receptors modulate thalamocortical plasticity and associative learning. Proc Natl Acad Sci U S A 113(10):E1382–E1391

94. Gewirtz JC, Marek GJ (2000) Behavioral evidence for interactions between a hallucinogenic drug and group II metabotropic glutamate receptors. Neuropsychopharmacology 23:569–576
95. Van der Kloot W (1991) The regulation of quantal size. Prog Neurobiol 36:93–130
96. Lovinger DM, McCool BA (1995) Metabotropic glutamate receptor-mediated presynaptic depression at corticostriatal synapses involves mGLuR2 or 3. J Neurophysiol 73:1076–1083
97. Maggio R, Vogel Z, Wess J (1993) Coexpression studies with mutant muscarinic/adrenergic receptors provide evidence for intermolecular "cross-talk" between G-protein-linked receptors. Proc Natl Acad Sci U S A 90:3103–3107
98. Kaupmann K, Huggel K, Heid J, Flor PJ, Bischoff S, Mickel SJ, McMaster G, Angst C, Bittiger H, Froestl W, Bettler B (1997) Expression cloning of GABA(B) receptors uncovers similarity to metabotropic glutamate receptors. Nature 386:239–246
99. Couve A, Filippov AK, Connolly CN, Bettler B, Brown DA, Moss SJ (1998) Intracellular retention of recombinant GABAB receptors. J Biol Chem 273:26361–26367
100. Jones KA, Borowsky B, Tamm JA, Craig DA, Durkin MM, Dai M, Yao WJ, Johnson M, Gunwaldsen C, Huang LY, Tang C, Shen Q, Salon JA, Morse K, Laz T, Smith KE, Nagarathnam D, Noble SA, Branchek TA, Gerald C (1998) GABA(B) receptors function as a heteromeric assembly of the subunits GABA(B)R1 and GABA(B)R2. Nature 396:674–679
101. Kaupmann K, Malitschek B, Schuler V, Heid J, Froestl W, Beck P, Mosbacher J, Bischoff S, Kulik A, Shigemoto R, Karschin A, Bettler B (1998) GABA(B)-receptor subtypes assemble into functional heteromeric complexes. Nature 396:683–687
102. White JH, Wise A, Main MJ, Green A, Fraser NJ, Disney GH, Barnes AA, Emson P, Foord SM, Marshall FH (1998) Heterodimerization is required for the formation of a functional GABA(B) receptor. Nature 396:679–682
103. Kuner R, Kohr G, Grunewald S, Eisenhardt G, Bach A, Kornau HC (1999) Role of heteromer formation in GABAB receptor function. Science 283:74–77
104. Margeta-Mitrovic M, Jan YN, Jan LY (2001) Function of GB1 and GB2 subunits in G protein coupling of GABA(B) receptors. Proc Natl Acad Sci U S A 98:14649–14654
105. Margeta-Mitrovic M, Jan YN, Jan LY (2001) Ligand-induced signal transduction within heterodimeric GABA(B) receptor. Proc Natl Acad Sci U S A 98:14643–14648
106. Margeta-Mitrovic M, Jan YN, Jan LY (2000) A trafficking checkpoint controls GABA(B) receptor heterodimerization. Neuron 27:97–106
107. Milligan G (2013) The prevalence, maintenance and relevance of GPCR oligomerization. Mol Pharmacol 84(1):158–169
108. Milligan G (2009) The role of dimerisation in the cellular trafficking of G-protein-coupled receptors. Curr Opin Pharmacol 10:23–29
109. Milligan G (2009) G protein-coupled receptor hetero-dimerization: contribution to pharmacology and function. Br J Pharmacol 158:5–14
110. Gonzalez-Maeso J (2011) GPCR oligomers in pharmacology and signaling. Mol Brain 4:20
111. Gonzalez-Maeso J (2014) Family a GPCR heteromers in animal models. Front Pharmacol 5:226
112. Han Y, Moreira IS, Urizar E, Weinstein H, Javitch JA (2009) Allosteric communication between protomers of dopamine class A GPCR dimers modulates activation. Nat Chem Biol 5:688–695
113. Guo W, Urizar E, Kralikova M, Mobarec JC, Shi L, Filizola M, Javitch JA (2008) Dopamine D2 receptors form higher order oligomers at physiological expression levels. Embo J 27:2293–2304
114. Guo W, Shi L, Javitch JA (2003) The fourth transmembrane segment forms the interface of the dopamine D2 receptor homodimer. J Biol Chem 278:4385–4388
115. Guo W, Shi L, Filizola M, Weinstein H, Javitch JA (2005) Crosstalk in G protein-coupled receptors: changes at the transmembrane homodimer interface determine activation. Proc Natl Acad Sci U S A 102:17495–17500
116. López-Giménez JF, Canals M, Pediani JD, Milligan G (2007) The alpha1b-adrenoceptor exists as a higher-order oligomer: effective oligomerization is required for receptor maturation, surface delivery, and function. Mol Pharmacol 71:1015–1029

117. Carrillo JJ, Lopez-Gimenez JF, Milligan G (2004) Multiple interactions between transmembrane helices generate the oligomeric alpha1b-adrenoceptor. Mol Pharmacol 66:1123–1137
118. Kniazeff J, Bessis A-S, Maurel D, Ansanay H, Prézeau L, Pin J-P (2004) Closed state of both binding domains of homodimeric mGlu receptors is required for full activity. Nat Struct Mol Biol 11:706–713
119. El Moustaine D, Granier S, Doumazane E, Scholler P, Rahmeh R, Bron P, Mouillac B, Baneres JL, Rondard P, Pin JP (2012) Distinct roles of metabotropic glutamate receptor dimerization in agonist activation and G-protein coupling. Proc Natl Acad Sci U S A 109:16342–16347
120. Xue L, Rovira X, Scholler P, Zhao H, Liu J, Pin JP, Rondard P (2015) Major ligand-induced rearrangement of the heptahelical domain interface in a GPCR dimer. Nat Chem Biol 11:134–140
121. Whorton MR, Jastrzebska B, Park PS, Fotiadis D, Engel A, Palczewski K, Sunahara RK (2008) Efficient coupling of transducin to monomeric rhodopsin in a phospholipid bilayer. J Biol Chem 283:4387–4394
122. Whorton MR, Bokoch MP, Rasmussen SG, Huang B, Zare RN, Kobilka B, Sunahara RK (2007) A monomeric G protein-coupled receptor isolated in a high-density lipoprotein particle efficiently activates its G protein. Proc Natl Acad Sci U S A 104:7682–7687
123. Kuszak AJ, Pitchiaya S, Anand JP, Mosberg HI, Walter NG, Sunahara RK (2009) Purification and functional reconstitution of monomeric mu-opioid receptors: allosteric modulation of agonist binding by Gi2. J Biol Chem 284:26732–26741
124. Rasmussen SGF, DeVree BT, Zou Y, Kruse AC, Chung KY, Kobilka TS, Thian FS, Chae PS, Pardon E, Calinski D, Mathiesen JM, Shah STA, Lyons JA, Caffrey M, Gellman SH, Steyaert J, Skiniotis G, Weis WI, Sunahara RK, Kobilka BK (2011) Crystal structure of the β2 adrenergic receptor-Gs protein complex. Nature 477(7366):549–555
125. Jordan BA, Devi LA (1999) G-protein-coupled receptor heterodimerization modulates receptor function. Nature 399:697–700
126. Gonzalez-Maeso J, Ang RL, Yuen T, Chan P, Weisstaub NV, Lopez-Gimenez JF, Zhou M, Okawa Y, Callado LF, Milligan G, Gingrich JA, Filizola M, Meana JJ, Sealfon SC (2008) Identification of a serotonin/glutamate receptor complex implicated in psychosis. Nature 452:93–97
127. Moreno JL, Muguruza C, Umali A, Mortillo S, Holloway T, Pilar-Cuellar F, Mocci G, Seto J, Callado LF, Neve RL, Milligan G, Sealfon SC, Lopez-Gimenez JF, Meana JJ, Benson DL, Gonzalez-Maeso J (2012) Identification of three residues essential for 5-HT2A-mGlu2 receptor heteromerization and its psychoactive behavioral function. J Biol Chem 287:44301–44319
128. Rives ML, Vol C, Fukazawa Y, Tinel N, Trinquet E, Ayoub MA, Shigemoto R, Pin JP, Prezeau L (2009) Crosstalk between GABAB and mGlu1a receptors reveals new insight into GPCR signal integration. Embo J 28:2195–2208
129. Delille HK, Becker JM, Burkhardt S, Bleher B, Terstappen GC, Schmidt M, Meyer AH, Unger L, Marek GJ, Mezler M (2012) Heterocomplex formation of 5-HT(2A)-mGlu(2) and its relevance for cellular signaling cascades. Neuropharmacology 62(7):2184–2191
130. Wright RA, Johnson BG, Zhang C, Salhoff C, Kingston AE, Calligaro DO, Monn JA, Schoepp DD, Marek GJ (2012) CNS distribution of metabotropic glutamate 2 and 3 receptors: transgenic mice and [(3)H]LY459477 autoradiography. Neuropharmacology 66:89–98
131. Chan P, Gonzalez-Maeso J, Ruf F, Bishop DF, Hof PR, Sealfon SC (2005) Epsilon-Sarcoglycan immunoreactivity and mRNA expression in mouse brain. J Comp Neurol 482:50–73
132. Chan P, Yuen T, Ruf F, Gonzalez-Maeso J, Sealfon SC (2005) Method for multiplex cellular detection of mRNAs using quantum dot fluorescent in situ hybridization. Nucleic Acids Res 33:e161
133. Fribourg M, Moreno JL, Holloway T, Provasi D, Baki L, Mahajan R, Park G, Adney SK, Hatcher C, Eltit JM, Ruta JD, Albizu L, Li Z, Umali A, Shim J, Fabiato A, Mackerell AD Jr, Brezina V, Sealfon SC, Filizola M, Gonzalez-Maeso J, Logothetis DE (2011) Decoding the signaling of a GPCR heteromeric complex reveals a unifying mechanism of action of antipsychotic drugs. Cell 147:1011–1023
134. Moreno JL, Miranda-Azpiazu P, Garcia-Bea A, Younkin J, Cui M, Kozlenkov A, Ben-Ezra A, Voloudakis G, Fakira AK, Baki L, Ge Y, Georgakopoulos A, Moron JA, Milligan G, Lopez-

Gimenez JF, Robakis NK, Logothetis DE, Meana JJ, Gonzalez-Maeso J (2016) Allosteric signaling through an mGlu2 and 5-HT2A heteromeric receptor complex and its potential contribution to schizophrenia. Sci Signal 9:ra5
135. Jastrzebska B, Fotiadis D, Jang GF, Stenkamp RE, Engel A, Palczewski K (2006) Functional and structural characterization of rhodopsin oligomers. J Biol Chem 281:11917–11922
136. Lopez-Gimenez JF, Canals M, Pediani JD, Milligan G (2007) The alpha1b-adrenoceptor exists as a higher-order oligomer: effective oligomerization is required for receptor maturation, surface delivery, and function. Mol Pharmacol 71:1015–1029
137. Bonaventura J, Navarro G, Casado-Anguera V, Azdad K, Rea W, Moreno E, Brugarolas M, Mallol J, Canela EI, Lluis C, Cortes A, Volkow ND, Schiffmann SN, Ferre S, Casado V (2015) Allosteric interactions between agonists and antagonists within the adenosine A2A receptor-dopamine D2 receptor heterotetramer. Proc Natl Acad Sci U S A 112:E3609–E3618
138. Wu B, Chien EYT, Mol CD, Fenalti G, Liu W, Katritch V, Abagyan R, Brooun A, Wells P, Bi FC, Hamel DJ, Kuhn P, Handel TM, Cherezov V, Stevens RC (2010) Structures of the CXCR4 chemokine GPCR with small-molecule and cyclic peptide antagonists. Science 330:1066–1071
139. Huang J, Chen S, Zhang JJ, Huang XY (2013) Crystal structure of oligomeric beta1-adrenergic G protein-coupled receptors in ligand-free basal state. Nat Struct Mol Biol 20:419–425
140. Manglik A, Kruse AC, Kobilka TS, Thian FS, Mathiesen JM, Sunahara RK, Pardo L, Weis WI, Kobilka BK, Granier S (2012) Crystal structure of the micro-opioid receptor bound to a morphinan antagonist. Nature 485:321–326
141. Gonzalez-Maeso J, Sealfon SC (2009) Psychedelics and schizophrenia. Trends Neurosci 32:225–232
142. Oquendo MA, Russo SA, Underwood MD, Kassir SA, Ellis SP, Mann JJ, Arango V (2006) Higher postmortem prefrontal 5-HT2A receptor binding correlates with lifetime aggression in suicide. Biol Psychiatry 59:235–243
143. Bekinschtein P, Renner MC, Gonzalez MC, Weisstaub N (2013) Role of medial prefrontal cortex serotonin 2A receptors in the control of retrieval of recognition memory in rats. J Neurosci 33:15716–15725
144. Slipczuk L, Tomaiuolo M, Garagoli F, Weisstaub N, Katche C, Bekinschtein P, Medina JH (2013) Attenuating the persistence of fear memory storage using a single dose of antidepressant. Mol Psychiatry 18:7–8
145. Morici JF, Ciccia L, Malleret G, Gingrich JA, Bekinschtein P, Weisstaub NV (2015) Serotonin 2a receptor and serotonin 1a receptor interact within the medial prefrontal cortex during recognition memory in mice. Front Pharmacol 6:298
146. Morris BJ, Cochran SM, Pratt JA (2005) PCP: from pharmacology to modelling schizophrenia. Curr Opin Pharmacol 5:101–106
147. Kristiansen LV, Huerta I, Beneyto M, Meador-Woodruff JH (2007) NMDA receptors and schizophrenia. Curr Opin Pharmacol 7:48–55
148. Umbricht D, Schmid L, Koller R, Vollenweider FX, Hell D, Javitt DC (2000) Ketamine-induced deficits in auditory and visual context-dependent processing in healthy volunteers: implications for models of cognitive deficits in schizophrenia. Arch. Gen. Psychiatry 57:1139–1147
149. Lahti AC, Weiler MA, Tamara Michaelidis BA, Parwani A, Tamminga CA (2001) Effects of ketamine in normal and schizophrenic volunteers. Neuropsychopharmacology 25:455–467
150. Anticevic A, Gancsos M, Murray JD, Repovs G, Driesen NR, Ennis DJ, Niciu MJ, Morgan PT, Surti TS, Bloch MH, Ramani R, Smith MA, Wang XJ, Krystal JH, Corlett PR (2012) NMDA receptor function in large-scale anticorrelated neural systems with implications for cognition and schizophrenia. Proc Natl Acad Sci U S A 109:16720–16725
151. Gouzoulis-Mayfrank E, Heekeren K, Neukirch A, Stoll M, Stock C, Daumann J, Obradovic M, Kovar KA (2006) Inhibition of return in the human 5HT(2A) agonist and NMDA antagonist model of psychosis. Neuropsychopharmacology 31(2):431–441
152. Gouzoulis-Mayfrank E, Heekeren K, Neukirch A, Stoll M, Stock C, Obradovic M, Kovar KA (2006) Psychological effects of (S)-ketamine and N,N-dimethyltryptamine (DMT): a double-blind, cross-over study in healthy volunteers. Pharmacopsychiatry 38:301–311

153. Reissig CJ, Carter LP, Johnson MW, Mintzer MZ, Klinedinst MA, Griffiths RR (2012) High doses of dextromethorphan, an NMDA antagonist, produce effects similar to classic hallucinogens. Psychopharmacology 223(1):1–15
154. Adler CM, Malhotra AK, Elman I, Goldberg T, Egan M, Pickar D, Breier A (1999) Comparison of ketamine-induced thought disorder in healthy volunteers and thought disorder in schizophrenia. Am J Psychiatry 156:1646–1649
155. Umbricht D, Koller R, Vollenweider FX, Schmid L (2002) Mismatch negativity predicts psychotic experiences induced by NMDA receptor antagonist in healthy volunteers. Biol Psychiatry 51:400–406
156. Vollenweider FX, Vollenweider-Scherpenhuyzen MF, Babler A, Vogel H, Hell D (1998) Psilocybin induces schizophrenia-like psychosis in humans via a serotonin-2 agonist action. Neuroreport 9:3897–3902
157. Young BG (1974) A phenomenological comparison of LSD and schizophrenic states. Br J Psychiatry 124:64–74
158. Hoch PH, Cattell JP, Pennes HH (1952) Effects of mescaline and lysergic acid (d-LSD-25). Am J Psychiatry 108:579–584
159. Anastasopoulos G, Photiades H (1962) Effects of LSD-25 on relatives of schizophrenic patients. J Ment Sci 108:95–98
160. Hermle L, Funfgeld M, Oepen G, Botsch H, Borchardt D, Gouzoulis E, Fehrenbach RA, Spitzer M (1992) Mescaline-induced psychopathological, neuropsychological, and neurometabolic effects in normal subjects: experimental psychosis as a tool for psychiatric research. Biol Psychiatry 32:976–991
161. Carhart-Harris RL, Erritzoe D, Williams T, Stone JM, Reed LJ, Colasanti A, Tyacke RJ, Leech R, Malizia AL, Murphy K, Hobden P, Evans J, Feilding A, Wise RG, Nutt DJ (2012) Neural correlates of the psychedelic state as determined by fMRI studies with psilocybin. Proc Natl Acad Sci U S A 109:2138–2143
162. Schmid Y, Enzler F, Gasser P, Grouzmann E, Preller KH, Vollenweider FX, Brenneisen R, Muller F, Borgwardt S, Liechti ME (2015) Acute effects of lysergic acid diethylamide in healthy subjects. Biol Psychiatry 78(8):544–553
163. Quednow BB, Kometer M, Geyer MA, Vollenweider FX (2011) Psilocybin-induced deficits in automatic and controlled inhibition are attenuated by ketanserin in healthy human volunteers. Neuropsychopharmacology 37(3):630–640
164. Bradford AM, Savage KM, Jones DN, Kalinichev M (2010) Validation and pharmacological characterisation of MK-801-induced locomotor hyperactivity in BALB/C mice as an assay for detection of novel antipsychotics. Psychopharmacology (Berl) 212:155–170
165. Yadav PN, Kroeze WK, Farrell MS, Roth BL (2011) Antagonist functional selectivity: 5-HT2A serotonin receptor antagonists differentially regulate 5-HT2A receptor protein level in vivo. J Pharmacol Exp Ther 339:99–105
166. Halberstadt AL, van der Heijden I, Ruderman MA, Risbrough VB, Gingrich JA, Geyer MA, Powell SB (2009) 5-HT(2A) and 5-HT(2C) receptors exert opposing effects on locomotor activity in mice. Neuropsychopharmacology 34:1958–1967
167. Halberstadt AL, Geyer MA (2010) LSD but not lisuride disrupts prepulse inhibition in rats by activating the 5-HT(2A) receptor. Psychopharmacology (Berl) 208:179–189
168. Halberstadt AL, Geyer MA (2011) Multiple receptors contribute to the behavioral effects of indoleamine hallucinogens. Neuropharmacology 61:364–381
169. Halberstadt AL, Geyer MA (2013) Characterization of the head-twitch response induced by hallucinogens in mice: detection of the behavior based on the dynamics of head movement. Psychopharmacology (Berl) 227:727–739
170. Halberstadt AL, Geyer MA (2014) Effects of the hallucinogen 2,5-dimethoxy-4-iodophenethylamine (2C-I) and superpotent N-benzyl derivatives on the head twitch response. Neuropharmacology 77:200–207
171. Halberstadt AL, Koedood L, Powell SB, Geyer MA (2011) Differential contributions of serotonin receptors to the behavioral effects of indoleamine hallucinogens in mice. J Psychopharmacol 25:1548–1561

172. Halberstadt AL, Powell SB, Geyer MA (2013) Role of the 5-HT(2)A receptor in the locomotor hyperactivity produced by phenylalkylamine hallucinogens in mice. Neuropharmacology 70:218–227
173. Halberstadt AL, Geyer MA (2010) LSD but not lisuride disrupts prepulse inhibition in rats by activating the 5-HT(2A) receptor. Psychopharmacology 208:179–189
174. Halberstadt AL, Koedood L, Powell SB, Geyer MA (2011) Differential contributions of serotonin receptors to the behavioral effects of indoleamine hallucinogens in mice. J Psychopharmacol (Oxford) 25:1548–1561
175. Marona-Lewicka D, Thisted RA, Nichols DE (2005) Distinct temporal phases in the behavioral pharmacology of LSD: dopamine D2 receptor-mediated effects in the rat and implications for psychosis. Psychopharmacology (Berl) 180:427–435
176. Marona-Lewicka D, Nichols DE (2007) Further evidence that the delayed temporal dopaminergic effects of LSD are mediated by a mechanism different than the first temporal phase of action. Pharmacol Biochem Behav 87:453–461
177. Nichols CD, Sanders-Bush E (2002) A single dose of lysergic acid diethylamide influences gene expression patterns within the mammalian brain. Neuropsychopharmacology 26:634–642
178. Nichols CD, Garcia EE, Sanders-Bush E (2003) Dynamic changes in prefrontal cortex gene expression following lysergic acid diethylamide administration. Brain Res Mol Brain Res 111:182–188
179. Abbas AI, Yadav PN, Yao W-D, Arbuckle MI, Grant SGN, Caron MG, Roth BL (2009) PSD-95 is essential for hallucinogen and atypical antipsychotic drug actions at serotonin receptors. J Neurosci 29:7124–7136
180. Canal CE, Olaghere da Silva UB, Gresch PJ, Watt EE, Sanders-Bush E, Airey DC (2010) The serotonin 2C receptor potently modulates the head-twitch response in mice induced by a phenethylamine hallucinogen. Psychopharmacology (Berl) 209:163–174
181. Wischhof L, Irrsack E, Dietz F, Koch M (2015) Maternal lipopolysaccharide treatment differentially affects 5-HT and mGlu2/3 receptor function in the adult male and female rat offspring. Neuropharmacology 97:275–288
182. Malkova NV, Gallagher JJ, Yu CZ, Jacobs RE, Patterson PH (2014) Manganese-enhanced magnetic resonance imaging reveals increased DOI-induced brain activity in a mouse model of schizophrenia. Proc Natl Acad Sci U S A 111:E2492–E2500
183. Moreno JL, Holloway T, Albizu L, Sealfon SC, Gonzalez-Maeso J (2011) Metabotropic glutamate mGlu2 receptor is necessary for the pharmacological and behavioral effects induced by hallucinogenic 5-HT2A receptor agonists. Neurosci Lett 493:76–79
184. Karaki S, Becamel C, Murat S, Mannoury la Cour C, Millan MJ, Prezeau L, Bockaert J, Marin P, Vandermoere F (2014) Quantitative phosphoproteomics unravels biased phosphorylation of serotonin 2A receptor at Ser280 by hallucinogenic versus nonhallucinogenic agonists. Mol Cell Proteomics 13:1273–1285
185. Garcia EE, Smith RL, Sanders-Bush E (2007) Role of G(q) protein in behavioral effects of the hallucinogenic drug 1-(2,5-dimethoxy-4-iodophenyl)-2-aminopropane. Neuropharmacology 52:1671–1677
186. Schmid CL, Raehal KM, Bohn LM (2008) Agonist-directed signaling of the serotonin 2A receptor depends on beta-arrestin-2 interactions in vivo. Proc Natl Acad Sci U S A 105:1079–1084
187. Zhai Y, George CA, Zhai J, Nisenbaum ES, Johnson MP, Nisenbaum LK (2003) Group II metabotropic glutamate receptor modulation of DOI-induced c-fos mRNA and excitatory responses in the cerebral cortex. Neuropsychopharmacology 28:45–52
188. Benneyworth MA, Xiang Z, Smith RL, Garcia EE, Conn PJ, Sanders-Bush E (2007) A selective positive allosteric modulator of metabotropic glutamate receptor subtype 2 blocks a hallucinogenic drug model of psychosis. Mol Pharmacol 72:477–484
189. Delille HK, Mezler M, Marek GJ (2013) The two faces of the pharmacological interaction of mGlu2 and 5-HT(2)A—relevance of receptor heterocomplexes and interaction through functional brain pathways. Neuropharmacology 70:296–305

190. Baki L, Fribourg M, Younkin J, Eltit JM, Moreno JL, Park G, Vysotskaya Z, Narahari A, Sealfon SC, Gonzalez-Maeso J, Logothetis DE (2016) Cross-signaling in metabotropic glutamate 2 and serotonin 2A receptor heteromers in mammalian cells. Pflugers Arch 468(5):775–793
191. Takahashi A, Camacho P, Lechleiter JD, Herman B (1999) Measurement of intracellular calcium. Physiol Rev 79:1089–1125
192. Francesconi A, Duvoisin RM (1998) Role of the second and third intracellular loops of metabotropic glutamate receptors in mediating dual signal transduction activation. J Biol Chem 273:5615–5624
193. Vidi P. A., Przybyla J. A., Hu C. D., Watts VJ (2010) Visualization of G protein-coupled receptor (GPCR) interactions in living cells using bimolecular fluorescence complementation (BiFC). Curr Protoc Neurosci Chapter 5:Unit 5.29
194. Medland SE, Jahanshad N, Neale BM, Thompson PM (2014) Whole-genome analyses of whole-brain data: working within an expanded search space. Nat Neurosci 17:791–800
195. Liu W, Downing AC, Munsie LM, Chen P, Reed MR, Ruble CL, Landschulz KT, Kinon BJ, Nisenbaum LK (2010) Pharmacogenetic analysis of the mGlu2/3 agonist LY2140023 monohydrate in the treatment of schizophrenia. Pharmacogenomics J 12(3):246–254
196. Nisenbaum LK, Downing AM, Zhao F, Millen BA, Munsie L, Kinon BJ, Adams DH, Gomez JC, Penny MA (2016) Serotonin 2A receptor SNP rs7330461 association with treatment response to pomaglumetad methionil in patients with schizophrenia. J Pers Med 6(1). https://doi.org/10.3390/jpm6010009
197. Patil ST, Zhang L, Martenyi F, Lowe SL, Jackson KA, Andreev BV, Avedisova AS, Bardenstein LM, Gurovich IY, Morozova MA, Mosolov SN, Neznanov NG, Reznik AM, Smulevich AB, Tochilov VA, Johnson BG, Monn JA, Schoepp DD (2007) Activation of mGlu2/3 receptors as a new approach to treat schizophrenia: a randomized Phase 2 clinical trial. Nat Med 13:1102–1107
198. Kinon BJ, Zhang L, Millen BA, Osuntokun OO, Williams JE, Kollack-Walker S, Jackson K, Kryzhanovskaya L, Jarkova N (2011) A multicenter, inpatient, phase 2, double-blind, placebo-controlled dose-ranging study of LY2140023 monohydrate in patients with DSM-IV schizophrenia. J Clin Psychopharmacol 31:349–355
199. Adams DH, Zhang L, Millen BA, Kinon BJ, Gomez JC (2014) Pomaglumetad methionil (LY2140023 Monohydrate) and aripiprazole in patients with schizophrenia: a Phase 3, multicenter, double-blind comparison. Schizophr Res Treatment 2014:758212
200. Adams DH, Kinon BJ, Baygani S, Millen BA, Velona I, Kollack-Walker S, Walling DP (2013) A long-term, phase 2, multicenter, randomized, open-label, comparative safety study of pomaglumetad methionil (LY2140023 monohydrate) versus atypical antipsychotic standard of care in patients with schizophrenia. BMC Psychiatry 13:143
201. Kurita M, Holloway T, Garcia-Bea A, Kozlenkov A, Friedman AK, Moreno JL, Heshmati M, Golden SA, Kennedy PJ, Takahashi N, Dietz DM, Mocci G, Gabilondo AM, Hanks J, Umali A, Callado LF, Gallitano AL, Neve RL, Shen L, Buxbaum JD, Han MH, Nestler EJ, Meana JJ, Russo SJ, Gonzalez-Maeso J (2012) HDAC2 regulates atypical antipsychotic responses through the modulation of mGlu2 promoter activity. Nat Neurosci 15:1245–1254
202. Kinon BJ, Millen BA, Zhang L, McKinzie DL (2015) Exploratory analysis for a targeted patient population responsive to the metabotropic glutamate 2/3 receptor agonist pomaglumetad methionil in schizophrenia. Biol Psychiatry 78(11):754–762
203. Deep-Soboslay A, Akil M, Martin CE, Bigelow LB, Herman MM, Hyde TM, Kleinman JE (2005) Reliability of psychiatric diagnosis in postmortem research. Biol Psychiatry 57:96–101
204. Deep-Soboslay A, Benes FM, Haroutunian V, Ellis JK, Kleinman JE, Hyde TM (2011) Psychiatric brain banking: three perspectives on current trends and future directions. Biol Psychiatry 69:104–112
205. Reynolds GP, Rossor MN, Iversen LL (1983) Preliminary studies of human cortical 5-HT2 receptors and their involvement in schizophrenia and neuroleptic drug action. J Neural Transm Suppl 18:273–277

206. Mita T, Hanada S, Nishino N, Kuno T, Nakai H, Yamadori T, Mizoi Y, Tanaka C (1986) Decreased serotonin S2 and increased dopamine D2 receptors in chronic schizophrenics. Biol Psychiatry 21:1407–1414
207. Laruelle M, Abi-Dargham A, Casanova MF, Toti R, Weinberger DR, Kleinman JE (1993) Selective abnormalities of prefrontal serotonergic receptors in schizophrenia. A postmortem study. Arch Gen Psychiatry 50:810–818
208. Burnet PW, Eastwood SL, Harrison PJ (1996) 5-HT1A and 5-HT2A receptor mRNAs and binding site densities are differentially altered in schizophrenia. Neuropsychopharmacology 15:442–455
209. Dean B, Hayes W (1996) Decreased frontal cortical serotonin2A receptors in schizophrenia. Schizophr Res 21:133–139
210. Dean B, Hayes W, Hill C, Copolov D (1998) Decreased serotonin2A receptors in Brodmann's area 9 from schizophrenic subjects. A pathological or pharmacological phenomenon? Mol Chem Neuropathol 34:133–145
211. Dean B, Hussain T, Hayes W, Scarr E, Kitsoulis S, Hill C, Opeskin K, Copolov DL (1999) Changes in serotonin2A and GABA(A) receptors in schizophrenia: studies on the human dorsolateral prefrontal cortex. J Neurochem 72:1593–1599
212. Matsumoto I, Inoue Y, Iwazaki T, Pavey G, Dean B (2005) 5-HT2A and muscarinic receptors in schizophrenia: a postmortem study. Neurosci Lett 379:164–168
213. Dean B, Crossland N, Boer S, Scarr E (2008) Evidence for altered post-receptor modulation of the serotonin 2a receptor in schizophrenia. Schizophr Res 104:185–197
214. Kang K, Huang XF, Wang Q, Deng C (2009) Decreased density of serotonin 2A receptors in the superior temporal gyrus in schizophrenia—a postmortem study. Prog Neuropsychopharmacol Biol Psychiatry 33:867–871
215. Dean B, Hayes W, Opeskin K, Naylor L, Pavey G, Hill C, Keks N, Copolov DL (1996) Serotonin2 receptors and the serotonin transporter in the schizophrenic brain. Behav Brain Res 73:169–175
216. Marazziti D, Giannaccini G, Giromella A, Betti L, Pesce D, Nardi I, Rossi A, Lucacchini A, Cassano GB (2003) [3H]-ketanserin binding sites in different psychiatric disorders. Neurochem Int 42:511–516
217. Muguruza C, Moreno JL, Umali A, Callado LF, Meana JJ, Gonzalez-Maeso J (2013) Dysregulated 5-HT(2A) receptor binding in postmortem frontal cortex of schizophrenic subjects. Eur Neuropsychopharmacol 23:852–864
218. Bennett JP Jr, Enna SJ, Bylund DB, Gillin JC, Wyatt RJ, Snyder SH (1979) Neurotransmitter receptors in frontal cortex of schizophrenics. Arch Gen Psychiatry 36:927–934
219. Whitaker PM, Crow TJ, Ferrier IN (1981) Tritiated LSD binding in frontal cortex in schizophrenia. Arch Gen Psychiatry 38:278–280
220. Joyce JN, Shane A, Lexow N, Winokur A, Casanova MF, Kleinman JE (1993) Serotonin uptake sites and serotonin receptors are altered in the limbic system of schizophrenics. Neuropsychopharmacology 8:315–336
221. Gurevich EV, Joyce JN (1997) Alterations in the cortical serotonergic system in schizophrenia: a postmortem study. Biol Psychiatry 42:529–545
222. Erritzoe D, Rasmussen H, Kristiansen KT, Frokjaer VG, Haugbol S, Pinborg L, Baare W, Svarer C, Madsen J, Lublin H, Knudsen GM, Glenthoj BY (2008) Cortical and subcortical 5-HT(2A) receptor binding in neuroleptic-naive first-episode schizophrenic patients. Neuropsychopharmacology 33:2435–2441
223. Rasmussen H, Erritzoe D, Andersen R, Ebdrup BH, Aggernaes B, Oranje B, Kalbitzer J, Madsen J, Pinborg LH, Baare W, Svarer C, Lublin H, Knudsen GM, Glenthoj B (2010) Decreased frontal serotonin2A receptor binding in antipsychotic-naive patients with first-episode schizophrenia. Arch Gen Psychiatry 67:9–16
224. Arora RC, Meltzer HY (1991) Serotonin2 (5-HT2) receptor binding in the frontal cortex of schizophrenic patients. J Neural Transm Gen Sect 85:19–29
225. Pralong D, Tomaskovic-Crook E, Opeskin K, Copolov D, Dean B (2000) Serotonin(2A) receptors are reduced in the planum temporale from subjects with schizophrenia. Schizophr Res 44:35–45

226. Huot P, Johnston TH, Darr T, Hazrati LN, Visanji NP, Pires D, Brotchie JM, Fox SH (2010) Increased 5-HT2A receptors in the temporal cortex of parkinsonian patients with visual hallucinations. Mov Disord 25:1399–1408
227. Ettrup A, da Cunha-Bang S, McMahon B, Lehel S, Dyssegaard A, Skibsted AW, Jorgensen LM, Hansen M, Baandrup AO, Bache S, Svarer C, Kristensen JL, Gillings N, Madsen J, Knudsen GM (2014) Serotonin 2A receptor agonist binding in the human brain with [(1)(1) C]Cimbi-36. J Cereb Blood Flow Metab 34:1188–1196
228. Ettrup A, Svarer C, McMahon B, da Cunha-Bang S, Lehel S, Moller K, Dyssegaard A, Ganz M, Beliveau V, Jorgensen LM, Gillings N, Knudsen GM (2016) Serotonin 2A receptor agonist binding in the human brain with [C]Cimbi-36: test-retest reproducibility and head-to-head comparison with the antagonist [F]altanserin. Neuroimage 130:167–174
229. Joiner TE Jr, Brown JS, Wingate LR (2005) The psychology and neurobiology of suicidal behavior. Annu Rev Psychol 56:287–314
230. Owen F, Cross AJ, Crow TJ, Deakin JF, Ferrier IN, Lofthouse R, Poulter M (1983) Brain 5-HT-2 receptors and suicide. Lancet 2:1256
231. Crow TJ, Cross AJ, Cooper SJ, Deakin JF, Ferrier IN, Johnson JA, Joseph MH, Owen F, Poulter M, Lofthouse R et al (1984) Neurotransmitter receptors and monoamine metabolites in the brains of patients with Alzheimer-type dementia and depression, and suicides. Neuropharmacology 23:1561–1569
232. Cheetham SC, Crompton MR, Katona CL, Horton RW (1988) Brain 5-HT2 receptor binding sites in depressed suicide victims. Brain Res 443:272–280
233. Arranz B, Eriksson A, Mellerup E, Plenge P, Marcusson J (1994) Brain 5-HT1A, 5-HT1D, and 5-HT2 receptors in suicide victims. Biol Psychiatry 35:457–463
234. Lowther S, De Paermentier F, Crompton MR, Katona CL, Horton RW (1994) Brain 5-HT2 receptors in suicide victims: violence of death, depression and effects of antidepressant treatment. Brain Res 642:281–289
235. Stockmeier CA, Dilley GE, Shapiro LA, Overholser JC, Thompson PA, Meltzer HY (1997) Serotonin receptors in suicide victims with major depression. Neuropsychopharmacology 16:162–173
236. Rosel P, Arranz B, San L, Vallejo J, Crespo JM, Urretavizcaya M, Navarro MA (2000) Altered 5-HT(2A) binding sites and second messenger inositol trisphosphate (IP(3)) levels in hippocampus but not in frontal cortex from depressed suicide victims. Psychiatry Res 99:173–181
237. Muguruza C, Miranda-Azpiazu P, Diez-Alarcia R, Morentin B, Gonzalez-Maeso J, Callado LF, Meana JJ (2014) Evaluation of 5-HT2A and mGlu2/3 receptors in postmortem prefrontal cortex of subjects with major depressive disorder: effect of antidepressant treatment. Neuropharmacology 86:311–318
238. Hrdina PD, Demeter E, Vu TB, Sotonyi P, Palkovits M (1993) 5-HT uptake sites and 5-HT2 receptors in brain of antidepressant-free suicide victims/depressives: increase in 5-HT2 sites in cortex and amygdala. Brain Res 614:37–44
239. Gross-Isseroff R, Salama D, Israeli M, Biegon A (1990) Autoradiographic analysis of [3H] ketanserin binding in the human brain postmortem: effect of suicide. Brain Res 507:208–215
240. Turecki G, Briere R, Dewar K, Antonetti T, Lesage AD, Seguin M, Chawky N, Vanier C, Alda M, Joober R, Benkelfat C, Rouleau GA (1999) Prediction of level of serotonin 2A receptor binding by serotonin receptor 2A genetic variation in postmortem brain samples from subjects who did or did not commit suicide. Am J Psychiatry 156:1456–1458
241. Moreno JL, Seto J, Hanks JB, Gonzalez-Maeso J (2015) Techniques for the study of GPCR heteromerization in living cells and animal models. Neuromethods 95:21–36
242. Strange PG (1998) Three-state and two-state models. Trends Pharmacol Sci 19:85–86
243. Strange PG (2008) Agonist binding, agonist affinity and agonist efficacy at G protein-coupled receptors. Br J Pharmacol 153:1353–1363
244. Stefansson H, Rujescu D, Cichon S, Pietilainen OP, Ingason A, Steinberg S, Fossdal R, Sigurdsson E, Sigmundsson T, Buizer-Voskamp JE, Hansen T, Jakobsen KD, Muglia P, Francks C, Matthews PM, Gylfason A, Halldorsson BV, Gudbjartsson D, Thorgeirsson

TE, Sigurdsson A, Jonasdottir A, Jonasdottir A, Bjornsson A, Mattiasdottir S, Blondal T, Haraldsson M, Magnusdottir BB, Giegling I, Moller HJ, Hartmann A, Shianna KV, Ge D, Need AC, Crombie C, Fraser G, Walker N, Lonnqvist J, Suvisaari J, Tuulio-Henriksson A, Paunio T, Toulopoulou T, Bramon E, Di Forti M, Murray R, Ruggeri M, Vassos E, Tosato S, Walshe M, Li T, Vasilescu C, Muhleisen TW, Wang AG, Ullum H, Djurovic S, Melle I, Olesen J, Kiemeney LA, Franke B, Sabatti C, Freimer NB, Gulcher JR, Thorsteinsdottir U, Kong A, Andreassen OA, Ophoff RA, Georgi A, Rietschel M, Werge T, Petursson H, Goldstein DB, Nothen MM, Peltonen L, Collier DA, St Clair D, Stefansson K, Kahn RS, Linszen DH, van Os J, Wiersma D, Bruggeman R, Cahn W, de Haan L, Krabbendam L, Myin-Germeys I (2008) Large recurrent microdeletions associated with schizophrenia. Nature 455:232–236

245. Stefansson H, Ophoff RA, Steinberg S, Andreassen OA, Cichon S, Rujescu D, Werge T, Pietilainen OP, Mors O, Mortensen PB, Sigurdsson E, Gustafsson O, Nyegaard M, Tuulio-Henriksson A, Ingason A, Hansen T, Suvisaari J, Lonnqvist J, Paunio T, Borglum AD, Hartmann A, Fink-Jensen A, Nordentoft M, Hougaard D, Norgaard-Pedersen B, Bottcher Y, Olesen J, Breuer R, Moller HJ, Giegling I, Rasmussen HB, Timm S, Mattheisen M, Bitter I, Rethelyi JM, Magnusdottir BB, Sigmundsson T, Olason P, Masson G, Gulcher JR, Haraldsson M, Fossdal R, Thorgeirsson TE, Thorsteinsdottir U, Ruggeri M, Tosato S, Franke B, Strengman E, Kiemeney LA, Melle I, Djurovic S, Abramova L, Kaleda V, Sanjuan J, de Frutos R, Bramon E, Vassos E, Fraser G, Ettinger U, Picchioni M, Walker N, Toulopoulou T, Need AC, Ge D, Yoon JL, Shianna KV, Freimer NB, Cantor RM, Murray R, Kong A, Golimbet V, Carracedo A, Arango C, Costas J, Jonsson EG, Terenius L, Agartz I, Petursson H, Nothen MM, Rietschel M, Matthews PM, Muglia P, Peltonen L, St Clair D, Goldstein DB, Stefansson K, Collier DA (2009) Common variants conferring risk of schizophrenia. Nature 460:744–747

246. Purcell SM, Wray NR, Stone JL, Visscher PM, O'Donovan MC, Sullivan PF, Sklar P (2009) Common polygenic variation contributes to risk of schizophrenia and bipolar disorder. Nature 460:748–752

247. Walsh T, McClellan JM, McCarthy SE, Addington AM, Pierce SB, Cooper GM, Nord AS, Kusenda M, Malhotra D, Bhandari A, Stray SM, Rippey CF, Roccanova P, Makarov V, Lakshmi B, Findling RL, Sikich L, Stromberg T, Merriman B, Gogtay N, Butler P, Eckstrand K, Noory L, Gochman P, Long R, Chen Z, Davis S, Baker C, Eichler EE, Meltzer PS, Nelson SF, Singleton AB, Lee MK, Rapoport JL, King MC, Sebat J (2008) Rare structural variants disrupt multiple genes in neurodevelopmental pathways in schizophrenia. Science 320:539–543

248. Consortium TIS (2008) Rare chromosomal deletions and duplications increase risk of schizophrenia. Nature 455:237–241

249. Schizophrenia Working Group of the Psychiatric Genomics Consortium (2014) Biological insights from 108 schizophrenia-associated genetic loci. Nature 511:421–427

250. Purcell SM, Moran JL, Fromer M, Ruderfer D, Solovieff N, Roussos P, O'Dushlaine C, Chambert K, Bergen SE, Kahler A, Duncan L, Stahl E, Genovese G, Fernandez E, Collins MO, Komiyama NH, Choudhary JS, Magnusson PK, Banks E, Shakir K, Garimella K, Fennell T, DePristo M, Grant SG, Haggarty SJ, Gabriel S, Scolnick EM, Lander ES, Hultman CM, Sullivan PF, McCarroll SA, Sklar P (2014) A polygenic burden of rare disruptive mutations in schizophrenia. Nature 506:185–190

251. Sekar A, Bialas AR, de Rivera H, Davis A, Hammond TR, Kamitaki N, Tooley K, Presumey J, Baum M, Van Doren V, Genovese G, Rose SA, Handsaker RE, Schizophrenia Working Group of the Psychiatric Genomics Consortium, Daly MJ, Carroll MC, Stevens B, McCarroll SA (2016) Schizophrenia risk from complex variation of complement component 4. Nature 530:177–183

252. Lewis DA, Levitt P (2002) Schizophrenia as a disorder of neurodevelopment. Annu Rev Neurosci 25:409–432

253. Cardno AG, Gottesman II (2000) Twin studies of schizophrenia: from bow-and-arrow concordances to star wars Mx and functional genomics. Am J Med Genet 97:12–17

254. Gottesman II, Erlenmeyer-Kimling L (2001) Family and twin strategies as a head start in defining prodromes and endophenotypes for hypothetical early-interventions in schizophrenia. Schizophr Res 51:93–102

255. Brown AS, Begg MD, Gravenstein S, Schaefer CA, Wyatt RJ, Bresnahan M, Babulas VP, Susser ES (2004) Serologic evidence of prenatal influenza in the etiology of schizophrenia. Arch Gen Psychiatry 61:774–780
256. Menninger KA (1919) Psychoses associated with influenza. J. Am. Med. Assoc. 72:235–241
257. Yudofsky SC (2009) Contracting schizophrenia: lessons from the influenza epidemic of 1918-1919. JAMA 301:324–326
258. Brown AS, Cohen P, Harkavy-Friedman J, Babulas V, Malaspina D, Gorman JM, Susser ESAE (2001) Bennett research award. Prenatal rubella, premorbid abnormalities, and adult schizophrenia. Biol Psychiatry 49:473–486
259. Sorensen HJ, Mortensen EL, Reinisch JM, Mednick SA (2009) Association between prenatal exposure to bacterial infection and risk of schizophrenia. Schizophr Bull 35:631–637
260. Brown AS, Schaefer CA, Quesenberry CP Jr, Liu L, Babulas VP, Susser ES (2005) Maternal exposure to toxoplasmosis and risk of schizophrenia in adult offspring. Am J Psychiatry 162:767–773
261. van Os J, Selten JP (1998) Prenatal exposure to maternal stress and subsequent schizophrenia. The May 1940 invasion of The Netherlands. Br J Psychiatry 172:324–326
262. Malaspina D, Corcoran C, Kleinhaus KR, Perrin MC, Fennig S, Nahon D, Friedlander Y, Harlap S (2008) Acute maternal stress in pregnancy and schizophrenia in offspring: a cohort prospective study. BMC Psychiatry 8:71
263. Susser E, St Clair D, He L (2008) Latent effects of prenatal malnutrition on adult health: the example of schizophrenia. Ann N Y Acad Sci 1136:185–192
264. Khashan AS, Abel KM, McNamee R, Pedersen MG, Webb RT, Baker PN, Kenny LC, Mortensen PB (2008) Higher risk of offspring schizophrenia following antenatal maternal exposure to severe adverse life events. Arch Gen Psychiatry 65:146–152
265. Moreno JL, Kurita M, Holloway T, Lopez J, Cadagan R, Martinez-Sobrido L, Garcia-Sastre A, Gonzalez-Maeso J (2011) Maternal influenza viral infection causes schizophrenia-like alterations of 5-HT2A and mGlu2 receptors in the adult offspring. J Neurosci 31:1863–1872
266. Akatsu S, Ishikawa C, Takemura K, Ohtani A, Shiga T (2015) Effects of prenatal stress and neonatal handling on anxiety, spatial learning and serotonergic system of male offspring mice. Neurosci Res 101:15–23
267. Wang Y, Ma Y, Hu J, Cheng W, Jiang H, Zhang X, Li M, Ren J, Li X (2015) Prenatal chronic mild stress induces depression-like behavior and sex-specific changes in regional glutamate receptor expression patterns in adult rats. Neuroscience 301:363–374
268. Holloway T, Moreno JL, Umali A, Rayannavar V, Hodes GE, Russo SJ, Gonzalez-Maeso J (2013) Prenatal stress induces schizophrenia-like alterations of serotonin 2A and metabotropic glutamate 2 receptors in the adult offspring: role of maternal immune system. J Neurosci 33:1088–1098
269. Pershing ML, Bortz DM, Pocivavsek A, Fredericks PJ, Jorgensen CV, Vunck SA, Leuner B, Schwarcz R, Bruno JP (2015) Elevated levels of kynurenic acid during gestation produce neurochemical, morphological, and cognitive deficits in adulthood: implications for schizophrenia. Neuropharmacology 90:33–41
270. Nasca C, Zelli D, Bigio B, Piccinin S, Scaccianoce S, Nistico R, McEwen BS (2015) Stress dynamically regulates behavior and glutamatergic gene expression in hippocampus by opening a window of epigenetic plasticity. Proc Natl Acad Sci U S A 112:14960–14965
271. Maple AM, Zhao X, Elizalde DI, McBride AK, Gallitano AL (2015) Htr2a expression responds rapidly to environmental stimuli in an Egr3-Dependent manner. ACS Chem Neurosci 6(7):1137–1142
272. Klein AB, Ultved L, Adamsen D, Santini MA, Tobena A, Fernandez-Teruel A, Flores P, Moreno M, Cardona D, Knudsen GM, Aznar S, Mikkelsen JD (2014) 5-HT(2A) and mGlu2 receptor binding levels are related to differences in impulsive behavior in the Roman Low- (RLA) and High- (RHA) avoidance rat strains. Neuroscience 263:36–45
273. Jorgensen CV, Jacobsen JP, Caron MG, Klein AB, Knudsen GM, Mikkelsen JD (2013) Cerebral 5-HT2A receptor binding, but not mGluR2, is increased in tryptophan hydroxylase 2 decrease-of-function mice. Neurosci Lett 555:118–122

274. Chiu HY, Chan MH, Lee MY, Chen ST, Zhan ZY, Chen HH (2014) Long-lasting alterations in 5-HT2A receptor after a binge regimen of methamphetamine in mice. Int J Neuropsychopharmacol 17(10):1647–1658
275. Chen YW, Lin HC, Ng MC, Hsiao YH, Wang CC, Gean PW, Chen PS (2014) Activation of mGluR2/3 underlies the effects of N-acetylcystein on amygdala-associated autism-like phenotypes in a valproate-induced rat model of autism. Front Behav Neurosci 8:219
276. Zhou R, Chen F, Feng X, Zhou L, Li Y, Chen L (2015) Perinatal exposure to low-dose of bisphenol a causes anxiety-like alteration in adrenal axis regulation and behaviors of rat offspring: a potential role for metabotropic glutamate 2/3 receptors. J Psychiatr Res 64:121–129
277. Holloway T, Gonzalez-Maeso J (2015) Epigenetic mechanisms of serotonin signaling. ACS Chem Neurosci 6:1099–1109

Schizophrenia

Anna Castañé and Albert Adell

Abstract The observations that hallucinogenic drugs are in fact serotonin 5-HT_{2A} receptor agonists and atypical antipsychotic drugs antagonize 5-HT_{2A} receptors in addition to dopamine D_2-like receptors, led to envisage a close relationship between schizophrenia and serotonin transmission. Post mortem studies have shown diminished 5-HT_{2A} receptor binding in brain tissue from people with schizophrenia. If these changes also occur in schizophrenia, the decreased 5-HT_{2A} binding might be a compensatory effect resulting from increased cortical serotonergic transmission. Therefore, schizophrenic personality was associated to an excess of brain serotonin. Overall, such data suggests that 5-HT systems may play a role in the etiology and therapy of some aspects of schizophrenia. Nevertheless, there is no simple hypothesis of schizophrenia involving a single transmitter dysfunction in the brain. Thus, although initial views suggest increased serotonergic and dopaminergic transmission over 5-HT_{2A} and D_2 receptors in schizophrenia, current hypotheses for both etiology and treatment of the disease also implicate 5-HT_{1A} and D_2 receptor partial agonism in the development of novel antipsychotics with a better therapeutic profile and fewer adverse effects. In this chapter we review the role that serotonin and 5-HT_{2A} receptors play in schizophrenia as well as the most relevant schizophrenia-related behavioral effects induced by hallucinogen drugs in rodents, focusing on those that involve a 5-HT_{2A} receptor mechanism.

Keywords Serotonin • 5-HT_{2A} receptors • Schizophrenia • Atypical antipsychotic • Prefrontal cortex

---

A. Castañé (✉)
CSIC-Institut d'Investigacions Biomèdiques de Barcelona-CIBERSAM-IDIBAPS, Barcelona, Spain
e-mail: acfnqi@iibb.csic.es

A. Adell
Institute of Biomedicine and Biotechnology of Cantabria, IBBTEC (CSIC-University of Cantabria), Santander, Spain
e-mail: albert.adell@unican.es

B.P. Guiard, G. Di Giovanni (eds.), *5-HT_{2A} Receptors in the Central Nervous System*, The Receptors 32, https://doi.org/10.1007/978-3-319-70474-6_8

The Serotonin Hypothesis of Schizophrenia

The initial serotonin (5-HT) hypothesis of schizophrenia emanates from early studies showing that the psychedelic compound lysergic acid diethylamide (LSD), a drug with structural similarities to serotonin and high affinity for 5-HT_{2A} receptors, showed hallucinogenic properties that resembled, at least in part, some of the symptoms of schizophrenia. However, although LSD was first postulated to model psychoses, the effects of hallucinogens are qualitatively different from the mental condition seen in schizophrenia [1]. Initially, LSD antagonized the effects of 5-HT in smooth muscle preparations. Hence, the initial hypothesis related schizophrenia to a brain 5-HT deficiency. Later work, however, suggested that indoleamine and phenethylamine classes of psychedelic hallucinogens might share a common mechanism of action, i.e. potent 5-HT_{2A} receptor agonism. Consequently, it was concluded for the first time that the schizophrenic personality resulted from an excess of brain 5-HT [2]. Moreover, few studies have shown that a number of direct and indirect 5-HT agonists can sometimes exacerbate the symptoms of schizophrenia [3, 4], see also [5] for review.

In 1963, Carlsson and Lindqvist [6] proposed that an overactive dopaminergic system was responsible for the positive symptoms of schizophrenia (hallucinations, delusions). This hypothesis was further supported by the correlation found between clinical effective doses of antipsychotic drugs and their potency to block dopamine D_2 receptors [7, 8]. Thus, by mid-seventies, the 5-HT hypothesis was completely eclipsed by the so-called dopamine hypothesis of schizophrenia. In later years, however, the 5-HT hypothesis regained momentum. One reason for this renewed interest in the role of 5-HT was the introduction of the first atypical antipsychotic drug, clozapine, and the realization that the most interesting advantage of this drug was its better therapeutic profile for positive and negative symptoms in comparison with chlorpromazine [9]. Clozapine is a weaker D_2-like blocker and more potent 5-HT_{2A} receptor blocker. Therefore, a highly active serotonergic transmission was again implicated in schizophrenia [10]. Overall, such data suggested that 5-HT systems may play a role in the etiology and therapy of some aspects of schizophrenia. Nevertheless, there is no simple hypothesis of schizophrenia involving a single transmitter dysfunction in the brain. Rather, it is now considered that this illness affects different transmitter pathways in different brain regions, which results in an impaired circuit function that is responsible for the multiple set of symptoms observed. Thus, a challenge will be to find treatments to target dysfunctional brain circuits without disrupting those that are functioning normally.

Role of 5-HT_{2A} Receptors in Schizophrenia

As mentioned above, the first hint that 5-HT_{2A} receptors have a role in schizophrenia comes from the observations that hallucinogenic drugs are in fact 5-HT_{2A} receptor agonists and atypical antipsychotic drugs have been designed to antagonize 5-HT_{2A}

receptors in addition to dopamine D_2-like receptors. However, 5-HT_{2A} receptor agonists usually evoke visual hallucinations, whereas those associated with schizophrenia are commonly auditory [11].

5-HT_{2A} receptors are particularly abundant in the pyramidal neurons from cortical layer V [12, 13], both in rats and rhesus monkeys (Fig. 1).

However, pyramidal cells are not the only cortical cell type enriched in 5-HT_{2A} receptors. In fact, 5-HT_{2A} receptor mRNA is also expressed in γ-aminobutyric acid (GABA) interneurons [14] (Fig. 2).

Post mortem studies have shown compelling, diminished 5-HT_{2A} receptor binding in brain tissue from people with schizophrenia [15–17] whereas more controversial results have been gathered from studies in living human brains. Intriguingly, the only two studies so far reporting significant 15–16% decreases in cortical 5-HT_{2A} receptors in schizophrenia [18, 19] were conducted in the youngest cohorts, which seems to suggest that decreases in 5-HT_{2A} receptors may be a potential prodromal marker [20]. A question that remains unanswered is, what causes this reduced 5-HT_{2A} binding? Although a hyperactive serotonergic transmission in the prefrontal cortex (PFC) has been implicated in schizophrenia [10], the precise role of cortical serotonin on this effect is not fully understood. It has been shown in an animal model of the illness that antagonists of the NMDA glutamate receptor increased serotonin release in PFC [21–23], an effect prevented by atypical but not typical antipsychotic drugs [23, 24]. If these changes also occur in schizophrenia, the decreased 5-HT_{2A} binding might be a compensatory effect resulting from increased cortical serotonergic transmission. Interestingly, although negative symptoms are usually thought to be associated with impaired serotonergic transmission in schizophrenia [25], the study by Rasmussen and colleagues [19] found a significant negative correlation between 5-HT_{2A} binding in the frontal cortex and positive symptoms in the group of male patients. However, similar correlations were not found between 5-HT_{2A} receptor levels and measures of working memory, problem-solving, or attention. Studies with larger populations are thus warranted to draw unequivocal conclusions on that matter. Also, an important point not to be missed is that PET measures were performed, for obvious reasons, in the resting state. Differences in serotonergic changes may thus exist, depending on the pathological phase of the illness.

From a pharmacological point of view, it is worth noting that blockade of 5-HT_{2A} receptors alone does not confer antipsychotic activity. Thus, clinical evaluation of the selective 5-HT_{2A} receptor antagonist, M100907, failed to demonstrate therapeutic efficacy [26], despite the high level of 5-HT_{2A} receptor occupancy achieved in frontal cortex [27]. Therefore, 5-HT_{2A} receptor antagonism might be a condition necessary, but not sufficient to achieve a clinical antipsychotic effect. However, it has been suggested that the improved clinical profile of atypical antipsychotic drugs is related to an increased 5-HT receptor antagonism [28]. In addition, recent years have witnessed a resurgence of interest in hallucinogenic drugs as models of schizophrenia, particularly its acute form [29]. Moreover, it has been reported that psilocybin, an indole chemically akin to serotonin produces schizophrenic-like symptoms

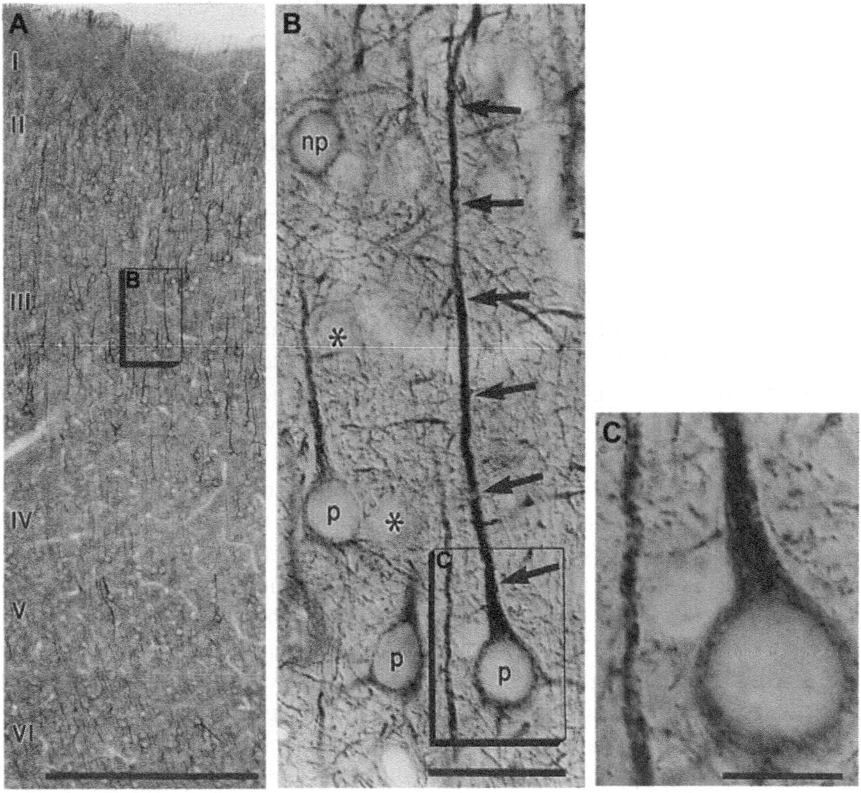

Fig. 1 Distribution of 5-HT$_{2A}$ receptors in prefrontal cortex of rhesus monkeys (*Macaca mulatta*). The receptor immunoreactivity in area 46 of the prefrontal cortex is showed in (**a**). 5-HT$_{2A}$ labeling is found in most (if not all) pyramidal neurons throughout cortical layers II and III and layers V and VI, including their dendritic branches in layer I. Receptor labeling is weak in layer IV because only moderately labeled interneurons and some *en passant* apical dendrites of layer V pyramids are present in this layer. The boxed area in layer III (enlarged in **b** and **c**) demonstrates receptor-labeled pyramidal cells (p), unlabeled (*asterisks*) and labeled nonpyramidal (np) cells, and receptor-positive fine processes. Scale bars: 0.5 mm in (**a**), 50 μm in (**b**), and 20 μm in (**c**). Taken with permission from RL Jakab and PS Goldman-Rakic' 5-Hydroxytryptamine2A serotonin receptors in the primate cerebral cortex: Possible site of action of hallucinogenic and antipsychotic drugs in pyramidal cell apical dendrites, published in Proceedings of the National Academy of Sciences USA, 95(2): 735–740 (1998). Copyright © 1998, National Academy of Sciences, USA

that are reversed by the serotonin antagonist ketanserin, but not by haloperidol, a dopamine antagonist [30]. Based on such observations, it has been suggested that serotonin receptors may be overactive in schizophrenia, an idea that is consonant with the serotonin hypothesis proposed by Woolley in 1962 [2].

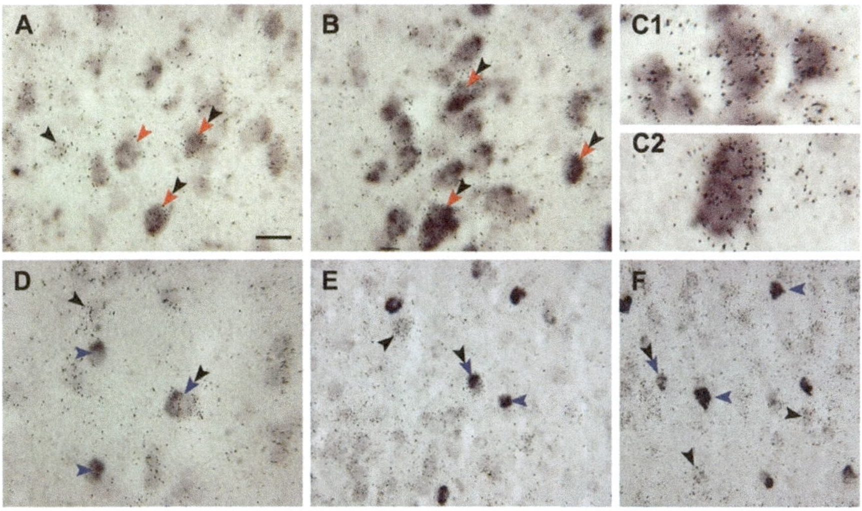

Fig. 2 Expression of 5-HT$_{2A}$ receptors in pyramidal and GABAergic cells of the rat cortex. Upper row (**a–c**): low and high level magnification photomicrographs showing the presence of 5-HT$_{2A}$ receptor mRNA (^{33}P–labeled oligonucleotides) in pyramidal cells, identified by the presence of vGluT1 mRNA (Dig-labeled oligonucleotides). **a** and **b** show, respectively, the presence of abundant cells expressing both transcripts in the prelimbic area and tenia tecta. *Red arrowheads* mark some cells positive for vGluT1 mRNA, *black arrowheads* mark cells positive for 5-HT$_{2A}$ receptor mRNA. Double labeled cells are marked by both *arrowheads*. A large number of glutamatergic cells expressed the 5-HT$_{2A}$ receptor mRNA, as denoted by the double labeling. Note also the presence of non-glutamatergic cells expressing the 5-HT$_{2A}$ receptor mRNA (*black arrowhead*). **c1** and **c2** show individual cells expressing both transcripts in the piriform cortex (**c1**) and prelimbic area (**c2**). Lower row (**d–f**): as opposed to pyramidal neurons, only a small percentage of GAD-containing cells (~20% on average) expressed the 5-HT$_{2A}$ receptor transcript. *Blue arrowheads* mark cells positive for GAD mRNA and *black arrowheads* mark cells positive for 5-HT$_{2A}$ receptor mRNA. Some double labeled cells are marked by both arrowheads. D shows a field in the prelimbic area containing a GABAergic neuron expressing the 5-HT$_{2A}$ receptor mRNA. **e** and **f** show two different fields, in the piriform cortex and prelimbic area, respectively, showing abundant non-GABAergic neurons expressing the 5-HT$_{2A}$ receptors. Occasional GABAergic cells expressing the 5-HT$_{2A}$ receptor were observed (*double arrowhead*). Scale bars: 20 μm (**a, b, d**); 50 μm (**e, f**); 10 μm (**c**). Taken with permission from N Santana, A Bortolozzi, J Serrats, G Mengod and F Artigas' Expression of Serotonin$_{1A}$ and Serotonin$_{2A}$ Receptors in Pyramidal and GABAergic Neurons of the Rat Prefrontal Cortex, published in Cerebral Cortex, 14:1100–1109 (2004). Copyright © 2004, Oxford University Press

Animal Models of Schizophrenia Impacting on 5-HT$_{2A}$ Receptors

Animal models of schizophrenia are very useful in basic, preclinical neuroscience research and in antipsychotic drug development. However, they have obvious limitations due to the unknown etiology of the disease, the high inter-individual variability of symptoms and the inability to reproduce subjective symptoms of the illness. Therefore, they are intended to assess specific endophenotypes (behavioral traits) instead of modeling the disease as a whole [31–33]. Moreover they are aimed at providing tools with high predictive validity.

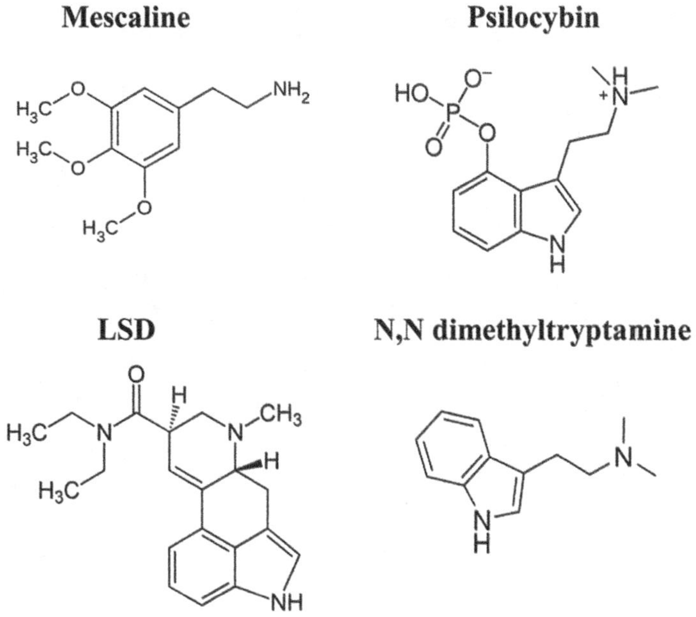

Fig. 3 Chemical structure of serotonergic hallucinogens

Serotonergic hallucinogens such as mescaline, psilocybin, lysergic acid diethylamide (LSD) and *N,N*-dimethyltryptamine (DMT) were one of the primary pharmacological agents proposed to produce effects bearing resemblance to some symptoms of schizophrenia. These compounds can produce important sensory distortions and visual hallucinations, but less frequently evoke severe delusions or auditory hallucinations, which are core features of the illness. From a pharmacological point of view, serotonergic hallucinogens, albeit having different chemical structures (Fig. 3), they all interact with 5-HT_{2A} receptors. There is strong evidence that agonist or partial agonist activity at 5-HT_{2A} receptors is necessary for the psychedelic effects, however it may be not sufficient to explain the differences in the behavioral effects of these drugs. For instance, the endogenous transmitter, serotonin, and other synthetic, exogenous compounds such as LSD might produce distinct ligand-receptor complexes when binding to 5-HT_{2A} receptors. These different conformations may lead to activation of different subsets of intracellular signaling pathways, which is known as ligand bias or functional selectivity. In this regard, Schmid et al. (2008) [34] demonstrated, in an elegant series of experiments, that the *in vivo* responses of serotonin and 2,5-dimethoxy-4-iodoamphetamine (DOI) at 5-HT_{2A} receptors are differentially altered by the presence of β-arrestin-2 scaffold [35] (Fig. 4). In addition, psychedelic drugs exhibit affinities for other targets such as serotonin 5-HT_{1A} and glutamate mGlu2 receptors, which further complicates the extent of their pharmacological effects (see [36] for review).

In this section we describe the most relevant schizophrenia-related behavioral effects induced by hallucinogenic drugs in rodents, focusing on those that involve a 5-HT_{2A} receptor mechanism:

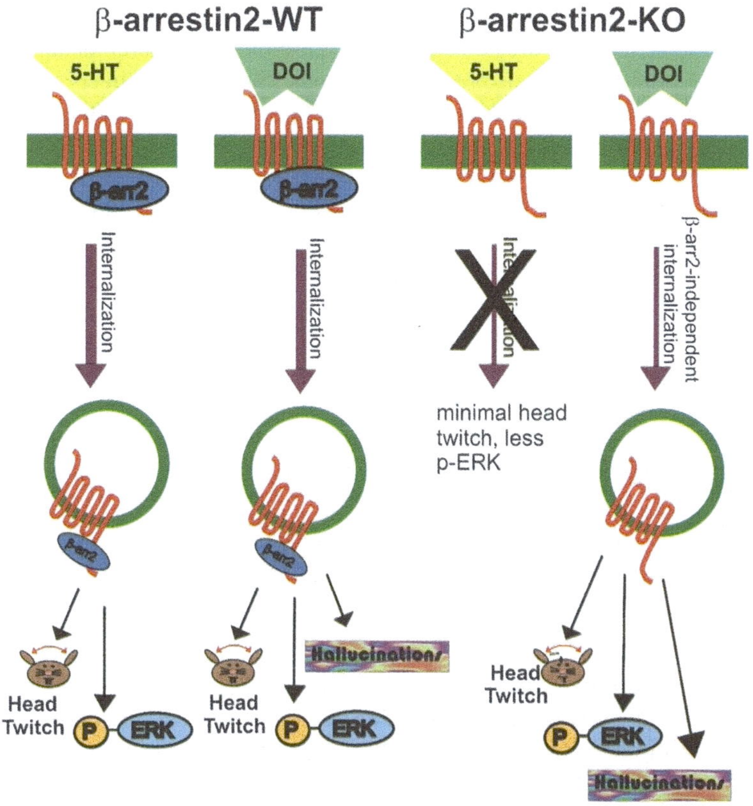

Fig. 4 Functional selectivity at the 5-HT$_{2A}$ receptor is mediated by β-arrestin. The absence of β-arrestin-2 abrogrates many serotonin-induced downstream events at 5-HT$_{2A}$ receptors, including internalization, head twitch, and p-ERK, but has little to no effect on those same signaling pathways when DOI is the ligand in question. Taken with permission from A Abbas and BL Roth' Arresting serotonin, published in Proceedings of the National Academy of Sciences USA, 105(3):831–832 (2008). Copyright © 2008, National Academy of Sciences, USA

(a) *Startle habituation.* Loud acoustic stimuli and tactile stimuli evoke a brief motor response (startle response) in both humans and animals, which decreases after repetitive stimuli exposure (habituation). Schizophrenia patients exhibit startle reflex habituation deficits [37–41], which suggests an inability to gate incoming stimuli. In rodents, serotonergic hallucinogens such as indoleamines (LSD; DMT; and psilocin, a psilocybin metabolite) and phenylalkylamines (mescaline; 2,5-dimethoxy-4-methylamphetamine, DOM; 2,5-dimethoxy-4-ethylamphetamine, DOET) increase startle response magnitudes and decreases startle habituation (mescaline, acute LSD). Moreover, mescaline-induced habituation effects are blocked by 5-HT$_{2A/2C}$ antagonists [42].

(b) *Prepulse inhibition (PPI).* The fundament of this procedure is that the presentation of a weak pre-stimulus prior to a startle-inducing stimulus attenuates the startle response [43, 44]. Prepulse inhibition (PPI) is a measure of sensorimotor gating across species and is disrupted in schizophrenia patients [38, 40, 45–49]. The hallucinogenic drugs LSD, DOI, 2,5-dimethoxy-4-bromoamphetamine (DOB) and mescaline also reduce PPI in rats [50–52]. The effects of DOI and LSD on PPI are prevented by selective 5-HT_{2A} antagonists M100907 and MDL 11939 but not by 5-HT_{1A} or 5-HT_{2C} antagonists. Moreover, the effects of DOI over PPI are blocked by atypical antipsychotics drugs [53–55].

(c) *Head twitch response (HTR).* HTR is a rapid lateral, side-to-side, head movement induced by a variety of psychedelic 5-HT_{2A} receptor agonists such as DOI, DOB, LSD and psilocin. LSD was the first drug used to study this behavioral response [56]. Currently, DOI is the prototype compound to investigate HTR in experimental animals (see [57] for review). The first study describing dose-dependent HTR of DOI in rats was published in 1989 by Arnt and Hyttel [58]. In 1990 Darmani and coworkers [59] demonstrated HTR of DOI in mice. Other species that show HTR after DOI are rabbit [60] and least shrew [61]. Activation of 5-HT_{2A} receptors is necessary for HTR of DOI in rodents since the response is blocked by fairly selective 5-HT_{2A} antagonists and 5-HT_{2A} KO mice do not show the response. Other serotonin receptors such as 5-HT_{2B}, 5-HT_{2C}, 5-HT_{1A} may have a modulatory role on HTR-induced by DOI [57].

(d) *Temporal processing.* Impaired timing and temporal processing is also a core feature in schizophrenia (see [55] for review). Similar deficits have been described after mescaline and LSD consumption in humans [62, 63]. Time perceptual disturbances are also reproduced in rodents after DOI administration. Thus, rats showed altered performance in the free-operant timing task [64–66] and the discrete-trials task [67]. The effects of DOI on interval timing are dependent on 5-HT_{2A} receptors since they are blocked by the selective 5-HT_{2A} receptor antagonist M100907 and ketanserin [65, 66].

(e) *Modulation of locomotor activity.* Locomotor activation has been used as a correlate of psychotic symptoms in experimental animals. Psychedelic drugs like DOI and mescaline have been shown to affect locomotor activity in mice and rats [68]. Low doses of DOI produce motor activation while high doses have and opposite effect. Genetic studies in 5-HT_{2A} KO mice and pharmacological studies involved 5-HT_{2A} receptors and 5-HT_{2C} in hyper- and hypo-locomotor effects induced by DOI, respectively [68, 69]. Non-psychedelic drugs did not produce motor activation. Therefore, psychedelic-induced motor activation has been proposed as an animal model of hallucinogenic potential in humans (see [70] for review).

(f) *Cognitive impairment.* Dysfunction of serotonergic PFC circuits may contribute to the cognitive deficits observed in schizophrenia, especially regarding flexible and impulsive behaviors [71]. DOI increases impulsive behavior in rodents which is prevented by ketanserin administration [72]. However, other cognitive domains also affected in schizophrenia such as working memory and attention have been less studied or they have been shown to be unaffected by 5-HT_{2A} ligands [73, 74].

(g) *Negative symptomatology.* Schizophrenia is also accompanied by negative symptoms such as depressed mood and social isolation among others.

Compelling evidence has demonstrated the role of the serotonergic system on the regulation of mood (for review see [75]). However, to date few studies have investigated the effects of 5-HT_{2A} ligands and the possible role of 5-HT_{2A} receptors on mood regulation in schizophrenia [76].

5-HT_{2A} Receptor Antagonism and Second Generation Antipsychotic Drugs

The discovery of atypical antipsychotics (a.k.a. second generation antipsychotics) was one of the most robust indications for the role of serotonin in schizophrenia. To date, several atypical antipsychotic are commercially available: clozapine, risperidone, olanzapine, quetiapine, ziprasidone, aripiprazole, zotepine, paliperidone, asenapine, lurasidone, iloperidone, amisulpride and sertindole. Atypical antipsychotics inhibit the actions of dopamine at D_2 receptors to a much lower extent than classical antipsychotics and additionally they interact with several serotonin receptors, predominantly through the blockade of 5-HT_{2A} and 5-HT_{2C}. With the exception of quetiapine and amisulpride, all currently approved atypical antipsychotic drugs display high affinity for 5-HT_{2A} receptors [77–80]. Some compounds are also agonists at 5-HT_{1A} receptors. Due to their pharmacodynamics, atypical antipsychotics display a more favorable hormonal and extrapyramidal side (EPS)-effects profile than first generation compounds [81]. Clozapine, which was discovered in 1958, was the first drug that showed antipsychotic action without EPS effects and high effectiveness in treatment-resistant schizophrenia [9]. Perhaps the most interesting aspect of atypical antipsychotics is that they possess multiple drug properties (agonist, partial agonist, antagonist, inverse agonist) at different receptors [82–85]. Indeed, the better therapeutic profile of clozapine (and possibly of other atypical drugs) could be accounted for by its affinity for a large number of receptors, thus being able to regulate different neurotransmitter systems and functions [78, 86–88] (Fig. 5).

However, atypical antipsychotics are not exempt from other adverse consequences such as agranulocytosis (clozapine), metabolic alterations, tardive dyskinesia or increased risk of stroke. Moreover, atypical antipsychotics do not provide clear advantages on relieving negative and cognitive symptomatology in schizophrenia [89]. Because, atypical antipsychotics have potent 5-HT_{2A} antagonist properties, agents directly targeting 5-HT_{2A} receptors such as ritanserin, M100907 and SR46349 (eplivanserin) have been tested as potential antipsychotics. Ritanserin induced a moderate reduction in core symptoms of schizophrenia [90]. The highly selective 5-HT_{2A} receptor antagonist M100907 showed superior efficacy over placebo but below haloperidol, which precipitated the discontinuation of phase III clinical trial in USA.

Similar intermediate efficacy between placebo and haloperidol was described for SR46349 [91]. Together, these clinical findings demonstrate that 5-HT_{2A} blockade alone results in some antipsychotic activity, yet optimal efficacy requests some degree of dopaminergic D_2 blockade.

	Classical		Atypical				
Receptor	Clorpromazine	Haloperidol	Clozapine	Olanzapine	Risperidone	Quetiapine	Ziprasidone
D_1	19.95	210	85	31	430	455	525
D_2	1.25	0.7	126	11	4	160	5
D_3		2	473	49	10		7
D_4		3	35	27	9		32
$5\text{-}HT_{1A}$		1100	875	>10 000	210	2800	3
$5\text{-}HT_{2A}$	7.94	45	16	4	0.5	295	0.4
$5\text{-}HT_{2C}$	12.59	>10 000	16	23	25		1
α_1	1	6	7	19	0.7	7	11
H_1		440	6	7	20	11	50
M_1		>1500	1.9	1.9	>10 000	120	>1000

Fig. 5 Main *in vitro* affinities (in nM) of some classical and atypical antipsychotic drugs. Data from human cloned receptors [78, 86–88]

Conclusions

- Hallucinogenic drugs are able to induce schizophrenia-related behavioral effects in experimental animals.
- The effects of hallucinogenic drugs in experimental animals are mainly mediated by $5\text{-}HT_{2A}$ receptors.
- $5\text{-}HT_{2A}$ receptors contribute to the pharmacological profile of atypical antypsychotics.

References

1. Lieberman JA, Mailman RB, Duncan G, Sikich L, Chakos M, Nichols DE et al (1998) Serotonergic basis of antipsychotic drug effects in schizophrenia. Biol Psychiatry 44(11):1099–1117
2. Wolley DW (1962) The biochemical bases of psychosis or the serotonin hypothesis of mental disease. Wiley, New York
3. Lerer B, Ran A, Blacker M, Silver H, Weller MP, Drummer D et al (1988) Neuroendocrine responses in chronic schizophrenia. Evidence for a serotonergic dysfunction. Schizophr Res 1(6):405–410
4. Maes M, Meltzer HY (1996) Effects of meta-chlorophenylpiperazine on neuroendocrine and behavioral responses in male schizophrenic patients and normal volunteers. Psychiatry Res 64(3):147–159
5. Roth BL, Meltzer HY (1995) The role of serotonin in schizophrenia. In: Bloom FE, Kupfer DJ (eds) Psychopharmacology, the 4th generation of progress. Raven Press, New York, pp 1215–1227

6. Carlsson A, Lindqvist M (1963) Effect of chlorpromazine or haloperidol on formation of 3-methoxytyramine and normetanephrine in mouse brain. Acta Pharmacol Toxicol 20:140–144
7. Seeman P, Lee T (1975) Antipsychotic drugs: direct correlation between clinical potency and presynaptic action on dopamine neurons. Science 188(4194):1217–1219
8. Creese I, Burt DR, Snyder SH (1976) Dopamine receptor binding predicts clinical and pharmacological potencies of antischizophrenic drugs. Science 192(4238):481–483
9. Kane J, Honigfeld G, Singer J, Meltzer H (1988) Clozapine for the treatment-resistant schizophrenic. A double-blind comparison with chlorpromazine. Arch Gen Psychiatry 45(9):789–796
10. Meltzer HY (1989) Clinical studies on the mechanism of action of clozapine: the dopamine-serotonin hypothesis of schizophrenia. Psychopharmacology 99(Suppl):S18–S27
11. Hollister LE (1962) Drug-induced psychoses and schizophrenic reactions: a critical comparison. Ann N Y Acad Sci 96:80–92
12. Willins DL, Deutch AY, Roth BL (1997) Serotonin 5-HT2A receptors are expressed on pyramidal cells and interneurons in the rat cortex. Synapse 27(1):79–82
13. Jakab RL, Goldman-Rakic PS (1998) 5-Hydroxytryptamine$_{2A}$ serotonin receptors in the primate cerebral cortex: possible site of action of hallucinogenic and antipsychotic drugs in pyramidal cell apical dendrites. Proc Natl Acad Sci U S A 95(2):735–740
14. Santana N, Bortolozzi A, Serrats J, Mengod G, Artigas F (2004) Expression of serotonin$_{1A}$ and serotonin$_{2A}$ receptors in pyramidal and GABAergic neurons of the rat prefrontal cortex. Cereb Cortex 14(10):1100–1109
15. Gurevich EV, Joyce JN (1997) Alterations in the cortical serotonergic system in schizophrenia: a postmortem study. Biol Psychiatry 42(7):529–545
16. Dean B (2003) The cortical serotonin$_{2A}$ receptor and the pathology of schizophrenia: a likely accomplice. J Neurochem 85(1):1–13
17. Selvaraj S, Arnone D, Cappai A, Howes O (2014) Alterations in the serotonin system in schizophrenia: a systematic review and meta-analysis of postmortem and molecular imaging studies. Neurosci Biobehav Rev 45:233–245
18. Ngan ET, Yatham LN, Ruth TJ, Liddle PF (2000) Decreased serotonin 2A receptor densities in neuroleptic-naive patients with schizophrenia: A PET study using [^{18}F]setoperone. Am J Psychiatry 157(6):1016–1018
19. Rasmussen H, Erritzoe D, Andersen R, Ebdrup BH, Aggernaes B, Oranje B et al (2010) Decreased frontal serotonin$_{2A}$ receptor binding in antipsychotic-naive patients with first-episode schizophrenia. Arch Gen Psychiatry 67(1):9–16
20. Dean B Serotonin$_{2A}$ Receptors and Schizophrenia: The Controversies Continue. Comment in Schizophrenia Research Forum [Posted 18 January 2010]. Available from: http://www.schizophreniaforum.org
21. Adams BW, Moghaddam B (2001) Effect of clozapine, haloperidol, or M100907 on phencyclidine-activated glutamate efflux in the prefrontal cortex. Biol Psychiatry 50(10):750–757
22. Amargós-Bosch M, López-Gil X, Artigas F, Adell A (2006) Clozapine and olanzapine, but not haloperidol, suppress serotonin efflux in the medial prefrontal cortex elicited by phencyclidine and ketamine. Int J Neuropsychopharmacol 9:565–573
23. López-Gil X, Babot Z, Amargós-Bosch M, Suñol C, Artigas F, Adell A (2007) Clozapine and haloperidol differently suppress the MK-801-increased glutamatergic and serotonergic transmission in the medial prefrontal cortex of the rat. Neuropsychopharmacology 32(10):2087–2097
24. López-Gil X, Artigas F, Adell A (2010) Unraveling monoamine receptors involved in the action of typical and atypical antipsychotics on glutamatergic and serotonergic transmission in prefrontal cortex. Curr Pharm Des 16(5):502–515
25. Reynolds GP (2004) Receptor mechanisms in the treatment of schizophrenia. J Psychopharmacol 18(3):340–345
26. De Paulis T (2001) M-100907 (Aventis). Curr Opin Investig Drugs 2(1):123–132
27. Offord SJ, Wong DF, Nyberg S (1999) The role of positron emission tomography in the drug development of M100907, a putative antipsychotic with a novel mechanism of action. J Clin Pharmacol 39:17S–24S

28. Owens MJ, Risch SC (2001) Atypical antipsychotics. In: Schatzberg AF, Nemeroff CB (eds) Essentials of clinical psychopharmacology. American Psychiatric Publishing, Inc, Washington DC, pp 125–154
29. Gouzoulis-Mayfrank E, Schneider F, Friedrich J, Spitzer M, Thelen B, Sass H (1998) Methodological issues of human experimental research with hallucinogens. Pharmacopsychiatry 31(Suppl 2):114–118
30. Vollenweider FX, Vollenweider-Scherpenhuyzen MF, Bäbler A, Vogel H, Hell D (1998) Psilocybin induces schizophrenia-like psychosis in humans via a serotonin-2 agonist action. Neuroreport 9(17):3897–3902
31. Amann LC, Gandal MJ, Halene TB, Ehrlichman RS, White SL, McCarren HS, Siegel SJ (2010) Mouse behavioral endophenotypes for schizophrenia. Brain Res Bull 83(3–4):147–161
32. Kaffman A, Krystal JH (2012) New frontiers in animal research of psychiatric illness. Methods Mol Biol 829:3–30
33. Robbins TW (2012) Animal models of neuropsychiatry revisited: a personal tribute to Teitelbaum. Behav Brain Res 231(2):337–342
34. Schmid CL, Raehal KM, Bohn LM (2008) Agonist-directed signaling of the serotonin 2A receptor depends on β-arrestin-2 interactions *in vivo*. Proc Natl Acad Sci U S A 105(3):1079–1084
35. Abbas A, Roth BL (2008) Arresting serotonin. Proc Natl Acad Sci U S A 105(3):831–832
36. Nichols DE (2016) Psychedelics. Pharmacol Rev 68(2):264–355
37. Bolino F, Manna V, Di Cicco L, Di Michele V, Daneluzzo E, Rossi A, Casacchia M (1992) Startle reflex habituation in functional psychoses: a controlled study. Neurosci Lett 145(2):126–128
38. Parwani A, Duncan EJ, Bartlett E, Madonick SH, Efferen TR, Rajan R, Sanfilipo M, Chappell PB, Chakravorty S, Gonzenbach S, Ko GN, Rotrosen JP (2000) Impaired prepulse inhibition of acoustic startle in schizophrenia. Biol Psychiatry 47(7):662–669
39. Taiminen T, Jääskeläinen S, Ilonen T, Meyer H, Karlsson H, Lauerma H, Leinonen KM, Wallenius E, Kaljonen A, Salokangas RK (2000) Habituation of the blink reflex in first-episode schizophrenia, psychotic depression and non-psychotic depression. Schizophr Res 44(1):69–79
40. Ludewig K, Geyer MA, Vollenweider FX (2003) Deficits in prepulse inhibition and habituation in never-medicated, first-episode schizophrenia. Biol Psychiatry 54(2):121–128
41. Meincke U, Light GA, Geyer MA, Braff DL, Gouzoulis-Mayfrank E (2004) Sensitization and habituation of the acoustic startle reflex in patients with schizophrenia. Psychiatry Res 126(1):51–61
42. Davis M (1987) Mescaline: excitatory effects on acoustic startle are blocked by serotonin2 antagonists. Psychopharmacology 93(3):286–291
43. Hoffman HS, Ison JR (1980) Reflex modification in the domain of startle: I. Some empirical findings and their implications for how the nervous system processes sensory input. Psychol Rev 87(2):175–189
44. Geyer MA, Ellenbroek B (2003) Animal behavior models of the mechanisms underlying antipsychotic atypicality. Prog Neuro-Psychopharmacol Biol Psychiatry 27(7):1071–1079
45. Braff D, Stone C, Callaway E, Geyer M, Glick I, Bali L (1978) Prestimulus effects on human startle reflex in normals and schizophrenics. Psychophysiology 15(4):339–343
46. Braff DL, Geyer MA (1990) Sensorimotor gating and schizophrenia. Human and animal model studies. Arch Gen Psychiatry 47(2):181–188
47. Bolino F, Di Michele V, Di Cicco L, Manna V, Daneluzzo E, Casacchia M (1994) Sensorimotor gating and habituation evoked by electro-cutaneous stimulation in schizophrenia. Biol Psychiatry 36(10):670–679
48. Braff DL, Geyer MA, Light GA, Sprock J, Perry W, Cadenhead KS, Swerdlow NR (2001) Impact of prepulse characteristics on the detection of sensorimotor gating deficits in schizophrenia. Schizophr Res 49(1–2):171–178
49. Quednow BB, Wagner M, Westheide J, Beckmann K, Bliesener N, Maier W, Kühn KU (2006) Sensorimotor gating and habituation of the startle response in schizophrenic patients randomly treated with amisulpride or olanzapine. Biol Psychiatry 59(6):536–545
50. Sipes TE, Geyer MA (1997) DOI disrupts prepulse inhibition of startle in rats via 5-HT2A receptors in the ventral pallidum. Brain Res 761(1):97–104

51. Pálenícek T, Balíková M, Bubeníková-Valesová V, Horácek J (2008) Mescaline effects on rat behavior and its time profile in serum and brain tissue after a single subcutaneous dose. Psychopharmacology 196(1):51–62
52. Halberstadt AL, Geyer MA (2010) LSD but not lisuride disrupts prepulse inhibition in rats by activating the 5-HT(2A) receptor. Psychopharmacology 208(2):179–189
53. Padich RA, McCloskey TC, Kehne JH (1996) 5-HT modulation of auditory and visual sensorimotor gating: II. Effects of the 5-HT2A antagonist MDL 100,907 on disruption of sound and light prepulse inhibition produced by 5-HT agonists in Wistar rats. Psychopharmacology 124(1–2):107–116
54. Kohnomi S, Suemaru K, Kawasaki H, Araki H (2008) Effect of aripiprazole on 5-HT2 receptor-mediated wet-dog shake responses and disruption of prepulse inhibition in rats. J Pharmacol Sci 106(4):645–650
55. Halberstadt AL, Geyer MA (2013) Serotonergic hallucinogens as translational models relevant to schizophrenia. Int J Neuropsychopharmacol 16(10):2165–2180
56. Keller DL, Umbreit WW (1956) Permanent alteration of behavior in mice by chemical and psychological means. Science 124(3225):723–724
57. Canal CE, Morgan D (2012) Head-twitch response in rodents induced by the hallucinogen 2,5-dimethoxy-4-iodoamphetamine: a comprehensive history, a re-evaluation of mechanisms, and its utility as a model. Drug Test Anal 4(7–8):556–576
58. Arnt J, Hyttel J (1989) Facilitation of 8-OHDPAT-induced forepaw treading of rats by the 5-HT2 agonist DOI. Eur J Pharmacol 161(1):45–51
59. Darmani NA, Martin BR, Pandey U, Glennon RA (1990) Pharmacological characterization of ear-scratch response in mice as a behavioral model for selective 5-HT2-receptor agonists and evidence for 5-HT1B- and 5-HT2-receptor interactions. Pharmacol Biochem Behav 37(1):95–99
60. Dave KD, Harvey JA, Aloyo VJ (2002) A novel behavioral model that discriminates between 5-HT2A and 5-HT2C receptor activation. Pharmacol Biochem Behav 72(1–2):371–378
61. Darmani NA, Mock OB, Towns LC, Gerdes CF (1994) The head-twitch response in the least shrew (Cryptotis parva) is a 5-HT2- and not a 5-HT1C-mediated phenomenon. Pharmacol Biochem Behav 48(2):383–396
62. Hoch PH, Cattell JP, Pennes HH (1952) Effects of mescaline and lysergic acid (d-LSD-25). Am J Psychiatry 108(8):579–584
63. Kenna JC, Sedman G (1964) The subjective experience of timing during lysergic acid diethylamide (LSD-25) intoxication. Psychopharmacologia 5:280–288
64. Stubbs DA (1980) Temporal discrimination and a free-operant psychophysical procedure. J Exp Anal Behav 33(2):167–185
65. Body S, Kheramin S, Ho MY, Miranda F, Bradshaw CM, Szabadi E (2003) Effects of a 5-HT2 receptor agonist, DOI (2,5-dimethoxy-4-iodoamphetamine), and antagonist, ketanserin, on the performance of rats on a free-operant timing schedule. Behav Pharmacol 14(8):599–607
66. Body S, Cheung TH, Bezzina G, Asgari K, Fone KC, Glennon JC, Bradshaw CM, Szabadi E (2006) Effects of d-amphetamine and DOI (2,5-dimethoxy-4-iodoamphetamine) on timing behavior: interaction between D1 and 5-HT2A receptors. Psychopharmacology 189(3):331–343
67. Hampson CL, Body S, den Boon FS, Cheung TH, Bezzina G, Langley RW, Fone KC, Bradshaw CM, Szabadi E (2010) Comparison of the effects of 2,5-dimethoxy-4-iodoamphetamine and D-amphetamine on the ability of rats to discriminate the durations and intensities of light stimuli. Behav Pharmacol 21(1):11–20
68. Halberstadt AL, Powell SB, Geyer MA (2013) Role of the 5-HT_2A receptor in the locomotor hyperactivity produced by phenylalkylamine hallucinogens in mice. Neuropharmacology 70:218–227
69. Halberstadt AL, van der Heijden I, Ruderman MA, Risbrough VB, Gingrich JA, Geyer MA, Powell SB (2009) 5-HT(2A) and 5-HT(2C) receptors exert opposing effects on locomotor activity in mice. Neuropsychopharmacology 34(8):1958–1967
70. Moreno JL, González-Maeso J (2013) Preclinical models of antipsychotic drug action. Int J Neuropsychopharmacol 16(10):2131–2144

71. Robbins TW (2005) Synthesizing schizophrenia: a bottom-up, symptomatic approach. Schizophr Bull 31(4):854–864
72. Hadamitzky M, Feja M, Becker T, Koch M (2009) Effects of acute systemic administration of serotonin2A/C receptor ligands in a delay-based decision-making task in rats. Behav Pharmacol 20(5–6):415–423
73. Ruotsalainen S, MacDonald E, Koivisto E, Stefanski R, Haapalinna A, Riekkinen P Jr, Sirviö J (1998) 5-HT1A receptor agonist (8-OH-DPAT) and 5-HT2 receptor agonist (DOI) disrupt the non-cognitive performance of rats in a working memory task. J Psychopharmacol 12(2):177–185
74. Rambousek L, Palenicek T, Vales K, Stuchlik A (2014) The effect of psilocin on memory acquisition, retrieval, and consolidation in the rat. Front Behav Neurosci 8:180
75. Baou M, Boumba VA, Petrikis P, Rallis G, Vougiouklakis T, Mavreas V (2016) A review of genetic alterations in the serotonin pathway and their correlation with psychotic diseases and response to atypical antipsychotics. Schizophr Res 170(1):18–29. https://doi.org/10.1016/j.schres.2015.11.003
76. Aghajanian GK, Marek GJ (1999) Serotonin and hallucinogens. Neuropsychopharmacology 21(2 Suppl):16S–23S
77. Meltzer HY, Matsubara S, Lee JC (1989) Classification of typical and atypical antipsychotic drugs on the basis of dopamine D-1, D-2 and serotonin2 pKi values. J Pharmacol Exp Ther 251(1):238–246
78. Arnt J, Skarsfeldt T (1998) Do novel antipsychotics have similar pharmacological characteristics? A review of the evidence. Neuropsychopharmacology 18(2):63–101
79. Miyamoto S, LaMantia AS, Duncan GE, Sullivan P, Gilmore JH, Lieberman JA (2003) Recent advances in the neurobiology of schizophrenia. Mol Interv 3(1):27–39
80. Farah A (2005) Atypicality of atypical antipsychotics. Prim Care Companion J Clin Psychiatry 7(6):268–274
81. Remington G, Kapur S (2000) Atypical antipsychotics: are some more atypical than others? Psychopharmacology 148(1):3–15
82. Clarke WP, Chavera TA, Silva M, Sullivan LC, Berg KA (2013) Signalling profile differences: paliperidone versus risperidone. Br J Pharmacol 170(3):532–545
83. Oyamada Y, Horiguchi M, Rajagopal L, Miyauchi M, Meltzer HY (2015) Combined serotonin $(5\text{-HT})_{1A}$ agonism, 5-HT_{2A} and dopamine D_2 receptor antagonism reproduces atypical antipsychotic drug effects on phencyclidine-impaired novel object recognition in rats. Behav Brain Res 285:165–175
84. Sullivan LC, Clarke WP, Berg KA (2015) Atypical antipsychotics and inverse agonism at 5-HT2 receptors. Curr Pharm Des 21(26):3732–3738
85. McCreary AC, Newman-Tancredi A (2015) Serotonin 5-HT1A receptors and antipsychotics—an update in light of new concepts and drugs. Curr Pharm Des 21(26):3725–3731
86. Bymaster FP, Calligaro DO, Falcone JF, Marsh RD, Moore NA, Tye NC, Seeman P, Wong DT (1996) Radioreceptor binding profile of the atypical antipsychotic olanzapine. Neuropsychopharmacology 14:87–96
87. Daniel DG, Zimbroff DL, Potkin SG, Reeves KR, Harrigan EP, Lakshminarayanan M (1999) Ziprasidone 80 mg/day and 160 mg/day in the acute exacerbation of schizophrenia and schizoaffective disorder: a 6-week placebo-controlled trial. Ziprasidone Study Group. Neuropsychopharmacology 20:491–505
88. Seeger TF, Seymour PA, Schmidt AW, Zorn SH, Schulz DW, Lebel LA, Heym J (1995) Ziprasidone (CP-88,059): a new antipsychotic with combined dopamine and serotonin receptor antagonist activity. J Pharmacol Exp Ther 275:101–113
89. Hill SK, Bishop JR, Palumbo D, Sweeney JA (2010) Effect of second-generation antipsychotics on cognition: current issues and future challenges. Expert Rev Neurother 10(1):43–57
90. Wiesel FA, Nordström AL, Farde L, Eriksson B (1994) An open clinical and biochemical study of ritanserin in acute patients with schizophrenia. Psychopharmacology 114(1):31–38
91. Marek G, Merchant K (2005) Developing therapeutics for schizophrenia and other psychotic disorders. NeuroRx 2(4):579–589

Role of Serotonin-2A Receptors in Pathophysiology and Treatment of Depression

Lucia Moravčíková, Kristína Csatlósová, Barbora Ďurišová, Katarína Ondáčová, Michaela Pavlovičová, Ľubica Lacinová, and Eliyahu Dremencov

Abstract This chapter aims to summarize the up-to-day evidence-based biomedical knowledge on serotonin-2A (5-HT_{2A}) receptors and their role in pathophysiology and treatment of central nervous system (CNS) disorders, with a primary focus on depression. The first paragraph provides a brief introduction to serotonin (5-HT) system and 5-HT receptors, focusing on serotonin-2 (5-HT_2) family and 5-HT_{2A} receptor specifically. The second paragraph is focused on molecular genetics of 5-HT_{2A} receptors, polymorphism of 5-HT_{2A} receptor (5HT2AR) gene, 5HT2AR gene epigenetic mechanisms, such as DNA methylation, and post-translational modifications of 5HT_{2A}R messenger ribonucleic acid (mRNA), such as alternative splicing. The molecular and cellular pharmacology and physiology of 5-HT_{2A} receptors in normal and pathological conditions are discussed in the third paragraph. The 5-HT_{2A} receptors-acting ligands are addresses. The fourth paragraph describes the role of 5-HT receptors in the interaction between 5-HT and other neurotransmitter systems in health and in CNS disorders. The fifth and the final paragraph specifically deals with the role of 5-HT_{2A} receptor in pathophysiology and treatment of depression, focusing on the 5-HT_{2A} receptor expressed in the hippocampus.

Keywords Serotonin-2A receptor (5HT_{2A}R) gene polymorphism • Deoxyribonucleic acid (DNA) methylation • Messenger ribonucleic acid (mRNA) alternative splicing

L. Moravčíková • K. Csatlósová • B. Ďurišová • K. Ondáčová • M. Pavlovičová • Ľ. Lacinová
Institute of Molecular Physiology and Genetics, Slovak Academy of Sciences, Bratislava, Slovakia

E. Dremencov, M.Med.Sc., Ph.D. (✉)
Institute of Molecular Physiology and Genetics, Slovak Academy of Sciences, Bratislava, Slovakia

Institute of Experimental Endocrinology, Biomedical Research Center, Slovak Academy of Sciences, Bratislava, Slovakia
e-mail: Eliyahu.Dremencov@savba.sk

B.P. Guiard, G. Di Giovanni (eds.), *5-HT_{2A} Receptors in the Central Nervous System*, The Receptors 32, https://doi.org/10.1007/978-3-319-70474-6_9

• G-protein coupled receptors (GPCR) • $G\alpha_{Q/Z}$-11 protein • Phospholipase C (PLC) • Inositol trisphosphate (IP_3) • Calcium signaling • Antidepressant drugs • Antipsychotic drugs • Hippocampus

Serotonin-2A Receptor: An Introduction

The 5-HT_{2A} receptors belong to the 5-HT_2 family consists of two more subtypes: 5-HT_{2B} and 5-HT_{2C} receptors. These subtypes have similar molecular structure, amino acid sequence, and signaling properties. The 5-HT_{2B} receptors have a restricted expression in CNS; they play an important role during the embryonic development [1]. The 5-HT_{2A} and 5-HT_{2C} receptors are widely distributed across the CNS and have multiple functions. All members of the 5-HT_2 receptor family primarily couple to PLC on activation. Like other G-protein coupled receptors (GPCRs), 5-HT_2 functional regulation also involves sensitization and desensitization-regulatory processes that help prevent overstimulation and allow recuperation of signaling competence, respectively [2].

Serotonin-2 receptor subtypes have been cloned from various species and tissues. The 5-HT_{2A} receptor from hamster, human, monkey, mouse, pig, rat, and sheep all have the same length of 471 amino acid. The 5-HT_{2B} receptor from human, mouse and rat have a length of 481, 504, and 479 amino acids and the 5-HT_{2C} receptor from human, mouse and rat have a length of 458, 459, and 460 amino acids, respectively [3]. The 5-HT_{2A} and 5-HT_{2C} receptors are glycosylated on multiple sites. The genes for the 5-HT_{2A} and 5-HT_{2B} receptor have 3 introns; the 5-HT_{2C} receptor gene has two introns. In humans, the genes are located on chromosome 13q14-q21 for the 5-HT_{2A} receptor, chromosome position 2q36.3–2q37.1 for the 5-HT_{2B} receptor, and chromosome X q24 for the 5-HT_{2C} receptor [1].

It has been shown that some GPCRs, including the 5-HT_{2A} receptor, exhibit critical differences in some aspects of functional regulation from those seen in conventionally studied model GPCRs such as the β_2-adrenergic receptor. This receptor couples to a number of intracellular signaling cascades, making it an important receptor to study. Therefore, the 5-HT_{2A} receptor could well serve as an important alternate paradigm in the study of GPCR function [2].

Though the receptor has been studied largely in relation to its multiple functions in the CNS, high levels of receptor expression in other areas such as the intestine, platelets, and endothelial cells suggest that it could play crucial roles in other aspects of physiology, as well. They mediate contractile responses in many vascular smooth muscle preparations (e.g. bronchial, uterine and urinary smooth muscle), and part of the contractile effects of 5-HT in the guinea pig ileum. In addition, platelet aggregation and increased capillary permeability following exposure to 5-HT have been attributed to 5-HT_{2A} receptor-mediated process. Moreover, 5-HT_2 receptor agonists, in addition to precursors of 5-HT and 5-HT releasing agents, mediate certain behavioral syndromes in vivo (e.g. head twitching in mice, and wet-dog shakes and back muscle contractions in rats) [4]. Centrally, these receptors are principally located in the cortex, claustrum and basal ganglia. 5-HT_{2A} receptor activation stimulates hormone secretion (e.g. ACTH, corticosterone, oxytocin, renin and prolactin) [5]. Considering the broad expression of 5-HT_{2A} receptors across the brain and their involvement in multiple CNS functions, it is expected that these receptors will play a role pathophysiology of brain disorders. Indeed, the CNS disorders in which the 5-HT_{2A} receptor seems to be involved range from schizophrenia, depression, obsessive compulsive disorder (OCD), and attention deficit–hyperactivity disorder (ADHD), to eating disorders such as anorexia nervosa, to autism spectrum disorders [2]. Implication of 5-HT_{2A} receptors in mental disorders with complex etiologies is still not clearly understood. There are a large number of drugs targeted to this receptor.

Molecular Genetics and Epigenetics of Serotonin-2A Receptor

Serotonin-2A Gene Polymorphism

The 5-HT_{2A} receptor, encoded by HTR2AR gene, is a widely-distributed post-synaptic target for 5-HT in the human brain. Serotonin-2A receptor heterogenity is affected by alternative polymorphisms and alternative splicing. The 5-HT_{2A} receptor is a target for atypical antipsychotics and antidepressants. The role of genetic variants of HTR2AR in signaling modulation remains unclear, despite positive clinical associations [6]. Methods for detecting genetic polymorphisms are advancing rapidly and now allow simultaneous genotyping of several nucleotide polymorphisms. The Genetic Association Database [7] reports 346 unique association studies between single nucleotide polymorphisms (SNPs) in HTR2AR gene and human phenotypes and more than half of these studies find positive genotype-phenotype associations. Most are related to cognition or risk for neuropsychiatric disorders, supporting the presence of functional genetic variants in HTR2AR gene. Some of SNPs (e.g., T102C, C516T, A1438G) are silent mutations and do not cause a change in the protein. Other SNPs (e.g., W25S, I197V, S421F, A447V, H452Y) result in a change in an amino acid. Although the A1438G mutation is silent and does not result in alteration of the amino acid sequence of 5-HT_{2A} receptor, it is located within promoter region of the gene. Thus was proposed that this mutation alters promoter activity and even so expression of 5-HT_{2A} receptors [8]. Lower 5-HT_{2A} receptor densities in some brain areas may cause another silent mutation, T102C [9]. On the other hand, mutation H452Y which caused change in protein has no effect on receptor expression, but reduces intracellular signaling capacity [10].

Numbers of studies have been conducted on the association between HTR2AR gene T102C polymorphism and major depressive disorder (MDD) [11–13]. To clarify the effects of HTR2AR gene T102C polymorphism on the risk of depression, Lin et al. [11] performed a meta-analysis in the Chinese population. Results have shown that HTR2AR gene T102C polymorphism is not associated with susceptibility to MDD in these population. Another study [14] demonstrated an association between T102C polymorphism of HTR2AR gene, lifespan, and the risk of age-related CNS disorders. Their results suggest that T102C is associated with mean life span, and thus this gene becomes a possible candidate for the group of adaptive genes to meat consumption proposed in the literature.

The $5HT_{2A}$ receptor gene polymorphisms rs7997012 and rs6311 has been suggested to be involved in major depressive disorder. Htr2a knock-out mice (Htr2a−/−) displayed an increase in depressive-like behavior, compared to wild type, thus suggesting, that lowered $5\text{-}HT_{2A}$ receptor transmission may favor the susceptibility and severity of major depressive episodes [15].

It is seems that genetic variants in the HTR2A gene affect the therapeutic effects of andtidepressant drugs but mechanism underlying the regulation of such response remains poorly described. According to study of Qesseveur et al. [16] the HTR2A gene may represent a relevant marker to predict the efficacy of antidepressant drugs. The effect of three HTR2A single nucleotide polymorphisms (SNPs- rs6313, rs6314 and rs7333412) was investigated. These three SNPs have potential functional consequences on $5\text{-}HT_{2A}$ receptor, on response and remission rates after 3 months of antidepressant treatments. Their clinical data indicated that GG patients for the rs7333412 SNP were less prone to respond to antidepressant drugs than AA/AG patients.

T102C and A1438G polymorphisms were associated with risk for schizophrenia [17–19]. The T102C polymorphism is also related to tobacco use [20] and the A1438G polymorphism of HTR2AR gene is involved in the development of alcohol dependence [21]. Polymorphisms of the HTR2AR gene are associated with hallucinatory symptoms and delusions in demented and non-demented cohorts. The study of Craig et al. [22] examined the role of the HTR2AR gene T102C polymorphism in influencing psychotic symptoms in a large Northern Ireland Alzheimer's disease (AD) population. No significant association was found either in frequency of genotype or allelic variation for either set of symptoms. On the other hand, Lam et al. [23] demonstrated significant association between neuropsychiatric symptoms in AD and HTR2AR gene polymorphisms.

Methylation

Differential DNA methylation has been suggested to contribute to differential activity of alleles C and T and thereby to genetic associations between the C/T(102) polymorphism in the HTR2AR gene and psychiatric disorders [24]. This study demonstrated methylation in two CpG sites, which are specific to allele C. The majority of allele C-specific CpG sites were methylated in human temporal cortex and peripheral leukocytes. Findings that methylation of allele C-specific CpG sites in the first exon correlated significantly with the expression of DNA methylase 1 but not S-adenosylhomocysteine hydrolase, support the hypothesis that allele-specific DNA methylation is involved in regulation of HTR2AR gene expression, influencing expression differences between alleles C and T.

De Luca et al. [25] developed an improved quantitative assay for the measurement of allele-specific methylation of the HTR2AR gene and genetic association between the HTR2AR gene T102C silent polymorphism and suicidality in patients with mood disorders and schizophrenia.

Falkenberg et al. [26] used functional and structural equation modeling (SEM) approaches to assess the contributions of the polymorphism (R6311S) to DNA methylation and HTR2AR gene expression in chronic fatigue syndrome (CFS) subjects from a population-based study. Their study suggests that the promoter polymorphism (rs6311) can affect both transcription factor binding and promoter methylation, and this along with an individual's stress response can impact the rate of HTR2A transcription in a genotype and methylation-dependent manner.

Alternative Splicing

The first alternatively spliced isoform of 5-HT_{2A} receptor was identified by Huang et al. [27] in the parasitic nematode species, *Ascaris Suum*. The 5-HT_{2A}-s1 and 5-HT_{2A}-s2 exhibited identical pharmacological profiles when stably expressed in human embryonic kidney (HEK) 293 cells. Both 5-HT_{2A}s isoforms had higher affinity for 5-HT than their closely related *Caenorhabditis Elegans* homolog (5-HT_{2C}-e).

Guest et al. [28] identified an alternatively spliced HTR2AR gene transcript by PCR of human brain cDNA using degenerate oligonucleotide primers to transmembrane domains. PCR analysis showed that truncated (5HT2ARtr) and native HTR2AR genes were co-expressed in most brain tissues, with the highest levels being found in hippocampus, corpus callosum, amygdala, and caudate nucleus. Western blot analysis of HEK-293 cells transfected transiently with a 5HT2ARtr construct showed that a 30-kDa protein was expressed in cell membranes. Co-transfection studies showed no effect of the 5HT2ARtr variant on 3H-ketanserin binding to the native HTR2AR or on functional coupling of the HTR2AR to 5-HT-stimulated calcium influx.

Molecular Pharmacology of and Serotonin-2A Receptors

Signal Transduction Pathways of Serotonin-2A Receptor

The activation of 5-HT_{2A} receptor leads to the dissociation of $G\alpha_{Q/Z}$ protein into α and $\beta\gamma$ subunits. The α subunit of $G\alpha_{Q/Z}$ protein activates the phospholipase C (PLC), which in turn catalyzes the dissociation of inositol 1,4,5-trisphosphate (IP_3)-diacylglycerol (DAG) complex into the IP_3 and DAG. The DAG activates protein kinase C (PKC), and IP_3 stimulates calcium (Ca^{2+}) release from endoplasmic reticulum (ER) into the cytoplasm, a characteristic activation signature of many GPCRs

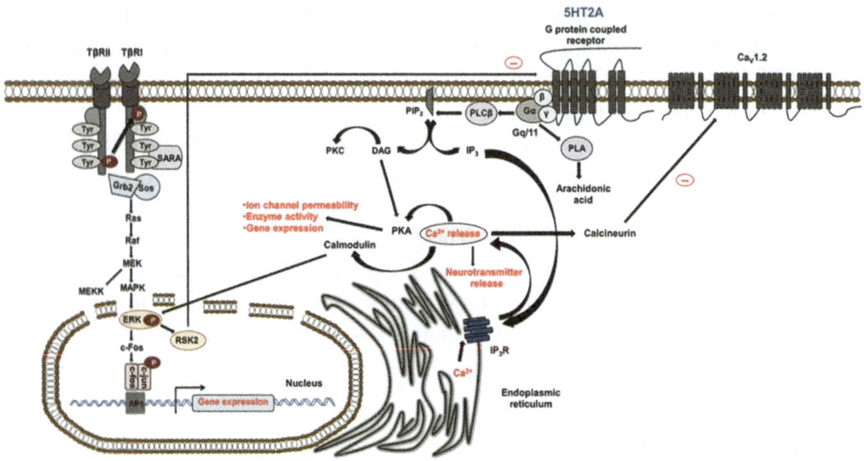

Fig. 1 Detailed signal transduction pathways of serotonin-2A receptors. Serotonin-2A (5-HT_{2A}) receptor activates protein kinase Cβ (PLCβ). Protein kinase Cβ hydrolysis phosphatidylinositol 4,5 bisphosphate (PIP_2) to diacylglycerol (DAG) which activates protein kinesis A (PKA) and inositol trisphosphate (IP_3) which acts through inositol trisphosphate receptors (IP_3R) localize on endoplasmic reticulum. Activation of this signaling pathway leads to increase in intracellular calcium concentration which affects ion channels, enzyme activity, and neurotransmission or gene expression. Intracellular calcium can also lead to activation of calmodulin which activates extracellular signal-regulated kinases (ERK) and activation of calcineurin leading to inhibition of voltage-dependent calcium channels. Activation of ERK signaling pathway suppresses 5-HT_{2A} receptor signaling through RSK2 kinase. Extracellular signal-regulated kinases can be activated by TGFβ receptor signaling pathway involving Ras GTP-ases interacting with Raf kinases and mitogen-activated protein kinase kinases (MEK) which phosphorylates mitogen-activated protein kinase (MAPK)

[29, 30]. This cascade has been the most extensively studied and is perhaps the most important signal transduction pathway regulated by this receptor (Fig. 1).

Stimulation of the 5-HT_{2A} receptor leads to the activation of at least three distinct signal transduction pathways: IP_3/DAG-, arachidonic acid (AA)-, and 2-arachidonylglycerol (2-AG)-mediated. In addition to PLC, 5-HT_{2A} receptors were also reported to activate phospholipase A2 (PLA2), so-called phospholipase B (PLB) [31].

Besides phospholipases-mediated calcium signaling, 5-HT_{2A} receptor activation also induces extracellular signal-regulated kinase (ERK) phosphorylation *via* diverse intracellular signaling mechanisms [32]. Src and calmodulin (CaM) promote 5-HT_{2A} receptor-mediated phosphorylation of ERK. In the PC12 cells, ERK phosphorylation by 5-HT_{2A} receptor may not depend on PLC/PKC signaling, and instead requires an increase in intracellular calcium, and the activation of CaM and Src [33]. The ERK target p90 ribosomal S6 kinase 2 (RSK2) directly acts on the third intracellular (i3) loop of 5-HT_{2A} receptor protein [34], leading to direct phosphorylation of the i3 loop at the conserved residue Ser-314 and to suppression of 5-HT_{2A} receptor signaling.

In addition, RSK2 is required for tyrosine kinases, such as the epidermal growth factor receptor and the platelet-derived growth factor receptor, both of which have been demonstrated to attenuate 5-HT_{2A} receptor functioning in primary cortical neurons [35, 36].

The 5-HT_{2A} receptors, like other members of 5-HT_2 family, couple preferentially via $G\alpha_{Q/Z}$-11 to the IP_3/PKC/Ca^{2+} pathway, although inhibition of cyclic adenosine monophosphate (cAMP) production has been reported [37].

The 5-HT_{2A} receptor also regulates the tyrosine kinase pathway activity [33]. Activation of neuronal 5-HT_{2A} receptor activates transglutaminase which leads to transamidation of Rac1, a small G protein, resulting in constitutive activation of Rac1 [38]. Chronic treatment with olanzapine, an atypical antipsychotic drug, causes the desensitization of 5-HT_{2A} receptor signaling. In rat frontal cortex, stimulation of the JAK-STAT pathway desensitizes the 5-HT_{2A} receptor-mediated PLC activation induced by olanzapine [39]. Furthermore, constitutive activation of 5-HT_{2A} receptor induces $G\alpha_{Q/Z}$-11 phosphorylation and desensitization (uncoupling) [40].

Functional Selectivity and Internalization of Serotonin-2A Receptors

Interestingly, different agonists of 5-HT_{2A} receptors vary in the efficacy with which they stimulate individual signal transduction pathways [2, 41]. This phenomena is called functional selectivity and the 5-HT_{2A} receptor was one of the first receptors for which this was described [29, 42]. This discovery was based of the observation that hallucinogenic effects of drugs such as LSD do not correlate with their activation of the IP_3/DAG pathway [2].

It has been suggested that hallucinogen, but not nonhallucinogen, 5-HT_{2A} receptor agonist induce phosphorylation of the 5-HT_{2A} receptor at S280 located in the third intracellular loop. Importantly, these authors also demonstrated that pretreating cells with pertussis toxin (PTX) decreased PLC activation induced by the hallucinogens 2,5-Dimethoxy-4-iodoamphetamine (DOI) and LSD, whereas PTX treatment did not affect lisuride and ergotamine responses [43]. Jones et al. [44] discovered, that application of the 5-HT_{2A} receptor agonist DOI to cultured cortical neurons induced phosphorylation of p21-activated kinase (PAK) via Rac guanine nucleotide exchange factor (RacGEF) kalirin-7 [44]. Taken together, these observations suggest that hallucinogens selectively activate $G\alpha_{I/O}$-dependent signaling, whereas non-hallucinogen 5-HT_{2A} receptor agonists do not [45].

Both *in vitro* and studies *in vivo* have shown receptor redistribution in response to exposure to antagonists. The 5-HT_{2A} receptor is internalized in response to both agonists and antagonists, adding a very interesting twist to its signaling properties [46, 47]. This feature of the 5-HT_{2A} receptor may play important roles in its signaling and in the actions of antipsychotic medications. The antagonist-mediated internalization of the rat 5-HT_{2A} receptor, unlike 5-HT-mediated internalization, is

Table 1 5-HT_{2A} ligands and their selectivity towards the 5-HT receptor family

Name of ligand	Effects of binging	Receptor affinity
Brexipiprazole	Antagonist	5-HT_{1A}, 5-HT_{2A}
Cyproheptadine	Antagonist/inverse agonist	5-HT_{1A}, 5-HT_{2A}, 5-HT_{2B}, 5-HT_{2C},5-HT_{3}, 5-HT_{6}, 5-HT_{7}
DOI	Agonist/partial agonist	5-HT_{2A}, 5-HT_{2B}, 5-HT_{2C}
MDL100907	Highly selective antagonist	5-HT_{2A}
Olanzapine	Agonist/inverse agonist	5-HT_{1A}, 5-HT_{3}, 5-HT_{6}, 5-HT_{7},5-HT_{2A}, 5-HT_{2B}, 5-HT_{2C}
Risperidone	Antagonist/inverse agonist/ irreversible antagonist	5-HT_{1A}, 5-HT_{1B}, 5-HT_{1D}, 5-HT_{5A}, 5-HT_{6}5-HT_{2A}, 5-HT_{2B}, 5-HT_{2C}5-HT_{7}
Ritanserin	Antagonist	5-HT_{2A}, 5-HT_{2C}
Seroquel	Antagonist	5-HT_{1A}, 5-HT_{2A}, 5-HT_{2C}, 5-HT_{7}
Spiperone	Antagonist	5-HT_{1A}, 5-HT_{1B}, 5-HT_{1D}, 5-HT_{1E}, 5-HT_{1F}, 5-HT_{2A}, 5-HT_{2B}, 5-HT_{2C}, 5-HT_{5A}, 5-HT_{6}, 5-HT_{7}
TCB-2	Agonist	5-HT_{2A}, 5-HT_{2C}
YM 992	Antagonist	5-HT_{2A}

independent of protein kinase C (PKC) activation [47]. Bhatnagar and colleagues [46] examined the internalization process of this receptor in detail, demonstrating that both agonist- and antagonist-induced internalization of the 5-HT_{2A} receptor were dynamin-dependent and via clathrin-mediated endocytosis. Activation of the 5-HT_{2A} receptor by agonists, but not antagonists, induced greater translocation of arrestin-3 than arrestin-2 to the plasma membrane, and resulted in differential sorting of arrestin-2, arrestin-3, and 5-HT_{2A} receptors into distinct plasma membrane and intracellular compartments. It is likely that these differences in distribution of the various signaling components induced by agonists and antagonists may be important in the "ligand-directed" of second messenger signals by the 5-HT_{2A} receptor, depending upon which ligand is used to stimulate the receptor. Authors discovered, that *in vitro* knockdown of Caveolin-1 (Cav-1, a scaffolding protein) nearly abolished 5-HT_{2A} receptor-mediated signal transduction as measured by calcium flux assays. Cav-1 appeared to modulate 5-HT_{2A} receptor signaling by facilitating the interaction of 5-HT_{2A} receptors with $G\alpha q$.

Serotonin-2A-Acting Drugs

Several drugs that have been developed for treatment of psychiatric disorders selectively bind to the 5-HT_{2A} receptor and modulate its signaling pathways (Table 1). The antipsychotic drugs spiperone and methiothepin with antipsychotic properties are nonselective antagonists of 5-HT_{1} and 5-HT_{2} receptors. Both prevent the 5-HT-dependent PLC activation at 10 μM concentration. However, cyproheptadine

(10 μM), another antagonist of 5-HT_1 and 5-HT_2 receptors, had no effect on PLC activity [48].

Brexpriprazole is an antagonist of 5-HT_{2A}, 5-HT_{1A} and D_2 receptors, is approved for the clinical use as a main pharmacotherapy in schizophrenia and as an adjunct in antidepressant-resistant depression. This drug demonstrated robust antipsychotic, antidepressant-like and anxiolytic activities, and limited extrapyramidal symptom liability with pro-cognitive efficacy in animal models [49]. Accumulating evidence suggests that antipsychotic drugs act by promoting neurite outgrowth. In the study of Ishima and colleagues [50] authors examined whether brexpiprazole can affect neurite outgrowth in cell culture. They found that brexpiprazole significantly potentiated nerve growth factor (NGF)-induced neurite outgrowth in PC12 cells, in a concentration dependent manner. Moreover, inhibitors of inositol IP_3 receptors, xestospongin C and 2-aminoethoxydiphenyl borate (2-APB), significantly blocked the effects of brexpiprazole. These findings suggest that brexpiprazole-induced neurite outgrowth is mediated through 5-HT_{1A} and 5-HT_{2A} receptors, and subsequent Ca^{2+} signaling via IP_3 receptors [50].

Role Serotonin-2A Receptors in the Regulation of CNS Circuits

Role of Serotonin-2A Receptors in the Interactions Between Serotonin and Glutamate and GABA Systems

DOI (1-[2,5-dimethoxy-4-iodophenyl-2-aminopropane]) is a hallucinogen acting as agonist of 5-HT_{2A} receptors, similarly to lysergic acid diethylamide (LSD). It was reported that DOI causes a dose-related inhibition of 5-HT neuronal activity, with the highest dose reducing firing rates by >80%. Pretreatment with the 5-HT_2 receptor antagonist ritanserin completely blocked the action of DOI [51]. Study of Quesseveur et al. [52] confirms this inhibitory effect of DOI on dorsal raphe (DR) nucleus 5-HT neuronal activity. DOI's response is dependent on 5-HT_{2A} receptors because it diminished in 5-HT_{2A} receptors lacking mice. Possible way of DOI inhibitory effect on DR 5-HT neuronal activity is via increasing of GABA release in DR. Other study shows that activation of 5-HT_{2A} receptors in the PFC by DOI increased the firing activity of DR 5-HT neurons. DOI administration also affected the firing rate of pyramidal neurons while most of them were excited, 11% were inhibited and rest was unaffected [53] In this case, excitatory and inhibitory actions of DOI on pyramidal cell firing are likely mediated by receptors located on pyramidal neurons and GABA interneurons, respectively. DOI also stimulates 5-HT release in the PFC, probably via a mechanism involving interaction between 5-HT_{2A} and AMPA (α-amino-3-hydroxy-5-methyl-4-isoxazolepropionic acid) receptors [54] (Fig. 2).

The PFC seems to play crucial role in depression. PFC is involved in higher brain functions and carries a control of brain functions through the processing and integration of signals from other brain areas, such as neocortex, several thalamic nuclei,

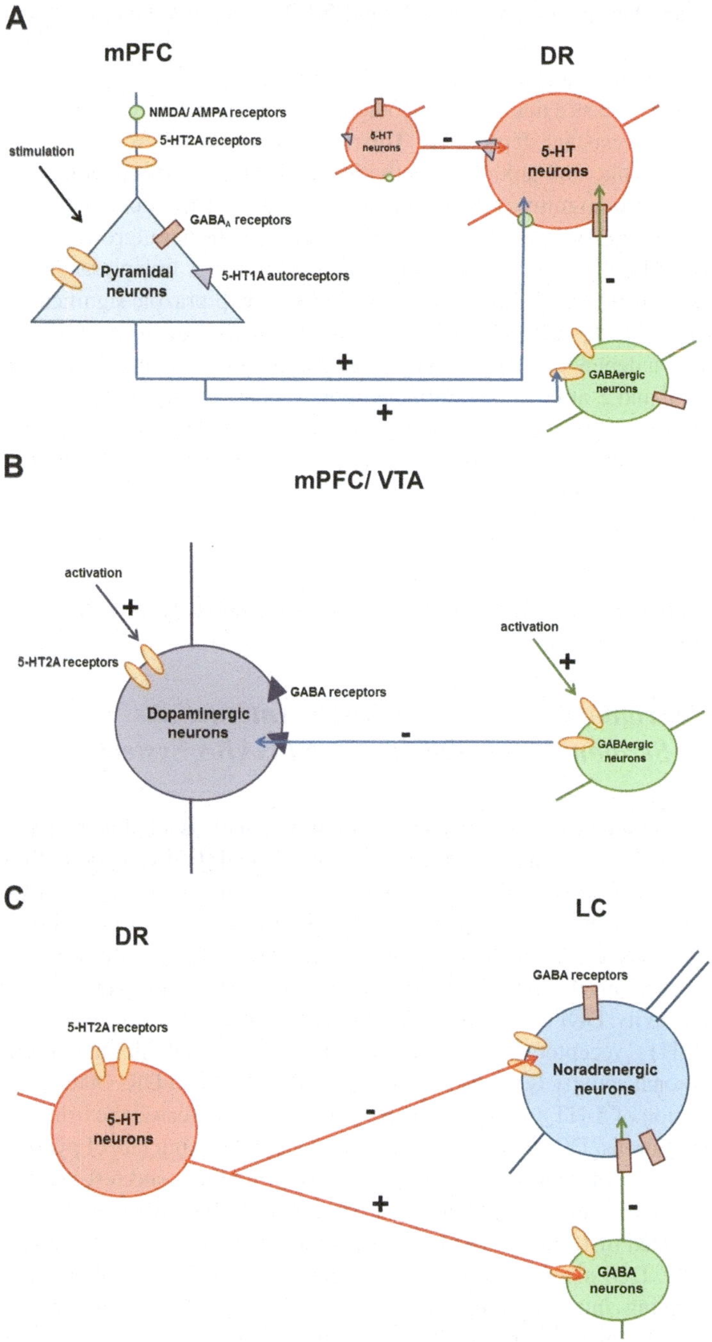

Fig. 2 Interactions between 5-HT_{2A} receptors and the other system. (**a**) Excitatory pyramidal neurons in the medial prefrontal cortex (mPFC) control activity of 5-HT neurons in dorsal raphe (DR)

and the brain stem. The apical and basal dendrites of pyramidal neurons of the PFC are highly enriched with 5-HT_{2A} receptors. These receptors are present also on large and medium-sized GABAergic interneurons that control the activity of local microcircuits [55]. The mPFC in rodents innervates via long glutamatergic axons various brain areas involved in depression, such as nucleus accumbens (NAcc), amygdala, and PFC [56]. As well, activity of dopaminergic neurons in ventral tegmental area (VTA) is under the excitatory control of 5-HT_{2A} receptors in mPFC. Neurons in mPFC excited through 5-HT_{2A} receptors increase the firing rate and burst firing of dopaminergic neuron and dopamine release in VTA [57].

The 5-HT_{2A} receptor activation located on thalamocortical afferents could increase glutamate release and increase spontaneous excitatory postsynaptic currents (EPSCs) through the activation of pyramidal AMPA receptors, however, this suggestion is based by the recent anatomical data indicating that the terminal 5-HT_{2A} receptors are not located on glutamate axons [58].

Role of Serotonin-2A Receptors in the Interactions Between Serotonin and Dopamine Systems

The 5-HT_{2A} receptor stimulation results in enhanced dopamine (DA) release in rat PFC, presumably via facilitation of 5-HT_{1A} receptor stimulation. Ability of clozapine to increase DA release may be boosted by antagonism of 5-HT_{2A} receptors [59].

The local infusion of DOI into the PFC dampened potassium (K^+)-mediated DA release in a dose-dependent manner. Regular intracortical administration of MDL 100907 caused an increase in cortical DA efflux, suggesting that cortical 5-HT_{2A} receptors potentiate the phasic release of DA [60]. The stimulatory effect of 5-HT on efflux of dopamine in the striatum is effective only when nigro-striatal DA transmission is elevated above basal levels [61]. Antagonism of 5-HT_{2A} receptors may modulate the activity of dopamine neurons in different areas. For the nigro-striatal dopaminergic pathway was suggested a model in which blockade of these receptors led to increased output of dopaminergic neurons into the striatum [62].

Brexpiprazole has higher affinity to D_2 than to the 5-HT_{2A} receptors. While other antipsychotic drugs act as D_2 antagonists, brexpiprazole is a partial agonist of the D_2

Fig. 2 (continued) through 3 different mechanisms: *N*-methyl-D-aspartate (NMDA) and 2-amino-3-(3-hydroxy-5-methylisoxazol-4-yl)propionate (AMPA) receptors- mediated excitation; $GABA_A$ receptors- mediated inhibition; and 5-HT_{1A} autoreceptors- mediated inhibition. (**b**) Regulation of the dopaminergic system through 5-HT_{2A} receptors. In the ventral tegmental area (VTA) or in medial prefrontal cortex (mPFC), 5-HT_{2A} receptors have also been identified in GABAergic interneurons. Their activation leads to the inhibition of dopaminergic activity. 5-HT_{2A} receptors might also be expressed in dopaminergic neurons in VTA region and their activation would stimulate dopaminergic activity. (**c**) Locus coeruleus (LC) receives dense 5-HT projections coming from dorsal raphe (DR), which have an inhibitory effect on noradrenergic neurons. Increased 5-HT levels act also on excitatory 5-HT_{2A} receptors on GABAergic neurons which lead to an inhibition of norepinephrine release

receptors [63, 64]. The D_2 receptor agonistic features could alter DA neurotransmission by stimulating D_2 receptors when the levels of DA are lowered, while decreasing their activation when DA levels are increased [65].

Increase in 5-HT levels inhibits dopaminergic neurons as the lesion of 5-HT neurons results in an increase of dopaminergic neuronal activity in the VTA [66]. Thus, an increase in the availability of 5-HT cause by SSRIs might result in attenuation of the firing of dopaminergic neurons. Neuronal activity of dopaminergic neurons has a critical role in the VTA in motivation, hedonia and reward, so the inhibition of this firing might contribute to SSRI resistance in some patients [67].

Role of Serotonin 2A Receptors in the Interactions Between Serotonin and Norepinephrine Systems

The 5-HT_{2A} receptor is likely to play an important role in the interaction between norepinephrine (NE) and serotonin (5-HT) systems [68]. Increased 5-HT levels act on excitatory 5-HT_{2A} receptors on GABA neurons, thus leading to an inhibition of NE release [69].

Acute brexpiprazole administration reduced inhibition of two important interaction nodes between the 5-HT and NE systems. The blockade of 5-HT_{2A} receptors revokes the tonic inhibition of NE neuronal firing activity, and the blocking of α_2-adrenergic receptors on the nerve terminals of NE neurons stimulates NE release [70].

YM992 [(S)-2-[[(7-fluoro-4-indanyl)oxy]methyl]morpholine monohydrochloride] is a selective serotonin reuptake inhibitor (SSRI) and a potent 5-HT_{2A} receptor antagonist. Acute injection of YM992 significantly decreased NE neuron firing activity and blocked the inhibitory effect of a subsequent injection of the 5-HT_2 receptor agonist DOI. After 2-day treatment the firing activity was elevated even more significantly, however after 7-day and 21-day treatment a partial recovery was observed. This NE activity may be a result of 5-HT reuptake inhibition plus 5-HT_{2A} receptor antagonism [69].

The activation of 5-HT_{2A} and 5-HT_{1A} receptors suppresses the firing of 5-HT and noradrenergic neurons of the locus coeruleus (LC). Serotoninergic neurons recover their firing rate with prolonged treatment, because of the desensitization of 5-HT_{1A} autoreceptors, but the firing rate of noradrenergic neurons does not recover over time [68].

Role of Serotonin-HT2A in the Response to Antidepressant and Mood Stabilizing Drugs

Selective serotonin reuptake inhibitors (SSRIs) induce inhibition of NE neuron firing [71]. It was reported in several open-label and blind studies that antagonists of 5-HT_{2A} receptors, such as atypical antipsychotic drugs, potentiate the therapeutic effect of SSRIs in patients with depression [72]. It is also reported that antidepressants induce down-regulation of 5-HT_{2A} receptors after repeated treatment [55].

Risperidone is 5-HT_{2A} and dopamine D_2 receptor antagonist which is the only antagonist known to saturate the 5-HT_{2A} receptors even at low doses (0.5–1 mg/day) [73]. It was reported that risperidone reverses SSRI-induced inhibition of NE neurons due to its 5-HT_{2A} receptor antagonistic property [71]. Co-administration of risperidone with venlafaxine or fluoxetine may enhance their antidepressant effects. Addition of yohimibine to the combination of risperidone with venlafaxine or fluoxetine augmented the antidepressant-like action proposing an interaction of α_2-adrenergic and 5-HT_{2A} receptor in mediating their action [74]. Palperidone is the main metabolite of risperidone. Although they share the same receptor binding profile, it seems that they have different effects on 5-HT and NE firing in vivo. Co-administration of paliperidone did not interfere with the effect of SSRIs, but still managed to inhibit the NE firing inhibition induced by the SSRIs which leads to assumption that it may be an effective enhancement of the treatment [75].

Amibegron (SR58611A)—selective β3 adrenergic agonist [76] interacts with serotonergic system in the brain resulting in an antidepressant effect [77]. It increases the synthesis of 5-HT and tryptophan levels in several brain areas, such as hippocampus, cortex, hypothalamus and striatum. Amibegron did not modify noradrenaline synthesis and metabolism, but it did increase its release [78]. A 5-HT_{2A} receptor antagonist ketanserin significantly reversed the effect of amibegron which leads to conclusion that these antidepressant-like effects are partially caused by the 5-HT_{2A} receptor activation, more precisely by interaction with 5-HT_{1A}, $5\text{-HT}_{2A/2C}$ and 5-HT_3 serotonin receptors [79, 80].

Function of cortical 5-HT_{2A} receptors has a specific role in the modulation of conflict anxiety. Weisstaub et al. [81] demonstrated that global disruption of 5-HT_{2A} receptor signaling in mice reduced inhibition in conflict anxiety paradigms without affecting fear-conditioned and depression-related behaviors. Selective restoration of 5HT_{2A} receptor signaling to the cortex normalized conflict anxiety behaviors.

The serotonergic system appears to play a role in episodic memory which is affected in pathologies such as schizophrenia, Alzheimer and depression. The 5-HT_{2A} receptors as one of the principal post-synaptic receptors for 5-HT in the brain are involved in neuropsychiatric and neurological disorders associated with memory deficits. Results of Morici et al. [82] showed that the 5-HT_{2A} and also 5-HT_{1A} receptors can be a novel target for drug development to improve episodic memory retrieval in psychiatric and neurological disorders.

Serotonin-2A Receptors in Pathophysiology and Treatment of Depression

Expression and Function of Serotonin-2A Receptors in the Hippocampus

Hippocampus is a brain structure which plays role in a spatial learning and declarative memory. It receives robust serotonergic innervation from medial and dorsal raphe nuclei. There is some evidence indicating role of 5-HT and its receptors in various aspects of cognitive functions including learning and memory. Nowadays, exact role of 5-HT in hippocampus is not fully understood. Results of functional studies are contradictory. One of possible explanation for these contradictory results is that 5-HT acts through different types of 5-HT receptors. The 5-HT_{2A} receptor subtype is related to memory disorders [83] and several neurological diseases like Alzheimer disease [84, 85] and schizophrenia [86–88].

The presence of 5-HT_{2A} receptors in hippocampus was demonstrated in different studies by multiple methods including immunohistochemistry, *in situ* hybridization, autoradiography and quantitative reverse transcription-polymerase chain reaction (RT-PCR). Results from these studies are quite different and depending on methodology which was used. Minimal levels of 5-HT_{2A} receptors were detected in human hippocampus by RT-PCR and autoradiography. They were barely detected in pyramidal cells in Cornu ammonis (CA) regions, and were not detected in dentate gyrus (DG) [89]. In rat hippocampus mRNA for 5-HT_{2A} receptors was detected in both CA regions and in DG [90]. In CA area of rat hippocampus low levels of 5-HT_{2A} receptors were detected by *in situ* hybridization and autoradiography methods. In ventral DG moderate levels of specific 5-HT_{2A} receptors binding were detected [91]. Immunohistochemistry studies showed that 5-HT_{2A} receptors expressed both excitatory glutamatergic and inhibitory GABAergic neurons [92–95]. Virtually all main hippocampal excitatory neurons (granular and pyramidal cells) expressed 5-HT_{2A} receptors. Strong expression is localized in apical dendrites of pyramidal cells, where 5-HT receptors can increase excitatory postsynaptic currents (EPSP) [92, 94]. Electrophysiological studies demonstrated that outward current induced by 5-HT and α-methyl-serotonin (5-HT_{2A} receptors agonist) in pyramidal cells of rat CA1 hippocampal area is blocked by ketanserin and spiperon (5-HT_{2A} receptors antagonist) in dose dependent manner [96]. The 5-HT_{2A} receptors are also expressed in mossy fiber in rat hippocampus [92]. Receptors localized on presynaptic side of mossy fibers could regulate excitatory neurotransmission and as result affect release of glutamate in hippocampus [97, 98]. On the other hand, colocalization analyses show that 5-HT_{2A} receptors are expressed in GABAergic neurons located in different rat hippocampal regions. This colocalization is similar in different hippocampal areas: in DG, CA1, CA2 and CA3 field. In hippocampal CA areas are 5-HT_{2A} receptors widespread in number of GABAergic interneurons distributed in pyramidal cell layer, in strata oriens, radiatum and lacunosum-moleculare.

The 5-HT_{2A} receptors are expressed on 90% of GABAergic neurons in hippocampus [92]. Electrophysiology studies showed that activation of 5-HT_{2A} receptors activate GABAergic neurons in rat DG [99] and in CA1 field [100]. High density of 5-HT_{2A} receptor in deeper layers of granular cell layer corresponds with study demonstrating that 5-HT receptors can regulate neurogenesis in subgranular zone of DG [101]. Because GABA regulates progenitor turnover and integration of newly synthetized neurons in DG [102], it can be assumed that GABA neurons distributed in subgranular zone can be involved in hippocampal progenitor proliferation mediated by 5-HT_{2A} receptors [103].

Function of 5-HT$_{2A}$ Receptors in Hippocampus in Health

Recent studies suggested that 5-HT$_{2A}$ receptors are included in several hippocampal functions although underlying mechanisms are still unclear. Activity of hippocampal pyramidal neurons can be modulated by 5-HT$_{2A}$ receptors in different ways: directly, by activation of 5-HT$_{2A}$ receptors in pyramidal cells, or indirectly, by activation of 5-HT$_{2A}$ receptors in GABA interneurons [96]. Serotonin 5-HT$_{2A}$ receptors can participate in information processing in hippocampus by participating in neurotransmission in different neuronal populations. Strong and widespread expression of 5-HT$_{2A}$ receptors in hippocampus is prerequisite for critical involvement of 5-HT receptors in number of brain functions including learning and memory [92]. It was shown that an application of M100907 (highly selective 5-HT$_{2A}$ receptors antagonist) to brain slices facilitates induction of long term potentiation (LTP) in CA1 field of rat hippocampus [104].

As a critical factor modulating brain plasticity is considered brain-derived neurotrophic factor (BDNF). Hippocampal BDNF mRNA expression was induced by physical activity which positively regulated neurogenesis and induced LTP [105]. This factor can acutely influence synaptic efficiency of neurons. Some electrophysiological studies demonstrate that application of BDNF on hippocampal slices results in increase of synaptic strength [106–110]. In hippocampus 5-HT$_{2A}$ receptors participate in regulation of BDNF levels as their agonist DOI decreased the expression of BDNF mRNA in granular cell layer in DG, but not in CA regions. Effect of agonist was blocked by pretreatment with selective antagonist of 5-HT$_{2A}$ receptors. Same decrease of BDNF expression in hippocampus is observed during stress and it is possible that this effect is mediated by 5-HT$_{2A}$ receptors. This hypothesis is supported by an observation that pretreatment with ketaserin significantly blocked stress induced decrease in BDNF expression [111].

Involvement of 5-HT$_{2A}$ receptors in process of learning and memory is supported by study where systematic activation of 5-HT$_{2A}$ receptors with agonist (TCB-2) enhanced the consolidation of both fear memory and object memory [112]. The memory strengthening effect of TCB-2 was blocked by pretreatment with 5-HT$_{2A}$ receptors antagonist (MDL11,939). Local microinfusion of TCB-2 into CA1 field of dorsal hippocampus had similar effect on memory consolidation observed after systemic treatment [113]. Postsynaptic 5-HT$_{2A}$ receptors can modulate memory storage associated with object also by influencing on *N*-Methyl-D-Aspartate (NMDA) receptors. It is supported by fact that hippocampal 5-HT$_{2A}$ receptors are predominantly expressed in dendritic part of pyramidal neurons [93, 114] and dendrites which expressed 5-HT$_{2A}$ receptors expressed also NMDAR subunit NR1 and GluR2 [114]. Activation of 5-HT$_{2A}$ receptors causes an increase of intracellular Ca^{2+} concentration which in combination with NMDA receptor-mediated calcium influx can strengthen the synaptic plasticity. These observations suggest that an activation of 5-HT$_{2A}$ receptors induces facilitation of object memory storage and can result from potentiating of glutamate release in hippocampus, temporal dynamics of pyramidal neurons and critical post-training period. These receptors may serve as a drug target for pharmacological intervention in the treatment of memory disorders [115]. It is known that new neurons are generated in mammal DG. These new neurons are later during life integrated into hippocampal circuit. Serotonin belongs to important factors influencing neurogenesis. Among others 5-HT receptor subtypes (5-HT_{1A}, 5-HT_{1B} and 5-HT_{2C}), activation of 5-HT_{2A} receptors is involved in the positive regulation of adult neurogenesis in DG caused by regulation of cell proliferation in this region [103]. It was reported that some animal models of depression produce decrease in hippocampal cell proliferation and neurogenesis. Unlike the depression, chronic treatment with antidepressants, such as SSRIs, seem to have the positive effect on neurogenesis which is sufficient to reduce anxiety and depression-related behavior [116].

Role of Hippocampal Serotonin-2A Receptors in Pathophysiology and Treatment of Depression

The main effect of antidepressants is increasing of synaptic 5-HT levels. There is some evidence suggesting that hippocampus can be influenced by depression. It is known that hypercorticosolemia, an animal model of depression, results in the death of hippocampal neurons [117]. Change of serotonergic function in hippocampus is likely to be involved in defects of mood regulation associated with the major depressive disorder (MDD). Serotonin 5-HT_{2A} receptors play role in these changes. Postmortem studies in depressed suicide completers documented changes in 5-HT_{2A} receptors binding in hippocampus [118, 119]. Magnetic resonance imaging (MRI) studies showed changes in 5-HT_{2A} receptors binding potential in hippocampus in patients with MDD [120, 121]. Magnetic resonance imaging studies also demonstrated decrease of hippocampal volume in patients with MDD which correlated with duration of depression [120, 121]. However, decrease in 5-HT_{2A} receptors binding potential is higher than volume loss and indicates that both conditions can coexist. Not only depression itself, but also the total number of days with depression inversely correlates with hippocampal volume [121, 122]. Serotonin 5-HT_{2A} receptor binding is not influenced by depression phase. However, patients not previously treated for depression have lower 5-HT_{2A} receptor binding than patients with previous medication treatment. It is possible that medication treatment provides compensatory upregulation of 5-HT_{2A} receptors [123]. It is well established that decreased 5-HT_{2A} receptor transmission is associated with depression [124]. It is also possible that decreased 5-HT_{2A} receptor-mediated neurotransmission has special importance. Indeed, decreased 5-HT_{2A} receptors binding was reported in patients with depression [123]. In addition, antidepressants treatment may cause changes in expression and binding of 5-HT_{2A} receptors and these changes can persist for a long time after treatment [1, 125–129].

Nowadays, the role of astrocytes in depression has been intensively studied [130]. 5-HT_{2A} receptors are expressed not only in hippocampal neurons, but also in astrocytes. This suggests the possibility that also 5-HT_{2A} receptors express in astrocyte have functional implications in psychiatric disorders [95]. Beside their housekeeping functions, astrocytes are dynamic regulators of synaptogenesis, synaptic strength and control neurogenesis in the adult DG [131]. Astrocytes synthesize and release many neurotrophic factors vital for neuronal health such as BDNF, glial-derived neurotrophic factor (GDNF), nerve growth factor (NGF), and neurotrophins 3 and 4/5 [132, 133]. Brain-derivated neurotrophic factor blocks neurogenesis in depression which is opposite to healthy condition. Its function has been implicated in the neurogenesis hypothesis of depression in which the antidepressants enhance neurogenesis, and BDNF is a key regulator of this mechanism. Antidepressants (including SSRIs) induce the CREB phosphorylation, CREB binds to the BDNF 13 promoter and induces BDNF transcription. Moreover, stress can reduce the expression of BDNF in the hippocampus and this reduction can be prevented by long-term chronic antidepressant treatment [134, 135]. *In vitro* studies reported that SSRIs stimulate the expression of BDNF, GDNF and vascular endothelial growth factor (VEGF) in primary culture of astrocytes [136–138]. *In vivo* data showed that the specific overexpression of BDNF in hippocampal astrocytes produced antidepressant-like effect accompanied by an increase in cell proliferation, maturation and survival of new neurons by generated cells in the DG of the hippocampus [139]. It is possible that astrocytes contribute to the enhancement in neurotrophic support and associated augmentation in synaptic plasticity that may form the basis for antidepressant efficacy. Several reports suggested that fluoxetine and other drugs can modulate the structural plasticity of astrocytes. Following chronic administration of lithium and some antipsychotic drugs, increased numbers of glia have been reported in the hippocampi of rats and nonhuman primates [140, 141]. In another study fluoxetine prevented the stress-induced decrease on a number of hippocampal astrocytes, but had no effect in nonstressed animals [142]. It demonstrates that fluoxetine, a prominent member of the SSRI family, can significantly modify the structural plasticity of astrocytes, and it is very likely that these morphological alterations either reflect or induce functional changes within the glial–neuronal interaction [142]. In particular, it is well accepted that SSRIs activate 5-HT_{2A} receptors and stimulate signaling intracellular cascades leading to the phosphorylation/activation of extracellular signal regulated kinases (ERK1/2). Hence, antidepressants may exert their therapeutic activity by stimulating this pathway. In the hippocampus ERK1/2 have been implicated in mood regulation [143] as suggested by their blunted activation and/or expression in both depressed patient [144] and animal models of depression [145].

Conclusion

The 5-HT_{2A} receptors belong to the 5-HT_2 receptor family, the only known group of 5-HT receptors which are coupled to $G\alpha_{Q/Z}$ proteins. The primary signal transduction mechanism of 5-HT_{2A} receptors involves activation of PLC and calcium signaling. However, 5-HT_{2A} receptor-mediated alteration of cAMP levels has also been reported. The 5-HT_{2A} receptor is a product of 5HT2AR gene. Genetic polymorphism of 5HT2AR gene, its epigenetic regulation, and post-translational modifications of 5HT2AR mRNA have been reported. Furthermore, pre- and post-translational 5HT2AR alterations correlate with certain CNS disorders, such as depression, schizophrenia, dementia, and alcohol and nicotine dependence. On the functional level, 5-HT_{2A} receptors play a central role in the interaction between 5-HT and norepinephrine systems and they are also involved in 5-HT-glutamate, 5-HT-GABA, and 5-HT-dopamine interactions. In addition, 5-HT_{2A} receptors are fundamental in the modulation of hippocampal neuronal circuits. These lines of evidence, taken together, indicate that 5-HT_{2A} receptors are one of the primary targets for antidepressant and mood stabilizing drugs and other CNS medications. And indeed, atypical antidepressant drugs act as antagonist of 5-HT_{2A} receptors.

Acknowledgements The authors of this chapter were supported by the Slovak Academy of Sciences (SAS; 2013 Scholarship Award), by the Scientific Grant Agency of Ministry of Education of Slovak Republic and SAS (grants VEGA-2/0019/15 and VEGA-2/0024/15) and by the Slovak Research and Development Agency (Grants APVV-15-0388).

References

1. Leysen JE (2004) 5-HT2 receptors. Curr Drug Targets CNS Neurol Disord 3(1):11–26
2. Ishier R, Bhattacharya A, Panicker MM (2007) Serotonin-2A (5-HT2A) receptor function: ligand-dependent mechanisms and pathways. In: Chattopadhyay A (ed) Serotonin receptors in neurobiology. CRC, Boca Raton, FL, pp 105–132
3. Kroeze WK, Kristiansen K, Roth BL (2002) Molecular biology of serotonin receptors structure and function at the molecular level. Curr Top Med Chem 2(6):507–528
4. Hoyer D, Hannon JP, Martin GR (2002) Molecular, pharmacological and functional diversity of 5-HT receptors. Pharmacol Biochem Behav 71(4):533–554. doi:S0091305701007468 [pii]
5. Van de Kar LD, Javed A, Zhang Y et al (2001) 5-HT2A receptors stimulate ACTH, corticosterone, oxytocin, renin, and prolactin release and activate hypothalamic CRF and oxytocin-expressing cells. J Neurosci 21(10):3572–3579. doi:21/10/3572 [pii]
6. Kleinrock M (2011) The use of medicines in the United States: review of 2010. IMS Institute for Healthcare Informatics, Danbury, CT
7. Becker KG, Barnes KC, Bright TJ et al (2004) The genetic association database. Nat Genet 36(5):431–432. https://doi.org/10.1038/ng0504-431
8. Parsons MJ, D'Souza UM, Arranz MJ, Kerwin RW, Makoff AJ (2004) The -1438A/Gpolymorphism in the 5-hydroxytryptamine type 2A receptor gene affects promoter activity. Biol Psychiatry 56(6):406–410
9. Serretti A, Drago A, De Ronchi D (2007) HTR2A gene variants and psychiatric disorders: a review of current literature and selection of SNPs for future studies. Curr Med Chem 14(19):2053–2069
10. Hazelwood LA, Sanders-Bush E (2004) His452Tyr polymorphism in the human 5-HT2A receptor destabilizes the signaling conformation. Mol Pharmacol 66(5):1293–1300
11. Lin CX, Hu Z, Yan ZM et al (2015) Association between HTR2A T102C polymorphism and major depressive disorder: a meta-analysis in the Chinese population. Int J Clin Exp Med 8(11):20897–20903
12. Tan J, Chen S, Su L et al (2014) Association of the T102C polymorphism in the HTR2A gene with major depressive disorder, bipolar disorder, and schizophrenia. Am J Med Genet B Neuropsychiatr Genet 165B(5):438–455. https://doi.org/10.1002/ajmg.b.32248

13. Zhao X, Sun L, Sun YH et al (2014) Association of HTR2A T102C and A-1438G polymorphisms with susceptibility to major depressive disorder: a meta-analysis. Neurol Sci 35(12):1857–1866. https://doi.org/10.1007/s10072-014-1970-7
14. Jobim PF, Prado-Lima PA, Schwanke CH et al (2008) The polymorphism of the serotonin-2A receptor T102C is associated with age. Braz J Med Biol Res 41(11):1018–1023. doi:S0100-879X2008005000045 [pii]
15. Petit AC, Quesseveur G, Gressier F et al (2014) Converging translational evidence for the involvement of the serotonin 2A receptor gene in major depressive disorder. Prog Neuro-Psychopharmacol Biol Psychiatry 54:76–82. https://doi.org/10.1016/j.pnpbp.2014.04.013
16. Qesseveur G, Petit AC, Nguyen HT, Dahan L, Colle R, Rotenberg S, Seif I, Robert P, David D, Guilloux JP, Gardier AM, Verstuyft C, Becquemont L, Corruble E, Guiard BP (2016) Genetic dysfunction of serotonin 2A receptor hampers response to antidepressant drugs: a translational approach. Neuropharmacology 105:142–153. https://doi.org/10.1016/j.neuropharm.2015.12.022
17. Abdolmaleky HM, Faraone SV, Glatt SJ et al (2004) Meta-analysis of association between the T102C polymorphism of the 5HT2a receptor gene and schizophrenia. Schizophr Res 67(1):53–62. doi:S092099640300183X [pii]
18. Joober R, Benkelfat C, Brisebois K et al (1999) T102C polymorphism in the 5HT2A gene and schizophrenia: relation to phenotype and drug response variability. J Psychiatry Neurosci 24(2):141–146
19. Serretti A, Benedetti F, Mandelli L et al (2008) Association between GSK-3beta -50T/C polymorphism and personality and psychotic symptoms in mood disorders. Psychiatry Res 158(2):132–140. doi:S0165-1781(07)00199-0 [pii]
20. do Prado-Lima PA, Chatkin JM, Taufer M et al (2004) Polymorphism of 5HT2A serotonin receptor gene is implicated in smoking addiction. Am J Med Genet B Neuropsychiatr Genet 128B(1):90–93. https://doi.org/10.1002/ajmg.b.30004
21. Nakamura T, Matsushita S, Nishiguchi N et al (1999) Association of a polymorphism of the 5HT2A receptor gene promoter region with alcohol dependence. Mol Psychiatry 4(1):85–88
22. Craig D, Donnelly C, Hart D et al (2007) Analysis of the 5HT-2A T102C receptor polymorphism and psychotic symptoms in Alzheimer's disease. Am J Med Genet B Neuropsychiatr Genet 144B(1):126–128. https://doi.org/10.1002/ajmg.b.30409
23. Lam LC, Tang NL, Ma SL et al (2004) 5-HT2A T102C receptor polymorphism and neuropsychiatric symptoms in Alzheimer's disease. Int J Geriatr Psychiatry 19(6):523–526. https://doi.org/10.1002/gps.1109
24. Polesskaya OO, Aston C, Sokolov BP (2006) Allele C-specific methylation of the 5-HT2A receptor gene: evidence for correlation with its expression and expression of DNA methylase DNMT1. J Neurosci Res 83(3):362–373. https://doi.org/10.1002/jnr.20732
25. De Luca V, Viggiano E, Dhoot R et al (2009) Methylation and QTDT analysis of the 5-HT2A receptor 102C allele: analysis of suicidality in major psychosis. J Psychiatr Res 43(5):532–537. https://doi.org/10.1016/j.jpsychires.2008.07.007
26. Falkenberg VR, Gurbaxani BM, Unger ER et al (2011) Functional genomics of serotonin receptor 2A (HTR2A): interaction of polymorphism, methylation, expression and disease association. Neuromol Med 13(1):66–76. https://doi.org/10.1007/s12017-010-8138-2
27. Huang X, Xiao H, Rex EB et al (2002) Functional characterization of alternatively spliced 5-HT2 receptor isoforms from the pharynx and muscle of the parasitic nematode, Ascaris suum. J Neurochem 83(2):249–258. doi:1067 [pii]
28. Guest PC, Salim K, Skynner HA et al (2000) Identification and characterization of a truncated variant of the 5-hydroxytryptamine(2A) receptor produced by alternative splicing. Brain Res 876(1–2):238–244. doi:S0006-8993(00)02664-0 [pii]
29. Berg KA, Maayani S, Goldfarb J et al (1998a) Pleiotropic behavior of 5-HT2A and 5-HT2C receptor agonists. Ann N Y Acad Sci 861:104–110
30. Hoyer D, Clarke DE, Fozard JR et al (1994) International Union of Pharmacology classification of receptors for 5-hydroxytryptamine (serotonin). Pharmacol Rev 46(2):157–203

31. Raymond JR, Mukhin YV, Gelasco A et al (2001) Multiplicity of mechanisms of serotonin receptor signal transduction. Pharmacol Ther 92(2-3):179–212. doi:S0163725801001693 [pii]
32. Gooz M, Gooz P, Luttrell LM et al (2006) 5-HT2A receptor induces ERK phosphorylation and proliferation through ADAM-17 tumor necrosis factor-alpha-converting enzyme (TACE) activation and heparin-bound epidermal growth factor-like growth factor (HB-EGF) shedding in mesangial cells. J Biol Chem 281(30):21004–21012. doi:M512096200 [pii]
33. Quinn JC, Johnson-Farley NN, Yoon J et al (2002) Activation of extracellular-regulated kinase by 5-hydroxytryptamine(2A) receptors in PC12 cells is protein kinase C-independent and requires calmodulin and tyrosine kinases. J Pharmacol Exp Ther 303(2):746–752. https://doi.org/10.1124/jpet.102.038083
34. Sheffler DJ, Kroeze WK, Garcia BG et al (2006) p90 ribosomal S6 kinase 2 exerts a tonic brake on G protein-coupled receptor signaling. Proc Natl Acad Sci U S A 103(12):4717–4722. doi:0600585103 [pii]
35. Strachan RT, Allen JA, Sheffler DJ et al (2010) p90 Ribosomal S6 kinase 2, a novel GPCR kinase, is required for growth factor-mediated attenuation of GPCR signaling. Biochemistry 49(12):2657-2671. https://doi.org/10.1021/bi901921k
36. Strachan RT, Sheffler DJ, Willard B et al (2009) Ribosomal S6 kinase 2 directly phosphorylates the 5-hydroxytryptamine 2A (5-HT2A) serotonin receptor, thereby modulating 5-HT2A signaling. J Biol Chem 284(9):5557–5573. https://doi.org/10.1074/jbc.M805705200
37. Arranz MJ, Munro J, Owen MJ et al (1998) Evidence for association between polymorphisms in the promoter and coding regions of the 5-HT2A receptor gene and response to clozapine. Mol Psychiatry 3(1):61–66
38. Dai Y, Dudek NL, Patel TB et al (2008) Transglutaminase-catalyzed transamidation: a novel mechanism for Rac1 activation by 5-hydroxytryptamine2A receptor stimulation. J Pharmacol Exp Ther 326(1):153–162. https://doi.org/10.1124/jpet.107.135046
39. Singh RK, Jia C, Garcia F et al (2010) Activation of the JAK-STAT pathway by olanzapine is necessary for desensitization of serotonin2A receptor-stimulated phospholipase C signaling in rat frontal cortex but not serotonin2A receptor-stimulated hormone release. J Psychopharmacol 24(7):1079–1088. https://doi.org/10.1177/0269881109103090
40. Shi J, Damjanoska KJ, Singh RK et al (2007) Agonist induced-phosphorylation of Galpha11 protein reduces coupling to 5-HT2A receptors. J Pharmacol Exp Ther 323(1):248–256. doi:jpet.107.122317 [pii]
41. Kurrasch-Orbaugh DM, Watts VJ, Barker EL et al (2003) Serotonin 5-hydroxytryptamine 2A receptor-coupled phospholipase C and phospholipase A2 signaling pathways have different receptor reserves. J Pharmacol Exp Ther 304(1):229–237. https://doi.org/10.1124/jpet.102.042184
42. Berg KA, Maayani S, Goldfarb J et al (1998b) Effector pathway-dependent relative efficacy at serotonin type 2A and 2C receptors: evidence for agonist-directed trafficking of receptor stimulus. Mol Pharmacol 54(1):94–104
43. Karaki S, Becamel C, Murat S et al (2014) Quantitative phosphoproteomics unravels biased phosphorylation of serotonin 2A receptor at Ser280 by hallucinogenic versus nonhallucinogenic agonists. Mol Cell Proteomics 13(5):1273–1285. https://doi.org/10.1074/mcp.M113.036558
44. Jones KA, Srivastava DP, Allen JA et al (2009) Rapid modulation of spine morphology by the 5-HT2A serotonin receptor through kalirin-7 signaling. Proc Natl Acad Sci U S A 106(46):19575–19580. https://doi.org/10.1073/pnas.0905884106
45. Preedy VR (2016) Neuropathology of drug addictions and substance misuse, 1st edn. Elsevier B.V., Amsterdam, The Netherlands
46. Bhatnagar A, Willins DL, Gray JA et al (2001) The dynamin-dependent, arrestin-independent internalization of 5-hydroxytryptamine 2A (5-HT2A) serotonin receptors reveals differential sorting of arrestins and 5-HT2A receptors during endocytosis. J Biol Chem 276(11):8269–8277. https://doi.org/10.1074/jbc.M006968200

47. Bhattacharyya S (2005) Internalization and recycling of the serotonin 2A receptor in non-neuronal and neuronal cells, in National Centre for Biological Sciences. Manipal Academy of Higher Education of Bangalore, Bangalore
48. Barbas D, DesGroseillers L, Castellucci VF et al (2003) Multiple serotonergic mechanisms contributing to sensitization in aplysia: evidence of diverse serotonin receptor subtypes. Learn Mem 10(5):373–386. https://doi.org/10.1101/lm.66103
49. Citrome L, Stensbol TB, Maeda K (2015) The preclinical profile of brexpiprazole: what is its clinical relevance for the treatment of psychiatric disorders? Expert Rev Neurother 15(10):1219–1229. https://doi.org/10.1586/14737175.2015.1086269
50. Ishima T, Futamura T, Ohgi Y et al (2015) Potentiation of neurite outgrowth by brexpiprazole, a novel serotonin-dopamine activity modulator: a role for serotonin 5-HT1A and 5-HT2A receptors. Eur Neuropsychopharmacol 25(4):505–511. https://doi.org/10.1016/j.euroneuro.2015.01.014
51. Boothman LJ, Allers KA, Rasmussen K, Sharp T. Evidence that central 5-HT2A and 5-HT2B/C receptors regulate 5-HT cell firing in the dorsal raphe nucleus of the anaesthetised rat. Br J Pharmacol. 2003;139(5):998-1004. Erratum in: Br J Pharmacol. 2003;140(1):227–8.
52. Quesseveur G, Repérant C, David DJ, Gardier AM, Sanchez C, Guiard BP (2013) 5-HT_{2A} receptor inactivation potentiates the acute antidepressant-like activity of escitalopram: involvement of the noradrenergic system. Exp Brain Res 226(2):285–295. https://doi.org/10.1007/s00221-013-3434-3
53. Puig MV, Celada P, Diaz-Mataix L et al (2003) In vivo modulation of the activity of pyramidal neurons in the rat medial prefrontal cortex by 5-HT2A receptors: relationship to thalamocortical afferents. Cereb Cortex 13(8):870–882
54. Bortolozzi A, Amargos-Bosch M, Adell A et al (2003) In vivo modulation of 5-hydroxytryptamine release in mouse prefrontal cortex by local 5-HT(2A) receptors: effect of antipsychotic drugs. Eur J Neurosci 18(5):1235–1246. doi:2829 [pii]
55. Celada P, Puig MV, Casanovas JM et al (2001) Control of dorsal raphe serotonergic neurons by the medial prefrontal cortex: involvement of serotonin-1A, GABA(A), and glutamate receptors. J Neurosci 21(24):9917–9929
56. Groenewegen HJ, Uylings HB (2000) The prefrontal cortex and the integration of sensory, limbic and autonomic information. Prog Brain Res 126:3–28. doi:S0079-6123(00)26003-2 [pii]
57. Bortolozzi A, Díaz-Mataix L, Scorza MC, Celada P, Artigas F (2005) The activation of 5-HT receptors in prefrontal cortex enhances dopaminergic activity. J Neurochem 95(6):1597–1607
58. Miner LA, Backstrom JR, Sanders-Bush E et al (2003) Ultrastructural localization of serotonin2A receptors in the middle layers of the rat prelimbic prefrontal cortex. Neuroscience 116(1):107–117. doi:S0306452202005808 [pii]
59. Ichikawa J, Dai J, Meltzer HY (2001) DOI, a 5-HT2A/2C receptor agonist, attenuates clozapine-induced cortical dopamine release. Brain Res 907(1-2):151–155. doi:S0006-8993(01)02596-3 [pii]
60. Pehek EA, McFarlane HG, Maguschak K et al (2001) M100,907, a selective 5-HT(2A) antagonist, attenuates dopamine release in the rat medial prefrontal cortex. Brain Res 888(1):51–59. doi:S0006-8993(00)03004-3 [pii]
61. Lucas G, Spampinato U (2000) Role of striatal serotonin2A and serotonin2C receptor subtypes in the control of in vivo dopamine outflow in the rat striatum. J Neurochem 74(2):693–701
62. Horacek J, Bubenikova-Valesova V, Kopecek M, Palenicek T, Dockery C, Mohr P, Höschl C (2006) Mechanism of action of atypical antipsychotic drugs and the neurobiology of schizophrenia. CNS Drugs 20(5):389–409
63. Maeda K, Lerdrup L, Sugino H et al (2014a) Brexpiprazole II: antipsychotic-like and procognitive effects of a novel serotonin-dopamine activity modulator. J Pharmacol Exp Ther 350(3):605–614. https://doi.org/10.1124/jpet.114.213819
64. Maeda K, Sugino H, Akazawa H et al (2014b) Brexpiprazole I: in vitro and in vivo characterization of a novel serotonin-dopamine activity modulator. J Pharmacol Exp Ther 350(3):589–604. https://doi.org/10.1124/jpet.114.213793

65. Burris KD, Molski TF, Xu C et al (2002) Aripiprazole, a novel antipsychotic, is a high-affinity partial agonist at human dopamine D2 receptors. J Pharmacol Exp Ther 302(1):381–389
66. Guiard BP, El Mansari M, Merali Z, Blier P (2008) Functional interactions between dopamine, serotonin and norepinephrine neurons: an in-vivo electrophysiological study in rats with monoaminergic lesions. Int J Neuropsychopharmacol 11(5):625–639. https://doi.org/10.1017/S1461145707008383
67. Dremencov E, El Mansari M, Blier P (2009) Effects of sustained serotonin reuptake inhibition on the firing of dopamine neurons in the rat ventral tegmental area. J Psychiatry Neurosci 34(3):223–229
68. Szabo ST, Blier P (2001) Effect of the selective noradrenergic reuptake inhibitor reboxetine on the firing activity of noradrenaline and serotonin neurons. Eur J Neurosci 13(11):2077–2087
69. Szabo ST, Blier P (2002) Effects of serotonin (5-hydroxytryptamine, 5-HT) reuptake inhibition plus 5-HT(2A) receptor antagonism on the firing activity of norepinephrine neurons. J Pharmacol Exp Ther 302(3):983–991
70. Oosterhof CA, El Mansari M, Blier P (2014) Acute effects of brexpiprazole on serotonin, dopamine, and norepinephrine systems: an in vivo electrophysiologic characterization. J Pharmacol Exp Ther 351(3):585–595. https://doi.org/10.1124/jpet.114.218578
71. Dremencov E, El Mansari M, Blier P (2007b) Noradrenergic augmentation of escitalopram response by risperidone: electrophysiologic studies in the rat brain. Biol Psychiatry 61(5):671–678. S0006-3223(06)00659-7 [pii]; https://doi.org/10.1016/j.biopsych.2006.05.015
72. Marek GJ, Carpenter LL, McDougle CJ et al (2003) Synergistic action of 5-HT2A antagonists and selective serotonin reuptake inhibitors in neuropsychiatric disorders. Neuropsychopharmacology 28(2):402–412. https://doi.org/10.1038/sj.npp.1300057
73. Schotte A, Janssen PF, Gommeren W et al (1996) Risperidone compared with new and reference antipsychotic drugs: in vitro and in vivo receptor binding. Psychopharmacology 124(1-2):57–73
74. Dhir A, Kulkarni SK (2008) Risperidone, an atypical antipsychotic enhances the antidepressant-like effect of venlafaxine or fluoxetine: possible involvement of alpha-2 adrenergic receptors. Neurosci Lett 445(1):83–88. https://doi.org/10.1016/j.neulet.2008.08.074
75. Dremencov E, El Mansari M, Blier P (2007a) Distinct electrophysiological effects of paliperidone and risperidone on the firing activity of rat serotonin and norepinephrine neurons. Psychopharmacology. https://doi.org/10.1007/s00213-007-0818-8
76. Bianchetti A, Manara L (1990) In vitro inhibition of intestinal motility by phenylethanolaminotetralines: evidence of atypical beta-adrenoceptors in rat colon. Br J Pharmacol 100(4):831–839
77. Overstreet DH, Stemmelin J, Griebel G (2008) Confirmation of antidepressantpotential of the selective beta3 adrenoceptor agonist amibegron in an animal model of depression. Pharmacol Biochem Behav 89(4):623–626. https://doi.org/10.1016/j.pbb.2008.02.020
78. Claustre Y, Leonetti M, Santucci V et al (2008) Effects of the beta3-adrenoceptor (Adrb3) agonist SR58611A (amibegron) on serotonergic and noradrenergic transmission in the rodent: relevance to its antidepressant/anxiolytic-like profile. Neuroscience 156(2):353–364. https://doi.org/10.1016/j.neuroscience.2008.07.011
79. Tanyeri P, Buyukokuroglu ME, Mutlu O et al (2013a) Evidence that the anxiolytic-like effects of the beta3 receptor agonist amibegron involve serotoninergic receptor activity. Pharmacol Biochem Behav 110:27–32. https://doi.org/10.1016/j.pbb.2013.05.017
80. Tanyeri P, Buyukokuroglu ME, Mutlu O et al (2013b) Involvement of serotonin receptor subtypes in the antidepressant-like effect of beta receptor agonist Amibegron (SR 58611A): an experimental study. Pharmacol Biochem Behav 105:12–16. https://doi.org/10.1016/j.pbb.2013.01.010
81. Weisstaub NV, Zhou M, Lira A, Lambe E, González-Maeso J, Hornung JP, Sibille E, Underwood M, Itohara S, Dauer WT, Ansorge MS, Morelli E, Mann JJ, Toth M, Aghajanian G, Sealfon SC, Hen R, Gingrich JA (2006) Cortical 5-HT2A receptor signaling modulates anxiety-like behaviors in mice. Science 313(5786):536–540

82. Morici JF, Ciccia L, Malleret G, Gingrich JA, Bekinschtein P, Weisstaub NV (2015) Serotonin 2a receptor and Serotonin 1a receptor interact within the medial prefrontal cortex during recognition memory in mice. Front Pharmacol 6:298. https://doi.org/10.3389/fphar.2015.00298. eCollection 2015
83. de Quervain DJ, Henke K, Aerni A et al (2003) A functional genetic variation of the 5-HT2a receptor affects human memory. Nat Neurosci 6(11):1141–1142. https://doi.org/10.1038/nn1146
84. Lai MK, Tsang SW, Alder JT et al (2005) Loss of serotonin 5-HT2A receptors in the postmortem temporal cortex correlates with rate of cognitive decline in Alzheimer's disease. Psychopharmacology 179(3):673–677. https://doi.org/10.1007/s00213-004-2077-2
85. Versijpt J, Van Laere KJ, Dumont F et al (2003) Imaging of the 5-HT2A system: age-, gender-, and Alzheimer's disease-related findings. Neurobiol Aging 24(4):553-561. doi:S0197458002001379 [pii].
86. Dean B (2003) The cortical serotonin2A receptor and the pathology of schizophrenia: a likely accomplice. J Neurochem 85(1):1–13. doi:1693 [pii]
87. Meltzer HY, Li Z, Kaneda Y et al (2003) Serotonin receptors: their key role in drugs to treat schizophrenia. Prog Neuro-Psychopharmacol Biol Psychiatry 27(7):1159–1172. doi:S0278-5846(03)00223-9 [pii]
88. Roth BL, Hanizavareh SM, Blum AE (2004) Serotonin receptors represent highly favorable molecular targets for cognitive enhancement in schizophrenia and other disorders. Psychopharmacology 174(1):17–24. https://doi.org/10.1007/s00213-003-1683-8
89. Burnet PW, Eastwood SL, Harrison PJ (1994) Detection and quantitation of 5-HT1A and 5-HT2A receptor mRNAs in human hippocampus using a reverse transcriptase-polymerase chain reaction (RT-PCR) technique and their correlation with binding site densities and age. Neurosci Lett 178(1):85–89. doi:0304-3940(94)90296-8 [pii]
90. Wright DE, Seroogy KB, Lundgren KH et al (1995) Comparative localization of serotonin1A, 1C, and 2 receptor subtype mRNAs in rat brain. J Comp Neurol 351(3):357–373. https://doi.org/10.1002/cne.903510304
91. Pazos A, Cortes R, Palacios JM (1985) Quantitative autoradiographic mapping of serotonin receptors in the rat brain. II. Serotonin-2 receptors. Brain Res 346(2):231–249. doi:0006-8993(85)90857-1 [pii]
92. Bombardi C (2012) Neuronal localization of 5-HT2A receptor immunoreactivity in the rat hippocampal region. Brain Res Bull 87(2-3):259–273. https://doi.org/10.1016/j.brainresbull.2011.11.006
93. Cornea-Hebert V, Riad M, Wu C et al (1999) Cellular and subcellular distribution of the serotonin 5-HT2A receptor in the central nervous system of adult rat. J Comp Neurol 409(2):187–209. https://doi.org/10.1002/(SICI)1096-9861(19990628)409:23.0.CO;2-P [pii].
94. Luttgen M, Ove Ogren S, Meister B (2004) Chemical identity of 5-HT2A receptor immunoreactive neurons of the rat septal complex and dorsal hippocampus. Brain Res 1010(1–2):156–165. https://doi.org/10.1016/j.brainres.2004.03.016
95. Xu T, Pandey SC (2000) Cellular localization of serotonin(2A) (5HT(2A)) receptors in the rat brain. Brain Res Bull 51(6):499–505. doi:S0361-9230(99)00278-6 [pii]
96. Uneyama H, Munakata M, Akaike N (1992) 5-HT response of rat hippocampal pyramidal cell bodies. Neuroreport 3(7):633–636
97. Guo JD, Rainnie DG (2010) Presynaptic 5-HT(1B) receptor-mediated serotonergic inhibition of glutamate transmission in the bed nucleus of the stria terminalis. Neuroscience 165(4):1390–1401. https://doi.org/10.1016/j.neuroscience.2009.11.071
98. Hashimoto K, Kita H (2008) Serotonin activates presynaptic and postsynaptic receptors in rat globus pallidus. J Neurophysiol 99(4):1723–1732. https://doi.org/10.1152/jn.01143.2007
99. Piguet P, Galvan M (1994) Transient and long-lasting actions of 5-HT on rat dentate gyrus neurones in vitro. J Physiol 481(Pt 3):629–639
100. Shen RY, Andrade R (1998) 5-Hydroxytryptamine2 receptor facilitates GABAergic neurotransmission in rat hippocampus. J Pharmacol Exp Ther 285(2):805–812

101. Kulkarni VA, Jha S, Vaidya VA (2002) Depletion of norepinephrine decreases the proliferation, but does not influence the survival and differentiation, of granule cell progenitors in the adult rat hippocampus. Eur J Neurosci 16(10):2008–2012
102. Ge S, Goh EL, Sailor KA et al (2006) GABA regulates synaptic integration of newly generated neurons in the adult brain. Nature 439(7076):589–593. doi:nature04404 [pii]
103. Banasr M, Hery M, Printemps R et al (2004) Serotonin-induced increases in adult cell proliferation and neurogenesis are mediated through different and common 5-HT receptor subtypes in the dentate gyrus and the subventricular zone. Neuropsychopharmacology 29(3):450–460. https://doi.org/10.1038/sj.npp.1300320
104. Wang RY, Arvanov VL (1998) M100907, a highly selective 5-HT2A receptor antagonist and a potential atypical antipsychotic drug, facilitates induction of long-term potentiation in area CA1 of the rat hippocampal slice. Brain Res 779(1-2):309–313. doi:S0006-8993(97)01174-8 [pii]
105. Farmer J, Zhao X, van Praag H, Wodtke K, Gage FH, Christie BR (2004) Effects of voluntary exercise on synaptic plasticity and gene expression in the dentate gyrus of adult male Sprague-Dawley rats in vivo. Neuroscience 124(1):71–79
106. Figurov A, Pozzo-Miller LD, Olafsson P et al (1996) Regulation of synaptic responses to high-frequency stimulation and LTP by neurotrophins in the hippocampus. Nature 381(6584):706–709. https://doi.org/10.1038/381706a0
107. Kang H, Schuman EM (1995) Long-lasting neurotrophin-induced enhancement of synaptic transmission in the adult hippocampus. Science 267(5204):1658–1662
108. Korte M, Carroll P, Wolf E et al (1995) Hippocampal long-term potentiation is impaired in mice lacking brain-derived neurotrophic factor. Proc Natl Acad Sci U S A 92(19):8856–8860
109. Levine ES, Dreyfus CF, Black IB et al (1995) Brain-derived neurotrophic factor rapidly enhances synaptic transmission in hippocampal neurons via postsynaptic tyrosine kinase receptors. Proc Natl Acad Sci U S A 92(17):8074–8077
110. Patterson SL, Abel T, Deuel TA et al (1996) Recombinant BDNF rescues deficits in basal synaptic transmission and hippocampal LTP in BDNF knockout mice. Neuron 16(6):1137–1145. doi:S0896-6273(00)80140-3 [pii]
111. Vaidya VA, Marek GJ, Aghajanian GK et al (1997) 5-HT2A receptor-mediated regulation of brain-derived neurotrophic factor mRNA in the hippocampus and the neocortex. J Neurosci 17(8):2785–2795
112. Zhang G, Asgeirsdottir HN, Cohen SJ et al (2013) Stimulation of serotonin 2A receptors facilitates consolidation and extinction of fear memory in C57BL/6J mice. Neuropharmacology 64:403–413. https://doi.org/10.1016/j.neuropharm.2012.06.007
113. Zhang G, Stackman RW Jr (2015) The role of serotonin 5-HT2A receptors in memory and cognition. Front Pharmacol 6:225. https://doi.org/10.3389/fphar.2015.00225
114. Peddie CJ, Davies HA, Colyer FM et al (2008) Colocalisation of serotonin2A receptors with the glutamate receptor subunits NR1 and GluR2 in the dentate gyrus: an ultrastructural study of a modulatory role. Exp Neurol 211(2):561–573. https://doi.org/10.1016/j.expneurol.2008.03.003
115. Aghajanian GK, Marek GJ (1999) Serotonin and hallucinogens. Neuropsychopharmacology 21(2 Suppl):16S-23S. https://doi.org/10.1016/S0893-133X(98)00135-3
116. Hill AS, Sahay A, Hen R (2015) Increasing Adult Hippocampal Neurogenesis is Sufficient to Reduce Anxiety and Depression-Like Behaviors. Neuropsychopharmacology. 40(10):2368–2378. https://doi.org/10.1038/npp.2015.85
117. Reagan LP, McEwen BS (1997) Controversies surrounding glucocorticoid-mediated cell death in the hippocampus. J Chem Neuroanat 13(3):149–167. doi:S0891061897000318 [pii]
118. Cheetham SC, Crompton MR, Katona CL et al (1988) Brain 5-HT2 receptor binding sites in depressed suicide victims. Brain Res 443(1-2):272–280. doi:0006-8993(88)91621-6 [pii]
119. Rosel P, Arranz B, Vallejo J et al (1998) Variations in [3H]imipramine and 5-HT2A but not [3H]paroxetine binding sites in suicide brains. Psychiatry Res 82(3):161–170
120. MacQueen GM, Campbell S, McEwen BS et al (2003) Course of illness, hippocampal function, and hippocampal volume in major depression. Proc Natl Acad Sci U S A 100(3):1387–1392. https://doi.org/10.1073/pnas.0337481100

121. Sheline YI, Sanghavi M, Mintun MA et al (1999) Depression duration but not age predicts hippocampal volume loss in medically healthy women with recurrent major depression. J Neurosci 19(12):5034–5043
122. Sheline YI, Wang PW, Gado MH et al (1996) Hippocampal atrophy in recurrent major depression. Proc Natl Acad Sci U S A 93(9):3908–3913
123. Mintun MA, Sheline YI, Moerlein SM et al (2004) Decreased hippocampal 5-HT2A receptor binding in major depressive disorder: in vivo measurement with [18F]altanserin positron emission tomography. Biol Psychiatry 55(3):217–224. doi:S000632230300920X [pii]
124. Meltzer HY, Maes M (1995) Effect of pindolol pretreatment on MK-212-induced plasma cortisol and prolactin responses in normal men. Biol Psychiatry. 38(5):310–318
125. Massou JM, Trichard C, Attar-Levy D et al (1997) Frontal 5-HT2A receptors studied in depressive patients during chronic treatment by selective serotonin reuptake inhibitors. Psychopharmacology 133(1):99–101
126. Meyer JH, Kapur S, Eisfeld B et al (2001) The effect of paroxetine on 5-HT(2A) receptors in depression: an [(18)F]setoperone PET imaging study. Am J Psychiatry 158(1):78–85. https://doi.org/10.1176/appi.ajp.158.1.78
127. Meyer JH, Kapur S, Houle S et al (1999) Prefrontal cortex 5-HT2 receptors in depression: an [18F]setoperone PET imaging study. Am J Psychiatry 156(7):1029–1034. https://doi.org/10.1176/ajp.156.7.1029
128. Yatham LN, Liddle PF, Shiah IS et al (2001) Effects of rapid tryptophan depletion on brain 5-HT(2) receptors: a PET study. Br J Psychiatry 178:448–453
129. Zanardi R, Artigas F, Moresco R et al (2001) Increased 5-hydroxytryptamine-2 receptor binding in the frontal cortex of depressed patients responding to paroxetine treatment: a positron emission tomography scan study. J Clin Psychopharmacol 21(1):53–58
130. Banasr M, Chowdhury GM, Terwilliger R et al (2010) Glial pathology in an animal model of depression: reversal of stress-induced cellular, metabolic and behavioral deficits by the glutamate-modulating drug riluzole. Mol Psychiatry 15(5):501–511. https://doi.org/10.1038/mp.2008.106
131. Nedergaard M, Ransom B, Goldman SA (2003) New roles for astrocytes: redefining the functional architecture of the brain. Trends Neurosci 26(10):523–530. doi:S0166-2236(03)00266-2 [pii]
132. Althaus HH, Richter-Landsberg C (2000) Glial cells as targets and producers of neurotrophins. Int Rev Cytol 197:203–277
133. Friedman WJ, Black IB, Kaplan DR (1998) Distribution of the neurotrophins brain-derived neurotrophic factor, neurotrophin-3, and neurotrophin-4/5 in the postnatal rat brain: an immunocytochemical study. Neuroscience 84(1):101–114. doi:S0306-4522(97)00526-5 [pii]
134. Duman RS, Heninger GR, Nestler EJ (1997) A molecular and cellular theory of depression. Arch Gen Psychiatry 54(7):597–606
135. Russo-Neustadt AA, Chen MJ (2005) Brain-derived neurotrophic factor and antidepressant activity. Curr Pharm Des 11(12):1495–1510
136. Allaman I, Belanger M, Magistretti PJ (2011) Astrocyte-neuron metabolic relationships: for better and for worse. Trends Neurosci 34(2):76–87. https://doi.org/10.1016/j.tins.2010.12.001
137. Mercier G, Lennon AM, Renouf B et al (2004) MAP kinase activation by fluoxetine and its relation to gene expression in cultured rat astrocytes. J Mol Neurosci 24(2):207–216. doi:JMN:24:2:207 [pii]
138. Tsuchioka M, Takebayashi M, Hisaoka K et al (2008) Serotonin (5-HT) induces glial cell line-derived neurotrophic factor (GDNF) mRNA expression via the transactivation of fibroblast growth factor receptor 2 (FGFR2) in rat C6 glioma cells. J Neurochem 106(1):244–257. https://doi.org/10.1111/j.1471-4159.2008.05357.x
139. David DJ, Wang J, Samuels BA et al (2010) Implications of the functional integration of adult-born hippocampal neurons in anxiety-depression disorders. Neuroscientist 16(5):578–591. https://doi.org/10.1177/1073858409360281
140. Rocha BA, Scearce-Levie K, Lucas JJ et al (1998) Increased vulnerability to cocaine in mice lacking the serotonin-1B receptor. Nature 393(6681):175–178. https://doi.org/10.1038/30259

141. Selemon LD, Lidow MS, Goldman-Rakic PS (1999) Increased volume and glial density in primate prefrontal cortex associated with chronic antipsychotic drug exposure. Biol Psychiatry 46(2):161–172. doi:S0006-3223(99)00113-4 [pii]
142. Czeh B, Simon M, Schmelting B et al (2006) Astroglial plasticity in the hippocampus is affected by chronic psychosocial stress and concomitant fluoxetine treatment. Neuropsychopharmacology 31(8):1616–1626. doi:1300982 [pii]
143. Galeotti N, Ghelardini C (2012) Regionally selective activation and differential regulation of ERK, JNK and p38 MAP kinase signalling pathway by protein kinase C in mood modulation. Int J Neuropsychopharmacol 15(6):781–793. https://doi.org/10.1017/S1461145711000897
144. Yuan P, Zhou R, Wang Y et al (2010) Altered levels of extracellular signal-regulated kinase signaling proteins in postmortem frontal cortex of individuals with mood disorders and schizophrenia. J Affect Disord 124(1–2):164–169. https://doi.org/10.1016/j.jad.2009.10.017
145. Gourley SL, Wu FJ, Kiraly DD et al (2008) Regionally specific regulation of ERK MAP kinase in a model of antidepressant-sensitive chronic depression. Biol Psychiatry 63(4):353–359. doi:S0006-3223(07)00713-5 [pii]

Brain Circuits Regulated by the 5-HT$_{2A}$ Receptor: Behavioural Consequences on Anxiety and Fear Memory

L. Moulédous, P. Roullet, and Bruno P. Guiard

Abstract Anxiety disorders including generalized anxiety disorder (GAD), panic disorder (PD), social anxiety disorder (SAD) or phobias are the most prevalent mental pathologies across the world with a median lifetime prevalence of approximately 15%. Anxiety imposes substantial economic costs which are among the highest of all mental disorders studied. Evidence is now accumulating that the serotonergic nervous system is involved in the pathology of anxiety and can provide benefits in the treatment of related disorders through its diverse functions, notably the modulation of stress, fear and memory. Among serotonin receptor subtypes, the 5-HT$_{2A}$ receptor arouses great interest. This receptor displays original pharmacological properties i.e., cooperation with β-arrestins and homo–/hetero-dimerization regulating its intracellular signaling and its ability to control the serotonergic system. The present chapter provides insight into the mechanisms by which the 5-HT$_{2A}$ receptor may alter the activity of 5-HT neurons but also of the brain regions receiving a dense serotonergic innervation (i.e, the amygdala, the hippocampus and the prefrontal cortex). An overview of the literature is proposed to recapitulate the pharmacological and genetic studies in patients or relevant animal models supporting a role of the 5-HT$_{2A}$ receptor on various forms of anxiety. Moreover, we envision the future directions that we might follow to develop new anxiolytic strategies based on the manipulation of 5-HT$_{2A}$-mediated signaling. Doing so, we also point some inconsistencies illustrating the difficulty to target this receptor as a valid alternative to benzodiazepines.

Keywords 5-HT$_{2A}$ receptor • Amygdala • Animal studies • Anxiety • Fear memory • Hippocampus • Monoaminergic circuits

---

L. Moulédous • P. Roullet
Research Center on Animal Cognition, Center for Integrative Biology, Université Paul Sabatier, UMR 5169 CNRS, Toulouse Cedex 9, France

B. P. Guiard (✉)
Faculté de Pharmacie, Université Paris Sud, Université Paris-Saclay, 92290 Chatenay-Malabry, France

Centre de Recherches sur la Cognition Animale (CRCA), Centre de Biologie Intégrative (CBI), Université de Toulouse; CNRS, UPS, Toulouse, France
e-mail: bruno.guiard@univ-tlse3.fr

B.P. Guiard, G. Di Giovanni (eds.), *5-HT$_{2A}$ Receptors in the Central Nervous System*, The Receptors 32, https://doi.org/10.1007/978-3-319-70474-6_10

Abbreviations

5-HT	Serotonin
5-HT_{2A}	5-Hydroxytryptamine 2A
AMY	Amygdala
BDNF	Brain-derived neurotrophic factor
BLA	Basolateral complex of amygdala
CeA	Central nucleus of amygdala
CRF	Corticotropin releasing factor
CRFR	Corticotropin releasing factor receptor
DA	Dopamine
DAG	Diacylglycerol
DCX	Doublecortin
DG	Dentate gyrus
DR	Dorsal raphe
EPM	Elevated plus maze
ERK	Extracellular signal-regulated kinase
ETM	Elevated T-maze
FPT	Four plate test
GAD	Generalized anxiety disorders
GC	Granule cell
GDNF	Glial cell line-derived neurotrophic factor
HP	Hippocampus
IP_3	Inositol Triphosphate
IPSCs	Inhibitory post-synaptic currents
LC	Locus coeruleus
LSD	Lysergic acid diethylamide
MeA	Medial amygdala
mPFCx	Medial prefrontal cortex
MR	Median raphe
NE	Norepinephrine
NSF	Novelty suppressed feeding
OF	Open field
OIC	Object in Context Recognition Task
PAG	Periaqueducal grey
PD	Panic disorders
PKC	Protein kinase C
PLC	Phospholipase C
PV	Parvalbumin
SAD	Social anxiety disorder
SGZ	Subgranular zone
SNOR	Spontaneous Novel Object Recognition task
SNP	Single nucleotide polymorphism
SOM	Somatostatin

SSRIs	Serotonin selective reuptake inhibitors
TI	Tonic immobility
TMOR	Temporal Order Recognition Task
Tph	Tryptophan hydroxylase
VEGF	Vascular endothelial growth factor
VTA	Ventral tegmental area

Introduction

The G-protein coupled serotonin 2A receptor (5-hydroxytryptamine 2A, or 5-HT_{2A}) subtype is the most important excitatory receptor of the serotonergic system. Its distribution in the brain has been extensively characterized using autoradiography, in situ hybridization, immunocytochemical techniques and in vivo imaging approaches [1–3]. In particular, the 5-HT_{2A} receptor has been detected in brain regions involved in cognition, perception, sensorimotor gating and mood such as the prefrontal cortex, the hippocampus, the amygdala but also the striatum and the thalamus. Moreover, this receptor has been identified in monoaminergic nuclei and functional studies demonstrated that it plays an important role in the regulation of serotonin (5-HT), norepinephrine (NE) and dopamine (DA) neuronal activities [4]. Also, it is now well accepted that the 5-HT_{2A} receptor is expressed both in neurons and glia although its role in the latter cell type remains poorly investigated. The 5-HT_{2A} receptor transduces signals primarily via the $G\alpha q$ signal cascade. Upon agonistic receptor stimulation, $G\alpha q$ and $\beta\gamma$ subunits of the G-protein dissociate and initiate downstream effector pathways. For example, the activity of phospholipase C (PLC) is stimulated, which subsequently promotes the release of diacylglycerol (DAG) and inositol triphosphate (IP_3). The stimulation of the 5-HT_{2A} receptor also contributes to the activation of protein kinase C (PKC) that ultimately affects the function of other proteins through their phosphorylation [5]. Findings in the last decade have identified some peculiarities of this receptor and notably its close interactions with β-arrestin proteins. As observed with other GPCRs, β-arrestin2 is involved in the downregulation/internalization of the 5-HT_{2A} receptor thereby leading to an attenuation of signaling pathways. This is the "arresting phase". However, depending on the nature of the agonist (endogenous vs. exogenous), the 5-HT_{2A} receptor may promote intracellular events as part of the so-called "signaling phases" [6, 7]. In support of this, it has been shown in β-arrestin-2 KO mice (β-Arr2−/−), in which 5-HT_{2A} receptor were predominantly localized to the cell surface, that 5-HT was no longer capable of inducing behavioral responses such as head-twitch. However, the authors found that the preferential 5-HT_{2A} receptor agonist DOI still produces the head-twitch in β-Arr2−/− mice thereby suggesting that this protein is not required for DOI-mediated response [6, 8]. These data emphasize the importance of the nature of the ligand in determining the receptor signaling pathway and, ultimately, the physiological responses induced by the compound. 5-HT_{2A} receptor binding to the intracellular scaffolding proteins β-arrestins can either dampen or facilitate GPCRs signaling, and therefore, represent a key point at which receptor signaling may diverge in response to particular ligands [9]. There is another mechanism by which the 5-HT_{2A} receptor can regulate signaling. Recent evidence demonstrates that this protein can form stable homo- [10] and heteromeric complexes with other types of GPCRs including the glutamatergic and dopaminergic mGluR2 and D2 receptors, respectively [11–13]. The in vivo functional consequences of such oligomerization of the 5-HT_{2A} receptor have yet to be determined but this process is likely responsible for changes in binding and coupling properties. Accordingly, it has been reported that head-twitch induced by the preferential 5-HT_{2A} receptor agonists lysergic acid diethylamide (LSD) and DOI is completely abolished in mGlu2 knock-out (mGlu2−/− KO) mice [13, 14]. Both examples illustrate the fact that the functional activity of the 5-HT_{2A} receptor is finely regulated. A better knowledge of the physiological relevance of such regulations may help identify new strategies aimed at modulating 5-HT_{2A} receptor-mediated signaling and related functions.

The present chapter synthesizes the current knowledge about the role of the 5-HT_{2A} receptor in the modulation of the 5-HT system itself and neuronal excitability of brain regions receiving serotonergic innervation. In this prospect, an emphasis will be given to the amygdala (AMY) and the hippocampus (HP). Moreover, because the 5-HT_{2A} receptor has long been associated with fear and anxiety [15], we will recapitulate the main preclinical and clinical data supporting a role of this receptor in the modulation of anxious behaviors but also in fear memory. So far, studies aimed at exploring 5-HT_{2A} receptor expression using positron emission tomography or post-mortem mRNA analysis in anxious patients remain somewhat equivocal. Indeed, some investigators reported decreased or increased expression of the 5-HT_{2A} receptor in patients with anxiety [16–19], others found no difference compared to controls [20]. Consequently, it is at this point uncertain whether a lower or a higher neurotransmission at the 5-HT_{2A} receptor has an impact on anxiety. Finally, we will envision the future directions based on 5-HT_{2A} receptor targeting to develop new anxiolytic strategies.

Regulation of the Presynaptic Serotonergic System by the 5-HT_{2A} Receptor

The main source of serotonergic neurons in the brain is located in the raphe nuclei notably the dorsal and median raphe nuclei (DR/MR) sending projections in areas such as the AMY, the HP and the median prefrontal cortex (mPFCx). In vitro recordings in the DR showed that local application of 5-HT produced hyperpolarization of tryptophan hydroxylase (Tph) positive neurons [21]. Similarly, in rat brain slices, the preferential 5-HT_{2A} receptor agonist DOI induced a concentration-dependent increase in the frequency of inhibitory postsynaptic currents (IPSCs). These studies suggested that endogenous 5-HT would act on 5-HT_{2A} receptor located on GABA

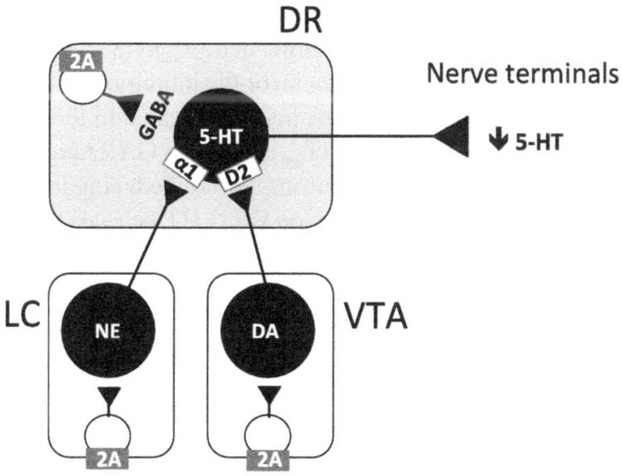

Fig. 1 Direct and indirect regulation of dorsal raphe (DR) serotonergic neurons by the 5-HT$_{2A}$ receptor. In the dorsal raphe (DR), local GABAergic interneurons express the 5-HT$_{2A}$ receptor (2A). Its activation increases GABA transmission leading to an inhibition of the firing rate of serotonin (5-HT) neurons. In the locus coeruleus (LC) and the ventral tegmental area (VTA), the 5-HT$_{2A}$ receptor is also expressed on GABAergic interneurons whose activation decreases the firing rate of norepinephrine (NE) and dopamine (DA) neurons. Both populations send projections to the DR in which serotonergic neurons express the excitatory alpha-1 (α1) and D2 receptors (for review, see [31]). Consequently, the increase in GABA levels resulting from the activation of the 5-HT$_{2A}$ in the LC and the VTA decreases these excitatory inputs thereby contributing to attenuate the firing rate of DR 5-HT neurons and likely the release of 5-HT at the nerve terminals

neurons to suppress the firing of DR 5-HT neurons. Although the levels of 5-HT$_{2A}$ receptor mRNA in the DR are low [22, 23], the latter findings are in agreement with the observation that systemic administration of DOI increased c-Fos immunoreactivity in the DR specifically in GABAergic interneurons [24]. Further pharmacological in vivo studies performed in rodents confirmed these data since it was reported that systemic or local administration of DOI in the DR reduced the discharge of 5-HT neurons [25–30] whereas these effects were reversed by ritanserin or MDL100907, 2 antagonists with a high affinity for the 5-HT$_{2A}$ receptor [25]. Interestingly, microdialysis experiments repeatedly found that the systemic administration of DOI in rats reduced the extracellular 5-HT concentrations at the nerve terminals such as the medial prefrontal cortex (mPFCx), an effect antagonized by MDL100907 [28]. Beyond the hypothesis that 5-HT$_{2A}$ receptor negatively regulates the activity of the serotonergic system through a local action in the DR (Fig. 1), evidence also indicates the recruitment of indirect mechanisms. In particular, the role of noradrenergic and dopaminergic neurons is highly suspected. The locus coeruleus (LC) which sends noradrenergic projections to the DR [31, 32], expresses 5-HT$_{2A}$ receptor. It is now well accepted that the enhancement of 5-HT transmission in the LC suppresses the firing activity of NE neurons through activation of excitatory 5-HT$_{2A}$ receptors also located on GABAergic interneurons [33, 34]. Given the excitatory influence of noradrenergic terminals on 5-HT neurons, notably via the alpha-1 heteroreceptor located on serotonergic neurons in the DR [32], activation of 5-HT_{2A} receptor in the LC could indirectly favor the inhibition of DR 5-HT neurons. Similar mechanisms could occur with DA neurons located in the ventral tegmental area (VTA). Indeed, the activation of 5-HT_{2A} located on GABAergic interneurons in the VTA inhibits the firing rate of DA neurons thereby reducing the excitatory influence exerted by these neuronal population on DR 5-HT neurons via the D2 receptor [32, 35, 36]. Indeed, although D2 receptor is a Gi coupled protein receptor, its ability to form heterodimers with D1 receptors may convert its signaling into excitatory responses [37]. Together, these findings strongly support the possibility that 5-HT_{2A} receptor also exerts indirect negative effects on the 5-HT system (Fig. 1).

However, the matter is more complex than it seems at first glance and the role of cortical 5-HT_{2A} receptor in the regulation of the 5-HT system should be carefully considered. Several studies from Dr. Artigas' group demonstrated that local application of DOI in the mPFCx increased the firing rate of DR 5-HT neurons [26, 28]. This might result from activation of excitatory cortical glutamatergic pyramidal neurons projecting to the DR. However, cortical glutamatergic neurons project to the DR mainly on GABAergic interneurons. Hence, the stimulation of the 5-HT_{2A} receptor in the cortex should reduce DR 5-HT neuronal activity. Beside these electrophysiological data, in microdialysis studies, local application of DOI in the mPFCx through reverse dialysis dose-dependently increased 5-HT local outflow and this effect was blocked by the application of MDL100907 [26, 28]. This elevation in cortical 5-HT tone could result from the activation of DR 5-HT neuronal activity but subsequent studies demonstrated that the neurochemical effects of DOI involved the local activation of 5-HT_{2A} receptor located on glutamatergic neurons whose post-synaptic receptors are directly expressed on 5-HT nerve endings [28, 38, 39]. In addition, it was also demonstrated that the 5-HT_{2A} receptor located in the mPFCx modulated the neuronal activity of VTA DA neurons [40]. Electrophysiological studies showed that local injection of DOI in the mPFCx increased VTA DA firing rate [41], an effect favorable to the enhancement of DR 5-HT neuron activity given the excitatory impact of DA in the DR as mentioned earlier [32, 35, 36].

Collectively, these results emphasize the fact that 5-HT_{2A} receptor activation elicits inhibitory or excitatory influences on DR 5-HT neuronal activity depending on the brain region where this receptor type is activated. Basically, the net effect of 5-HT_{2A} receptor activation after systemic administration of selective agonists is a decrease in 5-HT tone whereas the recruitment of 5-HT_{2A} receptors specifically in the cortex produces opposite responses. The hypothesis has been raised that changes in the expression/activity of the 5-HT_{2A} receptor switch the balance of activation/inhibition of the DR and then affect the release of endogenous 5-HT in projection areas [42].

Regulation of the Postsynaptic Serotonergic Areas by the 5-HT$_{2A}$ Receptor

5-HT$_{2A}$ and the Amygdala

The amygdala represents one of the most important brain region involved in the regulation of anxiety. It is endowed with a high density of 5-HT nerve terminals [43] and it is believed that part of 5-HT effects are mediated by the 5-HT$_{2A}$ receptor. The following section pays a specific attention to the role 5-HT$_{2A}$-mediated neurotransmission in the modulation of neuronal activity in different nuclei of the amygdala. First of all, it is now well known that anxiogenic stressors or fearful stimuli increase monoamine levels (including 5-HT) in the amygdala [44–46]. Conversely, the selective pharmacological or optogenetic manipulations of 5-HT in the amygdala have striking effects on fear and anxiety responses in experimental animals [47, 48]. For example, pharmacological depletion of serotonin in the basolateral amygdala (BLA) complex reduces anxiety and disrupts fear conditioning [49] whereas enhancement of serotonergic transmission, notably with serotonin selective reuptake inhibitors (SSRIs), induces opposite effects [50]. Overall, these studies emphasized the fact that 5-HT exerts an anxiogenic effect in this specific subdivision of the amygdala. One of the important mediators of serotonergic activity in response to anxiogenic stimuli is CRF. Indeed, it has been repeatedly shown that central infusion of CRF increases 5-HT levels in the amygdala [45, 51]. The DR receives CRF innervation, notably from the central nucleus of the amygdala (CeA), and expresses both CRF type 1 and 2 (CRF1 and CRF2) receptors [52, 53]. Direct infusion of CRF or CRF2 receptor agonists into the DR stimulates 5-HT release in both the BLA and the CeA [54–56] (Fig. 2). Interestingly, increased expression of CRF2 receptors occurs in the DR as a result of stress or in rat models of high anxiety [57] whereas CRF2 receptor antagonists infused directly into the DR reduce heightened anxiety-like behavior in rat models of early life stress [58].

In an attempt to determine the receptor type involved in the effect of 5-HT in the amygdala, different studies were conducted trying to dissociate the different subregions. Most of the studies focused their attention on the BLA and the medial amygdala (MeA) since both sites are endowed with 5-HT$_{2A}$ receptors [59] on pyramidal excitatory glutamatergic neurons and on GABAergic interneurons [e.g. Parvalbumin (PV) and somatostatin (SOM) positive cells] [60, 61]. Immunohistochemistry experiments reported increased c-Fos expression in PV-positive interneurons in response to the administration of anxiogenic drugs [60] or by subjecting rats to a novel open-field arena [62]. Strikingly, the numbers of c-Fos-immunoreactive (c-Fos-ir)/PV-ir GABAergic interneurons in the BLA were positively correlated with the numbers of c-Fos-ir 5-HT neurons in the DR and with a measure of anxiety-related behavior suggesting that the PV/5-HT$_{2A}$ receptor expressing GABAergic interneurons in the BLA are part of a DR-BLA neuronal circuit modulating anxiety-related behavior [60, 63, 64]. Evidence suggests that 5-HT depolarizes

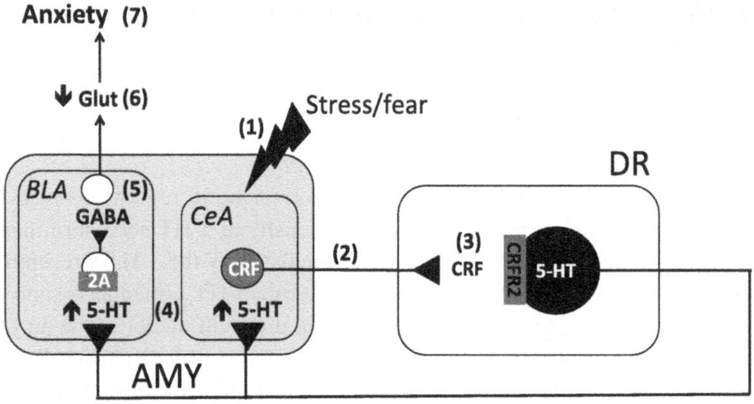

Fig. 2 Hypothetical role of the 5-HT_{2A} receptor in the amygdala (AMY). Stressful conditions and/or fear (1) activate CRF neurons in the central nucleus of the amygdala (CeA) (2). The subsequent release of CRF in the DR (3) is believed to activate CRFR2 receptors located on DR 5-HT neurons. Such activation favors the release of 5-HT at the nerve terminals and more particularly in the CeA and the basolateral nucleus of the amygdala (BLA) (4). The elevation of 5-HT in the BLA would activate 5-HT_{2A} receptor-expressing subpopulation of local parvalbumin and somatostatin inhibitory neurons which in turn release GABA in the synaptic cleft (5). This would reduce excitatory output (6), mediating anxiety-related behaviors (7)

PV-interneurons in the BLA, mainly via the 5-HT_{2A} receptor, leading to enhanced GABA release onto glutamatergic neurons [65–67] (Fig. 2). Accordingly, a recent study pointed that the stimulation of 5-HT axons in the BLA by the light activation of channelrhodopsin (ChR2), which was expressed selectively in DR 5-HT neurons, produced changes in the activity of microcircuits. Specifically, almost 50% of GABAergic neurons displayed a slow EPSC blocked by the 5-HT_{2A} receptor antagonist MDL 100907 [68] that, in turn, led to increased sIPSC frequency in glutamatergic neurons. The MeA also contains 5-HT_{2A} receptor immunoreactivity [69]. Although elusive, some studies showed that the pharmacological manipulation of this receptor subtype in this subregion of the AMY also influences anxious behaviors but in an opposite manner to that observed in the BLA (see chapter III-A). Collectively, these results indicate that anxiety is regulated in a subregion-dependent manner by the 5-HT_{2A} receptor in the amygdala. The possible explanation for the distinct consequences of the 5-HT_{2A} receptor activation in BLA and MeA could be the different cellular anatomy of these nuclei and the different neuronal population expressing this receptor.

5-HT_{2A} and the Hippocampus

The HP is another brain region that receives a 5-HT innervation and participates in the regulation of anxiety through its anatomical and functional interaction with the amygdala [70] but also through mechanisms that are specific to this area.

Adult Neurogenesis

Preclinical studies emphasized the role of granular cells (GCs) in the dentate gyrus (DG) of the HP in the pathophysiology of anxiety [71]. Indeed, using optogenetic techniques, it was demonstrated that elevating the activity of GCs in the ventral DG powerfully suppresses innate anxiety [72]. Therefore, adult hippocampal neurogenesis giving rise to new-born GCs might be an important process to control this kind of behavior. Interestingly, 5-HT is a potent regulator of adult hippocampal neurogenesis [73] and an increase in 5-HT tone resulting, for example, from the chronic administration of SSRIs favors this process [74]. In contrast, a decrease in 5-HT is believed to inhibit neurogenesis as suggested by the observation that 5-HT depletion in rats significantly reduced the number of newborn GCs in the subgranular zone (SGZ) of the HP [75]. It is now, well accepted that the beneficial effects of 5-HT on adult hippocampal neurogenesis rely, at least in part, from their ability to enhance the expression of brain-derived neurotrophic factor (BDNF) [76, 77] whereas evidence demonstrated that stress exerts opposite effects in the GCs layer of the HP [78]. Despite the extensive literature concerning the impact of 5-HT on adult neurogenesis, the precise effect of this complex cellular process on anxiety is not fully understood and various post-synaptic 5-HT receptors might have diverse, possibly opposing effects on different stages of neuronal development in the adult DG of the HP. A dense staining of the 5-HT_{2A} receptor in the hilus of the DG has been observed [79] and more precisely the receptor was proposed to be expressed at relatively high levels in GABAergic interneurons [22] suggesting a role in the modulation of adult hippocampal neurogenesis. Studies in rats pointed out that acute treatment with DOI had no effect on cell proliferation in the SGZ of the dorsal hippocampus [80, 81] whereas it led to a significantly decrease in brain-derived neurotrophic factor (BDNF) mRNA levels within the DG of the hippocampus [82]. It is possible the activation of the 5-HT_{2A} receptors located on inhibitory GABAergic interneurons in the DG could increase spontaneous GABA release [83, 84] and lead to a decrease in the production of BDNF by GCs themselves or neighboring cells. However, the 5-HT_{2A} receptor has been detected on GCs [85] prompting future research to determine to what extent the pharmacological manipulation of the 5-HT_{2A} receptor impacts adult hippocampal neurogenesis and dendritogenesis since it is at this point impossible to provide definitive conclusions. It is thus necessary to explore further the influence of 5-HT_{2A} receptor agonists/antagonists after their long-term administration since anatomical and morphological changes in the HP take several weeks to occur. In this regard, it was observed that the sustained administration of the non-selective 5-HT2 receptor antagonist ketanserin resulted in a robust increase in progenitor proliferation [80] without commensurate change in doublecortin (DCX)-positive immature neurons and dendritic maturation of DCX-positive newborn neurons [81].

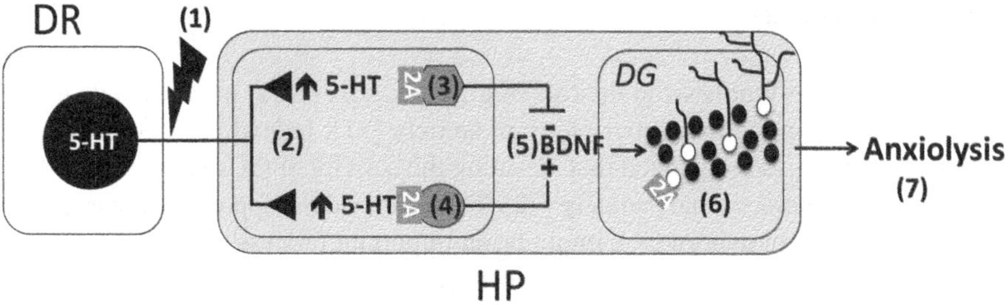

Fig. 3 Hypothetical role of the 5-HT$_{2A}$ receptor in the dentate gyrus (DG) of the hippocampus (HP). Pharmacological or optogenetic manipulation of the dorsal raphe (DR) serotonergic system (1) aimed at increasing extracellular 5-HT levels in the HP (2) favor the activation of the 5-HT$_{2A}$ receptors located on both astrocytes (3) and neurons (4). 5-HT$_{2A}$-mediated transmission is believed to enhance the synthesis and release of BDNF from astrocytes whereas opposite effects could occur in response to activation of 5-HT$_{2A}$ receptor located on GABAergic interneurons in the DG. A corollary of such mixed effects (5) is a lack of effect of acute stimulation of 5-HT$_{2A}$ receptors on adult hippocampal neurogenesis in the dentate gyrus (DG). However, the presence of the 5-HT$_{2A}$ on granule cells (6) might be a crucial mechanism to promote hippocampal plasticity such as dendritogenesis of granule cells themselves thereby facilitating their integration to the existing functional network. Evidence suggests that the enhancement of the latter process would favor anxiolysis (7)

Gliogenesis

The role of astrocytes in anxiety disorders is gaining growing interest [86, 87]. For example, it has been reported that the density of astrocytes was dramatically reduced in the brain of depressed patients [88] and in animal models of depression [89] for which anxious symptoms are strongly embedded. The 5-HT$_{2A}$ receptor was identified in primary cultures of glial cells and its activation enhanced the metabolic activity of astrocytes [90–92]. In particular, the pharmacological increase in 5-HT and the concomitant activation of the 5-HT$_{2A}$ receptor stimulates intracellular signaling cascades leading to the phosphorylation/activation of extracellular signal regulated kinases (ERK1/2) [93–95]. Indeed, by constituting a microenvironment permissive for neurogenesis and possibly for dendritogenesis [96], glia might dampen anxiety. Recent evidence also suggests that astrocytes could promote the synthesis and release of growth and neurotrophic factors [97, 98], a mechanism required for the neurogenesis-dependent activity of SSRIs. Accordingly, in vitro studies reported that SSRIs stimulate the expression of BDNF, Glial-derived neurotrophic factor (GDNF) and vascular endothelial growth factor (VEGF) in primary cultures of astrocytes and C6 glioma cells [98, 99]. In line with these data, we recently demonstrated that the specific over-expression of BDNF in hippocampal astrocytes produced anxiolytic-like effect in relation with an increase in cell proliferation, maturation and survival of new generated cells in the DG of the hippocampus [100]. Together, these data open a new avenue for a role of astroglial 5-HT$_{2A}$ receptors to promote the synthesis and release of factors that in turn, might positively reverberate on anxiety (Fig. 3).

Synaptic Plasticity

5-HT_{2A} receptors are present in a vast majority of the pyramidal and granule cells in the hippocampus and DG, notably in the dendrites and dendritic spines of DG and CA1 neurons, where glutamate NMDAR and AMPAR are assumed to be distributed [85, 101]. Of particular interest, the 5-HT_{2A} receptor has been found to directly interact with PSD-95 which regulates receptor trafficking and signal transduction [102]. It is possible that 5-HT_{2A} receptor activation, leading to a direct elevation of intracellular Ca^{2+}, combined with the recent or coincident elevation in intracellular Ca^{2+} due to NMDA receptor activation, would facilitate the induction of behaviourally triggered synaptic plasticity [103]. 5-HT_{2A} agonists could therefore act as memory enhancers. However, as in many other brain regions, the matter is complicated by the fact that 5-HT_{2A} receptors are also expressed in inhibitory interneurons and can therefore indirectly inhibit the activity of principal cells [70]. As a result, in certain situations, receptor antagonists could improve synaptic plasticity and memory. Accordingly, MDL100907, a highly selective 5-HT_{2A} receptor antagonist, has been shown to facilitate synaptic plasticity in area CA1 of the rat hippocampus [104]. Moreover, the 5-HT_{2A} receptor inverse agonist pimavanserin reversed object memory impairments induced by NMDA receptor antagonism [105], suggesting a complex modulatory influence of 5-HT_{2A} receptor on NMDA receptor-dependent memory mechanisms.

The Role of the 5-HT_{2A} Receptor in Anxiety

Serotonin regulates a variety of brain functions and is strongly implicated in the etiology and drug treatment of mood and anxiety disorders [106]. These behavioral effects arise from the 5-HT modulation of multiple neuronal circuits, but 5-HT transmission in the amygdala, the hippocampus and the prefrontal cortex is likely to play a critical role. In this chapter, we will recapitulate the genetic and pharmacological studies implemented to shed some lights on the role of the 5-HT_{2A} receptor in the regulation of anxiety.

Pharmacological Studies

In Rodents

The role of the 5-HT_{2A} receptor in the regulation of anxious behavior arose great interest in preclinical pharmacology studies. Initial investigations reported that the 5-HT_{2A} receptor antagonist MDL100907 (Volinanserin) alone lacked consistent activity in selected rodent models of anxiety [107, 108] while other studies found, on the contrary, that the systemic administration of 5-HT_{2A} receptor antagonists

Table 1 Effect of acute administration of preferential 5-HT_{2A} receptor agonists on anxiety

5-HT_{2A} receptor agonist	Route of administration	Test	Specie	Anxiety	References
DOI	Systemic	EPM	Mouse	↓	[110]
DOI	Systemic	FPT	Mouse	↓	[113]
DOI	Systemic	EPM/FPT	Mouse	↓	[111, 112]
DOI	Intra-PAG	EPM	Mouse	↓	[120]
DOI	Intra-HP (CA2)	FPT	Mouse	↓	[119]
mCPP	Intra-AMY	EPM	Mouse	↑	[115]
α-methyl-5-HT	Intra-AMY (BLA)	TI	Guinea pig	↓	[116]
TCB-2	Intra-AMY (BLA)	OF	Hamster	↑	[114]
α-methyl-5-HT	Intra-AMY (MeA)	TI	Guinea pig	↑	[118]

PAG periaqueducal grey; *HP* hippocampus; *AMY* amygdala; *BLA* basolateral nucleus of amygdala; *MeA* medial amygdala; *EPM* elevated plus maze; *FPT* four plate test; *OF* open-field; *TI* tonic immobility; ↑ and ↓ increased and decreased anxiety; respectively.

such as SR46949B elicited anxiety in the elevated plus maze (EPM) [109]. As regards the activation of the 5-HT_{2A} receptor, the systemic administration of the preferential 5-HT_{2A} receptor agonist DOI in mouse produced anxiolytic-like activity in the EPM whereas this effect was attenuated by mianserin, ketanserin or the 5-HT_{2A} receptor antagonist SR46949B [110]. The observation that the 5-HT_{2C} or 5-$HT_{2B/2C}$ receptor antagonists RS10-2221 and SB206553, respectively, failed to block DOI-induced anxiolysis highlighted the specific contribution of the 5-HT_{2A} receptor in such anxiolytic-like property [111, 112]. Interestingly, the lesion of the 5-HT neurons did not affect the anxiolytic-like effect of DOI evaluated in the four plate test (FPT) [113] thereby suggesting a post-synaptic action. In order to determine the brain region involved in this behavior, various experiments evaluated the pharmacological effects of local administration of DOI or other agonists.

In the amygdala, despite the anxiolytic effect of the systemic administration of DOI, it seems that the activation of the 5-HT_{2A} receptor would favour anxiety. Indeed, it was shown, for example, that the local injection of agonists displaying affinity for the 5-HT_{2A} receptor such TCB-2 or mCPP increased anxiety behavior in the open-field [114] or the EPM [115]. Nevertheless, it is important to remind that the amydgala is organized in distinct nuclei with specific anatomical and functional features. In the BLA, the activation of excitatory $5HT_2$ receptors is usually associated with increased anxiety-like behaviors (Table 1). Moreover, in agreement with this negative effect, it was reported that three consecutive days of immobilization with tail-shocks elicited anxiety and promoted the concomitant down-regulation of the 5-HT_{2A} receptor in the BLA [66] as a possible consequence of an over-stimulation of this receptor. Surprisingly, the microinjection of α-methyl-5-HT into the BLA reduces tonic immobility (TI) duration in guinea pigs [116] i.e., an innate fear behavior associated with intensely dangerous situations and considered a last resort aimed at the survival of the animal [117]. Although these findings defy the theory of an anxiogenic effect of the 5-HT_{2A} receptor in the BLA, they have to be interpreted with caution given the poor selectivity of α-methyl-5-HT towards the 5-HT_{2A} receptor. In the MeA, the activation of 5-HT_{2A} receptors was recently shown to promote an increase in TI duration [118], most likely due to an anxiogenic-like effect, while the blocking of this 5-HT receptor by the microinjection of ketanserin reduced TI. Future work should determine whether 5-HT and more particularly the 5-HT_{2A} exerts anxiolytic- or anxiogenic-like effect in the MeA.

Finally, it is important to note that a limited number of studies explored the role of the 5-HT_{2A} in relation with anxiety in other brain regions. Evidence demonstrated that the activation of 5-HT_{2A} receptors in response to the microinjection of DOI in the CA2 of the hippocampus (HP) [119] or the periaqueducal grey (PAG) [120] elicited anxiolysis. Unexpectedly, the role of cortical 5-HT_{2A} receptors in the regulation of anxiety has been poorly explored using pharmacological approaches although recent data showed that anxiety induced by the SSRI fluoxetine, as well as specific gene expression changes in the prefrontal cortex, were prevented by 5-HT_{2A}/C receptor blockade [121].

In Human

There is now accumulating support for the therapeutic interest to target the 5-HT_{2A} receptor for the relief of anxious symptoms. Indeed, the 5-$HT_{2A/2C}$ receptor antagonists ritanserin and mianserin exert anxiolytic effects in patients and effectively block the anxiogenic effects of m-chlorophenylpiperazine [122, 123]. Moreover, the antidepressant nefazodone possesses antagonistic activity at the 5-HT_{2A} receptor (along with 5-HT and norepinephrine reuptake inhibition properties) and is more effective than imipramine in the treatment of anxiety disorders [124]. Mirtazapine is an antidepressant with anxiolytic activity and its ability to block the 5-HT_{2A} receptor [125] is possibly important for promoting therapeutic activity. Obviously, atypical antipsychotics with prominent 5-HT_{2A} receptor antagonistic profile are being studied for their efficacy in anxiety disorders [126]. The first highly selective 5-HT_{2A} receptor antagonist to be developed was glemanserin (MDL11939), and it was initially studied for GAD without success [127]. Since this initial study, novel atypical antipsychotics such as asenapine that possesses high affinity for the 5-HT_{2A} receptor, have been shown to provide additional benefit in Veterans with PTSD who had not responded to an adequate course of treatment with an SSRI, venlafaxine, or mirtazapine [128]. Other drugs with a 5-HT_{2A} receptor antagonistic activity are in Phase III studies for the treatment of PTSD such as brexpiprazole [129, 130]. Moreover, since anxiety is frequently observed in schizophrenia [131], notably social phobia [132] and PTSD [133], the effects of atypical antipsychotics in these populations of patients have been explored. A recent review identified clinical trials in schizophrenia in which anxiety was a primary or secondary outcome measure [134]. It reports that amisulpiride, lurasidone, and asenapine show anxiolytic effects in patients with schizophrenia. It is worth noting that, in contrast to results obtained from preclinical investigations, human studies strongly suggest the necessity to block rather than to stimulate the 5-HT_{2A} receptor to obtain therapeutic effects. These discrepancies are of particular concern for the development of drugs with therapeutic potential in preclinical research and therefore for the identification of innovative strategies. Several reasons might explain these conflicting effects such as different anatomical distribution of the 5-HT_{2A} receptors between species. Technical consideration should also be taken into account. Indeed, in animal studies the potential of 5-HT_{2A} agonists/antagonists has mainly been tested after their acute administration and in behavioral paradigms (EPM, OF, FPT) that do not model chronic human disorders.

Genetic Studies

In Rodents

In constitutive 5-HT_{2A} receptor knock-out mice (5-$HT_{2A}^{-/-}$ KO), no modification of anxiety was observed using the novelty suppressed feeding paradigm or the elevated plus maze [135, 136]. Contrasting results had however been reported a few years before since it was shown that 5-$HT_{2A}^{-/-}$ KO mice display a low-anxiety behavioral phenotype in the open field (OF) and the light-dark test [137]. Although the discrepancies between these studies remain somewhat unexplained, it is possible that the choice of the behavioral paradigm is crucial to unveil the involvement of the 5-HT_{2A} receptor in anxiety. Hence, the development of treatment strategies for the various anxiety disorders should take into consideration the degree of anxiety-like and fear-like symptomology. Interestingly, the reduced level of anxiety detected in 5-$HT_{2A}^{-/-}$ KO mice was completely reversed in response to the selective re-introduction of the receptor in the cortex [137]. Despite these promising results underlying an anxiogenic effect of cortical 5-HT_{2A} receptors, genetic studies have yet to be conducted to determine whether other brain regions such as the AMY and the HP are involved in this specific behavior.

Importantly, 5-HT is a core neurotransmitter in the physiopathology of anxiety disorders and SSRIs are the first-line treatment [138]. We recently reported the effects of antidepressant drugs in 5-$HT_{2A}^{-/-}$ KO mice. Our results indicated that the ability of chronic administration of the SSRI fluoxetine to produce anxiolytic-like effects in the novelty suppressed feeding (NSF) was completely dampened in mutants [136]. Several hypotheses can be proposed to explain how the disruption of this receptor gene impaired behavioral responses to SSRIs. We raised the possibility that the constitutive loss of the 5-HT_{2A} receptor lead to developmental and compensatory effects such as hypersensitization/upregulation of the 5-HT_{1A} receptor [136], processes known to mitigate the therapeutic activity of SSRI in clinical [139] and preclinical [140] studies.

In Human

Given the dysfunction of the serotonergic system in psychiatric disorders [15], genetic studies focused on the association between genetic variants at the gene encoding for the 5-HT_{2A} receptor (HTR_{2A}) and anxiety. The HTR_{2A} gene is located on chromosome 13q14-q21 and consists of three exons and two introns, spanning over 63 kb. The single nucleotides polymorphism (SNP) databases contain 230 SNPs within this gene region. The HTR_{2A} 102 T/C polymorphism, one of the most studied, is a mutation that is defined by a T to C transition at position 102 that does not alter the amino acid composition of the receptor protein [141]. A polymorphism in the promoter region of the HTR_{2A} gene, 1438A/G, is in linkage disequilibrium with the 102 T/C polymorphism [142] and it has been suggested that it alters promoter activity and expression of the 5-HT_{2A} receptor. Interestingly, evidence suggests that the 102 T/C polymorphism might be associated with anxiety disorders including social anxiety disorder (SAD), panic disorders (PD) or generalized anxiety disorders (GAD) [143, 144]. SAD is a common, disabling condition, characterized by a significant amount of fear in one or more social situations causing considerable distress and impaired ability to function. Public speaking, or eating/drinking in front of others represent different forms of SAD [145]. The hypothesis of a genetic vulnerability for SAD is supported by recent findings showing that individuals carrying at least one T allele of the 5-HT_{2A} T102C polymorphism are less dependent on external stimuli for pleasure or reward or have a greater tendency to form pessimistic (rather than optimistic) attributions and expectations of socially-related events compared to those homozygous for the C-allele [146]. Hence, given that 5-HT_{2A} T102C polymorphism would result in increased expression of 5-HT_{2A} [147, 148], the latter study suggested that 5-HT_{2A}-mediated transmission would favor anxious traits.

The HTR_{2A} has also been implicated in PD, another form of anxiety characterized by an acute, intense attack of anxiety accompanied by feelings of impending doom [149]. In a recent work, the differences in genotype and allele frequencies between PD patients and controls, but also the association of the polymorphisms with symptom severity among PD patients in the Korean population were investigated. Although no genotype or allele distribution differences between PD patients and controls were unveiled, it was proposed that the C allele of HTR_{2A} 102T/C and G allele of HTR_{2A} 1438A/G are associated with the severity of symptoms of PD [150]. Interestingly, this significant association was also shown in a Japanese sample [151] but other studies failed to replicate such positive results [152, 153] reinforcing the idea that this question warrants further investigation. Because there is a striking comorbidity between anxiety disorders and major depressive disorders (i.e., anxiety represents a general risk factor for emotional disorders) [154], the question of common genetic markers has been raised. In particular, the influence of the $5HTR_{2A}$ 102T/C and 1438A/G SNPs was carefully investigated. Inconsistent results for the C allele of 102T/C SNP [association: [155–157], no association [158–160] or the A allele of 1438A/G SNP [association: 143, 161, 162, no association: 158, 159, 163] were observed. Although interesting, it is difficult to provide definitive conclusions and valuable insights into the role of the 5-HT_{2A} receptor in anxiety as long as the influence of a specific genotype on the receptor expression or function is not firmly established.

Particularly interesting in the context of this review, variations in the gene encoding for the 5-HT_{2A} receptor have been associated with clinical outcome after antidepressant treatment. For example, a recent meta-analysis pointed out an association between the C allele of the 102T/C SNP and antidepressant drugs response [164]. Naturalistic studies, in which all classes of antidepressant drugs were administered using "real-world" treatment options in patients with anxious symptoms revealed significant genotypic associations with remission for different SNPs of HTR_{2A} [165–171]. Unfortunately, the latter studies did not emphasize whether these SNPs specifically impact individual anxious symptoms.

The Role of the 5-HT_{2A} Receptor in Learning and Memory

As mentioned earlier, the 5-HT_{2A} receptor is present in brain regions involved in learning and memory processes [70]. Reduced binding capacity of 5-HT_{2A} receptor has been observed in aging subjects [172] but also in patients suffering from Alzheimer's disease [173, 174] thereby strengthening the hypothesis that this receptor subtype might play a role in learning and memory. Accordingly, Weisstaub and collaborators explored recognition memory in 5-$HT_{2A}{}^{-/-}$ KO mice. Results indicated that the constitutive loss of the 5-HT_{2A} receptor produced deficits in the Object in Context Recognition Task (OIC) and the Temporal Order Recognition Task (TMOR) task while the performance of mutants were normal in the Spontaneous Novel Object Recognition task (SNOR) [175]. From these observations, the authors suggested that the 5-HT_{2A} receptor might be necessary to control the expression of the relevant memory traces when complex representations must be used for successful retrieval. Results from two working memory tasks also suggested that 5-HT_{2A} receptor signalling is helpful to perform correctly only when the interference load is high. Specific role of 5-HT_{2A} receptor in the different memory phases was previously described using a pharmacological approach. Indeed, systemic administration of the 5-HT_{2A} receptor agonist TCB-2 improved memory consolidation in the SNOR task [176]. On the contrary, the pharmacological blockade of 5-HT_{2A} receptors with MDL11939 in the mPFCx of rats during the test phase impaired memory retrieval only in OIC and SNOR, two tasks that cannot be solved by a single item strategy [177]. Thus, the 5-HT_{2A} receptors seem implicated both in memory consolidation and memory retrieval in some non-aversive memories.

Even more interesting in the context of the present review, a recent study also emphasized a role of the 5-HT_{2A} receptor in fear memory. It has been shown in mice that the systemic administration of the 5-HT_{2A} receptor agonist TCB-2 right after conditioning enhanced the freezing to both cue and context during subsequent test sessions in a trace conditioning paradigm which depends on the hippocampus [176]. Conversely, the memory enhancing effect was absent in a hippocampus-independent delay fear conditioning paradigm. These data suggest that the activation of the 5-HT_{2A} receptor facilitates not all the fear memory consolidation but only the consolidation of episodic-like fear memory. Moreover, the authors demonstrated that TCB-2 was also able to significantly enhance the acquisition of extinction of cued trace fear memory and extinction of cued delay fear memory [176] whereas blockade of the 5-HT_{2A} receptors with MDL11939 exerted opposite effects in both procedures. In light of the plastic changes in neural circuitry that occur during the acquisition of fear extinction, it is possible that TCB-2 influences either the infralimbic cortical neurons or the "extinction neurons" of the BLA to facilitate fear extinction.

Further studies are now needed to clarify the neurophysiological influences of 5-HT, and the 5-HT_{2A} receptor on the neural circuitry supporting fear memory encoding, consolidation, retrieval, reconsolidation and, extinction. A number of neuropsychological disorders display fear memory symptoms, and may involve maladaptive processes in brain areas expressing the 5-HT_{2A} receptor, such as the AMY, the HP and the mPFCx. Thus, the observation of a facilitating influence of the 5-HT_{2A} receptor on the extinction of fear memory may have significant impact on the development of therapeutic approaches for subjects with fear memory-related disorders, such as phobias and PTSD [178]. Another possibility to block fear memory in these two disorders is to act during memory reconsolidation i.e. when memory is recalled and when this memory again becomes labile and sensitive to disruption. Serotonin receptors, especially the 5-HT_{5A}, 5-HT_6 and 5-HT_7, participate in fear memory reconsolidation measured in the contextual fear conditioning task [179]. Whether or not the 5-HT_{2A} receptor is involved in this process has yet to be demonstrated.

To conclude, activation of this receptor in association with exposure therapy may thus either facilitate extinction or block reconsolidation of pathological fear-memory in these patients by accelerating the building of new circuits and/or the reorganization of existing pathways to alleviate fear memory.

Concluding Remarks

The present review of the literature highlights the role of the 5-HT_{2A} receptor in the regulation of anxiety. The neurobiological mechanisms underlying such property is more complex than it appears at first sight. Indeed, depending on the brain region studied, the activation of this receptor subtype may elicit anxiolytic or anxiogenic-like effects as demonstrated in preclinical studies. Considering that an excess of 5-HT could be responsible for the development of anxiety, as repeatedly shown after short-term administration of SSRIs, the ability of the 5-HT_{2A} receptor to dampen the activity of the serotonergic system by inhibiting DR 5-HT neuronal activity should favour anxiolysis. However, in the AMY and more particularly in the BLA, the activation of this receptor produces opposite effects. Collectively, these results emphasize and justify the necessity to block specifically the post-synaptic

5-HT_{2A} receptor in order to relieve symptoms of anxiety. On this background, one might expect beneficial effects of atypical antipsychotics with 5-HT_{2A} antagonistic profile (even if these pharmacological agents are susceptible to enhance 5-HT neuronal activity). We can also question the interest to use antipsychotics as an add-on strategy with benzodiazepines to improve the therapeutic activity of the latter drugs. However, beneficial effects of such combination are poorly documented. In an old survey, it was reported that patients treated with atypical antipsychotics and benzodiazepines required a lower dose of antipsychotics to manage the negative symptoms of schizophrenia [180], a mental disorders characterized with a high level of anxiety. A more appropriate use of scales evaluating anxiety symptoms might have yielded more definitive data. Nevertheless, the beneficial effects of adjunctive benzodiazepines is suspected to be helpful in managing this type of disorder during the acute phase of treatment, until the antipsychotics has had time to be therapeutically effective [134]. It is also conceivable that the development of biased agonists would be of particular interest. Indeed, taking into consideration the fact that the activation of the 5-HT_{2A} produces distinct effects in the AMY or the HP, targeting a subpopulation of 5-HT_{2A} receptors with a specific coupling property might represent an alternative pharmacological strategy to optimize treatments outcomes. The relevance of such approach has been recently demonstrated with 5-HT_{1A} receptor agonist that exhibits biased agonism for postsynaptic receptors (preferentially coupled to $G\alpha i3$ protein subunits in the cortex), and with potential for selectively reducing aggression [181], depression or cognitive deficits in schizophrenia [182]. Clearly, there is a strong potential for all these avenues in the field of mental disorders and they should gain high attention in the near future.

References

1. Bombardi C (2012) Neuronal localization of 5-HT2A receptor immunoreactivity in the rat hippocampal region. Brain Res Bull 87:259–273
2. Bombardi C (2014) Neuronal localization of the 5-HT2 receptor family in the amygdaloid complex. Front Pharmacol 5:68
3. Kroll T, Elmenhorst D, Matusch A, Celik AA, Wedekind F, Weisshaupt A, Beer S, Bauer A (2014) [(1)(8)F]Altanserin and small animal PET: impact of multidrug efflux transporters on ligand brain uptake and subsequent quantification of 5-HT(2)A receptor densities in the rat brain. Nucl Med Biol 41:1–9
4. Quesseveur G, Nguyen HT, Gardier AM, Guiard BP (2012) 5-HT2 ligands in the treatment of anxiety and depression. Expert Opin Investig Drugs 21:1701–1725
5. Roth BL (2011) Irving Page Lecture: 5-HT(2A) serotonin receptor biology: interacting proteins, kinases and paradoxical regulation. Neuropharmacology 61:348–354
6. Abbas A, Roth BL (2008) Arresting serotonin. Proc Natl Acad Sci U S A 105:831–832
7. Guiard BP, Di Giovanni G (2015) Central serotonin-2A (5-HT2A) receptor dysfunction in depression and epilepsy: the missing link? Front Pharmacol 6:46
8. Schmid CL, Raehal KM, Bohn LM (2008) Agonist-directed signaling of the serotonin 2A receptor depends on beta-arrestin-2 interactions in vivo. Proc Natl Acad Sci U S A 105:1079–1084

9. Raote I, Bhattacharya A, Panicker MM (2007) Serotonin 2A (5-HT2A) receptor function: ligand-dependent mechanisms and pathways. CRC, Boca Raton, FL
10. Brea J, Castro M, Giraldo J, Lopez-Gimenez JF, Padin JF, Quintian F, Cadavid MI, Vilaro MT, Mengod G, Berg KA, Clarke WP, Vilardaga JP, Milligan G, Loza MI (2009) Evidence for distinct antagonist-revealed functional states of 5-hydroxytryptamine(2A) receptor homodimers. Mol Pharmacol 75:1380–1391
11. Delille HK, Becker JM, Burkhardt S, Bleher B, Terstappen GC, Schmidt M, Meyer AH, Unger L, Marek GJ, Mezler M (2012) Heterocomplex formation of 5-HT2A-mGlu2 and its relevance for cellular signaling cascades. Neuropharmacology 62:2184–2191
12. Gonzalez-Maeso J, Ang RL, Yuen T, Chan P, Weisstaub NV, Lopez-Gimenez JF, Zhou M, Okawa Y, Callado LF, Milligan G, Gingrich JA, Filizola M, Meana JJ, Sealfon SC (2008) Identification of a serotonin/glutamate receptor complex implicated in psychosis. Nature 452:93–97
13. Moreno JL, Holloway T, Albizu L, Sealfon SC, Gonzalez-Maeso J (2011) Metabotropic glutamate mGlu2 receptor is necessary for the pharmacological and behavioral effects induced by hallucinogenic 5-HT2A receptor agonists. Neurosci Lett 493:76–79
14. Hanks JB, Gonzalez-Maeso J (2013) Animal models of serotonergic psychedelics. ACS Chem Neurosci 4:33–42
15. Lowry CA, Johnson PL, Hay-Schmidt A, Mikkelsen J, Shekhar A (2005) Modulation of anxiety circuits by serotonergic systems. Stress 8:233–246
16. Dean B, Tawadros N, Seo MS, Jeon WJ, Everall I, Scarr E, Gibbons A (2014) Lower cortical serotonin 2A receptors in major depressive disorder, suicide and in rats after administration of imipramine. Int J Neuropsychopharmacol 17:895–906
17. Massou JM, Trichard C, Attar-Levy D, Feline A, Corruble E, Beaufils B, Martinot JL (1997) Frontal 5-HT2A receptors studied in depressive patients during chronic treatment by selective serotonin reuptake inhibitors. Psychopharmacology 133:99–101
18. Muguruza C, Miranda-Azpiazu P, Diez-Alarcia R, Morentin B, Gonzalez-Maeso J, Callado LF, Meana JJ (2014) Evaluation of 5-HT2A and mGlu2/3 receptors in postmortem prefrontal cortex of subjects with major depressive disorder: effect of antidepressant treatment. Neuropharmacology 86:311–318
19. Zanardi R, Artigas F, Moresco R, Colombo C, Messa C, Gobbo C, Smeraldi E, Fazio F (2001) Increased 5-hydroxytryptamine-2 receptor binding in the frontal cortex of depressed patients responding to paroxetine treatment: a positron emission tomography scan study. J Clin Psychopharmacol 21:53–58
20. Messa C, Colombo C, Moresco RM, Gobbo C, Galli L, Lucignani G, Gilardi MC, Rizzo G, Smeraldi E, Zanardi R, Artigas F, Fazio F (2003) 5-HT(2A) receptor binding is reduced in drug-naive and unchanged in SSRI-responder depressed patients compared to healthy controls: a PET study. Psychopharmacology 167:72–78
21. Bayliss DA, Li YW, Talley EM (1997) Effects of serotonin on caudal raphe neurons: activation of an inwardly rectifying potassium conductance. J Neurophysiol 77:1349–1361
22. Pompeiano M, Palacios JM, Mengod G (1994) Distribution of the serotonin 5-HT2 receptor family mRNAs: comparison between 5-HT2A and 5-HT2C receptors. Brain Res Mol Brain Res 23:163–178
23. Xie H, Ma F, Zhang YQ, Gao X, Wu GC (2002) Expression of 5-HT(2A) receptor mRNA in some nuclei of brain stem enhanced in monoarthritic rats. Brain Res 954:94–99
24. Boothman LJ, Sharp T (2005) A role for midbrain raphe gamma aminobutyric acid neurons in 5-hydroxytryptamine feedback control. Neuroreport 16:891–896
25. Boothman LJ, Allers KA, Rasmussen K, Sharp T (2003) Evidence that central 5-HT2A and 5-HT2B/C receptors regulate 5-HT cell firing in the dorsal raphe nucleus of the anaesthetised rat. Br J Pharmacol 139:998–1004
26. Bortolozzi A, Amargos-Bosch M, Adell A, Diaz-Mataix L, Serrats J, Pons S, Artigas F (2003) In vivo modulation of 5-hydroxytryptamine release in mouse prefrontal cortex by local 5-HT(2A) receptors: effect of antipsychotic drugs. Eur J Neurosci 18:1235–1246

27. Garratt JC, Kidd EJ, Wright IK, Marsden CA (1991) Inhibition of 5-hydroxytryptamine neuronal activity by the 5-HT agonist, DOI. Eur J Pharmacol 199:349–355
28. Martin-Ruiz R, Puig MV, Celada P, Shapiro DA, Roth BL, Mengod G, Artigas F (2001) Control of serotonergic function in medial prefrontal cortex by serotonin-2A receptors through a glutamate-dependent mechanism. J Neurosci 21:9856–9866
29. Quesseveur G, Reperant C, David DJ, Gardier AM, Sanchez C, Guiard BP (2013a) 5-HT(2) A receptor inactivation potentiates the acute antidepressant-like activity of escitalopram: involvement of the noradrenergic system. Exp Brain Res 226:285–295
30. Wright IK, Garratt JC, Marsden CA (1990) Effects of a selective 5-HT2 agonist, DOI, on 5-HT neuronal firing in the dorsal raphe nucleus and 5-HT release and metabolism in the frontal cortex. Br J Pharmacol 99:221–222
31. Guiard BP, El Mansari M, Blier P (2009) Prospect of a dopamine contribution in the next generation of antidepressant drugs: the triple reuptake inhibitors. Curr Drug Targets 10:1069–1084
32. Guiard BP, El Mansari M, Merali Z, Blier P (2008) Functional interactions between dopamine, serotonin and norepinephrine neurons: an in-vivo electrophysiological study in rats with monoaminergic lesions. Int J Neuropsychopharmacol 11:625–639
33. Szabo ST, Blier P (2001) Functional and pharmacological characterization of the modulatory role of serotonin on the firing activity of locus coeruleus norepinephrine neurons. Brain Res 922:9–20
34. Szabo ST, de Montigny C, Blier P (1999) Modulation of noradrenergic neuronal firing by selective serotonin reuptake blockers. Br J Pharmacol 126:568–571
35. Aman TK, Shen RY, Haj-Dahmane S (2007) D2-like dopamine receptors depolarize dorsal raphe serotonin neurons through the activation of nonselective cationic conductance. J Pharmacol Exp Ther 320:376–385
36. Haj-Dahmane S (2001) D2-like dopamine receptor activation excites rat dorsal raphe 5-HT neurons in vitro. Eur J Neurosci 14:125–134
37. Dziedzicka-Wasylewska M, Faron-Gorecka A, Gorecki A, Kusemider M (2008) Mechanism of action of clozapine in the context of dopamine D1-D2 receptor hetero-dimerization--a working hypothesis. Pharmacol Rep 60:581–587
38. Celada P, Puig MV, Casanovas JM, Guillazo G, Artigas F (2001) Control of dorsal raphe serotonergic neurons by the medial prefrontal cortex: involvement of serotonin-1A, GABA(A), and glutamate receptors. J Neurosci 21:9917–9929
39. Puig MV, Celada P, Diaz-Mataix L, Artigas F (2003) In vivo modulation of the activity of pyramidal neurons in the rat medial prefrontal cortex by 5-HT2A receptors: relationship to thalamocortical afferents. Cereb Cortex 13:870–882
40. Vazquez-Borsetti P, Celada P, Cortes R, Artigas F (2011) Simultaneous projections from prefrontal cortex to dopaminergic and serotonergic nuclei. Int J Neuropsychopharmacol 14:289–302
41. Bortolozzi A, Diaz-Mataix L, Scorza MC, Celada P, Artigas F (2005) The activation of 5-HT receptors in prefrontal cortex enhances dopaminergic activity. J Neurochem 95:1597–1607
42. Celada P, Puig MV, Artigas F (2013) Serotonin modulation of cortical neurons and networks. Front Integr Neurosci 7:25
43. Asan E, Steinke M, Lesch KP (2013) Serotonergic innervation of the amygdala: targets, receptors, and implications for stress and anxiety. Histochem Cell Biol 139:785–813
44. Inoue T, Koyama T, Yamashita I (1993) Effect of conditioned fear stress on serotonin metabolism in the rat brain. Pharmacol Biochem Behav 44:371–374
45. Mo B, Feng N, Renner K, Forster G (2008) Restraint stress increases serotonin release in the central nucleus of the amygdala via activation of corticotropin-releasing factor receptors. Brain Res Bull 76:493–498
46. Spannuth BM, Hale MW, Evans AK, Lukkes JL, Campeau S, Lowry CA (2011) Investigation of a central nucleus of the amygdala/dorsal raphe nucleus serotonergic circuit implicated in fear-potentiated startle. Neuroscience 179:104–119

47. Baratta MV, Kodandaramaiah SB, Monahan PE, Yao J, Weber MD, Lin PA, Gisabella B, Petrossian N, Amat J, Kim K, Yang A, Forest CR, Boyden ES, Goosens KA (2016) Stress enables reinforcement-elicited serotonergic consolidation of fear memory. Biol Psychiatry 79:814–822
48. Christianson JP, Ragole T, Amat J, Greenwood BN, Strong PV, Paul ED, Fleshner M, Watkins LR, Maier SF (2010) 5-hydroxytryptamine 2C receptors in the basolateral amygdala are involved in the expression of anxiety after uncontrollable traumatic stress. Biol Psychiatry 67:339–345
49. Johnson PL, Molosh A, Fitz SD, Arendt D, Deehan GA, Federici LM, Bernabe C, Engleman EA, Rodd ZA, Lowry CA, Shekhar A (2015) Pharmacological depletion of serotonin in the basolateral amygdala complex reduces anxiety and disrupts fear conditioning. Pharmacol Biochem Behav 138:174–179
50. Hamon M (1994) Neuropharmacology of anxiety: perspectives and prospects. Trends Pharmacol Sci 15:36–39
51. Matsuzaki I, Takamatsu Y, Moroji T (1989) The effects of intracerebroventricularly injected corticotropin-releasing factor (CRF) on the central nervous system: behavioural and biochemical studies. Neuropeptides 13:147–155
52. Day HE, Greenwood BN, Hammack SE, Watkins LR, Fleshner M, Maier SF, Campeau S (2004) Differential expression of 5HT-1A, alpha 1b adrenergic, CRF-R1, and CRF-R2 receptor mRNA in serotonergic, gamma-aminobutyric acidergic, and catecholaminergic cells of the rat dorsal raphe nucleus. J Comp Neurol 474:364–378
53. Pringle RB, Mouw NJ, Lukkes JL, Forster GL (2008) Amphetamine treatment increases corticotropin-releasing factor receptors in the dorsal raphe nucleus. Neurosci Res 62:62–65
54. Amat J, Tamblyn JP, Paul ED, Bland ST, Amat P, Foster AC, Watkins LR, Maier SF (2004) Microinjection of urocortin 2 into the dorsal raphe nucleus activates serotonergic neurons and increases extracellular serotonin in the basolateral amygdala. Neuroscience 129:509–519
55. Pernar L, Curtis AL, Vale WW, Rivier JE, Valentino RJ (2004) Selective activation of corticotropin-releasing factor-2 receptors on neurochemically identified neurons in the rat dorsal raphe nucleus reveals dual actions. J Neurosci 24:1305–1311
56. Scholl JL, Vuong SM, Forster GL (2010) Chronic amphetamine treatment enhances corticotropin-releasing factor-induced serotonin release in the amygdala. Eur J Pharmacol 644:80–87
57. Lukkes J, Vuong S, Scholl J, Oliver H, Forster G (2009) Corticotropin-releasing factor receptor antagonism within the dorsal raphe nucleus reduces social anxiety-like behavior after early-life social isolation. J Neurosci 29:9955–9960
58. Bledsoe AC, Oliver KM, Scholl JL, Forster GL (2011) Anxiety states induced by post-weaning social isolation are mediated by CRF receptors in the dorsal raphe nucleus. Brain Res Bull 85:117–122
59. Bombardi C (2011) Distribution of 5-HT2A receptor immunoreactivity in the rat amygdaloid complex and colocalization with gamma-aminobutyric acid. Brain Res 1370:112–128
60. Hale MW, Johnson PL, Westerman AM, Abrams JK, Shekhar A, Lowry CA (2010) Multiple anxiogenic drugs recruit a parvalbumin-containing subpopulation of GABAergic interneurons in the basolateral amygdala. Prog Neuro-Psychopharmacol Biol Psychiatry 34:1285–1293
61. McDonald AJ, Mascagni F (2007) Neuronal localization of 5-HT type 2A receptor immunoreactivity in the rat basolateral amygdala. Neuroscience 146:306–320
62. Hale MW, Bouwknecht JA, Spiga F, Shekhar A, Lowry CA (2006) Exposure to high- and low-light conditions in an open-field test of anxiety increases c-Fos expression in specific subdivisions of the rat basolateral amygdaloid complex. Brain Res Bull 71:174–182
63. Hale MW, Hay-Schmidt A, Mikkelsen JD, Poulsen B, Bouwknecht JA, Evans AK, Stamper CE, Shekhar A, Lowry CA (2008a) Exposure to an open-field arena increases c-Fos expression in a subpopulation of neurons in the dorsal raphe nucleus, including neurons projecting to the basolateral amygdaloid complex. Neuroscience 157:733–748

64. Hale MW, Hay-Schmidt A, Mikkelsen JD, Poulsen B, Shekhar A, Lowry CA (2008b) Exposure to an open-field arena increases c-Fos expression in a distributed anxiety-related system projecting to the basolateral amygdaloid complex. Neuroscience 155:659–672
65. Bocchio M, Fucsina G, Oikonomidis L, McHugh SB, Bannerman DM, Sharp T, Capogna M (2015) Increased serotonin transporter expression reduces fear and recruitment of parvalbumin interneurons of the amygdala. Neuropsychopharmacology 40:3015–3026
66. Jiang X, Xing G, Yang C, Verma A, Zhang L, Li H (2009) Stress impairs 5-HT2A receptor-mediated serotonergic facilitation of GABA release in juvenile rat basolateral amygdala. Neuropsychopharmacology 34:410–423
67. Rainnie DG (1999) Serotonergic modulation of neurotransmission in the rat basolateral amygdala. J Neurophysiol 82:69–85
68. Sengupta A, Bocchio M, Bannerman DM, Sharp T, Capogna M (2017) Control of amygdala circuits by 5-HT neurons via 5-HT and glutamate cotransmission. J Neurosci 37:1785–1796
69. Cornea-Hebert V, Riad M, Wu C, Singh SK, Descarries L (1999) Cellular and subcellular distribution of the serotonin 5-HT2A receptor in the central nervous system of adult rat. J Comp Neurol 409:187–209
70. Bombardi C, Di Giovanni G (2013) Functional anatomy of 5-HT2A receptors in the amygdala and hippocampal complex: relevance to memory functions. Exp Brain Res 230:427–439
71. Miller BR, Hen R (2015) The current state of the neurogenic theory of depression and anxiety. Curr Opin Neurobiol 30:51–58
72. Kheirbek MA, Drew LJ, Burghardt NS, Costantini DO, Tannenholz L, Ahmari SE, Zeng H, Fenton AA, Hen R (2013) Differential control of learning and anxiety along the dorsoventral axis of the dentate gyrus. Neuron 77:955–968
73. Samuels BA, Hen R (2011) Neurogenesis and affective disorders. Eur J Neurosci 33: 1152–1159
74. Malberg JE, Eisch AJ, Nestler EJ, Duman RS (2000) Chronic antidepressant treatment increases neurogenesis in adult rat hippocampus. J Neurosci 20:9104–9110
75. Brezun JM, Daszuta A (2000) Serotonin may stimulate granule cell proliferation in the adult hippocampus, as observed in rats grafted with foetal raphe neurons. Eur J Neurosci 12:391–396
76. Nibuya M, Morinobu S, Duman RS (1995) Regulation of BDNF and trkB mRNA in rat brain by chronic electroconvulsive seizure and antidepressant drug treatments. J Neurosci 15:7539–7547
77. Sairanen M, Lucas G, Ernfors P, Castren M, Castren E (2005) Brain-derived neurotrophic factor and antidepressant drugs have different but coordinated effects on neuronal turnover, proliferation, and survival in the adult dentate gyrus. J Neurosci 25:1089–1094
78. Smith MA, Makino S, Kvetnansky R, Post RM (1995) Stress and glucocorticoids affect the expression of brain-derived neurotrophic factor and neurotrophin-3 mRNAs in the hippocampus. J Neurosci 15:1768–1777
79. Klempin F, Babu H, De Pietri Tonelli D, Alarcon E, Fabel K, Kempermann G (2010) Oppositional effects of serotonin receptors 5-HT1a, 2, and 2c in the regulation of adult hippocampal neurogenesis. Front Mol Neurosci 3
80. Banasr M, Hery M, Printemps R, Daszuta A (2004) Serotonin-induced increases in adult cell proliferation and neurogenesis are mediated through different and common 5-HT receptor subtypes in the dentate gyrus and the subventricular zone. Neuropsychopharmacology 29:450–460
81. Jha S, Rajendran R, Fernandes KA, Vaidya VA (2008) 5-HT2A/2C receptor blockade regulates progenitor cell proliferation in the adult rat hippocampus. Neurosci Lett 441:210–214
82. Vaidya VA, Marek GJ, Aghajanian GK, Duman RS (1997) 5-HT2A receptor-mediated regulation of brain-derived neurotrophic factor mRNA in the hippocampus and the neocortex. J Neurosci 17:2785–2795
83. Freund TF, Gulyas AI, Acsady L, Gorcs T, Toth K (1990) Serotonergic control of the hippocampus via local inhibitory interneurons. Proc Natl Acad Sci U S A 87:8501–8505

84. Piguet P, Galvan M (1994) Transient and long-lasting actions of 5-HT on rat dentate gyrus neurones in vitro. J Physiol 481(Pt 3):629–639
85. Peddie CJ, Davies HA, Colyer FM, Stewart MG, Rodriguez JJ (2008) Colocalisation of serotonin2A receptors with the glutamate receptor subunits NR1 and GluR2 in the dentate gyrus: an ultrastructural study of a modulatory role. Exp Neurol 211:561–573
86. Banasr M, Chowdhury GM, Terwilliger R, Newton SS, Duman RS, Behar KL, Sanacora G (2010) Glial pathology in an animal model of depression: reversal of stress-induced cellular, metabolic and behavioral deficits by the glutamate-modulating drug riluzole. Mol Psychiatry 15:501–511
87. Quesseveur G, Gardier AM, Guiard BP (2013b) The monoaminergic tripartite synapse: a putative target for currently available antidepressant drugs. Curr Drug Targets 14:1277–1294
88. Rajkowska G (2000) Postmortem studies in mood disorders indicate altered numbers of neurons and glial cells. Biol Psychiatry 48:766–777
89. Czeh B, Simon M, Schmelting B, Hiemke C, Fuchs E (2006) Astroglial plasticity in the hippocampus is affected by chronic psychosocial stress and concomitant fluoxetine treatment. Neuropsychopharmacology 31:1616–1626
90. Hansson E, Simonsson P, Alling C (1987) 5-Hydroxytryptamine stimulates the formation of inositol phosphate in astrocytes from different regions of the brain. Neuropharmacology 26:1377–1382
91. Hirst WD, Price GW, Rattray M, Wilkin GP (1997) Identification of 5-hydroxytryptamine receptors positively coupled to adenylyl cyclase in rat cultured astrocytes. Br J Pharmacol 120:509–515
92. Zhang S, Li B, Lovatt D, Xu J, Song D, Goldman SA, Nedergaard M, Hertz L, Peng L (2010) 5-HT2B receptors are expressed on astrocytes from brain and in culture and are a chronic target for all five conventional 'serotonin-specific reuptake inhibitors'. Neuron Glia Biol 6:113–125
93. Li B, Zhang S, Li M, Hertz L, Peng L (2009) Chronic treatment of astrocytes with therapeutically relevant fluoxetine concentrations enhances cPLA2 expression secondary to 5-HT2B-induced, transactivation-mediated ERK1/2 phosphorylation. Psychopharmacology 207:1–12
94. Musazzi L, Mallei A, Tardito D, Gruber SH, El Khoury A, Racagni G, Mathe AA, Popoli M (2010) Early-life stress and antidepressant treatment involve synaptic signaling and Erk kinases in a gene-environment model of depression. J Psychiatr Res 44:511–520
95. Qi X, Lin W, Li J, Li H, Wang W, Wang D, Sun M (2008) Fluoxetine increases the activity of the ERK-CREB signal system and alleviates the depressive-like behavior in rats exposed to chronic forced swim stress. Neurobiol Dis 31:278–285
96. Morrens J, Van Den Broeck W, Kempermann G (2011) Glial cells in adult neurogenesis. Glia
97. Li B, Zhang S, Zhang H, Nu W, Cai L, Hertz L, Peng L (2008) Fluoxetine-mediated 5-HT2B receptor stimulation in astrocytes causes EGF receptor transactivation and ERK phosphorylation. Psychopharmacology 201:443–458
98. Tsuchioka M, Takebayashi M, Hisaoka K, Maeda N, Nakata Y (2008) Serotonin (5-HT) induces glial cell line-derived neurotrophic factor (GDNF) mRNA expression via the transactivation of fibroblast growth factor receptor 2 (FGFR2) in rat C6 glioma cells. J Neurochem 106:244–257
99. Meller R, Babity JM, Grahame-Smith DG (2002) 5-HT2A receptor activation leads to increased BDNF mRNA expression in C6 glioma cells. NeuroMolecular Med 1:197–205
100. Quesseveur G, David DJ, Gaillard MC, Pla P, Wu MV, Nguyen HT, Nicolas V, Auregan G, David I, Dranovsky A, Hantraye P, Hen R, Gardier AM, Deglon N, Guiard BP (2013c) BDNF overexpression in mouse hippocampal astrocytes promotes local neurogenesis and elicits anxiolytic-like activities. Transl Psychiatry 3:e253
101. Zhang G, Stackman RW (2015) The role of serotonin 5-HT2A receptors in memory and cognition. Front Pharmacol 6:225
102. Xia Z, Gray JA, Compton-Toth BA, Roth BL (2003) A direct interaction of PSD-95 with 5-HT2A serotonin receptors regulates receptor trafficking and signal transduction. J Biol Chem 278:21901–21908

103. Arvanov VL, Liang X, Russo A, Wang RY (1999) LSD and DOB: interaction with 5-HT2A receptors to inhibit NMDA receptor-mediated transmission in the rat prefrontal cortex. Eur J Neurosci 11:3064–3072
104. Wang RY, Arvanov VL (1998) M100907, a highly selective 5-HT2A receptor antagonist and a potential atypical antipsychotic drug, facilitates induction of long-term potentiation in area CA1 of the rat hippocampal slice. Brain Res 779:309–313
105. Snigdha S, Horiguchi M, Huang M, Li Z, Shahid M, Neill JC, Meltzer HY (2010) Attenuation of phencyclidine-induced object recognition deficits by the combination of atypical antipsychotic drugs and pimavanserin (ACP 103), a 5-hydroxytryptamine(2A) receptor inverse agonist. J Pharmacol Exp Ther 332:622–631
106. Nemeroff CB (2002) Recent advances in the neurobiology of depression. Psychopharmacol Bull 36(Suppl 2):6–23
107. Griebel G, Perrault G, Sanger DJ (1997) A comparative study of the effects of selective and non-selective 5-HT2 receptor subtype antagonists in rat and mouse models of anxiety. Neuropharmacology 36:793–802
108. Kehne JH, Baron BM, Carr AA, Chaney SF, Elands J, Feldman DJ, Frank RA, van Giersbergen PL, McCloskey TC, Johnson MP, McCarty DR, Poirot M, Senyah Y, Siegel BW, Widmaier C (1996) Preclinical characterization of the potential of the putative atypical antipsychotic MDL 100,907 as a potent 5-HT2A antagonist with a favorable CNS safety profile. J Pharmacol Exp Ther 277:968–981
109. Setem J, Pinheiro AP, Motta VA, Morato S, Cruz AP (1999) Ethopharmacological analysis of 5-HT ligands on the rat elevated plus-maze. Pharmacol Biochem Behav 62:515–521
110. Onaivi ES, Bishop-Robinson C, Darmani NA, Sanders-Bush E (1995) Behavioral effects of (+/−)-1-(2,5-dimethoxy-4-iodophenyl)-2-aminopropane, (DOI) in the elevated plus-maze test. Life Sci 57:2455–2466
111. Nic Dhonnchadha BA, Bourin M, Hascoet M (2003a) Anxiolytic-like effects of 5-HT2 ligands on three mouse models of anxiety. Behav Brain Res 140:203–214
112. Nic Dhonnchadha BA, Hascoet M, Jolliet P, Bourin M (2003b) Evidence for a 5-HT2A receptor mode of action in the anxiolytic-like properties of DOI in mice. Behav Brain Res 147:175–184
113. Masse F, Hascoet M, Dailly E, Bourin M (2006) Effect of noradrenergic system on the anxiolytic-like effect of DOI (5-HT2A/2C agonists) in the four-plate test. Psychopharmacology 183:471–481
114. Clinard CT, Bader LR, Sullivan MA, Cooper MA (2015) Activation of 5-HT2a receptors in the basolateral amygdala promotes defeat-induced anxiety and the acquisition of conditioned defeat in Syrian hamsters. Neuropharmacology 90:102–112
115. Cornelio AM, Nunes-de-Souza RL (2007) Anxiogenic-like effects of mCPP microinfusions into the amygdala (but not dorsal or ventral hippocampus) in mice exposed to elevated plus-maze. Behav Brain Res 178:82–89
116. Leite-Panissi CR, Ferrarese AA, Terzian AL, Menescal-de-Oliveira L (2006) Serotoninergic activation of the basolateral amygdala and modulation of tonic immobility in guinea pig. Brain Res Bull 69:356–364
117. Klemm WR (1971) Neurophysiologic studies of the immobility reflex ("animal hypnosis"). Neurosci Res (N Y) 4:165–212
118. de Paula BB, Leite-Panissi CR (2016) Distinct effect of 5-HT1A and 5-HT2A receptors in the medial nucleus of the amygdala on tonic immobility behavior. Brain Res 1643:152–158
119. Petit-Demouliere B, Masse F, Cogrel N, Hascoet M, Bourin M (2009) Brain structures implicated in the four-plate test in naive and experienced Swiss mice using injection of diazepam and the 5-HT2A agonist DOI. Behav Brain Res 204:200–205
120. Gomes KS, Nunes-De-Souza RL (2009) Implication of the 5-HT2A and 5-HT2C (but not 5HT1A) receptors located within the periaqueductal gray in the elevated plus-maze test-retest paradigm in mice. Prog Neuro-Psychopharmacol Biol Psychiatry 33:1261–1269

121. Sarkar A, Chachra P, Vaidya VA (2014) Postnatal fluoxetine-evoked anxiety is prevented by concomitant 5-HT2A/C receptor blockade and mimicked by postnatal 5-HT2A/C receptor stimulation. Biol Psychiatry 76:858–868
122. Ceulemans DL, Hoppenbrouwers ML, Gelders YG, Reyntjens AJ (1985) The influence of ritanserin, a serotonin antagonist, in anxiety disorders: a double-blind placebo-controlled study versus lorazepam. Pharmacopsychiatry 18:303–305
123. Pigott TA, Zohar J, Hill JL, Bernstein SE, Grover GN, Zohar-Kadouch RC, Murphy DL (1991) Metergoline blocks the behavioral and neuroendocrine effects of orally administered m-chlorophenylpiperazine in patients with obsessive-compulsive disorder. Biol Psychiatry 29:418–426
124. Bystritsky A, Rosen R, Suri R, Vapnik T (1999) Pilot open-label study of nefazodone in panic disorder. Depress Anxiety 10:137–139
125. Ribeiro L, Busnello JV, Kauer-Sant'Anna M, Madruga M, Quevedo J, Busnello EA, Kapczinski F (2001) Mirtazapine versus fluoxetine in the treatment of panic disorder. Braz J Med Biol Res 34:1303–1307
126. Blier P, Szabo ST (2005) Potential mechanisms of action of atypical antipsychotic medications in treatment-resistant depression and anxiety. J Clin Psychiatry 66(Suppl 8):30–40
127. Sramek JJ, Robinson RE, Suri A, Cutler NR (1995) Efficacy trial of the 5-HT2 antagonist MDL 11,939 in patients with generalized anxiety disorder. J Clin Psychopharmacol 15:20–22
128. Pilkinton P, Berry C, Norrholm S, Bartolucci A, Birur B, Davis LL (2016) An open label pilot study of adjunctive asenapine for the treatment of posttraumatic stress disorder. Psychopharmacol Bull 46:8–17
129. Citrome L, Stensbol TB, Maeda K (2015) The preclinical profile of brexpiprazole: what is its clinical relevance for the treatment of psychiatric disorders? Expert Rev Neurother 15:1219–1229
130. Greig SL (2015) Brexpiprazole: First Global Approval. Drugs 75:1687–1697
131. Llorca PM, Lancon C, Blanc O, de Chazeron I, Samalin L, Caci H, Lesturgeon JA, Bayle FJ (2014) A composite scale applied to evaluate anxiety in schizophrenic patients (SAES). Eur Arch Psychiatry Clin Neurosci 264:171–178
132. Pallanti S, Cantisani A, Grassi G (2013) Anxiety as a core aspect of schizophrenia. Curr Psychiatry Rep 15:354
133. Braga RJ, Reynolds GP, Siris SG (2013) Anxiety comorbidity in schizophrenia. Psychiatry Res 210:1–7
134. Garay RP, Samalin L, Hameg A, Llorca PM (2015) Investigational drugs for anxiety in patients with schizophrenia. Expert Opin Investig Drugs 24:507–517
135. Petit AC, Quesseveur G, Gressier F, Colle R, David DJ, Gardier AM, Ferreri F, Lepine JP, Falissard B, Verstuyft C, Guiard BP, Corruble E (2014) Converging translational evidence for the involvement of the serotonin 2A receptor gene in major depressive disorder. Prog Neuro-Psychopharmacol Biol Psychiatry 54:76–82
136. Qesseveur G, Petit AC, Nguyen HT, Dahan L, Colle R, Rotenberg S, Seif I, Robert P, David D, Guilloux JP, Gardier AM, Verstuyft C, Becquemont L, Corruble E, Guiard BP (2016) Genetic dysfunction of serotonin 2A receptor hampers response to antidepressant drugs: a translational approach. Neuropharmacology 105:142–153
137. Weisstaub NV, Zhou M, Lira A, Lambe E, Gonzalez-Maeso J, Hornung JP, Sibille E, Underwood M, Itohara S, Dauer WT, Ansorge MS, Morelli E, Mann JJ, Toth M, Aghajanian G, Sealfon SC, Hen R, Gingrich JA (2006) Cortical 5-HT2A receptor signaling modulates anxiety-like behaviors in mice. Science 313:536–540
138. Bandelow B (2008) The medical treatment of obsessive-compulsive disorder and anxiety. CNS Spectr 13:37–46
139. Lemonde S, Du L, Bakish D, Hrdina P, Albert PR (2004) Association of the C(−1019) G 5-HT1A functional promoter polymorphism with antidepressant response. Int J Neuropsychopharmacol 7:501–506

140. Richardson-Jones JW, Craige CP, Guiard BP, Stephen A, Metzger KL, Kung HF, Gardier AM, Dranovsky A, David DJ, Beck SG, Hen R, Leonardo ED (2010) 5-HT1A autoreceptor levels determine vulnerability to stress and response to antidepressants. Neuron 65:40–52
141. Bondy B, Spaeth M, Offenbaecher M, Glatzeder K, Stratz T, Schwarz M, de Jonge S, Kruger M, Engel RR, Farber L, Pongratz DE, Ackenheil M (1999) The T102C polymorphism of the 5-HT2A-receptor gene in fibromyalgia. Neurobiol Dis 6:433–439
142. Arranz MJ, Munro J, Owen MJ, Spurlock G, Sham PC, Zhao J, Kirov G, Collier DA, Kerwin RW (1998) Evidence for association between polymorphisms in the promoter and coding regions of the 5-HT2A receptor gene and response to clozapine. Mol Psychiatry 3:61–66
143. Enoch MA, Kaye WH, Rotondo A, Greenberg BD, Murphy DL, Goldman D (1998) 5-HT2A promoter polymorphism -1438G/A, anorexia nervosa, and obsessive-compulsive disorder. Lancet 351:1785–1786
144. Walitza S, Wewetzer C, Warnke A, Gerlach M, Geller F, Gerber G, Gorg T, Herpertz-Dahlmann B, Schulz E, Remschmidt H, Hebebrand J, Hinney A (2002) 5-HT2A promoter polymorphism -1438G/A in children and adolescents with obsessive-compulsive disorders. Mol Psychiatry 7:1054–1057
145. D'Avanzato C, Dalrymple KL (2016) Recent insight into the subtypes of social anxiety disorder. Curr Psychiatry Rep 18:50
146. Lochner C, Hemmings S, Seedat S, Kinnear C, Schoeman R, Annerbrink K, Olsson M, Eriksson E, Moolman-Smook J, Allgulander C, Stein DJ (2007) Genetics and personality traits in patients with social anxiety disorder: a case-control study in South Africa. Eur Neuropsychopharmacol 17:321–327
147. Khait VD, Huang YY, Zalsman G, Oquendo MA, Brent DA, Harkavy-Friedman JM, Mann JJ (2005) Association of serotonin 5-HT2A receptor binding and the T102C polymorphism in depressed and healthy Caucasian subjects. Neuropsychopharmacology 30:166–172
148. Polesskaya OO, Sokolov BP (2002) Differential expression of the "C" and "T" alleles of the 5-HT2A receptor gene in the temporal cortex of normal individuals and schizophrenics. J Neurosci Res 67:812–822
149. Kogan CS, Stein DJ, Maj M, First MB, Emmelkamp PM, Reed GM (2016) The classification of anxiety and fear-related disorders in the ICD-11. Depress Anxiety 33:1141–1154
150. Yoon HK, Yang JC, Lee HJ, Kim YK (2008) The association between serotonin-related gene polymorphisms and panic disorder. J Anxiety Disord 22:1529–1534
151. Inada Y, Yoneda H, Koh J, Sakai J, Himei A, Kinoshita Y, Akabame K, Hiraoka Y, Sakai T (2003) Positive association between panic disorder and polymorphism of the serotonin 2A receptor gene. Psychiatry Res 118:25–31
152. Martinez-Barrondo S, Saiz PA, Morales B, Garcia-Portilla MP, Coto E, Alvarez V, Bobes J (2005) Serotonin gene polymorphisms in patients with panic disorder. Actas Esp Psiquiatr 33:210–215
153. Rothe C, Koszycki D, Bradwejn J, King N, De Luca V, Shaikh S, Franke P, Garritsen H, Fritze J, Deckert J, Kennedy JL (2004) Association study of serotonin-2A receptor gene polymorphism and panic disorder in patients from Canada and Germany. Neurosci Lett 363:276–279
154. Pollack MH (2005) Comorbid anxiety and depression. J Clin Psychiatry 66(Suppl 8):22–29
155. Arias B, Gasto C, Catalan R, Gutierrez B, Pintor L, Fananas L (2001) The 5-HT(2A) receptor gene 102T/C polymorphism is associated with suicidal behavior in depressed patients. Am J Med Genet 105:801–804
156. Du L, Bakish D, Lapierre YD, Ravindran AV, Hrdina PD (2000) Association of polymorphism of serotonin 2A receptor gene with suicidal ideation in major depressive disorder. Am J Med Genet 96:56–60
157. Zhang HY, Ishigaki T, Tani K, Chen K, Shih JC, Miyasato K, Ohara K (1997) Serotonin2A receptor gene polymorphism in mood disorders. Biol Psychiatry 41:768–773
158. Illi A, Setala-Soikkeli E, Viikki M, Poutanen O, Huhtala H, Mononen N, Lehtimaki T, Leinonen E, Kampman O (2009) 5-HTR1A, 5-HTR2A, 5-HTR6, TPH1 and TPH2 polymorphisms and major depression. Neuroreport 20:1125–1128

159. Kishi T, Kitajima T, Tsunoka T, Ikeda M, Yamanouchi Y, Kinoshita Y, Kawashima K, Okochi T, Okumura T, Inada T, Ozaki N, Iwata N (2009) Genetic association analysis of serotonin 2A receptor gene (HTR2A) with bipolar disorder and major depressive disorder in the Japanese population. Neurosci Res 64:231–234
160. Minov C, Baghai TC, Schule C, Zwanzger P, Schwarz MJ, Zill P, Rupprecht R, Bondy B (2001) Serotonin-2A-receptor and -transporter polymorphisms: lack of association in patients with major depression. Neurosci Lett 303:119–122
161. Christiansen L, Tan Q, Iachina M, Bathum L, Kruse TA, McGue M, Christensen K (2007) Candidate gene polymorphisms in the serotonergic pathway: influence on depression symptomatology in an elderly population. Biol Psychiatry 61:223–230
162. Kamata M, Suzuki A, Yoshida K, Takahashi H, Higuchi H, Otani K (2011) Genetic polymorphisms in the serotonergic system and symptom clusters of major depressive disorder. J Affect Disord 135:374–376
163. Tencomnao T, Thongrakard V, Phuchana W, Sritharathikhun T, Suttirat S (2010) No relationship found between -1438A/G polymorphism of the serotonin 2A receptor gene (rs6311) and major depression susceptibility in a northeastern Thai population. Genet Mol Res 9:1171–1176
164. McMahon FJ, Buervenich S, Charney D, Lipsky R, Rush AJ, Wilson AF, Sorant AJ, Papanicolaou GJ, Laje G, Fava M, Trivedi MH, Wisniewski SR, Manji H (2006) Variation in the gene encoding the serotonin 2A receptor is associated with outcome of antidepressant treatment. Am J Hum Genet 78:804–814
165. Fabbri C, Souery D, Calati R, Crisafulli C, Chierchia A, Albani D, Forloni G, Chiesa A, Martines R, Sentissi O, Mendlewicz J, De Girolamo G, Serretti A (2015) Genetics of psychotropic medication induced side effects in two independent samples of bipolar patients. J Neural Transm (Vienna) 122:43–58
166. Horstmann S, Lucae S, Menke A, Hennings JM, Ising M, Roeske D, Muller-Myhsok B, Holsboer F, Binder EB (2010) Polymorphisms in GRIK4, HTR2A, and FKBP5 show interactive effects in predicting remission to antidepressant treatment. Neuropsychopharmacology 35:727–740
167. Kato M, Zanardi R, Rossini D, De Ronchi D, Okugawa G, Kinoshita T, Colombo C, Serretti A (2009) 5-HT2A gene variants influence specific and different aspects of antidepressant response in Japanese and Italian mood disorder patients. Psychiatry Res 167:97–105
168. Kishi T, Yoshimura R, Kitajima T, Okochi T, Okumura T, Tsunoka T, Yamanouchi Y, Kinoshita Y, Kawashima K, Naitoh H, Nakamura J, Ozaki N, Iwata N (2010) HTR2A is associated with SSRI response in major depressive disorder in a japanese cohort. NeuroMolecular Med 12:237–242
169. Noro M, Antonijevic I, Forray C, Kasper S, Kocabas NA, Lecrubier Y, Linotte S, Mendlewicz J, Montgomery S, Snyder L, Souery D, Verbanck P, Zohar J, Massat I (2010) 5HT1A and 5HT2A receptor genes in treatment response phenotypes in major depressive disorder. Int Clin Psychopharmacol 25:228–231
170. Viikki M, Huuhka K, Leinonen E, Illi A, Setala-Soikkeli E, Huuhka M, Mononen N, Lehtimaki T, Kampman O (2011) Interaction between two HTR2A polymorphisms and gender is associated with treatment response in MDD. Neurosci Lett 501:20–24
171. Wilkie MJ, Smith G, Day RK, Matthews K, Smith D, Blackwood D, Reid IC, Wolf CR (2009) Polymorphisms in the SLC6A4 and HTR2A genes influence treatment outcome following antidepressant therapy. Pharmacogenomics J 9:61–70
172. Meltzer CC, Smith G, DeKosky ST, Pollock BG, Mathis CA, Moore RY, Kupfer DJ, Reynolds CF III (1998a) Serotonin in aging, late-life depression, and Alzheimer's disease: the emerging role of functional imaging. Neuropsychopharmacology 18:407–430
173. Marner L, Frokjaer VG, Kalbitzer J, Lehel S, Madsen K, Baare WF, Knudsen GM, Hasselbalch SG (2012) Loss of serotonin 2A receptors exceeds loss of serotonergic projections in early Alzheimer's disease: a combined [11C]DASB and [18F]altanserin-PET study. Neurobiol Aging 33:479–487

174. Meltzer CC, Smith G, Price JC, Reynolds CF III, Mathis CA, Greer P, Lopresti B, Mintun MA, Pollock BG, Ben-Eliezer D, Cantwell MN, Kaye W, DeKosky ST (1998b) Reduced binding of [18F]altanserin to serotonin type 2A receptors in aging: persistence of effect after partial volume correction. Brain Res 813:167–171
175. Morici JF, Ciccia L, Malleret G, Gingrich JA, Bekinschtein P, Weisstaub NV (2015) Serotonin 2a receptor and serotonin 1a receptor interact within the medial prefrontal cortex during recognition memory in mice. Front Pharmacol 6:298
176. Zhang G, Asgeirsdottir HN, Cohen SJ, Munchow AH, Barrera MP, Stackman RW Jr (2013) Stimulation of serotonin 2A receptors facilitates consolidation and extinction of fear memory in C57BL/6J mice. Neuropharmacology 64:403–413
177. Bekinschtein P, Renner MC, Gonzalez MC, Weisstaub N (2013) Role of medial prefrontal cortex serotonin 2A receptors in the control of retrieval of recognition memory in rats. J Neurosci 33:15716–15725
178. Quirk GJ, Pare D, Richardson R, Herry C, Monfils MH, Schiller D, Vicentic A (2010) Erasing fear memories with extinction training. J Neurosci 30:14993–14997
179. Schmidt SD, Furini CR, Zinn CG, Cavalcante LE, Ferreira FF, Behling JA, Myskiw JC, Izquierdo I (2016) Modulation of the consolidation and reconsolidation of fear memory by three different serotonin receptors in hippocampus. Neurobiol Learn Mem 142:48–54
180. Pecknold JC (1993) Survey of the adjuvant use of benzodiazepines for treating outpatients with schizophrenia. J Psychiatry Neurosci 18:82–84
181. Stein DJ, Miczek KA, Lucion AB, de Almeida RM (2013) Aggression-reducing effects of F15599, a novel selective 5-HT1A receptor agonist, after microinjection into the ventral orbital prefrontal cortex, but not in infralimbic cortex in male mice. Psychopharmacology 230:375–387
182. Llado-Pelfort L, Assie MB, Newman-Tancredi A, Artigas F, Celada P (2010) Preferential in vivo action of F15599, a novel 5-HT(1A) receptor agonist, at postsynaptic 5-HT(1A) receptors. Br J Pharmacol 160:1929–1940

$5\text{-HT}_{2A/2B/2C}$ Receptors, Memory and Therapeutic Targets

Alfredo Meneses, Rossana Nieto-Vera, and Rosa María Anaya-Jiménez

Abstract The 5-HT_2 receptors subdivision into the $5\text{-HT}_{2A/2B/2C}$ subtypes along with the advent of the selective antagonists has allowed a more detailed investigation on the role and therapeutic significance of these subtypes in cognitive functions. It is suggested that, $5\text{-HT}_{2B/2C}$ receptors might be involved on memory formation probably mediating a suppressive or constraining action. Whether the drug-induced learning impairments in this study are explained by simple agonism, antagonism or inverse agonism at 5-HT_2 receptors remains unclear at this time. Notably, the 5-HT_2 receptor subtypes blockade may provide some benefit to reverse poor memory consolidation conditions associated with decreased cholinergic, glutamatergic, and/or serotonergic neurotransmission.

Keywords Autoshaping • $5\text{-HT}_{2A/2B/2C}$ • Receptors • Memory consolidation • Serotonin • Rats

Introduction

Evidence from aplysia to human studies indicates that serotonin (5-hydroxytryptamine; 5-HT) systems mediate learning and memory processes [1]. Even though the precise receptors and mechanisms have not been elucidated yet. 5-HT receptors characterized so far in mammals, include 5-HT_1 through 5-HT_7 subfamilies [2–6], showing a regional distribution in brain areas implicated in learning and memory, such as hippocampus, amygdala and cortex (see [5, 6], for reviews). 5-HT receptors participate in different learning and memory tasks, using different schedules for drug administration, doses [5]. Serotonergic inverse agonists [6–9] have been documented, raising the possibility that, inverse agonism may actually have physiological implications and even, a possible impact in drug development [10]. Certainly, whether or not

A. Meneses (✉)
Department of Pharmacobiology, CINVESTAV, México City, Mexico
e-mail: ameneses@msn.com

R. Nieto-Vera • R.M. Anaya-Jiménez
Facultad de Ciencias de la Salud, Universidad Anáhuac México Norte, México City, Mexico

B.P. Guiard, G. Di Giovanni (eds.), *5-HT$_{2A}$ Receptors in the Central Nervous System*, The Receptors 32, https://doi.org/10.1007/978-3-319-70474-6_11

Table 1 Affinities (pK_i) of several 5HT receptor compounds for various 5-HT receptor (sub)types

	$5HT_{1A}$	$5HT_{1B}$	$5HT_{1D}$	$5HT_{2A}$	$5HT_{2B}$	5-HT	$5\text{-}HT_3$	$5HT_7$
mCPP	6.5	6.6	5.1	6.7	7.6	6.9	7.0	6.6
1-NP	7.2	6.6	8.1	7.2	8.3	8.2	–	7.7
Mesulergine	6.2	4.9	5.2	8.4	7.4	8.7	–	7.7
TFMPP	6.3	6.4	6.2	6.6	7.1	6.5	5.7	6.6
DOI	5.3	5.9	5.9	7.8	7.6	7.7	–	–
Ketanserin	<5.0	<5.0	5.7	8.9	5.4	7.1	<5.0	6.6
SB200646	<5.0	–	<5.0	5.2	7.5	6.9	<5.0	–
LY215840	7.4	6.0	6.2	7.7	8.7	8.4	–	7.8
MDL100907	6.0	5.2	–	9.4	–	6.9	–	–

References in Meneses 2002 [8]

different stages of the learning process have a link with changes in constitutive activity of 5-HT receptors remains open for investigation and speculative in the light of the available evidence, mainly $5\text{-}HT_{2A/2B/2C}$ receptors. Interestingly, the $5\text{-}HT_2$ receptor subdivision into the $5\text{-}HT_{2A/2B/2C}$ categories [3, 11], along with the advent of the selective antagonists MDL100907 ($5\text{-}HT_{2A}$) and SB-200646 ($5\text{-}HT_{2B/2C}$) (see Table 1 for some affinities) or agonist (−)-MBP ($5\text{-}HT_{2C}$; [12]) has allowed or will allow a more detailed investigation on the role and (potential) therapeutic significance of these subtypes in cognitive functions. The amino acid sequences of $5\text{-}HT_2$ receptors [2, 3, 13] have a high degree of homology within the seven transmembrane domains, being structurally distinct from other 5-HT receptors, sharing a characteristic of all genes in having either two introns ($5\text{-}HT_{2A}$ and $5\text{-}HT_{2B}$ receptors) or three ($5\text{-}HT_{2C}$) in the coding sequence. All three are coupled positively, via G_q, to phospholipase C and increased accumulation of inositol phosphates and mobilize intracellular Ca^{2+}; though, $5\text{-}HT_{2B}$ receptor in the gut is not associated with G_q and the phospholipase C pathway. As above-mentioned, whether $5\text{-}HT_{2A/2B/2C}$ an agonistic, antagonistic and/or inverse agonistic action modulates learning and memory is unclear.

$5\text{-}HT_{2A/2B/2C}$ Receptors

The $5\text{-}HT_{2A}$ receptor has also been shown to be particularly involved in the action of hallucinogens such as lysergic acid diethylamide (LSD) in the cortex as well as in the therapeutic efficacy of antipsychotic medications (see [14]). Also, m-chlorophenylpiperazine (displays affinity for $5\text{-}HT_2$ receptors) induces migraine attacks [15]. Orban et al. [16] reported that $5\text{-}HT_{2C}$ receptor is involved in the development of epilepsy and $5HT_{2C}$ receptor inverse agonism in antipsychotic effects [17]. The present analysis of $5\text{-}HT_2$ receptor regarding memory tasks, aims illustrating the subject. For instance, $5\text{-}HT_2$ receptors role on memory consolidation have, e.g., revealed that the SB-200646 (a selective $5\text{-}HT_{2B/2C}$ receptor antagonist) and

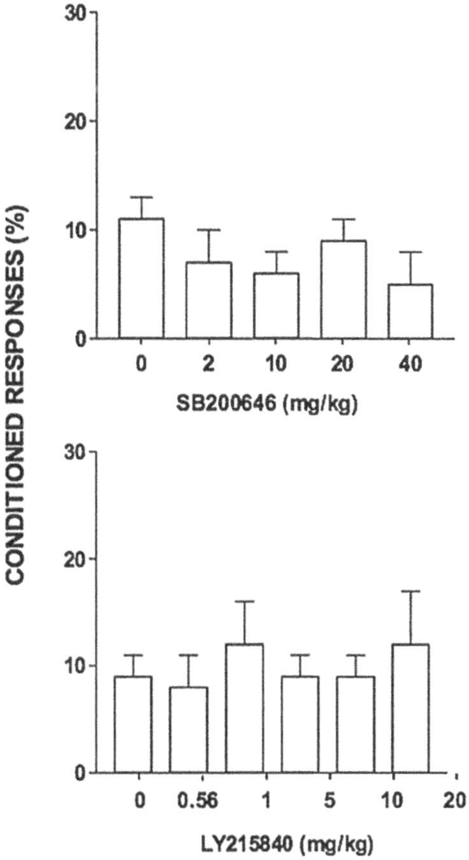

Fig. 1 The effect of acute posttraining administration (i.p) of (**a**) SB-200646 and (**b**) LY215840 in an autoshaping learning task in fasted animals. Data are plotted as percentage of conditioned responses. All rats received an injection immediately after the first training session and data correspond to session carried out 24 h later. Top bar values represent the mean, and vertical lines denote the s.e. mean of eight different animals. Neither compounds significantly altered performance. (Reproduced with permission from [45])

LY215840 (a nonselective $5\text{-}HT_{2/7}$ receptor antagonist) post-training administration had no effect on an autoshaped (Fig. 6) memory consolidation; however, both drugs significantly and differentially antagonized the memory impairments induced by serotonergic drugs as 1-(3-chlorophenyl)piperazine (mCPP), 1-naphtyl-piperazine (1-NP), mesulergine, or N-(3-trifluoromethylphenyl) piperazine (TFMPP). In contrast, SB-200646 failed to modify the facilitatory procognitive effect produced by (+/−)-2.5-dimethoxy-4-iodoamphetamine (DOI) or ketanserin, which were sensitive to MDL100907 (selective $5\text{-}HT_{2A}$ receptor antagonist) and to a LY215840 high dose. Finally, SB-200646 reversed the memory deficit induced by dizocilpine (glutamatergic antagonist), but not that by scopolamine (cholinergic antagonist): while SB-200646 and MDL100907 coadministration reversed memory deficits induced by both drugs (Figs. 1, 2, 3, 4 and 5). Hence, $5\text{-}HT_{2B/2C}$ receptors might be involved on memory formation probably mediating a suppressive or constraining action; certainly, whether the drug-induced memory impairments in the above study are explained by simple agonism, antagonism, or inverse agonism at $5\text{-}HT_2$ receptors, remains unclear [9]. Similar situation occurs regarding $5\text{-}HT_{1A}$, $5\text{-}HT_6$ and $5\text{-}HT_7$

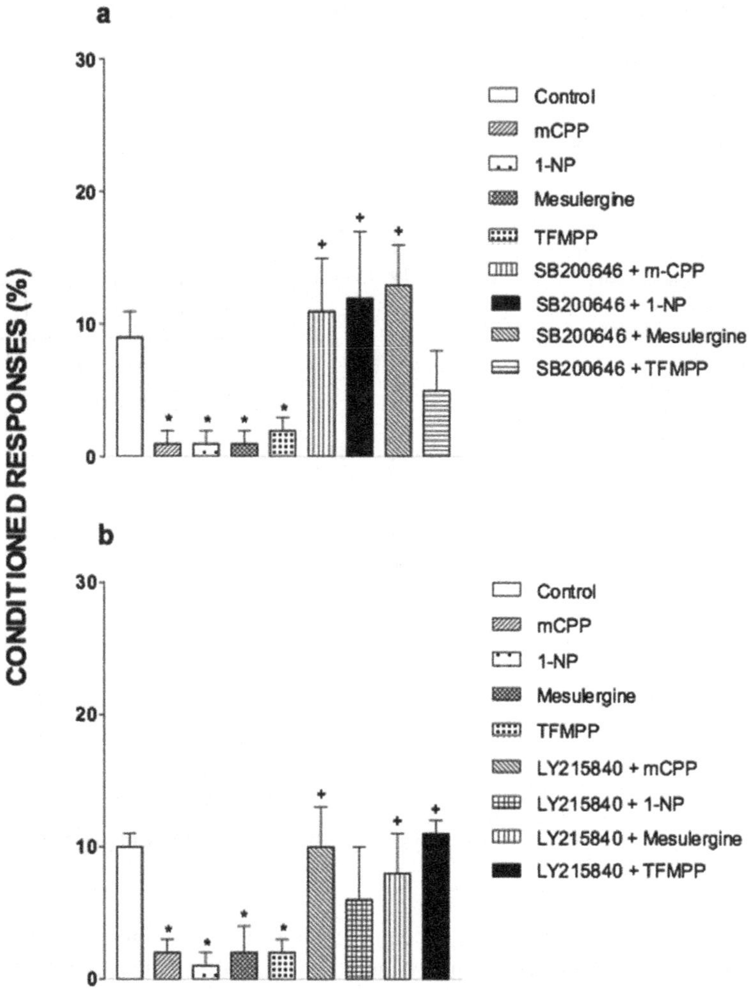

Fig. 2 The effect of acute posttraining administration (i.p) of (**a**) SB-200646 (2 mg/kg) and (**b**) LY215840 (0.56 mg/kg) on the responses induced by mCPP (10 mg/kg), 1-NP (1 mg/kg), mesulergine (0.4 mg/kg) and TFMPP (10 mg/kg) in an autoshaping learning task in fasted animals. Data are plotted as percentage of conditioned responses. All rats received an injection immediately after the first training session and data correspond to session carried out 24 h later. Top bar values represent the mean, and vertical lines denote the s.e. mean of eight different animals. $^*P < 0.05$ vs. vehicletreated rats. $+P < 0.05$ vs. initial drug treatments. (Reproduced with permission from [45])

receptors [5, 9]. Until now, 5-HT_2 receptor subtypes blockade may provide some benefit to reverse poor memory consolidation associated with decreased cholinergic, glutamatergic, and/or serotonergic neurotransmission [7].

Vanover et al. [18] reported that AC-90179 (a selective 5-HT_{2A} receptor inverse agonist) had no effect on acquisition of a (Pavlovian/instrumental autoshaping; (Fig. 6) nose-poke response until the highest dose (30 mg/kg s.c.). Haloperidol significantly reduced the number of reinforcers earned and clozapine dose-dependently

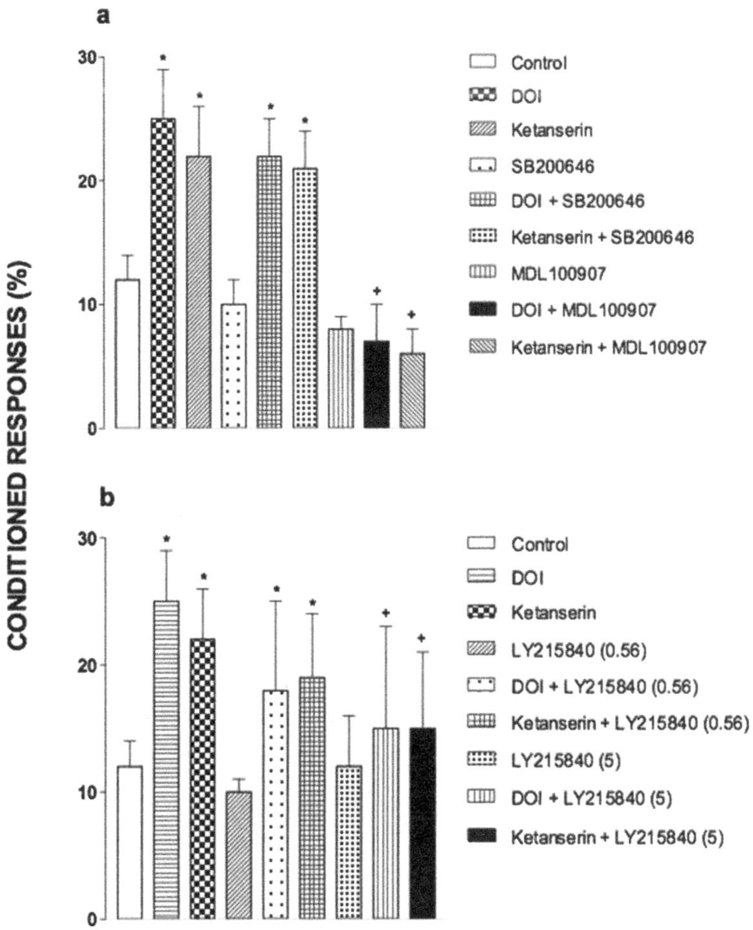

Fig. 3 The effect of acute posttraining administration (i.p) of (**a**) SB-200646 (2 mg/kg) or MDL100907 (1 mg/kg); and (**b**) LY215849 (0.56 and 5 mg/kg) on the responses induced by DOI (0.1 mg/kg) and Ketanserin (0.1 mg/kg) in an autoshaping learning task in fasted animals. Data are plotted as percentage of conditioned responses. All rats received an injection immediately after the first training session and data correspond to session carried out 24 h later. Top bar values represent the mean, and vertical lines denote the s.e. mean of eight different animals. $^{*}P < 0.05$ vs. vehicle-treated rats. $+P < 0.05$ vs. initial drug treatments. (Reproduced with permission from [45])

decreased the number of reinforcers earned, but the effect failed to reach statistical significance due to individual variability [18]. Regarding individual variability in autoshaping tasks (see [19, 20]). Notably, some of these data are comparable to those observed in other memory tasks [20].

Very importantly, the status of inverse agonism at serotonin2A (5-HT_{2A}) and 5-HT_{2C} receptors was recently revised (see [21]). Indeed, contemporary receptor theory was developed to account for the existence of constitutive activity, as defined by the presence of receptor signaling in the absence of any ligand; thus, ligands acting at a constitutively active receptor, can act as agonists, antagonists, and inverse

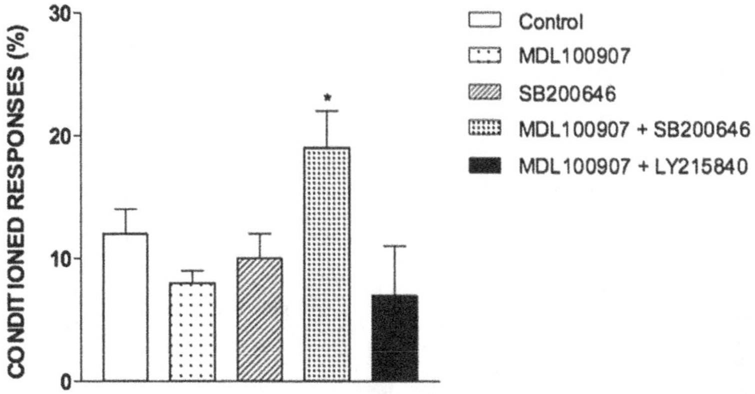

Fig. 4 The effect of acute posttraining administration (i.p) of SB-200646 (2 mg/kg), MDL100907 (1 mg/kg), and SB-200646 (2 mg/kg) combined with MDL100907 (1 mg/kg) or LY215840 (0.56 mg/kg) in an autoshaping task in fasted animals. Data are plotted as percentage of conditioned responses. All rats received an injection immediately after the first training session and data correspond to session carried out 24 h later. Top bar values represent the mean, and vertical lines denote the s.e. mean of eight different animals. *P < 0.05 vs. vehicle treated rats. (Reproduced with permission from [45])

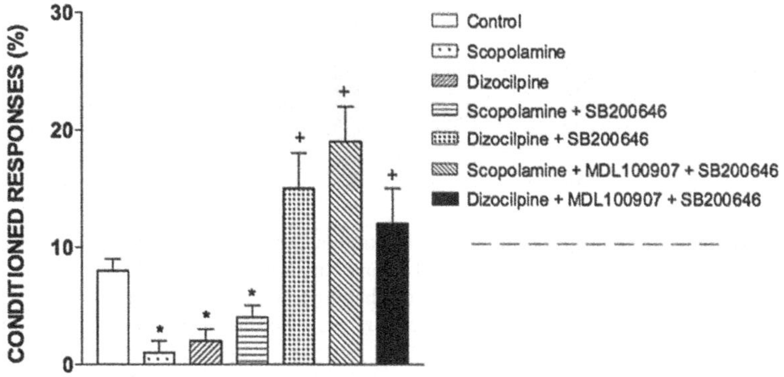

Fig. 5 The effect of acute posttraining administration (i.p) of SB-200646 (2 mg/kg) and the combination of SB-200646 (2 mg/kg) plus MDL100907 (1 mg/kg) on the impairing responses induced by scopolamine (0.17 mg/kg) and dizocilpine (0.1 mg/kg) in an autoshaping task in fasted animals. Data are plotted as percentage of conditioned responses. All rats received an injection immediately after the first training session and data correspond to session carried out 24 h later. Top bar values represent the mean, and vertical lines denote the s.e. mean of eight different animals. *P < 0.05 vs. vehicletreated rats. +P < 0.05 vs. values obtained with scopolamine and dizocilpine alone. (Reproduced with permission from [45])

Fig. 6 Illustration of behavioral memory tasks, specifically autoshaping memory task for rats

agonists [21]. *In-vitro* and *ex-vivo* studies have also revealed the complexity of ligand/receptor interactions including agonist-directed stimulus trafficking, a finding that has led to multi-active state models of receptor function (e.g., [5, 21]). Studies with a variety of cell types have established that the serotonin $5\text{-}HT_{2A}$ and $5\text{-}HT_{2C}$ receptors also demonstrate constitutive activity and inverse agonism [21]; however, until recently, there has been no evidence to suggest that these receptors also demonstrate constitutive activity and hence reveal inverse agonist properties of ligands *in vivo*. Aloyo et al. [21] describe the current knowledge of constitutive activity *in-vitro* and then examine the evidence for constitutive activity *in-vivo*. According with Aloyo et al. [21], both $5\text{-}HT_{2A}$ and $5\text{-}HT_{2C}$ receptors are involved in a number of physiological and behavioral functions and are the targets for treatment of schizophrenia, anxiety, weight control, Parkinsonism, and other disorders; the existence of constitutive activity at these receptors *in-vivo*, along with the possibility of inverse agonism, provides new avenues for drug development. In the context of memory, $5\text{-}HT_{2A}$ receptor inverse agonists seem to be important. Notably, regarding, $5\text{-}HT_{2C}$ receptor-specific agonist and $5\text{-}HT_{2A}$ competitive antagonist/$5\text{-}HT_{2B}$ inverse agonist with preclinical efficacy for psychoses are interesting [11].

It should be noted, the inverse agonism and its therapeutic significance postulate that a large number of G-protein-coupled receptors (GPCRs) show varying degrees of basal or constitutive activity. This constitutive activity is usually minimal in natural receptors but is markedly observed in wild type and mutated (naturally or induced) receptors [22]. Conventional two-state drug receptor interaction model, binding of a ligand may initiate activity (agonist with varying degrees of positive intrinsic activity) or prevent the effect of an agonist (antagonist with zero intrinsic activity) [21]. Inverse agonists bind with the constitutively active receptors, stabilize them, and thus reduce the activity (negative intrinsic activity). According with Khilnani and Khilanani [21] receptors of many classes (α-and β-adrenergic, histaminergic, GABAergic, serotonergic, opiate, and angiotensin receptors) have shown basal activity in suitable *in-vitro* models. For instance, drugs that have been conventionally classified as antagonists (e.g., β-blockers, antihistaminics) have shown inverse agonist effects on corresponding constitutively active receptors and some drugs have significant inverse agonistic activity that contributes partly or wholly to their therapeutic value [21]. Inverse agonism may also help explain the underlying mechanism of beneficial effects of e.g., clozapine in psychosis (see below). Notably, understanding inverse agonisms has paved a way for newer drug development [21], which have only desired therapeutic value and are devoid of unwanted (or reduced) adverse effect (e.g., anxiety, antinociceptive, obesity, chronic asthma). According with Khilnani and Khilanani [21] pimavanserin (ACP-103), a highly selective 5-HT_{2A} inverse agonist, attenuates psychosis in patients with Parkinson's disease with psychosis and is devoid of extrapyramidal side effects; therefore, inverse agonism is an important aspect of drug-receptor interaction and has immense untapped therapeutic potential [21].

Navailles et al. [23] updating the growing number of studies showing (by means of pharmacological tools) the participation of the constitutive activity of 5-HT_{2C} receptors in the control of various biochemical and behavioral functions *in-vivo*; and emphasizing the functional organization of this constitutive control together with the phasic and tonic (involving the spontaneous release of 5-HT) modalities of the 5-HT_{2C} receptor in the brain [23]. Moreover, functional anatomy of 5-HT_{2A} receptors in the amygdala and hippocampal complex revealed relevance to memory functions [24]. And investigation about 5-HT_{2A} receptor role in memory had showed (e.g., [14]) that memory fields of putative pyramidal cells were attenuated by iontophoresis of 5-HT_{2A} antagonists, which primarily produced a reduction in delay activity for preferred target localizations. Conversely, 5-HT_{2A} stimulation by alpha-methyl-5-HT or 5-HT itself, accentuated the spatial tuning of these neurons by producing a modest increase in activity for preferred target locations and/or a reduction in activity for nonpreferred locations. The agonist effects could be reversed by the selective antagonist MDL100907 and were dose-dependent, such that high levels attenuated spatial tuning by profoundly reducing delay activity. A role for feed-forward inhibitory circuitry in these effects was supported by the finding that 5-HT_{2A} blockade also attenuated the memory fields of putative interneurons. Williams et al. [14] conclude that prefrontal 5-HT_{2A} receptors have a hitherto unrecognized role in the cognitive function of working memory, which involves actions at both excitatory and inhibitory elements within local circuitry (see also [25]). Notably, Dougherty and Oristaglio [26] hypothesize that long-term drug treatments resulting in 5-HT_{2A} receptor up-regulation may be useful in enhancing recall of learned behaviors and thus may have potential for treating cognitive impairment associated with neurodegenerative disorders. Likewise, Dougherty and Oristaglio [26] point out that their observations suggest a widespread modulatory role of 5-HT_{2A} and 5-HT_{2C} receptors in learning and memory, with the net effect being dependent on task requirements and the specific mnemonic systems recruited. It should be noted that the methodology employed by Dougherty and Oristaglio [26] is a first step in better characterizing drug effects on goal-directed behavior and identifying quantifiable factors that can underlie changes in response latency; they suggest using more elaborate methodologies, such as video tracking, could add resolution to this analysis and provide a more complete profile of motor variability under baseline and drug-influenced conditions. In addition, such analyses could be an important consideration for evaluating the behavioral performance of different strains of mice, particularly in aged or neurodegenerative models where latencies in choice behavior and motor variability might be considerably higher [26]. For instance, increases in both reaction time and reaction time variability on cognitive tasks are associated with aging and cognitive decline in humans [26].

Moreover, 5-HT_{2B2C} receptors and memory investigation (e.g., [27]) had showed that systemic administration of 5-HT_{2C} and 5-HT_{2A} receptor antagonists significantly enhanced and impaired spatial reversal learning, respectively (e.g., [28]). Indeed, the role of these receptors in the mediation of these opposing effects was further investigated regarding neuroanatomical specificity within some of the brain regions implicated in cognitive flexibility [27], namely the orbitofrontal cortex (OFC), medial prefrontal cortex (mPFC), and nucleus accumbens (NAc), by means of targeted infusions of selective 5-HT_{2C} and 5-HT_{2A} receptor antagonists (SB-242084 and M100907, respectively). Intra-OFC 5-HT_{2C} receptor antagonism produced dose-dependent effects similar to those of systemic administration, i.e., improved spatial reversal learning by reducing the number of trials (doses: 0.1, 0.3, and 1.0 μg) and perseverative errors to criterion (0.3 and 1.0 μg) compared with controls [27]. However, the highest dose (1.0 μg) showed a nonselective effect, as it also affected retention preceding the reversal phase and decreased learning errors. Intracerebral infusions of SB-242084 into the mPFC or NAc produced no significant effects on any behavioral measures. Similarly, no significant differences were observed with intra-OFC, −mPFC, or -NAc infusions of M100907. According with Boulougouris and Robbins [27], these data suggest that the improved performance in reversal learning observed after 5-HT_{2C} receptor antagonism is mediated within the OFC. Also, the data also bear on the issue of whether 5-HT_{2C} receptor antagonism within the OFC might help elucidate the biological substrate of obsessive-compulsive disorder, offering the potential for therapeutic application. Moreover, novel 5-HT_{2A}/5-HT_{2C} receptor agonists with pro-cognitive effects have been reported [28]. For further evidence see also Meneses [7] and Puig and Gulledge [29].

According with Hanks and Gonzalez-Maeso [30], psychedelic 5-HT_{2A} receptor agonists LSD and DOI, but not lisuride, enhance trace conditioning of the nictitating membrane response in rabbits (a simple associative learning of a motor response; an effect reversed by $5\text{-HT}_{2A/2C}$ receptor antagonists. Fear memory in a trace conditioning paradigm was also affected by activation of the 5-HT_{2A} receptor in rats, and post-training administration of the 5-HT_{2A} receptor agonist (4-bromo-3,6-dimethoxybenzocyclobuten-1-yl)methylamine hydrobromide (TCB-2)115 enhanced subsequent freezing in a trace fear conditioning test [30].

Although the role of $5\text{-HT}_{2A/2B/2C}$ receptors in memory is unclear the above and other studies suggest new insights. For instance, evidence reported by Blasi et al. [31] suggest that HTR2A affects 5-HT_{2A} receptor expression and functionally contributes to genetic modulation of known endophenotypes of schizophrenia-like higher-level cognitive behaviors and related prefrontal activity, as well as response to treatment with olanzapine. Moreover, true but not false memories seem to be associated with the HTR2A gene [32].

Genetic Delation and 5-HT_2 Receptors

Interestingly, new horizons are emerging. For instance, acute pharmacological blockade and constitutive loss of 5-HT_{2C} receptor activity provide insights into the serotonergic regulation of executive control processes (part of cognition; e.g., [33]) and suggest that impaired 5-HT_{2C} receptor signaling during development may predispose to executive function disorders [34]. The 5-HT_{2C} receptor gene mutation revealed abnormal performance in a spatial learning task and altered exploratory behavior, associated with perturbed long-term potentiation (supposedly physiological model of memory) restricted to the dentate gyrus perforant path synapse [35]. Certainly, 5-HT_{2C} receptor activation inhibits appetitive and consummatory components of feeding and increases brain c-fos immunoreactivity in mice [36]. Del'guidice et al. [37] reported that stimulation of 5-HT_{2C} receptor improves cognitive deficits induced by human tryptophan hydroxylase 2 loss of function mutation. Dissociable effects of 5-HT_{2C} receptor antagonism and genetic inactivation on perseverance and learned non-reward in an egocentric spatial reversal task revealed that 5-HT_{2C} receptor reducing activity facilitates reversal learning in the mouse by reducing the influence of previously non-rewarded associations [38]. Finally, 5-HT_{2C} receptor inactivation potentiates consequences of serotonin reuptake blockade [39].

5-HT_2 Receptors and Addiction

Volkow et al. [40] have noted that during addiction, the enhanced value of the drug in the reward, motivation, and memory circuits overcomes the inhibitory control exerted by the prefrontal cortex, thereby favoring a positive-feedback loop initiated by the consumption of the drug and perpetuated by the enhanced activation of the motivation/drive and memory circuits. Moreover, Cunningham and colleagues [41] reported synergism between 5-HT_{2A} receptor antagonism and 5-HT_{2C} receptor agonism pharmacotherapeutics for cocaine addiction. Importantly, selective 5-HT_{2C} receptor agonism and 5-HT2A receptor antagonism may be potential targets for therapies to treat some aspects of nicotine dependence [42]. Chronic fluvoxamine (an antidepressant) treatment changes $5\text{-HT}_{2A/2C}$ receptor-mediated behavior in olfactory bulbectomized mice (a model of depression) [43]. In addition, 5-HT_{2A}

receptor in the orbitofrontal cortex facilitates reversal learning and contributes to the beneficial cognitive effects of chronic citalopram (antidepressant) treatment [44]. In conclusion, all together the above data allow illustrating the therapeutic diversity and utility of $5\text{-HT}_{2A/2B/2C}$ receptors.

Acknowledgments The author also wants to thank Sofia Meneses for the language review and the technical assistance of Roberto Gonzalez.

References

1. Barbas D, Zappulla JP, Angers S, Bouvier M, Castellucci VF, DesGroseillers L (2002) Functional characterization of a novel serotonin receptor (5-HT_{ap2}) expressed in the CNS of Aplysia Californica. J Neurochem 80:335–345
2. Barnes NM, Sharp T (1999) A review of central 5-HT receptors and their function. Neuropharmacology 38:1083–1152
3. Hoyer D, Clarke DE, Fozard JR, Hartig PR, Martin GR, Mylecharane EJ, Saxena PR, Humphrey PPA (1994) International Union of Pharmacology classification of receptors for 5hydroxytryptamine (serotonin). Pharmacol Rev 46:157203
4. Hoyer D, Hannon JP, Martin GR (2002) Molecular, pharmacological and functional diversity of 5-HT receptors. Pharmacol Biochem Behav 71:533–554
5. Meneses A (1999) 5-HT system and cognition. Neurosci Biobehav Rev 23:1111–1125
6. Meneses A (2013) 5-HT systems: emergent targets for memory formation and memory alterations. Rev Neurosci 24(6):629–664
7. Harvey JA (1996) Serotonergic regulation of associative learning. Behav Brain Res 73:47–50
8. Meneses A (2002) Involvement of 5-HT(2A/2B/2C) receptors on memory formation: simple agonism, antagonism, or inverse agonism? Cell Mol Neurobiol 22(5–6):675–688
9. Meneses A (2014) Memory formation and memory alterations: 5-HT6 and 5-HT7 receptors, novel alternative. Rev Neurosci 25(3):325–356
10. De Ligt RAF, Kourounakis AP, Ijzerman AP (2000) Inverse agonism at G protein-coupled receptors: (patho)physiological relevance and implications for drug discovery. Br J Pharmacol 130:1–12
11. Martin GR, Eglen RM, Hamblin MW, Hoyer D, Yocca F (1998) The structure and signaling properties of 5-HT receptors: an endless diversity. Trends Pharmacol Sci 19:2–4
12. Canal CE, Morgan D, Felsing D, Kondabolu K, Rowland NE, Robertson KL, Sakhuja R, Booth RG (2014) A novel aminotetralin-type serotonin (5-HT) 2C receptor-specific agonist and 5-HT2A competitive antagonist/5-HT2B inverse agonist with preclinical efficacy for psychoses. J Pharmacol Exp Ther 349:310–318
13. Williams GV, Rao SG, Goldman-Rakic PS (2002) The physiological role of 5-HT2A receptors in working memory. J Neurosci 22:2843–2854
14. Leone M, Attanasio A, Croci D, Filippini G, D'Amico D, Grazzi L, Nespolo A, Bussone G (2000) The serotonergic agentm- chlorophenylpiperazine induces migraine attacks: a controlled study. Neurology 55:169–139
15. Orban G, Bombardi C, Marino Gammazza A, Colangeli R, Pierucci M, Pomara C, Pessia M, Bucchieri F, Arcangelo B, Smolders I, De Deurwaerdère P, Di Giovanni G (2014) Role(s) of the 5-HT2C receptor in the development of maximal dentate activation in the hippocampus of anesthetized rats. CNS Neurosci Ther 20:651–661
16. Rauser L, Savage JE, Meltzer HY, Roth BL (2001) Inverse agonist actions of typical and typical and atypical antipsychotic drugs at the human $5\text{-hydroxytryptamine}_{2C}$ receptor. J Pharmacol Exp Ther 299:83–89

17. Vanover KE, Harvey SC, Son T, Bradley SR, Kold H, Makhay M, Veinbergs I, Spalding TA, Weiner DM, Andersson CM, Tolf BR, Brann MR, Hacksell U, Davis RE (2004) Pharmacological characterization of AC-90179 [2-(4-methoxyphenyl)-N-(4-methyl-benzyl)-N-(1-methyl-piperidin-4-yl)-acetamide hydrochloride]: a selective serotonin 2A receptor inverse agonist. J Pharmacol Exp Ther 310(3):943–951
18. Meneses A, Perez-Garcia G, Liy-Salmeron G, Ponce-López T, Lacivita E, Leopoldo M (2015) 5-HT_7 receptor activation by LP-211 increases LTM and cAMP production and attenuates memory impairment. Psychopharmacology 232(3):595–603
19. Perez-Garcia G, Meneses A (2008) Memory formation, amnesia, improved memory and reversed amnesia: 5-HT role. Behav Brain Res 195(1):17–29
20. Aloyo VJ, Berg KA, Spampinato U, Clarke WP, Harvey JA (2009) Current status of inverse agonism at serotonin2A (5-HT2A) and 5-HT2C receptors. Pharmacol Ther 121(2):160–173
21. Khilnani G, Khilnani AK (2011) Inverse agonism and its therapeutic significance. Indian J Pharmacol 43(5):492–501
22. Navailles S, Lagière M, Le Moine C, De Deurwaerdère P (2013) Role of 5-HT2C receptors in the enhancement of c-Fos expression induced by a 5-HT2B/2C inverse agonist and 5-HT 2 agonists in the rat basal ganglia. Exp Brain Res 230(4):525–535
23. Bombardi C, Di Giovanni G (2013) Functional anatomy of 5-HT2A receptors in the amygdala and hippocampal complex: relevance to memory functions. Exp Brain Res 230(4):427–439
24. Zhang G, Ásgeirsdóttir HN, Cohen SJ, Munchow AH, Barrera MP, Stackman RW Jr (2013) Stimulation of serotonin 2A receptors facilitates consolidation and extinction of fear memory in C57BL/6J mice. Neuropharmacology 64:403–413
25. Dougherty JP, Oristaglio J (2013) Chronic treatment with the serotonin 2A/2C receptor antagonist SR 46349B enhances the retention and efficiency of rule-guided behavior in mice. Neurobiol Learn Mem 103:50–63
26. Boulougouris V, Robbins TW (2010) Enhancement of spatial reversal learning by 5-HT2C receptor antagonism is neuroanatomically specific. J Neurosci 30(3):930–938
27. López-Vázquez MA, Gutiérrez-Guzmán BE, Cervantes M, Olvera-Cortés ME (2011) 5-HT2C receptors in learning. In: Di Giovanni G, et al. (eds) 5-HT 2C receptors in the pathophysiology of CNS disease. Springer Science + Business Media, New York: 461-507
28. Jensen AA, Plath N, Pedersen MH, Isberg V, Krall J, Wellendorph P, Stensbøl TB, Gloriam DE, Krogsgaard-Larsen P, Frølund B (2013) Design, synthesis, and pharmacological characterization of N- and O-substituted 5,6,7,8-tetrahydro-4H-isoxazolo[4,5-d]azepin-3-ol analogues: novel 5-HT(2A)/5-HT(2C) receptor agonists with pro-cognitive properties. J Med Chem 56(3):1211–1227
29. Puig MV, Gulledge AT (2011) Serotonin and prefrontal cortex function: neurons, networks, and circuits. Mol Neurobiol 44(3):449–464
30. Hanks JB, González-Maeso J (2013) Animal models of serotonergic psychedelics. ACS Chem Neurosci 4(1):33–42
31. Blasi G, De Virgilio C, Papazacharias A, Taurisano P, Gelao B, Fazio L, Ursini G, Sinibaldi L, Andriola I, Masellis R, Romano R, Rampino A, Di Giorgio A, Lo Bianco L, Caforio G, Piva F, Popolizio T, Bellantuono C, Todarello O, Kleinman JE, Gadaleta G, Weinberger DR, Bertolino A (2013) Converging evidence for the association of functional genetic variation in the serotonin receptor 2a gene with prefrontal function and olanzapine treatment. JAMA Psychiat 70(9):921–930
32. Zhu B, Chen C, Loftus EF, Moyzis RK, Dong Q, Lin C (2013) True but not false memories are associated with the HTR2A gene. Neurobiol Learn Mem 106:204–209
33. Ruan L, Lau BW, Wang J, Huang L, Zhuge Q, Wang B, Jin K, So KF (2014) Neurogenesis in neurological and psychiatric diseases and brain injury: from bench to bedside. Prog Neurobiol 115C:116–137
34. Pennanen L, Van der Hart M, Yu L, Tecott LH (2013) Impact of serotonin (5-HT)2C receptors on executive control processes. Neuropsychopharmacology 38:957–967

35. Heisler LK, Tecott LH (2009) Knockout Corner: neurobehavioural consequences of a serotonin 5-HT(2C) receptor gene mutation. Int J Neuropsychopharmacol 2(1):67–69
36. Somerville EM, Horwood JM, Lee MD, Kennett GA, Clifton PG (2007) 5-HT(2C) receptor activation inhibits appetitive and consummatory components of feeding and increases brain c-fos immunoreactivity in mice. Eur J Neurosci 25:3115–3124
37. Del'guidice T, Lemay F, Lemasson M, Levasseur-Moreau J, Manta S, Etievant A, Escoffier G, Doré FY, Roman FS, Beaulieu JM (2014) Stimulation of 5-HT2C receptors improves cognitive deficits induced by human tryptophan hydroxylase 2 loss of function mutation. Neuropsychopharmacology 39:1125–1134
38. Nilsson SR, Somerville EM, Clifton PG (2013) Dissociable effects of 5-HT2C receptor antagonism and genetic inactivation on perseverance and learned non-reward in an egocentric spatial reversal task. PLoS One 30:e77762
39. Cremer TI, Giorgetti M, Bosker FJ, Hogg S, Arnt J, Mork A, Honig G, Bogeso KP, Westerink BH, Den Boer H, Wikstrom HV, Tecott LH (2004) Inactivation of 5-HT(2C) receptors potentiates consequences of serotonin reuptake blockade. Neuropsychopharmacology 29:1782–1789
40. Volkow ND, Baler RD, Goldstein RZ (2011) Addiction: pulling at the neural threads of social behaviors. Neuron 69:599–602
41. Cunningham KA, Anastasio NC, Fox RG, Stutz SJ, Bubar MJ, Swinford SE, Watson CS, Gilbertson SR, Rice KC, Rosenzweig-Lipson S, Moeller FG (2013) Synergism between a serotonin 5-HT2A receptor (5-HT2AR) antagonist and 5-HT2CR agonist suggests new pharmacotherapeutics for cocaine addiction. ACS Chem Neurosci 4:110–121
42. Fletcher PJ, Rizos Z, Noble K, Soko AD, Silenieks LB, Lê AD, Higgins GA (2012) Effects of the 5-HT2C receptor agonist Ro60-0175 and the 5-HT2A receptor antagonist M100907 on nicotine self-administration and reinstatement. Neuropharmacology 62:2288–2298
43. Oba A, Nakagawasai O, Onogi H, Nemoto W, Yaoita F, Arai Y, Tan-No K, Tadano T (2013) Chronic fluvoxamine treatment changes 5-HT(2A/2C) receptor-mediated behavior in olfactory bulbectomized mice. Life Sci 92:119–124
44. Furr A, Lapiz-Bluhm MD, Morilak DA (2012) 5-HT_{2A} receptors in the orbitofrontal cortex facilitate reversal learning and contribute to the beneficial cognitive effects of chronic citalopram treatment in rats. Int J Neuropsychopharmacol 15:1295–1305
45. Meneses A (2002) Involvement of 5-HT2A/2B/2C receptors on memory formation: simple agonism, antagonism, or inverse agonism? Cell Mol Neurobiol 22:675–688

$5\text{-}HT_{2A}$ Receptors in the Basal Ganglia

Cristina Miguelez, Teresa Morera-Herreras, and Philippe De Deurwaerdère

Abstract The serotonin$_{2A}$ ($5\text{-}HT_{2A}$) receptor is present in the basal ganglia (BG), a group of subcortical structures involved in the control of motor behaviours. It is one of the numerous serotonin (5-HT) G-protein coupled receptors responding to the release of 5-HT from neurons of the dorsal raphe nucleus. The interest brought to the function of $5\text{-}HT_{2A}$ receptors in the BG is related to the possible implication of $5\text{-}HT_2$ receptors in the regulation of mesencephalic dopaminergic neurons and the deleterious side effects of long-term treatment with antipsychotic medication.

The $5\text{-}HT_{2A}$ receptors are mostly expressed in the cortex and to a lesser extent in the BG, where other 5-HT receptor subtypes show stronger expression. Nonetheless, numerous cells including dopaminergic, GABAergic, glutamatergic or cholinergic neurons express $5\text{-}HT_{2A}$ receptors brain-wide. Correspondingly, $5\text{-}HT_{2A}$ receptors modulate the metabolic and electrophysiological activity of some neuronal populations including dopaminergic and GABAergic neurons. This control involves $5\text{-}HT_{2A}$ receptors in the BG and is specific and state-dependent, in particular with regard to the level of dopaminergic transmission. Behavioural data have also shown that $5\text{-}HT_{2A}$ receptor agents modulate the effects of a variety of psychotropic agents including drugs of abuse and antipsychotic drugs. Moreover, the $5\text{-}HT_{2A}$ receptor-mediated modulation is altered in animal models of Parkinson's disease, tardive dyskinesia, L-DOPA-induced dyskinesia and drug addiction.

This chapter summarizes data exploring the role of $5\text{-}HT_{2A}$ receptors in the BG, which remains an important topic for research aimed at ameliorating current treatments of schizophrenia, Parkinson's disease and addiction.

C. Miguelez • T. Morera-Herreras
Department of Pharmacology, Faculty of Medicine and Nursing, University of the Basque Country (UPV/EHU), Leioa, Spain

P. De Deurwaerdère (✉)
Université de Bordeaux, Bordeaux, France

Centre National de la Recherche Scientifique, UMR 5287, Bordeaux, France
e-mail: deurwaer@u-bordeaux.fr

B.P. Guiard, G. Di Giovanni (eds.), *5-HT$_{2A}$ Receptors in the Central Nervous System*, The Receptors 32, https://doi.org/10.1007/978-3-319-70474-6_12

Keywords Striatum • Cortex • Substantia Nigra • Serotonin • Dopamine • Drug of Abuse • Electrophysiology • Behaviour • Intracerebral Microdialysis • Pharmacology

Abbreviations

5-HT	Serotonin
5-HT_{2A} receptor	Serotonin 2A receptor subtype
6-OHDA	6-Hydroxydopamine
BG	Basal Ganglia
DA	Dopamine
DOI	1-(2,5-dimethoxy-4-iodophenyl)-2-aminopropane
DRN	Dorsal raphe nucleus
EP	Entopeduncular nucleus
EPS	Extrapyramidal side effects
GPe	External globus pallidus
GPi	Internal globus pallidus
IHC	Immunohistochemical studies
LSD	Lysergic acid diethylamide
m-CPP	Metachlorophenylpiperazine
MDL 100,907	(R-(+)-a-(2,3-dimethoxyphenyl)-1-[2-(4-fluorophenyl) ethyl]-4-piperidinemethanol)
MPTP	1-Methyl-4-phenyl-1,2,3,6-tetrahydropyridine
MRN	Medial raphe nucleus
mRNA ISH	mRNA *in situ* hybridization
NAc	Nucleus Accumbens
OCD	Obsessive Compulsive Disorders
PCP	Phencyclidine
pCPA	*para*-chlorophenylalanine
PD	Parkinson's disease
PET	Positron emission tomography
PPE	Preproenkephalin
PPT	Preprotachykinin
Ro 60–0175	S-2-(6-chloro-5-fluoroindol-1-yl)-1-methylethylamine
RT-PCR	Reverse transcription polymerase chain reaction
RU-29469	5-Methoxy-3-(1,2,5,6-tetrahydro-4-pyridinyl)-1H–indole
SB 228357	1–5[−fluoro-3-(3-pyridyl)phenyl-carbamoyl]-5-methoxy-6-trifluoromethylindoline
SNc	Substantia nigra *pars* compacta
SNr	Substantia nigra *pars* reticulata
SPNs	Spiny projecting neurons
STN	Subthalamic nucleus
TFMPP	Trifluoromethylphenylpiperazine
TH	Tyrosine hydroxylase
VTA	Ventral tegmental area

Introduction

The research on serotonin$_{2A}$ (5-HT_{2A}) receptor subtypes within the basal ganglia (BG) started during the 70s. The BG encompass a group of subcortical and intermingled regions involved in the control of motor behaviours [1]. They house mesencephalic dopamine (DA) neurons, which play a critical role in the function of these brain regions, as suggested by their involvement in numerous diseases including Parkinson's disease (PD), schizophrenia and drug abuse [2–6]. Soon, it was suggested that 5-HT_2 receptors subtypes were able to control DA release, and notably to exert an inhibitory control on DA neuron activity [7–9]. The interest was growing when the therapeutic benefit of atypical antipsychotic drugs, marked by the lower incidence of extrapyramidal and vegetative side effects compared to typical antipsychotics, was proposed to rely on their higher affinity for 5-HT_2 compared to DA D_2 receptors [10–12]. At that moment, the lack of selective 5-HT_{2A}, 5-HT_{2B} and 5-HT_{2C} receptor compounds delayed the understanding of the role of each of these subtypes [13, 14]. However, during the 90s, the synthesis of selective ligands permitted to elucidate the action exerted by each receptor subtype in the BG, demonstrating distinct and even opposite effects on DA activity [13].

Beyond the complex regulation of DA neurons, anatomical, electrophysiological and neurochemical data provide evidence for the influence of 5-HT_{2A} receptors on various neuronal populations in the BG. Their role is further stressed when looking at the behavioural data related to BG function. Yet, the role of 5-HT_{2A} receptors specifically expressed in the BG is often unclear. Thus, the aim of this chapter is to summarize data indicating that 5-HT_{2A} receptors expressed in the BG play a role in modulating neuronal function and control of motor behaviours.

Basal Ganglia Contains Serotonergic Fibers and 5-HT_{2A} Receptors

Functional Organization of the Basal Ganglia

The BG are a highly-organized network formed by different subcortical nuclei that connects the cortex with the thalamus, creating the cortico-BG-thalamo-cortical loop. Based on topographical organization, this network is divided in sensorimotor, associative and limbic circuits and is therefore involved in motor and cognitive functions [1, 15, 16]. The BG comprise the striatum, the external and internal segments of the globus pallidus (GPe and GPi or entopenduncular nucleus, EP, in rodents), the substantia nigra (*pars* compacta, SNc, and *pars* reticulata, SNr) and the subthalamic nucleus (STN) [17]. Other limbic structures, such as the nucleus accumbens (NAc) situated in the ventral striatum, or the ventral pallidum are important to take into account because they integrate information from the BG and limbic systems [18]. The STN and the striatum are considered the input stations of the BG, and receive direct afferences from the cortex, limbic structures and the thalamus, while the SNr and the GPi are the output nuclei that send the information back to premotor and motor cortical areas by means of the thalamus. Based on the cortico-nigral projections, three parallel pathways have been described: the hyperdirect pathway, which arises from cortical structures and projects to the output nuclei *via* the STN; the direct pathway that conveys cortical inputs to the SNr through specific striatal projecting neurons; and the indirect pathway, which connects the cortex with the SNr passing first through the striatum, GPe and STN [19].

Although GABAergic and glutamatergic connections govern the flow of information in the BG, the proper activity of these nuclei depends on the functional integrity of the DA neurons. DA cell bodies are located in the SNc and the ventral tegmental area (VTA), and innervate all BG nuclei at different levels. Degeneration of DA neurons and subsequent loss of DA control produces an imbalance in the BG pathways that leads to dramatic molecular, cellular, and behavioural modifications [2, 20, 21].

Serotonergic Innervation of the Basal Ganglia

The topographical organization of serotonergic projections in the brain is similar across different species [22–25], with the midbrain raphe nuclei being the origin of the ascending innervation. Both the medial (MRN) and the dorsal raphe nucleus (DRN) send projections to common areas implicated in motor control, such as the thalamus and the BG. Although the MRN innervates areas as the SN or the NAc, the nature of these projections is non-serotonergic and therefore, the DRN is the main responsible nucleus for the serotonergic control of the BG (reviewed in [26, 27]). Serotonergic fibres primarily within the medial forebrain bundle ascend to innervate all areas of the BG, including the striatum, GP, SNc, SNr and STN [28–32]. In general, the SN is the most densely innervated nucleus followed by the striatum, STN, GPi and GPe. The DRN also provides most of the 5-HT innervation in the VTA, NAc, ventral pallidum and prefrontal cortex [31, 33, 34].

Distribution of 5-HT_{2A} Receptors in the Basal Ganglia

General Overview

Serotonin (5-HT) physiologically modulates BG nuclei activity by acting on 5-HT receptors (reviewed in [26, 35]). Among this large family of receptors, the 5-HT_{2A} receptor (originally termed 5-HT_2 [36]) is coupled to the activation of protein kinase C via G proteins of the Gq/G11 subtype. The distribution of these receptors is summarized in Table 1.

Table 1 Distribution of 5-HT_{2A} receptors in the basal ganglia.

	Caudate/ Putamen	NAc	GPe/ GPi	EP/ SNr	SN/ VTA	STN	References
Rodents							
Autoradiography	++	++	+	0/+	+	+	[37, 38]
mRNA ISH	+	++	0/+	0/+	+	0	[48, 51, 52, 55, 237]
IHC	++	++	+	+	+	+	[61–65, 68–71]
Non-human primates							
Autoradiography	++	++	+	+	+	+	[39, 40, 42]
mRNA ISH	0	0	0	0	0	/	[41, 53, 233]
IHC	+	/	/	/	/	/	
Humans							
Autoradiography	++	++	+	/	+	+	[40, 43, 45]
mRNA ISH	0	/	/	/	/	/	[41, 53, 233]
IHC	/	/	/	/	+	/	[66]

Information in the table reflects data from autoradiographical, mRNA *in situ* hybridization (mRNA ISH) and immunohistochemical (IHC) studies performed in tissue from rodents, non-human primates and humans. The intensity of the expression is graded as low (+), moderate (++) or absent (0), debated (0/+) or unstudied (/). For abbreviations, see list.

Autoradiographic binding studies using 5-HT_{2A} receptor radioligands, as [^{3}H]-ketanserin, [^{3}H]-mesulergine, [^{3}H]-LSD, [^{3}H]-spiperone, or [^{3}H]-MDL 100907, have found high levels of 5-HT_{2A} receptor binding sites in neocortical areas, caudate-putamen and NAc and lower levels (sometimes barely detectable) in the SN, GP or STN in rodents [37, 38], monkeys [39–42] and humans [43–45]. In human brain tissue, the use of more specific ligands confirmed the presence of the receptor in striosomes in the caudate-putamen [44]. The striosomes represent the part of the striatum with low levels of acetylcholinesterase, rich in mu opioid receptors, and containing the cell bodies of the GABAergic neurons projecting to the SNc [46, 47]. In the rat, in contrast, there is reportedly an inverse distribution of 5-HT_{2A} receptor mRNA with predominant expression in the matrix compared to the striosomes [48].

Numerous studies using *in situ* hybridization have demonstrated widespread and similar mapping of 5-HT_{2A} receptor mRNA among species [41, 49–52]. In the rodent, monkey and human, the highest expression has been reported in the cortex, which has been extensively characterized. In the rat brain, strong hybridization signal has been observed in the deeper laminae of the prefrontal cingulate and the retrosplenial cortex while in the human brain, intense expression was found in the neocortex, where pyramidal neurons and interneurons were labelled [49, 53]. The olfactory bulb also expresses high levels of mRNA, while other nuclei as the raphe nuclei or hippocampus (except the CA3) show barely detectable or very low expression of 5-HT_{2A} receptor mRNA [49, 51, 54]. In the rodent BG, some studies, but not all [49], have found intermediate levels of expression in the SNc or the caudate-putamen, and higher expression in the ventral striatum, including the NAc or the olfactory tubercle [51, 52, 55]. In the human and non-human primate brain, however, no expression was reported in the BG components [41, 53]. In general, receptor autoradiography studies agree with the results from *in situ* hybridization. However, some areas lacking 5-HT_{2A} receptor mRNA were confirmed to express low but detectable receptor binding [41]. These differences between the density of receptors and the presence of the correspondent mRNA may rely on different issues. On the one hand, the detection limits of *in situ* hybridization techniques may not be adequate for detecting low expression of mRNA or simply, 5-HT_{2A} receptors may be heteroceptors in some areas where they would be located in non-5HT afferent axons. On the other hand, the disparity of radioligand specificities used among the autoradiography studies may also contribute to the overestimation of 5-HT_{2A} receptor expression in areas where correspondent mRNA was not detected.

Results from positron emission tomography (PET) with specific antagonist radioligands as [^{11}C]-MDL100907 or [^{18}F]-altanserin are in good accordance with the distribution of 5-HT_{2A} receptors in the brain defined by autoradiography studies. Both in rats and humans, high radioactive signal was observed in the neocortex (occipital, temporal and frontal cortex), with lower expression in the cerebellum, thalamus, caudate-putamen, pons and very weak expression in the cerebellum [56–60]. Supporting the 5-HT_{2A} receptor mapping described by the previous techniques, immunohistochemical assays largely confirmed the location of 5-HT_{2A} receptors in the olfactory system, hippocampus, amygdala, and cerebral cortex [61, 62]. Regarding the pattern of 5-HT_{2A} receptor distribution in the rat BG nuclei, detectable levels have been shown in the caudate-putamen, NAc and ventral pallidum in most studies [61–65]. In addition, others have found modest receptor expression also in the GP, EP, SNr, SNc, VTA and STN in rats [61] and in the SNc and VTA in humans [66].

Cell Types Expressing the 5-HT_{2A} Receptors

The use of single or double *in situ* hybridization, reverse transcription coupled to polymerase chain reaction (RT-PCR) and immunohistochemistry has helped to identify some of the cell types that express 5-HT_{2A} receptors. Brain areas with high expression have been extensively characterized, including the prefrontal cortex where it is known that pyramidal cells expressing 5-HT_{2A} receptors send direct projections to the NAc (for review see [67]). In the BG, however, the phenotypical description of 5-HT_{2A} receptor-containing cells has been poor and mainly focused on the striatum, which shows relatively moderate (ventral striatum) or low (dorsal striatum) levels of expression. In the caudate-putamen and shell of the NAc, the 5-HT_{2A} receptor has been found in the cell-bodies, dendrites, dendritic spines, axons and axon terminals [65] and, as in the other BG nuclei, the location seems to be cytoplasmic [61, 65].

In the NAc, 5-HT_{2A} receptors are mostly located in spiny projecting neurons (SPNs) [61], although one study has also found labelled glial cells in the core and shell regions [62]. Further phenotypic cellular characterization using double-labelling *in situ* hybridization demonstrated that in the ventral striatum, 5-HT_{2A} receptor mRNA is present in both enkephalin- and dynorphin- containing SPNs [48, 55], although not all cells were found to express the 5-HT receptor. In the caudate-putamen, 5-HT_{2A} receptors are expressed by various neuronal populations including parvalbumin-positive interneurons and SPNs [61–63, 68]. In agreement with the pattern observed in the ventral area, neurons containing 5-HT_{2A} receptors in the posterior part of the striatum seem to be positive for dynorphin, enkephalin or substance P [48], which suggests that neither striatopallidal nor striatonigral pathways are specifically influenced by 5-HT_{2A} receptor control (see below). Additionally, one study has reported that some cells in the caudate-putamen and the NAc co-express 5-HT_{2A} and glutamatergic NMDA receptors [65].

In the midbrain, DA neurons in the VTA and SNc express both tyrosine hydroxylase (TH) and 5-HT_{2A} receptors [66, 69, 70] and interestingly, the existence of $5\text{-HT}_{2A}/D_2$ heterodimers has been recently suggested in the SNr by some authors [71].

Conclusions

5-HT_{2A} receptors are unevenly expressed in the BG. Although low expression has been observed in most BG nuclei, the NAc, ventral pallidum and medial striatum have higher expression. In those regions, mainly SPNs contain the receptor, but the pattern of expression is not restricted to a specific pathway classification. Taking into account that the NAc is the structure of the BG with higher expression, one can speculate that the functional control of 5-HT_{2A} receptors may mainly influence limbic or associative rather than sensorimotor circuits.

5-HT_{2A} Receptor-Mediated Electrophysiological Control of the Basal Ganglia

5-HT_{2A} Receptor Mediated Modulation of Neuron Activity in the Striatum

Reports about the serotonergic modulation of striatal function are inconsistent. The local injection of 5-HT into the striatum or the electrical stimulation of the DRN induce inhibition of the vast majority of SPNs [72–74]. However, other studies based on intracellular recordings have shown, first, that DRN stimulation evokes striatal excitatory postsynaptic potentials and second, that 5-HT administration excites SPNs by reducing voltage-dependent potassium currents [75–78].

The 5-HT_1 receptor family controls striatal activity by regulating the release of different neurotransmitters. Stimulation of 5-HT_{1A} or 5-HT_{1B} receptors leads to an inhibition of 5-HT release [79, 80]. Moreover, 5-HT_{1A} receptor agonists decrease glutamate release from corticostriatal projections [81–83], while the 5-HT_{1A} receptor antagonist, WAY100135, increases it [84]. On the other hand, striatal DA tone is increased after 5-HT_{1B} receptor stimulation through a mechanism that implicates a decrease of GABA release from the SNr [79].

Regarding 5-HT_2 receptor family, both 5-HT_{2A} and 5-HT_{2C} receptor agonists evoke inhibitory responses in the caudate nucleus [85, 86]. This inhibition may, however, depend on the striatal area analyzed [78]. Indeed, other studies stress that 5-HT_2 receptor stimulation (mainly 5-HT_{2C}), also induces an activation of cholinergic and fast-spiking interneurons, leading to an increased release of acetylcholine and subsequent GABAergic postsynaptic inhibition, which inhibits SPNs [87, 88].

In the NAc, although the vast majority of the studies have determined the serotonergic modulation by microdialysis, intracellular recordings in rat brain slices have shown that 5-HT depolarizes the neurons by closing inwardly rectifying potassium channels. This effect is antagonized by the preferential 5-HT_{2A} antagonist ketanserin, suggesting that it is mediated by this receptor subtype [89]. On the other hand, presynaptic 5-HT_{1B} receptor stimulation reduces frequency of synaptic glutamate input in this nucleus [90].

5-HT_{2A} Receptor Modulation of Neuron Activity in the Substantia Nigra

Considering the variety of 5-HT receptors within the BG nuclei, it is not surprising that endogenous 5-HT exerts mixed effects on the electrical activity of SNr neurons. *In vivo* studies suggest that 5-HT can induce inhibitory or excitatory responses in the SNr, with 5-HT_{2C} receptors responsible for the excitatory component [91–94]. Furthermore, 5-HT depletion produces an inhibitory effect on SNr neurons, whose firing pattern becomes more bursty [95]. *In vitro* studies performed in brain slices also report complex effects. On the one hand, 5-HT excites SNr neurons via direct stimulation of 5-HT_{2C} receptors or indirect activation of 5-HT_{1B} receptors on striatonigral GABAergic terminals [96–99]. On the other hand, 5-HT_{1B} receptor activation in the STN leads to an indirect inhibition of SNr cells [100]. So far, the few studies that have investigated the role of 5-HT_{2A} receptors on the activity of SNr neurons have not reported an effect of 5-HT_{2A} receptor agonists [97] or participation of this receptor in the inhibitory effect produced by 5-HT [96].

In the SNc, the role of 5-HT is still unclear. Generalized 5-HT depletion seems to decrease, or not to have any effect, on SNc neuron excitability [101, 102]. DRN electrical stimulation evokes mainly inhibitory responses, but in some occasions excitation, probably due to the release of neurotransmitters other than 5-HT [101, 103]. *In vivo* or *in vitro* local administration of 5-HT does not modify SNc neuron activity [91, 104]. Regarding 5-HT_2 receptors, non-selective antagonists like ritanserin can produce diverse effects on SNc neurons, including stimulation [105] or no effect [106]. In general, *in vivo* or *in vitro* acute administration of 5-HT_{2A} antagonists (SR 46349 or MDL 100907) does not alter the firing rate of VTA and SNc DA neurons [104, 106–109] but can influence the number of active cells or firing pattern [110]. As we describe later, 5-HT_{2A} receptors regulate the firing activity of DA neurons in case of drastic changes of activity. Thus, it is known that drugs able to block 5-HT_{2A} receptors counteract the ability of the psychostimulant drug amphetamine to inhibit the activity of DA neurons in the SNc and the VTA [109]. This effect seems specific of amphetamine. Indeed, 5-HT_{2A} receptor antagonists are unable to modify the inhibition produced by D_2 receptor agonists on DA neurons [108]. Similarly, unpublished results of Dr Ennio Esposito indicated that low doses of the preferential 5-HT_{2A} receptor antagonist SR 46349 reduced the increase in firing rate triggered by cumulative doses of haloperidol. On the other hand, SR 46349 was not able to counteract the increase in DA neuron firing frequency induced by cumulative doses of morphine [107]. This pattern of response is complex and could be related to the ability of amphetamine, haloperidol, but not morphine, to stimulate intraneuronal DA synthesis [26].

5-HT_{2A} Receptor Modulation of Neuron Activity in the Globus Pallidus

The presence of different 5-HT receptors predicts the role of 5-HT in controlling neuron activity both in the GPe and GPi/EP.

In the GPe, several studies have confirmed the tonic excitatory tone exerted by 5-HT. Acute 5-HT depletion decreases basal firing rate and the number of regular neurons in the GPe [111]. Moreover, local application of 5-HT both *in vivo* and *in vitro*, triggers excitatory responses in the GP, and occasionally some inhibitory responses [112–116]. The stimulatory effect may depend on pre- and postsynaptic mechanisms involving various 5-HT receptors (5-HT_{1B}, 5-HT_{1A}, 5-HT_4 or 5-HT_7 receptor subtypes) but not 5-HT_{2A} receptors [112–114, 117].

In the ventral part of the pallidum, 5-HT also controls neuron activity. Bath application of 5-HT excites GABAergic neurons and inhibits cholinergic cells by mechanisms that rule out the participation of 5-HT_{2A} receptors [118] but probably require 5-HT_{1A} receptor activation [119].

5-HT_{2A} Receptor Modulation of STN Neuron Activity

5-HT exerts a complex effect on STN neuron activity, which is considered an important centre of BG motor circuits and receives serotonergic innervation mainly from the DRN (see above). *In vivo* electrophysiological studies have shown that STN firing rate and burst activity are increased after DRN pharmacological lesion or 5-HT depletion induced by pretreatment with *para*-chlorophenylalanine (pCPA, a 5-HT synthesis inhibitor) [120, 121]. Both *in vitro* and *in vivo* studies have reported how the excitability of STN neurons is controlled by various 5-HT receptors including 5-HT_{1A}, 5-HT_{1B}, 5-HT_{2C} or 5-HT_4 receptors [122, 123]. Although studies using selective compounds are still lacking, 5-HT_{2A} receptors seem not to participate in STN excitability [124].

Regulation of Neuroanatomical Markers by 5-HT_{2A} Receptors in the Basal Ganglia

Although nowadays few agonists with a selective profile for 5-HT_{2A} receptors are available, less specific drugs have been traditionally used for analysing the impact of 5-HT_{2A} receptor stimulation on the cellular activity in the BG. Among them, the lead compound to study 5-HT_{2A} receptor function is DOI, which also displays strong affinity toward 5-HT_{2B} and 5-HT_{2C} receptor subtypes [14, 125]. Regarding the protooncogene c-Fos or other markers, most data arise from studies performed in the striatum while the impact of 5-HT_{2A} receptors in other brain regions is scarce.

Some immunohistochemical studies have reported that systemic administration of DOI enhances the expression of c-Fos in the striatum [126–129], mainly in the medial striatum and the NAc [126, 130]. This c-Fos expression is suppressed by the unspecific 5-HT_2 antagonist ritanserin or reduced by the 5-HT_{2A}/D_2 antagonist spiperone [126], suggesting a participation of 5-HT_{2A} receptors in this effect. Other drugs like m-chlorophenylpiperazine (m-CPP), 1-(3-trifluoromethylphenyl) piperazine (TFMPP), RU-29469 or Ro 60–0175 also enhance the expression of c-Fos in the striatum [131–134]. However, their effect is not necessarily related to 5-HT_{2A} or even 5-HT_{2C} receptors and the pattern of expression is often different from that produced by DOI. Indeed, moderate to high doses of mCPP, RU-29469 and TFMPP substantially enhance striatal c-Fos expression in all striatal areas, presumably by stimulating 5-HT_{1B} receptors [135]. The preferential 5-HT_{2C} receptor agonist Ro 60–0175 and lower doses of mCPP (1 mg/kg) induces lower expression of c-Fos compared to the above situation and a pattern of effect similar to that reported for DOI [132, 133, 136]. Since 5-HT_{2C} receptors are unlikely to be involved in the effect of these compounds [136], the affinity toward 5-HT_{2A} receptors might contribute to c-Fos expression. It is noteworthy that the control exerted by 5-HT_{2A} receptors is probably very peculiar. In fact, as happens with DOI [137], the non-selective 5-HT agonist and hallucinogenic compound lysergic acid diethylamide (LSD) enhances c-Fos expression in the cortex through the stimulation of 5-HT_{2A} receptors without affecting the expression in the striatum or the NAc [138]. The overall picture indicates that the overexpression of immediate early genes by 5-HT_{2A} receptor stimulation has a low magnitude in the striatum compared to the cortex, paralleling the lower density of 5-HT_{2A} receptor expression. In contrast to 5-HT_{2C} receptors [139], selective or mixed 5-HT_2 receptor antagonists do not induce c-Fos expression in the striatum [138, 140–142].

5-HT_{2A} receptors have also been proposed to regulate, at least in part, the c-Fos expression induced by various psychotropic drugs. Thus, cocaine administration induces the expression of c-Fos and zif248 by a 5-HT-dependent mechanism, as this effect is reduced after 5-HT depletion by *p*-chloroamphetamine [143]. Specifically, cocaine-induced c-Fos expression is reduced by the co-administration of a low dose of MDL 100,907 (0.125 mg/kg) plus an agonist of 5-HT_{2C} receptors [144], while cocaine-induced zif248 expression is reduced by MDL 100,907 (0.5 mg/kg) administration [145]. The non-selective NMDA channel blocker phencyclidine (PCP) enhances c-Fos expression in the NAc, an effect that is suppressed in rats pretreated by MDL 100,907 [141]. Amphetamine also enhances c-Fos expression but its effect is not reduced by ritanserin [146]. The 5-HT releasers fenfluramine/d-fenfluramine enhance c-Fos in the striatum [128, 147–149], however, the contribution of 5-HT_2 receptors to this effect is rather poor or inexistent [128, 147–149].

Several researches have analysed the profile of c-Fos expression induced by typical and atypical antipsychotic drugs, the latter often displaying a higher affinity for 5-HT_{2A} compared to D_2 receptors [10, 150]. All studies have reported that the profile of c-Fos expression induced by haloperidol (typical) was different compared to that induced by clozapine (atypical). Haloperidol induces strong c-Fos expression in the striatum and the NAc with a higher response in the dorsolateral quadrant, whereas clozapine preferentially enhances c-Fos in the shell of the NAc and the cortex [129, 142, 151]. The distinction occurs for numerous antipsychotics [129, 142, 151–153], although not as straightforward as initially thought [153]. Interestingly, the administration of the agonist DOI reduces the effect of haloperidol, particularly in the ventrolateral striatum [129, 154]. Conversely, blockade of 5-HT_{2A} receptors with MDL 100,907 potentiates c-Fos expression stimulated by haloperidol in the NAc, but not in the striatum [155].

As described above, few data are available regarding the possible influence of 5-HT_{2A} receptors on the activity of cell populations in other parts of the BG. The increase in c-Fos produced by preferential 5-HT_{2C} receptor agonists such as mCPP and Ro 60–0175 has been reported in the STN while a faint labelling is sometimes reported in the SNr or the EP and no labelling is observed in the GP [131–134, 136]. Again, the role of 5-HT_{2C} receptors does not appear to be exclusive in these effects [136], likely suggesting the participation of additional receptors including 5-HT_{2A} ones.

Effect of 5-HT_{2A} Receptors on Neurotransmitter Levels in the Basal Ganglia

Some evidence supports that 5-HT_{2A} receptors are able to regulate the neurochemical activity of numerous transmitter systems in the BG. Again, the great majority of the data concern the striatum and the NAc, while very little information is available about other brain regions. Moreover, measurement of transmitter levels in the brain is a complicated task. Most data refer to tissue and extracellular levels of DA using electrochemical detection or, to a lesser extent, GABA or glutamate using fluorometric or electrochemical detection. Levels of neuropeptides are indirectly evaluated by *in situ* hybridization.

Regulation of Extracellular DA Levels

The control exerted by 5-HT_{2A} receptors on subcortical DA release evokes different situations, indicating that these receptors exert a state-dependent facilitating influence on DA release. Indeed, this regulation is complex, first because it does not strictly follow the electrical activity of DA neurons of the SNc and VTA and, second, because it depends on various conditions. It has been further complicated by some data reporting *in vitro* that exogenously applied 5-HT inhibits H^+-stimulated $[^3H]$-DA release in rat striatal slices [156]. These results should be carefully interpreted due to the concomitant amphetamine-like effect triggered by exogenous 5-HT [157].

Thus, in contrast to the prefrontal cortex [158–161], 5-HT_{2A} receptor stimulation by DOI does not increase basal DA release in the striatum or the NAc [162–164]. Similarly, blockade of 5-HT_2 receptors with non-selective antagonists or with preferential 5-HT_{2A} antagonists does not alter extracellular DA levels in those regions [106, 107, 109, 162, 165–171]. Of note, SR46349 has been shown to enhance striatal DA release when administered at high dose (10 mg/kg) [172]. At this regimen, SR46349 behaves as a 5-HT_{2C} receptor antagonist/inverse agonist [165], known to disinhibit DA neuron activity [13]. It is generally accepted that basal and subcortical DA release is not under the tonic or phasic regulation exerted by central 5-HT_{2A} receptors.

Conversely, 5-HT_{2A} receptors regulate DA release through mechanisms involving control of either DA synthesis or DA neuron firing rate [162, 170, 173, 174] (Table 2). Waldmeier and Delini-Stula [175] pioneered the first studies in the field by analysing the effect of non-selective 5-HT_2 antagonists on the increase in DA metabolism induced by haloperidol in post-mortem tissue [175]. Years after, the real concept of state-dependent regulation of DA neuron function came from microdialysis studies aimed at studying the influence of 5-HT_{2A} receptors on the mechanism of action [176] and excitotoxicity of MDMA [169, 177, 178]. Briefly, preferential or selective 5-HT_{2A} antagonists reduced the ability of MDMA to enhance DA release in the NAc and the striatum [162, 170, 171, 176]. The excitatory control exerted by 5-HT_{2A} receptors unmasked with selective antagonists has also been reported in the mechanism of action of amphetamine [107, 179–182], haloperidol [167, 168], d-fenfluramine [183], the non-selective NMDA channel blocker dizocilpine (or MK-801) [174]) or DRN electrical stimulation [166] (Table 2). Moreover, amphetamine and MDMA-stimulated DA efflux is further enhanced by systemic administration of DOI in the NAc and striatum [162, 163]. The excitatory influence of 5-HT_{2A} receptors on DA release is not observed in all situations and the involvement of these receptors appears to be restricted to specific conditions.

The involvement of 5-HT_{2A} receptors on DA release is complex and slightly different between the striatum and the NAc (Table 2). Thus, morphine enhances striatal and accumbal DA release, but this effect is not modified by 5-HT_{2A} receptor blockade [107]. Cocaine-induced raise of extracellular levels of DA in the NAc is not further modified by the 5-HT_{2A} agonist DOI [164]. In rhesus macaques, the increase

Table 2 Involvement of 5-HT$_{2A}$ receptors in the control of DA release in the nucleus accumbens and the striatum.

	DA released			5-HT	5-HT$_{2A}$			Selected
	Exocyt	Synth.	Firing	tone	agents	NAc	ST	references[a]
Basal conditions								
DOI	Na	Na	0	–/0		↔	↔	[162, 164]
SR46349	Na	Na	0	0		↔	↔/↑	[166]
MDL 100,907	Na	Na	0	0		↔	↔	[303]
Ritanserin	Na	Na	0	0		↔	↔	[106]
Ketanserin	Na	Na	0	0		Nd	↔	[176]
Activated conditions								
Amphetamine	No	+	–	0	Agonist	↑	↑	[163]
					Antagonist	↓	↓	[107]
MDMA	No/yes	+	–	+	Agonist	↑	↑	[162]
					Antagonist	↓	↓	[170]
Cocaine	Yes	–	–	+	Agonist	↔	Nd	[164]
					Antagonist	↔	↓	[184]
Haloperidol	Yes	+	+	–	Agonist	Nd	↑	[191]
					Antagonist	↓	↓/↔	[167, 168]
					Inverse ag.	↓	Nd	[186]
Morphine	Yes	+	+	0	Agonist	↓	Nd	[164]
					Antagonist	↔	↔	[107]
D-fenfluramine	Yes	+	–	+	Antagonist	Nd	↓	[183]
Mk-801	Yes	0	+	+	Antagonist	↓	Na	[174]
DRN stim (20 Hz)	Yes	+	–	+	Antagonist	↓	Na	[166]
Local applications								
Exogenous 5-HT	No/yes	0	Na	+	Antagonist	↓	↔	[157]
MDMA	No/yes	0	+	+	Antagonist	Nd	↔	[193]
D-fenfluramine	No/yes	0	+	+	Antagonist	Nd	↔	[194]

This table illustrates the state-dependent nature of the regulation exerted by 5-HT$_{2A}$ receptors on DA release in basal or activated conditions triggered by pharmacological drugs. Those drugs enhance DA release in an exocytotic (Exocyt.) (vesicular), non-exocytotic manner or both (indicated by no/yes) and their effect is associated with changes (+, – or 0 for no change) in DA synthesis (Synth.), DA neuron firing rate (Firing) and/or 5-HT tone. Depending on the drug, 5-HT$_{2A}$ receptor agonists (usually DOI) or antagonists (often MDL100907 and also non-selective compounds) modulate DA release in the nucleus accumbens (NAc) and/or the striatum (STR). Local applications of exogenous 5-HT, MDMA or D-fenfluramine are associated with strong enhancement of DA release (activated conditions). The arrows indicate a potentiation (↑), a reduction (↓) or no effect (↔) of DA release.

[a]Additional references are indicated in the text. *Na* not applicable; *Nd* not determined.

in DA overflow induced by the intravenously administration of cocaine was reduced by the 5-HT$_{2A}$ antagonist MDL 100,907 in the striatum, but not in the NAc [184]. A closer look at the available data shows that 5-HT$_{2A}$ antagonists counteract the increase in DA release induced by 2.5 but not 1 mg/kg amphetamine [107, 185] or 0.01–0.1 but not 1 mg/kg haloperidol in the striatum [168].

At least four factors have been proposed to account for the involvement of 5-HT_{2A} receptors in the regulation of subcortical DA release (Table 2). First, it has been suggested that the effect mediated by 5-HT_{2A} receptors is conditioned by the presence of enhanced DA transmission, i.e. an increased DA tone at post-synaptic cells. This hypothesis is unclear and could reveal region-dependent effects. Indeed, 5-HT_{2A} receptors lose their ability to modulate enhanced DA release at high doses of haloperidol in the striatum but not in the NAc [167, 168, 186]. Second, together with the change in DA transmission, it has been suggested that the involvement of 5-HT_{2A} receptors depends on additional 5-HT release. Nonetheless, although this hypothesis is compatible with the mechanism of action of MDMA, cocaine or d-fenfluramine, it is less evident with amphetamine and haloperidol, which do not increase extracellular levels of 5-HT [168, 187, 188] or do it poorly [189]. A third hypothesis concerns the nature of DA release especially if this release is sustained by DA synthesis. This hypothesis was proposed for the mechanism of action of MDMA. In this case, the increase in DA synthesis is required for the permissive role of 5-HT_{2A} receptors on maintaining the outflow of DA [169, 177, 178, 190]. This is in line with the involvement of 5-HT_{2A} receptors in the effect of amphetamine and haloperidol, or their non-involvement in the effect of cocaine (NAc only) or morphine, which do not increase DA synthesis. This hypothesis is not exclusive (see cocaine and the situation in the striatum) since some drugs, as MK-801, are also able to regulate DA release in the NAc regardless the increase in DA synthesis [174]. The last criterion of the state-dependent control exerted by 5-HT_{2A} receptors on DA release is probably the less understood and could be the most important. It concerns an anatomic-functional mechanism distinct from the striatum and capable of triggering the excitatory influence on DA release. Indeed, the increase in striatal DA release induced by MDMA, haloperidol or d-fenfluramine *in vivo* is reduced by the striatal application of non-selective and/or selective 5-HT_{2A} antagonists [171, 183, 188, 191]. This indicates that part of this excitatory control is located within the striatum. Schmidt et al. [171] pursued the idea that 5-HT_{2A} receptor activation induced DA synthesis, thereby sustaining the increase in DA release induced by MDMA. Surprisingly, the increase in DA release induced by the application of MDMA on striatal slices was not altered by the 5-HT_{2A} receptor antagonist MDL 100,907 despite the alteration of DA synthesis [171]. Yet, this result is consistent with all the available data *in vitro* and *in vivo* (Table 2). Thus, the systemic injection of ketanserin does not modify the increase in striatal DA release induced by the intrastriatal application of MDMA [192]. The effect of MDMA on striatal DA release is reduced by 5-HT depletion upon its systemic, but not upon its intrastriatal application [193]. The intrastriatal application of the non-selective antagonist methiotepin reduces the DA effect triggered by the systemic but not the intrastriatal application of d-fenfluramine [183, 194].

The whole picture is consistent with the excitatory effect of exogenous 5-HT on striatal DA release [157]. Indeed, the enhancement of striatal DA release induced by exogenous 5-HT itself *in vitro* and *in vivo* is not modified by a variety of 5-HT_2 receptor antagonists [193, 195–200]. The data obtained with the sole application of DOI on DA terminals led to discrepant results. DOI (10–300 μM) increases accumbal DA release and this effect is blocked by the co-application of high concentration (50 μM) of the non-selective $5\text{-}HT_2$ antagonists LY-53,857 or ketanserin [201]. Bowers and collaborators have reported that the increase in DA release induced by DOI (25–250 μM) occurs in the posterior area of the NAc [202]. The data in the striatum are more controversial because the local infusion of DOI has been shown to inhibit [203], produce no effect [196] or enhance [191] DA release. The application of DOI in slices does not alter [^{3}H]-DA release in the NAc and the striatum [200]. Moreover, $5\text{-}HT_{2A}$ receptors are unlikely to be involved in the excitatory effect of DOI on basal DA release *in vivo* because the co-application of the $5\text{-}HT_{2A}$ receptor antagonist SR 46349 failed to affect the effect induced by DOI [191].

In conclusion, the involvement of $5\text{-}HT_{2A}$ receptors in the regulation of striatal DA function requires extra-striatal factors and changes in DA neuron activity. One may wonder whether the differences observed in the $5\text{-}HT_{2A}$ receptor-dependent control of DA release in the striatum and the NAc represent interconnected neurobiological loops hypothesized by Haber, Fudge and McFacland [204]. Indeed, the full blockade of DA transmission by haloperidol impaired the ability of $5\text{-}HT_{2A}$ antagonists (and more generally 5-HT drugs) to control striatal DA release, but not accumbal DA release. Conversely, cocaine would trigger this control in the striatum and not in the NAc. It does not discard the idea that specific conditions are required for $5\text{-}HT_{2A}$ receptors to modulate DA release and would explain also how $5\text{-}HT_{2A}$ receptors can modulate behaviours independently from changes in DA release [26].

Regulation of Extracellular GABA Levels

The data regarding the control exerted by $5\text{-}HT_{2A}$ receptors on GABA release are scarce. In line with the data above, MDMA has been shown to reduce extracellular GABA levels in the SNr. This inhibition was mediated by $5\text{-}HT_2$ receptors [188] and other authors later reported that, GABA release in the SNr specifically depends on $5\text{-}HT_{2C}$ receptors rather than $5\text{-}HT_{2A}$ receptors [94]. A similar study has also been published at the level of the NAc with similar results [205]. In this case, the effect was blocked by application of the $5\text{-}HT_{2B/2C}$ antagonist SB206553 into the VTA. The role of $5\text{-}HT_{2A}$ receptors is still not well known, but may be not very relevant. It is noteworthy that DOI (7 mg/kg for 9 consecutive days) did not modify the level of expression of GAD65 and GAD67 mRNA in the NAc, striatum or GP [206].

Regulation of Extracellular Glutamate Levels

Recent studies suggest that $5\text{-}HT_{2A}$ receptors may control extracellular levels of glutamate, mainly in the striatum. Indeed, $5\text{-}HT_{2A}$ receptors are expressed by pyramidal cells projecting to the VTA and the NAc [207] and local application of DOI into the prefrontal cortex enhances glutamate release in both the VTA and NAc [207–209]. The control exerted by cortical $5\text{-}HT_{2A}$ receptors via glutamatergic fibres occurs also at the level of the DRN (see below). In the striatum, several studies failed to report an increase in basal glutamate release after the lesion of the nigrostriatal DA system in rats [210–212], although an increase in d-aspartate concentrations has been reported after extended, but not partial, lesion [210]. In one recent study, 1-methyl-4-phenyl-1,2,3,6-tetrahydropyridine (MPTP) lesion in mice enhanced striatal glutamate extracellular levels three weeks after the neurotoxin administration and the intrastriatal application of MDL 100,907 reduced these higher extracellular levels [213]. However, the increase in glutamate was observed only after the acute administration of MPTP [214].

Regulation of Extracellular 5-HT Levels

The regulation of 5-HT release by 5-HT_{2A} receptors is probable in the BG. It is likely because it is well known that DOI inhibits the electrical activity of 5-HT-like neurons in the DRN [215–217]. The authors argued that this effect, which depends on a cortical decrease of 5-HT release, was associated with an alteration of the autoregulatory control that 5-HT_{1A} receptors produce in the DRN [216, 217]. This has been confirmed and extended later [218]. Indeed, systemic administration of DOI inhibits cortical 5-HT release probably via its inhibitory effect on 5-HT-like neurons in the DRN and this effect is suppressed by the selective 5-HT_{2A} receptor antagonist MDL 100,907. Conversely, local cortical administration of DOI does not inhibit cortical 5-HT release and even increases it through a 5-HT_{2A} receptor-dependent mechanism. The excitation of pyramidal cells by cortical 5-HT_{2A} receptors likely alters the activity of cortical-raphe fibres [219].

Overall, one may predict some decrease in 5-HT release after DOI administration in the BG, but this effect might not be related to 5-HT_{2A} receptors expressed within the BG.

Regulation of Neuropeptides in the Striatum

Several pieces of evidence demonstrate that 5-HT_{2A} receptors regulate the level of expression of peptide mRNA in the striatum. Thus, DOI (infused by minipump for 9 consecutive days) has been shown to enhance preprotachykinin (PPT) mRNA evaluated by northern blot and SP immunomaterial by western blot [220]. A single injection of DOI (1, 2 or 7 mg/kg depending on the studies) increases PPT and/or preproenkephalin (PPE) mRNA in the dorsolateral and posterior striatum, but not within the anterior ventromedial striatum [221–223]. The regulation exerted by 5-HT_{2A} receptors appears to be more prominent on PPT mRNA levels compared to PPE ones [221]. Interestingly, a 9-days treatment with DOI (7 mg/kg) inhibited preprodynorphin mRNA expression, another marker of the direct striatofugal pathway [206].

The effect of DOI on PPT mRNA expression is likely related to 5-HT_{2A} receptors, since it can be reversed by ritanserin and the preferential 5-HT_{2A} antagonist ketanserin [221]. The authors report also that the decrease in PPT mRNA produced by a lesion of DA neurons is reversed by systemic administration of DOI via 5-HT_{2A} receptors [221]. Similarly, the acute administration of the 5-HT/DA releaser *p*-chloroamphetamine enhanced PPT mRNA levels in the striatum, which was blocked by a pretreatment with the mixed 5-HT_2 receptor antagonist ritanserin [224]. All these data underscore a link between 5-HT_{2A} receptors and the regulation of D_1 receptor-dependent pathways, mostly the direct pathway of the BG.

Regulation of 5-HT_{2A} Receptor Expression and Transmission in BG

The expression of 5-HT_{2A} receptors in the BG can be modulated by various neurotransmitter systems, drugs and physiological conditions. This is important as the changes in 5-HT_{2A} receptor expression condition the magnitude of the 5-HT_{2A} receptor-dependent responses. Here, we discuss how changes in DA transmission can alter the expression of 5-HT_{2A} receptors and describe some data in humans obtained with PET ligands. We also discuss possible changes of 5-HT_{2A} receptor transmission related to heterodimerization.

The tight and complex connection between 5-HT_{2A} receptors and DA neuron activity delineated above is mirrored in the fact that changes in DA transmission are also associated with drastic changes in 5-HT_{2A} receptor expression. It has been reported that lesion of DA neurons in adult rats enhances the striatal binding of $[^{125}I]$-DOI on 5-HT_{2A} receptors [225]. Consistently, a lesion of SNc DA neurons induces an upregulation of the mRNA encoding 5-HT_{2A} receptors in the striatum [52, 226, 227]. This effect is paralleled by an increase in $[^{125}I]$-DOI binding [226]. Qualitatively, from a medial distribution, the labelling for both mRNA and protein reaches lateral parts of the striatum after the lesion. The expression of the mRNA is observed in both PPE and non-PPE neurons, being enhanced specifically in non-PPE neurons [226]. These findings are consistent with the ability of DOI to enhance PPT mRNA and restore its levels after DA lesion by stimulating 5-HT_{2A} receptors [221]. An upregulation of 5-HT_{2A} receptors assessed by western blot and caused by MPTP treatment has been reported in mice [213]. In the nigrostriatal DA-deficient Pitx3 mutant mice, 5-HT hyperinnervation and increase in 5-HT_{2A} receptor mRNA, have been observed in the dorsal striatum [228]. The increase in 5-HT_{2A} receptor expression would not be due to the 5-HT hyperinnervation that takes place in some parkinsonian models [229]. Indeed, in tryptophan hydroxylase 2 deficient mice (with very low levels of endogenous 5-HT) the binding of $[^3H]$-MDL 100,907 is enhanced both in the striatum and the SN [230]. In monkeys, however, MPTP treatment did not alter the labelling of $[^3H]$-ketanserin in the caudate-putamen or GP [42, 231]. Although a species-related issue could be hypothesized, it has been reported, also in rodents, that $[^3H]$-ketanserin was slightly or drastically decreased three weeks after the injection of 6-hydroxydopamine (6-OHDA) [203, 232]. Rather than a species issue, one may question the pharmacological agents used for the labelling. Among the differences, DOI is an agonist which could label distinct conformations of the receptor compared to ketanserin which is an antagonist/inverse agonist [233]. Yet, it is difficult to imagine that an increase in 5-HT_{2A} receptor mRNA observed in three studies, and the increase in DOI binding, can be associated with a decrease in ketanserin binding. In fact, ketanserin has been shown for many years to bind the vesicular transporter (a tetrabenazine site), a site associated with monoamine vesicles. It has a high affinity ranging from 4 nM to 45 nM depending on the temperature [234–236]. This additional binding site of ketanserin is responsible for a strong non-specific binding in the striatum that is reduced in case of DA lesion [40, 235, 237]. The above studies reporting a decrease in [^{3}H]-ketanserin after DA neuron lesion did not take into account this property, though the studies performed in monkeys discussed the point that at the temperature used for their binding experiments, ketanserin should be dissociated from the "tetrabenazine" site [231]. Yet, it cannot be excluded that a small contribution of this site to the whole labelling, presumably altered after MPTP treatment, could have diminished an expected increase in [^{3}H]-ketanserin binding. To the best of our knowledge, [^{3}H]-MDL 100,907, which present a much higher favourable binding profile compared to ketanserin in this respect [37, 40, 44], has not been used to label 5-HT_{2A} receptors after DA lesion. In humans, a recent neuroimaging study using single-photon emission computed tomography (SPECT) reported a decrease in the premotor cortex and the anterior striatum in *de novo* patients with PD [238].

DA treatments also modulate the expression of 5-HT_{2A} receptors. In animal models of PD, the injection of the non-selective DA agonist apomorphine or the preferential DA D_1 agonist SKF-38393 reverses the increase in striatal 5-HT_{2A} receptor expression [226]. On the other hand, L-DOPA enhances the binding of [^{3}H]-ketanserin in MPTP-treated monkeys [42, 231]. These data might not be controversial as L-DOPA is supposed to create a low DA tone in the striatum of DA neurons-deficient animals compared to the physiological situation [211, 239]. Nonetheless, additional studies are warranted in this field of research.

Studies with antipsychotic drugs have also found modifications of 5-HT_{2A} receptor expression. One of the most recent publications indicated that chronic and continuous administration of haloperidol with osmotic minipumps increases and decreases 5-HT_{2A} receptor density assessed by [^{3}H]-ketanserin in the striatum and NAc shell, respectively [240]. On the other hand, numerous studies have failed to report any changes regarding 5-HT_{2A} receptor mRNA/binding density after systemic injections of haloperidol [155, 237, 241], while clozapine reduced it in some cortical areas and in the striatum in some [241] but not all [237] studies. A transient decrease in 5-HT_{2A} receptor expression has also been noticed after olanzapine injection in the striatum [242].

In humans, the findings concerning subcortical expression of 5-HT_{2A} receptors in psychiatric diseases are unclear. Obsessive-compulsive disorder (OCD) is regularly considered as a cortico-subcortical loop disturbance involving DA and 5-HT systems. Patients may receive either atypical antipsychotic and/or antidepressant drugs [243]. In drug-naïve OCD patients [^{18}F]-altanserin increases binding in the caudate nuclei in comparison with control individuals [244]. The binding of [^{18}F]-altanserin in the caudate is not correlated with the severity of symptoms in patients. In other studies in drug-naïve OCD patients, decreased cortical binding of [^{11}C]-MDL100907 was correlated to the severity of symptoms or the age onset, and no differences of binding were reported in the caudate nuclei [245, 246]. Similarly, the difference of 5-HT_{2A} receptor expression in schizophrenic patients would converge toward a decrease in prefrontal cortex expression whereas no clear effects are reported in the BG [247].

The last point we would like to evoke concerns possible changes of 5-HT_{2A} receptor transmission through molecular interactions with other receptors. It has been well established that 5-HT receptors, including 5-HT_{2A} receptors, function as oligo- and heterodimers *in vitro* [248, 249] and that the process of oligomerization might alter the receptor function [249, 250]. Regarding 5-HT_{2A} receptors, it has been reported by two separate groups that 5-HT_{2A} receptors physically interact with D_2 receptors [71, 251]. Colocalization of these receptors has been reported, at least, in the medial prefrontal cortex, the striatum and the SN [71]. In the ventral striatum, the binding of [^{3}H]-raclopride is sensitive to the heterodimerization of 5-HT_{2A} and D_2 receptors and the heterodimerization with D_2 receptors in HEK293 cells is sensitive to the nature of 5-HT_{2A} ligands [252]. The functionality of these molecular interactions is not fully understood, but it definitely constitutes an important field of research in the treatment of schizophrenia and PD [253]. In the striatum, these specific interactions might explain the different reactivity of 5-HT_{2A} receptor expression in the direct (mostly D_1 receptor) versus the indirect pathway (mostly D_2 receptors) [254, 255], occurring after a lesion of DA neurons [226]. Also, heterodimers formed by 5-HT_{2C} and 5-HT_{2A} receptors have been noticed in heterologous cell expression systems [248] and additional data are needed to explore the pertinence of these heterodimers *in vivo*.

5-HT_{2A} Receptors and Behaviour Related to Basal Ganglia Functions

Numerous studies have pointed to a role of 5-HT_{2A} receptors in the regulation of learning, automatisms, executive and cognitive functions [256, 257]. In most cases, these functions are likely located in the cortex or, as recently exemplified, along the cortico-thalamic pathway [258]. The involvement of 5-HT_{2A} receptors in the BG has been reported in the control of motor behaviours mostly associated with changes in DA transmission. Here, we will go through the studies reporting interaction of 5-HT_{2A} receptor ligands with drugs of abuse, antiparkinsonian and antipsychotic drugs.

5-HT_{2A} Receptors and the Regulation of Locomotor Behaviour

The implication of 5-HT_{2A} receptors in the control of motor behaviours has been well documented. Earlier studies have reported that injection of 5-HT induces contralateral turning through the activation of 5-HT_2 receptors, suggesting a permissive role of 5-HT_2 receptors in motor behaviour [259]. More recently, the 5-HT_{2A}

receptor antagonist MDL 100,907 has been shown to reduce the hyperlocomotion induced by amphetamine, cocaine or MDMA [260–263]. Conversely, the 5-HT_{2A} receptor agonist, DOI, is able to increase it [264]. These findings suggest that compounds that increase locomotor activity by enhancing striatal/accumbal DA tone trigger a permissive role of 5-HT_{2A} receptors on locomotor activity. The relationship with the control of DA release is not established [26]. The reduction of hyperlocomotive effect of cocaine by 5-HT_{2A} receptor blockade would be related to 5-HT_{2A} receptors located in the VTA, not the NAc [262], and would occur independently from measurable changes of extracellular levels of DA (see Table 2). In any case, 5-HT_{2A} receptor antagonism emerges as a new therapeutic tool to treat drug abuse, particularly cocaine abuse [265].

5-HT_{2A} Receptor Role in Parkinson's Disease and L-DOPA-Induced Dyskinesia Treatment

PD is a neurodegenerative disease characterized by a progressive loss of DA neurons in the SNc. The drop in DA concentration in the striatum drastically alters functionality of the BG circuitry and impacts also 5-HT receptor expression (see above regarding 5-HT_{2A} receptors and, for review, see [35]). It has been reported that the 5-HT_{2A} receptor agonist DOI or antagonist MDL 100,907 enhances and decreases, D_1 receptor-induced locomotor activity in DA-depleted rats, respectively [266–268]. These results suggest that, in DA depletion conditions, 5-HT_{2A} receptors more likely influence striatal D1 receptor function and the direct pathway output [266]. Although 5-HT_{2A} agonists have been proposed to facilitate D_1 receptor-dependent responses, their hallucinogenic properties prevent such a strategy.

Conversely, 5-HT_{2A} antagonist could be interesting to limit excessive D_1 receptor tone. Indeed, DA replacement therapy with the DA precursor, L-DOPA, is the most effective and common pharmacological treatment in PD. However, its efficacy after long-term treatment is impaired by motor complications, such as L-DOPA-induced dyskinesia and L-DOPA-induced psychosis [269]. The development of L-DOPA-induced dyskinesia is still poorly understood, although numerous studies have implicated abnormal D1 receptor signalling in the striatum [269–271]. In line with this, 5-HT_{2A} antagonists could be effective in alleviating L-DOPA-induced dyskinesia. Thus, the atypical antipsychotics clozapine or quetiapine, which are non-selective 5-HT_{2A} receptor antagonists, reduce L-DOPA-induced dyskinesia in animal models of PD [272] and PD patients [273, 274]. However, these drugs compromise L-DOPA antiparkinsonian effects [275, 276]. Pimavanserin (ACP-103), a selective 5-HT_{2A} receptor inverse agonist tested to treat L-DOPA induced psychosis [277], alleviates L-DOPA-induced dyskinesia in MPTP-lesioned monkeys [278] and in PD patients without worsening L-DOPA antiparkinsonian effect [279]. However, additional data are required because in rats MDL 100,907 was not efficacious against L-DOPA-induced dyskinesia [280].

$5\text{-}HT_{2A}$ Receptor in Antipsychotic-Induced Extrapyramidal Side Effects

The chronic use of antipsychotics in the treatment of schizophrenia is associated with the induction of extrapyramidal side effects (EPS) characterized by PD-like syndrome and acute or tardive dyskinesia [281]. Whereas all antipsychotics exert their clinical benefit by blocking D_2 receptors likely along the mesoaccumbal DA pathway [282, 283], the chronic or excessive blockade of D_2 transmission in the nigrostriatal DA pathway is responsible for the development of EPS [8, 284]. The superiority of atypical antipsychotic drugs over the typical ones in terms of lower induction of EPS has been related to their 5-HT properties and notably their ability to block 5-HT_{2A} receptors [11, 12, 285]. The rodent models that are classically used to address the incidence of EPS by antipsychotics are the catalepsy, a model of parkinsonian syndrome, and oral dyskinesia induced by long-term injection of typical antipsychotics.

The first behavioural studies in rats showed that increasing or inhibiting central 5-HT transmission worsened or ameliorated haloperidol-induced catalepsy, respectively [175, 286–288]. The $5\text{-HT}_{2A/2C}$ receptors could be implicated in these effects. Indeed, administration of the non-selective 5-HT_2 receptor antagonist ritanserin is able to attenuate EPS in schizophrenic patients treated with classical antipsychotics [289, 290] and haloperidol-induced catalepsy in rats [286, 291]. In line with this, atypical antipsychotics such as clozapine, displaying strong 5-HT_{2A} receptor antagonists and 5-HT_{2C} receptor inverse agonist properties, are not capable of inducing catalepsy [292–296]. Yet, the relative contribution of 5-HT_{2A} *vs* 5-HT_{2C} receptor antagonism is not completely clear. Haloperidol- and risperidone-induced catalepsy has been shown to be reduced by the 5-HT_{2A} inverse agonist ACP 103 or by the $5\text{-HT}_{2B/2C}$ antagonist SB 228357 [297, 298]. Nevertheless, MDL 100,907 and the 5-HT_{2B} antagonist SB 215505 have no effect on this motor behaviour [297, 298]. It could reflect the preferential involvement of 5-HT_{2C} receptors over 5-HT_{2A} receptors [133, 299–302]. Similarly, the oral dyskinesia arising from chronic administration of neuroleptics is modulated by the 5-HT system and 5-HT_2 receptor subtypes, but it seems that the responses are more related to 5-HT_{2C} receptors [13, 133]. Thus, the therapeutic benefit associated with 5-HT_{2A} receptor blockade in terms of lower incidence of EPS is still not a reality.

Concluding Remarks

Numerous data indicate that the 5-HT_{2A} receptor plays an important role in the BG and their action appears to be intimately linked to DA transmission. Under some circumstances, 5-HT_{2A} receptor can exert a state-dependent excitatory regulation of striatal and NAc DA release. Despite the progress made over the years, their influence in the BG is still largely misunderstood, in part due to the confounding and powerful controls they exert in the cortex. The main body of data suggests that

5-HT_{2A} receptors exert a subtle regulation of corticostriatal afferences and loops. Additional data are needed to elucidate more precisely their role in the BG, notably in relation to the therapeutic potential of 5-HT_{2A} receptor blockade in the treatment of addiction to drugs of abuse and schizophrenia.

References

1. Mink JW (1996) The basal ganglia: focused selection and inhibition of competing motor programs. Prog Neurobiol 50(4):381–425
2. Albin RL, Young AB, Penney JB (1989) The functional anatomy of basal ganglia disorders. Trends Neurosci 12(10):366–375
3. Di Chiara G (1995) The role of dopamine in drug abuse viewed from the perspective of its role in motivation. Drug Alcohol Depend 38(2):95–137
4. Kapur S, Agid O, Mizrahi R, Li M (2006) How antipsychotics work-from receptors to reality. NeuroRx 3(1):10–21. https://doi.org/10.1016/j.nurx.2005.12.003
5. Koob GF (1992) Drugs of abuse: anatomy, pharmacology and function of reward pathways. Trends Pharmacol Sci 13(5):177–184
6. Mink JW (2006) Neurobiology of basal ganglia and Tourette syndrome: basal ganglia circuits and thalamocortical outputs. Adv Neurol 99:89–98
7. Dray A, Davies J, Oakley NR, Tongroach P, Vellucci S (1978) The dorsal and medial raphe projections to the substantia nigra in the rat: electrophysiological, biochemical and behavioural observations. Brain Res 151(3):431–442
8. Kapur S, Remington G (1996) Serotonin-dopamine interaction and its relevance to schizophrenia. Am J Psychiatry 153(4):466–476
9. Westfall TC, Tittermary V (1982) Inhibition of the electrically induced release of [3H]dopamine by serotonin from superfused rat striatal slices. Neurosci Lett 28(2):205–209
10. Meltzer HY, Li Z, Kaneda Y, Ichikawa J (2003) Serotonin receptors: their key role in drugs to treat schizophrenia. Prog Neuro-Psychopharmacol Biol Psychiatry 27(7):1159–1172. https://doi.org/10.1016/j.pnpbp.2003.09.010
11. Meltzer HY, Matsubara S, Lee JC (1989a) Classification of typical and atypical antipsychotic drugs on the basis of dopamine D-1, D-2 and serotonin2 pKi values. J Pharmacol Exp Ther 251(1):238–246
12. Meltzer HY, Matsubara S, Lee JC (1989b) The ratios of serotonin2 and dopamine2 affinities differentiate atypical and typical antipsychotic drugs. Psychopharmacol Bull 25(3):390–392
13. Di Giovanni G, De Deurwaerdere P (2016) New therapeutic opportunities for 5-HT2C receptor ligands in neuropsychiatric disorders. Pharmacol Ther 157:125–162. https://doi.org/10.1016/j.pharmthera.2015.11.009
14. Knight AR, Misra A, Quirk K, Benwell K, Revell D, Kennett G et al (2004) Pharmacological characterisation of the agonist radioligand binding site of 5-HT(2A), 5-HT(2B) and 5-HT(2C) receptors. Naunyn Schmiedeberg's Arch Pharmacol 370(2):114–123. https://doi.org/10.1007/s00210-004-0951-4
15. Graybiel AM (2005) The basal ganglia: learning new tricks and loving it. Curr Opin Neurobiol 15(6):638–644. https://doi.org/10.1016/j.conb.2005.10.006
16. Obeso JA, Rodriguez-Oroz MC, Benitez-Temino B, Blesa FJ, Guridi J, Marin C et al (2008) Functional organization of the basal ganglia: therapeutic implications for Parkinson's disease. Mov Disord 23(Suppl 3):S548–S559
17. Alexander GE, Crutcher MD (1990) Functional architecture of basal ganglia circuits: neural substrates of parallel processing. Trends Neurosci 13(7):266–271
18. Mogenson GJ, Yang CR (1991) The contribution of basal forebrain to limbic-motor integration and the mediation of motivation to action. Adv Exp Med Biol 295:267–290

19. Kolomiets BP, Deniau JM, Glowinski J, Thierry AM (2003) Basal ganglia and processing of cortical information: functional interactions between trans-striatal and trans-subthalamic circuits in the substantia nigra pars reticulata. Neuroscience 117(4):931–938
20. DeLong MR (1990) Primate models of movement disorders of basal ganglia origin. Trends Neurosci 13(7):281–285
21. Galvan A, Wichmann T (2008) Pathophysiology of parkinsonism. Clin Neurophysiol 119(7):1459–1474. doi: S1388-2457(08)00198-3 [pii]. https://doi.org/10.1016/j.clinph.2008.03.017
22. Azmitia EC, Segal M (1978) An autoradiographic analysis of the differential ascending projections of the dorsal and median raphe nuclei in the rat. J Comp Neurol 179(3):641–667
23. Lavoie B, Parent A (1990) Immunohistochemical study of the serotoninergic innervation of the basal ganglia in the squirrel monkey. J Comp Neurol 299(1):1–16. https://doi.org/10.1002/cne.902990102
24. Parent A, Descarries L, Beaudet A (1981) Organization of ascending serotonin systems in the adult rat brain. A radioautographic study after intraventricular administration of [3H]5-hydroxytryptamine. Neuroscience 6(2):115–138
25. Parent M, Wallman MJ, Gagnon D, Parent A (2011) Serotonin innervation of basal ganglia in monkeys and humans. J Chem Neuroanat 41(4):256–265. doi: S0891-0618(11)00048-2 [pii]. https://doi.org/10.1016/j.jchemneu.2011.04.005
26. De Deurwaerdere P, Di Giovanni G (2017) Serotonergic modulation of the activity of mesencephalic dopaminergic systems: therapeutic implications. Prog Neurobiol 151:175–236. https://doi.org/10.1016/j.pneurobio.2016.03.004
27. Di Matteo V, Pierucci M, Esposito E, Crescimanno G, Benigno A, Di Giovanni G (2008) Serotonin modulation of the basal ganglia circuitry: therapeutic implication for Parkinson's disease and other motor disorders. Prog Brain Res 172:423–463
28. Moukhles H, Bosler O, Bolam JP, Vallee A, Umbriaco D, Geffard M et al (1997) Quantitative and morphometric data indicate precise cellular interactions between serotonin terminals and postsynaptic targets in rat substantia nigra. Neuroscience 76(4):1159–1171
29. Soghomonian JJ, Descarries L, Watkins KC (1989) Serotonin innervation in adult rat neostriatum. II. Ultrastructural features: a radioautographic and immunocytochemical study. Brain Res 481(1):67–86
30. Steinbusch HW (1981) Distribution of serotonin-immunoreactivity in the central nervous system of the rat-cell bodies and terminals. Neuroscience 6(4):557–618. doi: 0306-4522(81)90146-9 [pii]
31. Sur C, Betz H, Schloss P (1996) Immunocytochemical detection of the serotonin transporter in rat brain. Neuroscience 73(1):217–231. doi: 0306452296000309 [pii]
32. Wallman MJ, Gagnon D, Parent M (2011) Serotonin innervation of human basal ganglia. Eur J Neurosci 33(8):1519–1532. https://doi.org/10.1111/j.1460-9568.2011.07621.x
33. Herve D, Pickel VM, Joh TH, Beaudet A (1987) Serotonin axon terminals in the ventral tegmental area of the rat: fine structure and synaptic input to dopaminergic neurons. Brain Res 435(1–2):71–83
34. Van Bockstaele EJ, Biswas A, Pickel VM (1993) Topography of serotonin neurons in the dorsal raphe nucleus that send axon collaterals to the rat prefrontal cortex and nucleus accumbens. Brain Res 624(1–2):188–198
35. Miguelez C, Morera-Herreras T, Torrecilla M, Ruiz-Ortega JA, Ugedo L (2014) Interaction between the 5-HT system and the basal ganglia: functional implication and therapeutic perspective in Parkinson's disease. Front Neural Circuits 8:21. https://doi.org/10.3389/fncir.2014.00021
36. Hoyer D, Clarke DE, Fozard JR, Hartig PR, Martin GR, Mylecharane EJ et al (1994) International Union of Pharmacology classification of receptors for 5-hydroxytryptamine (serotonin). Pharmacol Rev 46(2):157–203
37. Lopez-Gimenez JF, Mengod G, Palacios JM, Vilaro MT (1997) Selective visualization of rat brain 5-HT2A receptors by autoradiography with [3H]MDL 100,907. Naunyn Schmiedeberg's Arch Pharmacol 356(4):446–454

38. Pazos A, Cortes R, Palacios JM (1985) Quantitative autoradiographic mapping of serotonin receptors in the rat brain. II. Serotonin-2 receptors. Brain Res 346(2):231–249
39. Huot P, Johnston TH, Winkelmolen L, Fox SH, Brotchie JM (2012) 5-HT2A receptor levels increase in MPTP-lesioned macaques treated chronically with L-DOPA. Neurobiol Aging 33(1):194 e195–115. https://doi.org/10.1016/j.neurobiolaging.2010.04.035
40. Lopez-Gimenez JF, Vilaro MT, Palacios JM, Mengod G (1998) [3H]MDL 100,907 labels 5-HT2A serotonin receptors selectively in primate brain. Neuropharmacology 37(9): 1147–1158
41. Lopez-Gimenez JF, Vilaro MT, Palacios JM, Mengod G (2001a) Mapping of 5-HT2A receptors and their mRNA in monkey brain: [3H]MDL100,907 autoradiography and in situ hybridization studies. J Comp Neurol 429(4):571–589
42. Riahi G, Morissette M, Parent M, Di Paolo T (2011) Brain 5-HT(2A) receptors in MPTP monkeys and levodopa-induced dyskinesias. Eur J Neurosci 33(10):1823–1831. https://doi.org/10.1111/j.1460-9568.2011.07675.x
43. Hall H, Farde L, Halldin C, Lundkvist C, Sedvall G (2000) Autoradiographic localization of 5-HT(2A) receptors in the human brain using [(3)H]M100907 and [(11)C] M100907. Synapse 38(4):421–431. https://doi.org/10.1002/1098-2396(20001215)38:4<421::AID-SYN7>3.0.CO;2-X
44. Lopez-Gimenez JF, Mengod G, Palacios JM, Vilaro MT (1999) Human striosomes are enriched in 5-HT2A receptors: autoradiographical visualization with [3H]MDL100,907,[125I](+/−) DOI and [3H]ketanserin. Eur J Neurosci 11(10):3761–3765
45. Pazos A, Probst A, Palacios JM (1987) Serotonin receptors in the human brain--IV. Autoradiographic mapping of serotonin-2 receptors. Neuroscience 21(1):123–139
46. Gerfen CR (1984) The neostriatal mosaic: compartmentalization of corticostriatal input and striatonigral output systems. Nature 311(5985):461–464
47. Gerfen CR (1985) The neostriatal mosaic. I. Compartmental organization of projections from the striatum to the substantia nigra in the rat. J Comp Neurol 236(4):454–476. https://doi.org/10.1002/cne.902360404
48. Ward RP, Dorsa DM (1996) Colocalization of serotonin receptor subtypes 5-HT2A, 5-HT2C, and 5-HT6 with neuropeptides in rat striatum. J Comp Neurol 370(3):405–414
49. Burnet PW, Eastwood SL, Lacey K, Harrison PJ (1995) The distribution of 5-HT1A and 5-HT2A receptor mRNA in human brain. Brain Res 676(1):157–168. doi: 0006-8993(95)00104-X [pii]
50. Mengod G, Pompeiano M, Martinez-Mir MI, Palacios JM (1990) Localization of the mRNA for the 5-HT2 receptor by in situ hybridization histochemistry. Correlation with the distribution of receptor sites. Brain Res 524(1):139–143
51. Pompeiano M, Palacios JM, Mengod G (1994) Distribution of the serotonin 5-HT2 receptor family mRNAs: comparison between 5-HT2A and 5-HT2C receptors. Brain Res Mol Brain Res 23(1–2):163–178
52. Zhang X, Andren PE, Svenningsson P (2007) Changes on 5-HT2 receptor mRNAs in striatum and subthalamic nucleus in Parkinson's disease model. Physiol Behav 92(1–2):29–33. https://doi.org/10.1016/j.physbeh.2007.05.033
53. Pasqualetti M, Nardi I, Ladinsky H, Marazziti D, Cassano GB (1996) Comparative anatomical distribution of serotonin 1A, 1D alpha and 2A receptor mRNAs in human brain postmortem. Brain Res Mol Brain Res 39(1–2):223–233
54. Wright DE, Seroogy KB, Lundgren KH, Davis BM, Jennes L (1995) Comparative localization of serotonin1A, 1C, and 2 receptor subtype mRNAs in rat brain. J Comp Neurol 351(3):357–373. https://doi.org/10.1002/cne.903510304
55. Mijnster MJ, Raimundo AG, Koskuba K, Klop H, Docter GJ, Groenewegen HJ et al (1997) Regional and cellular distribution of serotonin 5-hydroxytryptamine2a receptor mRNA in the nucleus accumbens, olfactory tubercle, and caudate putamen of the rat. J Comp Neurol 389(1):1–11. https://doi.org/10.1002/(SICI)1096-9861(19971208)389:1<1::AID-CNE1>3.0.CO;2-6. [pii]

56. Biver F, Goldman S, Luxen A, Monclus M, Forestini M, Mendlewicz J et al (1994) Multicompartmental study of fluorine-18 altanserin binding to brain 5HT2 receptors in humans using positron emission tomography. Eur J Nucl Med 21(9):937–946
57. Haugbol S, Pinborg LH, Arfan HM, Frokjaer VM, Madsen J, Dyrby TB et al (2007) Reproducibility of 5-HT2A receptor measurements and sample size estimations with [18F] altanserin PET using a bolus/infusion approach. Eur J Nucl Med Mol Imaging 34(6):910–915. https://doi.org/10.1007/s00259-006-0296-y
58. Ito H, Nyberg S, Halldin C, Lundkvist C, Farde L (1998) PET imaging of central 5-HT2A receptors with carbon-11-MDL 100,907. J Nucl Med 39(1):208–214
59. Pinborg LH, Adams KH, Svarer C, Holm S, Hasselbalch SG, Haugbol S et al (2003) Quantification of 5-HT2A receptors in the human brain using [18F]altanserin-PET and the bolus/infusion approach. J Cereb Blood Flow Metab 23(8):985–996. https://doi.org/10.1097/01.WCB.0000074092.59115.23
60. Riss PJ, Hong YT, Williamson D, Caprioli D, Sitnikov S, Ferrari V et al (2011) Validation and quantification of [18F]altanserin binding in the rat brain using blood input and reference tissue modeling. J Cereb Blood Flow Metab 31(12):2334–2342. https://doi.org/10.1038/jcbfm.2011.94
61. Cornea-Hebert V, Riad M, Wu C, Singh SK, Descarries L (1999) Cellular and subcellular distribution of the serotonin 5-HT2A receptor in the central nervous system of adult rat. J Comp Neurol 409(2):187–209. https://doi.org/10.1002/(SICI)1096-9861(19990628)409:2<187::AID-CNE2>3.0.CO;2-P. [pii]
62. Xu T, Pandey SC (2000) Cellular localization of serotonin(2A) (5HT(2A)) receptors in the rat brain. Brain Res Bull 51(6):499–505. doi: S0361-9230(99)00278-6 [pii]
63. Morilak DA, Garlow SJ, Ciaranello RD (1993) Immunocytochemical localization and description of neurons expressing serotonin2 receptors in the rat brain. Neuroscience 54(3):701–717. doi: 0306-4522(93)90241-7 [pii]
64. Morilak DA, Somogyi P, Lujan-Miras R, Ciaranello RD (1994) Neurons expressing 5-HT2 receptors in the rat brain: neurochemical identification of cell types by immunocytochemistry. Neuropsychopharmacology 11(3):157–166. https://doi.org/10.1038/sj.npp.1380102.1380102. [pii]
65. Rodriguez JJ, Garcia DR, Pickel VM (1999) Subcellular distribution of 5-hydroxytryptamine2A and N-methyl-D-aspartate receptors within single neurons in rat motor and limbic striatum. J Comp Neurol 413(2):219–231. https://doi.org/10.1002/(SICI)1096-9861(19991018)413:2<219::AID-CNE4>3.0.CO;2-F
66. Ikemoto K, Nishimura A, Okado N, Mikuni M, Nishi K, Nagatsu I (2000) Human midbrain dopamine neurons express serotonin 2A receptor: an immunohistochemical demonstration. Brain Res 853(2):377–380
67. Mengod G, Palacios JM, Cortes R (2015) Cartography of 5-HT1A and 5-HT2A receptor subtypes in prefrontal cortex and its projections. ACS Chem Neurosci 6(7):1089–1098. https://doi.org/10.1021/acschemneuro.5b00023
68. Bubser M, Backstrom JR, Sanders-Bush E, Roth BL, Deutch AY (2001) Distribution of serotonin 5-HT(2A) receptors in afferents of the rat striatum. Synapse 39(4):297–304. https://doi.org/10.1002/1098-2396(20010315)39:4<297::AID-SYN1012>3.0.CO;2-Q
69. Doherty MD, Pickel VM (2000) Ultrastructural localization of the serotonin 2A receptor in dopaminergic neurons in the ventral tegmental area. Brain Res 864(2):176–185
70. Nocjar C, Roth BL, Pehek EA (2002) Localization of 5-HT(2A) receptors on dopamine cells in subnuclei of the midbrain A10 cell group. Neuroscience 111(1):163–176
71. Lukasiewicz S, Polit A, Kedracka-Krok S, Wedzony K, Mackowiak M, Dziedzicka-Wasylewska M (2010) Hetero-dimerization of serotonin 5-HT(2A) and dopamine D(2) receptors. Biochim Biophys Acta 1803(12):1347–1358. doi: S0167-4889(10)00226-0 [pii]. https://doi.org/10.1016/j.bbamcr.2010.08.010
72. Davies J, Tongroach P (1978) Neuropharmacological studies on the nigro-striatal and raphe-striatal system in the rat. Eur J Pharmacol 51(2):91–100

73. Olpe HR, Koella WP (1977) The response of striatal cells upon stimulation of the dorsal and median raphe nuclei. Brain Res 122(2):357–360
74. Yakel JL, Trussell LO, Jackson MB (1988) Three serotonin responses in cultured mouse hippocampal and striatal neurons. J Neurosci 8(4):1273–1285
75. Park MR, Gonzales-Vegas JA, Kitai ST (1982) Serotonergic excitation from dorsal raphe stimulation recorded intracellularly from rat caudate-putamen. Brain Res 243(1):49–58
76. Stefani A, Surmeier DJ, Kitai ST (1990) Serotonin enhances excitability in neostriatal neurons by reducing voltage-dependent potassium currents. Brain Res 529(1–2):354–357
77. Vandermaelen CP, Bonduki AC, Kitai ST (1979) Excitation of caudate-putamen neurons following stimulation of the dorsal raphe nucleus in the rat. Brain Res 175(2):356–361. doi: 0006-8993(79)91016-3 [pii]
78. Wilms K, Vierig G, Davidowa H (2001) Interactive effects of cholecystokinin-8S and various serotonin receptor agonists on the firing activity of neostriatal neuronesin rats. Neuropeptides 35(5–6):257–270. https://doi.org/10.1054/npep.2001.0875
79. Gerber R, Altar CA, Liebman JM (1988) Rotational behavior induced by 8-hydroxy-DPAT, a putative 5HT-1A agonist, in 6-hydroxydopamine-lesioned rats. Psychopharmacology 94(2):178–182
80. Knobelman DA, Kung HF, Lucki I (2000) Regulation of extracellular concentrations of 5-hydroxytryptamine (5-HT) in mouse striatum by 5-HT(1A) and 5-HT(1B) receptors. J Pharmacol Exp Ther 292(3):1111–1117
81. Antonelli T, Fuxe K, Tomasini MC, Bartoszyk GD, Seyfried CA, Tanganelli S et al (2005) Effects of sarizotan on the corticostriatal glutamate pathways. Synapse 58(3):193–199. https://doi.org/10.1002/syn.20195
82. Mignon L, Wolf WA (2007) Postsynaptic 5-HT1A receptor stimulation increases motor activity in the 6-hydroxydopamine-lesioned rat: implications for treating Parkinson's disease. Psychopharmacology 192(1):49–59. https://doi.org/10.1007/s00213-006-0680-0
83. Mignon LJ, Wolf WA (2005) 8-hydroxy-2-(di-n-propylamino)tetralin reduces striatal glutamate in an animal model of Parkinson's disease. Neuroreport 16(7):699–703. doi: 00001756-200505120-00009 [pii]
84. Dijk SN, Francis PT, Stratmann GC, Bowen DM (1995) NMDA-induced glutamate and aspartate release from rat cortical pyramidal neurones: evidence for modulation by a 5-HT1A antagonist. Br J Pharmacol 115(7):1169–1174
85. el Mansari M, Blier P (1997) In vivo electrophysiological characterization of 5-HT receptors in the guinea pig head of caudate nucleus and orbitofrontal cortex. Neuropharmacology 36(4–5):577–588
86. el Mansari M, Radja F, Ferron A, Reader TA, Molina-Holgado E, Descarries L (1994) Hypersensitivity to serotonin and its agonists in serotonin-hyperinnervated neostriatum after neonatal dopamine denervation. Eur J Pharmacol 261(1–2):171–178
87. Blomeley CP, Bracci E (2009) Serotonin excites fast-spiking interneurons in the striatum. Eur J Neurosci 29(8):1604–1614. https://doi.org/10.1111/j.1460-9568.2009.06725.x
88. Bonsi P, Cuomo D, Ding J, Sciamanna G, Ulrich S, Tscherter A et al (2007) Endogenous serotonin excites striatal cholinergic interneurons via the activation of 5-HT 2C, 5-HT6, and 5-HT7 serotonin receptors: implications for extrapyramidal side effects of serotonin reuptake inhibitors. Neuropsychopharmacology 32(8):1840–1854. https://doi.org/10.1038/sj.npp.1301294
89. North RA, Uchimura N (1989) 5-Hydroxytryptamine acts at 5-HT2 receptors to decrease potassium conductance in rat nucleus accumbens neurones. J Physiol 417:1–12
90. Muramatsu M, Lapiz MD, Tanaka E, Grenhoff J (1998) Serotonin inhibits synaptic glutamate currents in rat nucleus accumbens neurons via presynaptic 5-HT1B receptors. Eur J Neurosci 10(7):2371–2379
91. Collingridge GL, Davies J (1981) The influence of striatal stimulation and putative neurotransmitters on identified neurones in the rat substantia nigra. Brain Res 212(2):345–359. doi: 0006-8993(81)90467-4 [pii]

92. Di Giovanni G, Di Matteo V, La Grutta V, Esposito E (2001) M-Chlorophenylpiperazine excites non-dopaminergic neurons in the rat substantia nigra and ventral tegmental area by activating serotonin-2C receptors. Neuroscience 103(1):111–116
93. Dray A (1981) Serotonin in the basal ganglia: functions and interactions with other neuronal pathways. J Physiol Paris 77(2–3):393–403
94. Invernizzi RW, Pierucci M, Calcagno E, Di Giovanni G, Di Matteo V, Benigno A et al (2007) Selective activation of 5-HT(2C) receptors stimulates GABA-ergic function in the rat substantia nigra pars reticulata: a combined in vivo electrophysiological and neurochemical study. Neuroscience 144(4):1523–1535
95. Delaville C, Chetrit J, Abdallah K, Morin S, Cardoit L, De Deurwaerdere P et al (2012a) Emerging dysfunctions consequent to combined monoaminergic depletions in parkinsonism. Neurobiol Dis 45(2):763–773. https://doi.org/10.1016/j.nbd.2011.10.023
96. Gongora-Alfaro JL, Hernandez-Lopez S, Flores-Hernandez J, Galarraga E (1997) Firing frequency modulation of substantia nigra reticulata neurons by 5-hydroxytryptamine. Neurosci Res 29(3):225–231
97. Rick CE, Stanford IM, Lacey MG (1995) Excitation of rat substantia nigra pars reticulata neurons by 5-hydroxytryptamine in vitro: evidence for a direct action mediated by 5-hydroxytryptamine2C receptors. Neuroscience 69(3):903–913
98. Stanford IM, Kantaria MA, Chahal HS, Loucif KC, Wilson CL (2005) 5-Hydroxytryptamine induced excitation and inhibition in the subthalamic nucleus: action at 5-HT(2C), 5-HT(4) and 5-HT(1A) receptors. Neuropharmacology 49(8):1228–1234
99. Stanford IM, Lacey MG (1996) Differential actions of serotonin, mediated by 5-HT1B and 5-HT2C receptors, on GABA-mediated synaptic input to rat substantia nigra pars reticulata neurons in vitro. J Neurosci 16(23):7566–7573
100. Ding S, Li L, Zhou FM (2013) Presynaptic serotonergic gating of the subthalamonigral glutamatergic projection. J Neurosci 33(11):4875–4885. https://doi.org/10.1523/JNEUROSCI.4111-12.2013
101. Kelland MD, Freeman AS, Chiodo LA (1990) Serotonergic afferent regulation of the basic physiology and pharmacological responsiveness of nigrostriatal dopamine neurons. J Pharmacol Exp Ther 253(2):803–811
102. Minabe Y, Emori K, Ashby CR Jr (1996) The depletion of brain serotonin levels by para-chlorophenylalanine administration significantly alters the activity of midbrain dopamine cells in rats: an extracellular single cell recording study. Synapse 22(1):46–53
103. Gervais J, Rouillard C (2000) Dorsal raphe stimulation differentially modulates dopaminergic neurons in the ventral tegmental area and substantia nigra. Synapse 35(4):281–291
104. Olijslagers JE, Werkman TR, McCreary AC, Siarey R, Kruse CG, Wadman WJ (2004) 5-HT2 receptors differentially modulate dopamine-mediated auto-inhibition in A9 and A10 midbrain areas of the rat. Neuropharmacology 46(4):504–510
105. Ugedo L, Grenhoff J, Svensson TH (1989) Ritanserin, a 5-HT2 receptor antagonist, activates midbrain dopamine neurons by blocking serotonergic inhibition. Psychopharmacology 98(1):45–50
106. Di Giovanni G, De Deurwaerdere P, Di Mascio M, Di Matteo V, Esposito E, Spampinato U (1999) Selective blockade of serotonin-2C/2B receptors enhances mesolimbic and mesostriatal dopaminergic function: a combined in vivo electrophysiological and microdialysis study. Neuroscience 91(2):587–597
107. Porras G, Di Matteo V, Fracasso C, Lucas G, De Deurwaerdere P, Caccia S et al (2002) 5-HT2A and 5-HT2C/2B receptor subtypes modulate dopamine release induced in vivo by amphetamine and morphine in both the rat nucleus accumbens and striatum. Neuropsychopharmacology 26(3):311–324. https://doi.org/10.1016/S0893-133X(01)00333-5
108. Shi WX, Nathaniel P, Bunney BS (1995) Ritanserin, a 5-HT2A/2C antagonist, reverses direct dopamine agonist-induced inhibition of midbrain dopamine neurons. J Pharmacol Exp Ther 274(2):735–740

109. Sorensen SM, Kehne JH, Fadayel GM, Humphreys TM, Ketteler HJ, Sullivan CK et al (1993) Characterization of the 5-HT2 receptor antagonist MDL 100907 as a putative atypical antipsychotic: behavioral, electrophysiological and neurochemical studies. J Pharmacol Exp Ther 266(2):684–691
110. Minabe Y, Hashimoto K, Watanabe KI, Ashby CR Jr (2001) Acute and repeated administration of the selective 5-HT(2A) receptor antagonist M100907 significantly alters the activity of midbrain dopamine neurons: an in vivo electrophysiological study. Synapse 40(2):102–112. https://doi.org/10.1002/syn.1031
111. Delaville C, Navailles S, Benazzouz A (2012b) Effects of noradrenaline and serotonin depletions on the neuronal activity of globus pallidus and substantia nigra pars reticulata in experimental parkinsonism. Neuroscience 202:424–433. https://doi.org/10.1016/j.neuroscience.2011.11.024
112. Chen L, Yung KK, Chan YS, Yung WH (2008) 5-HT excites globus pallidus neurons by multiple receptor mechanisms. Neuroscience 151(2):439–451. https://doi.org/10.1016/j.neuroscience.2007.11.003
113. Kita H, Chiken S, Tachibana Y, Nambu A (2007) Serotonin modulates pallidal neuronal activity in the awake monkey. J Neurosci 27(1):75–83. https://doi.org/10.1523/JNEUROSCI.4058-06.2007
114. Querejeta E, Oviedo-Chavez A, Araujo-Alvarez JM, Quinones-Cardenas AR, Delgado A (2005) In vivo effects of local activation and blockade of 5-HT1B receptors on globus pallidus neuronal spiking. Brain Res 1043(1–2):186–194. doi: S0006-8993(05)00342-2 [pii]. https://doi.org/10.1016/j.brainres.2005.02.055
115. Wang H, Chen XY, Chen WF, Xue Y, Wei L, Chen L (2013) Anticataleptic effects of 5-HT(1B) receptors in the globus pallidus. Neurosci Res 77(3):162–169. doi: S0168-0102(13)00199-5 [pii]. https://doi.org/10.1016/j.neures.2013.09.002
116. Zhang SJ, Wang H, Xue Y, Yung WH, Chen L (2010) Behavioral and electrophysiological effects of 5-HT in globus pallidus of 6-hydroxydopamine lesioned rats. J Neurosci Res 88(7):1549–1556. https://doi.org/10.1002/jnr.22325.
117. Hashimoto K, Kita H (2008) Serotonin activates presynaptic and postsynaptic receptors in rat globus pallidus. J Neurophysiol 99(4):1723–1732. https://doi.org/10.1152/jn.01143.2007
118. Bengtson CP, Lee DJ, Osborne PB (2004) Opposing electrophysiological actions of 5-HT on noncholinergic and cholinergic neurons in the rat ventral pallidum in vitro. J Neurophysiol 92(1):433–443. https://doi.org/10.1152/jn.00543.2003
119. Heidenreich BA, Napier TC (2000) Effects of serotonergic 5-HT1A and 5-HT1B ligands on ventral pallidal neuronal activity. Neuroreport 11(13):2849–2853
120. Aristieta A, Morera-Herreras T, Ruiz-Ortega JA, Miguelez C, Vidaurrazaga I, Arrue A et al (2013) Modulation of the subthalamic nucleus activity by serotonergic agents and fluoxetine administration. Psychopharmacology 231:1913. https://doi.org/10.1007/s00213-013-3333-0
121. Liu J, Chu YX, Zhang QJ, Wang S, Feng J, Li Q (2007) 5,7-dihydroxytryptamine lesion of the dorsal raphe nucleus alters neuronal activity of the subthalamic nucleus in normal and 6-hydroxydopamine-lesioned rats. Brain Res 1149:216–222. doi: S0006-8993(07)00485-4 [pii]. https://doi.org/10.1016/j.brainres.2007.02.052
122. Barwick VS, Jones DH, Richter JT, Hicks PB, Young KA (2000) Subthalamic nucleus microinjections of 5-HT2 receptor antagonists suppress stereotypy in rats. Neuroreport 11(2):267–270
123. Shen KZ, Johnson SW (2008) 5-HT inhibits synaptic transmission in rat subthalamic nucleus neurons in vitro. Neuroscience 151(4):1029–1033. https://doi.org/10.1016/j.neuroscience.2007.12.001
124. Xiang Z, Wang L, Kitai ST (2005) Modulation of spontaneous firing in rat subthalamic neurons by 5-HT receptor subtypes. J Neurophysiol 93(3):1145–1157
125. Zifa E, Fillion G (1992) 5-Hydroxytryptamine receptors. Pharmacol Rev 44(3):401–458
126. Leslie RA, Moorman JM, Coulson A, Grahame-Smith DG (1993) Serotonin2/1C receptor activation causes a localized expression of the immediate-early gene c-fos in rat brain: evidence for involvement of dorsal raphe nucleus projection fibres. Neuroscience 53(2):457–463

127. Moorman JM, Leslie RA (1996) P-chloroamphetamine induces c-fos in rat brain: a study of serotonin2A/2C receptor function. Neuroscience 72(1):129–139
128. Rouillard C, Bovetto S, Gervais J, Richard D (1996) Fenfluramine-induced activation of the immediate-early gene c-fos in the striatum: possible interaction between serotonin and dopamine. Brain Res Mol Brain Res 37(1–2):105–115
129. Tremblay PO, Gervais J, Rouillard C (1998) Modification of haloperidol-induced pattern of c-fos expression by serotonin agonists. Eur J Neurosci 10(11):3546–3555
130. Sebens JB, Kuipers SD, Koch T, Ter Horst GJ, Korf J (2000) Limited participation of 5-HT(1A) and 5-HT(2A/2C) receptors in the clozapine-induced Fos-protein expression in rat forebrain regions. Eur J Pharmacol 408(1):11–17
131. Beyeler A, Kadiri N, Navailles S, Boujema MB, Gonon F, Moine CL et al (2010) Stimulation of serotonin2C receptors elicits abnormal oral movements by acting on pathways other than the sensorimotor one in the rat basal ganglia. Neuroscience 169(1):158–170
132. De Deurwaerdere P, Chesselet MF (2000) Nigrostriatal lesions alter oral dyskinesia and c-Fos expression induced by the serotonin agonist 1-(m-chlorophenyl)piperazine in adult rats. J Neurosci 20(13):5170–5178
133. De Deurwaerdere P, Lagiere M, Bosc M, Navailles S (2013) Multiple controls exerted by 5-HT2C receptors upon basal ganglia function: from physiology to pathophysiology. Exp Brain Res 230(4):477–511. https://doi.org/10.1007/s00221-013-3508-2
134. Stark JA, Davies KE, Williams SR, Luckman SM (2006) Functional magnetic resonance imaging and c-Fos mapping in rats following an anorectic dose of m-chlorophenylpiperazine. NeuroImage 31(3):1228–1237
135. Lucas JJ, Segu L, Hen R (1997b) 5-Hydroxytryptamine1B receptors modulate the effect of cocaine on c-fos expression: converging evidence using 5-hydroxytryptamine1B knockout mice and the 5-hydroxytryptamine1B/1D antagonist GR127935. Mol Pharmacol 51(5):755–763
136. Navailles S, Lagiere M, Le Moine C, De Deurwaerdere P (2013) Role of 5-HT2C receptors in the enhancement of c-Fos expression induced by a 5-HT2B/2C inverse agonist and 5-HT 2 agonists in the rat basal ganglia. Exp Brain Res 230(4):525–535. https://doi.org/10.1007/s00221-013-3562-9
137. Tilakaratne N, Friedman E (1996) Genomic responses to 5-HT1A or 5-HT2A/2C receptor activation is differentially regulated in four regions of rat brain. Eur J Pharmacol 307(2):211–217
138. Gresch PJ, Strickland LV, Sanders-Bush E (2002) Lysergic acid diethylamide-induced Fos expression in rat brain: role of serotonin-2A receptors. Neuroscience 114(3):707–713
139. Kadiri N, Lagiere M, Le Moine C, Millan MJ, De Deurwaerdere P, Navailles S (2012) Diverse effects of 5-HT2C receptor blocking agents on c-Fos expression in the rat basal ganglia. Eur J Pharmacol 689(1–3):8–16. https://doi.org/10.1016/j.ejphar.2012.05.022
140. De Deurwaerdere P, Le Moine C, Chesselet MF (2010) Selective blockade of serotonin 2C receptor enhances Fos expression specifically in the striatum and the subthalamic nucleus within the basal ganglia. Neurosci Lett 469(2):251–255
141. Habara T, Hamamura T, Miki M, Ohashi K, Kuroda S (2001) M100907, a selective 5-HT(2A) receptor antagonist, attenuates phencyclidine-induced Fos expression in discrete regions of rat brain. Eur J Pharmacol 417(3):189–194
142. Wan W, Ennulat DJ, Cohen BM (1995) Acute administration of typical and atypical antipsychotic drugs induces distinctive patterns of Fos expression in the rat forebrain. Brain Res 688(1–2):95–104
143. Bhat RV, Baraban JM (1993) Activation of transcription factor genes in striatum by cocaine: role of both serotonin and dopamine systems. J Pharmacol Exp Ther 267(1):496–505
144. Pockros LA, Pentkowski NS, Conway SM, Ullman TE, Zwick KR, Neisewander JL (2012) 5-HT(2A) receptor blockade and 5-HT(2C) receptor activation interact to reduce cocaine hyperlocomotion and Fos protein expression in the caudate-putamen. Synapse 66(12):989–1001. https://doi.org/10.1002/syn.21592

145. Burton CL, Rizos Z, Diwan M, Nobrega JN, Fletcher PJ (2013) Antagonizing 5-HT(2) a receptors with M100907 and stimulating 5-HT(2)C receptors with Ro60-0175 blocks cocaine-induced locomotion and zif268 mRNA expression in Sprague-Dawley rats. Behav Brain Res 240:171–181. https://doi.org/10.1016/j.bbr.2012.11.030
146. Genova LM, Hyman SE (1998) 5-HT3 receptor activation is required for induction of striatal c-Fos and phosphorylation of ATF-1 by amphetamine. Synapse 30(1):71–78. https://doi.org/10.1002/(sici)1098-2396(199809)30:1<71::aid-syn9>3.0.co;2-h
147. Gardier AM, Moratalla R, Cuellar B, Sacerdote M, Guibert B, Lebrec H et al (2000) Interaction between the serotoninergic and dopaminergic systems in d-fenfluramine-induced activation of c-fos and jun B genes in rat striatal neurons. J Neurochem 74(4):1363–1373
148. Guerra MJ, Liste I, Labandeira-Garcia JL (1998) Interaction between the serotonergic, dopaminergic, and glutamatergic systems in fenfluramine-induced Fos expression in striatal neurons. Synapse 28(1):71–82. https://doi.org/10.1002/(sici)1098-2396(199801)28:1<71::aid-syn9>3.0.co;2-9
149. Javed A, Van de Kar LD, Gray TS (1998) The 5-HT1A and 5-HT2A/2C receptor antagonists WAY-100635 and ritanserin do not attenuate D-fenfluramine-induced fos expression in the brain. Brain Res 791(1–2):67–74
150. Meltzer HY (1993) Serotonin receptors and antipsychotic drug action. Psychopharmacol Ser 10:70–81
151. Robertson GS, Fibiger HC (1992) Neuroleptics increase c-fos expression in the forebrain: contrasting effects of haloperidol and clozapine. Neuroscience 46(2):315–328
152. Ishibashi T, Tagashira R, Nakamura M, Noguchi H, Ohno Y (1999) Effects of perospirone, a novel 5-HT2 and D2 receptor antagonist, on Fos protein expression in the rat forebrain. Pharmacol Biochem Behav 63(4):535–541
153. Oka T, Hamamura T, Lee Y, Miyata S, Habara T, Endo S et al (2004) Atypical properties of several classes of antipsychotic drugs on the basis of differential induction of Fos-like immunoreactivity in the rat brain. Life Sci 76(2):225–237. https://doi.org/10.1016/j.lfs.2004.08.009
154. Zhao C, Li M (2010) C-Fos identification of neuroanatomical sites associated with haloperidol and clozapine disruption of maternal behavior in the rat. Neuroscience 166(4):1043–1055. https://doi.org/10.1016/j.neuroscience.2010.01.023
155. Creed-Carson M, Oraha A, Nobrega JN (2011) Effects of 5-HT(2A) and 5-HT(2C) receptor antagonists on acute and chronic dyskinetic effects induced by haloperidol in rats. Behav Brain Res 219(2):273–279. https://doi.org/10.1016/j.bbr.2011.01.025
156. Muramatsu M, Tamaki-Ohashi J, Usuki C, Araki H, Chaki S, Aihara H (1988) 5-HT2 antagonists and minaprine block the 5-HT-induced inhibition of dopamine release from rat brain striatal slices. Eur J Pharmacol 153(1):89–95
157. Navailles S, De Deurwaerdere P (2011) Presynaptic control of serotonin on striatal dopamine function. Psychopharmacology 213(2–3):213–242
158. Gobert A, Millan MJ (1999) Serotonin (5-HT)2A receptor activation enhances dialysate levels of dopamine and noradrenaline, but not 5-HT, in the frontal cortex of freely-moving rats. Neuropharmacology 38(2):315–317
159. Ichikawa J, Dai J, Meltzer HY (2001) DOI, a 5-HT2A/2C receptor agonist, attenuates clozapine-induced cortical dopamine release. Brain Res 907(1–2):151–155
160. Pehek EA, McFarlane HG, Maguschak K, Price B, Pluto CP (2001) M100,907, a selective 5-HT(2A) antagonist, attenuates dopamine release in the rat medial prefrontal cortex. Brain Res 888(1):51–59
161. Pehek EA, Nocjar C, Roth BL, Byrd TA, Mabrouk OS (2006) Evidence for the preferential involvement of 5-HT2A serotonin receptors in stress- and drug-induced dopamine release in the rat medial prefrontal cortex. Neuropsychopharmacology 31(2):265–277. https://doi.org/10.1038/sj.npp.1300819
162. Gudelsky GA, Yamamoto BK, Nash JF (1994) Potentiation of 3,4-methylenedioxymethamphetamine-induced dopamine release and serotonin neurotoxicity by 5-HT2 receptor agonists. Eur J Pharmacol 264(3):325–330
163. Ichikawa J, Meltzer HY (1995) DOI, a 5-HT2A/2C receptor agonist, potentiates amphetamine-induced dopamine release in rat striatum. Brain Res 698(1–2):204–208

164. Willins DL, Meltzer HY (1998) Serotonin 5-HT2C agonists selectively inhibit morphine-induced dopamine efflux in the nucleus accumbens. Brain Res 781(1–2):291–299
165. Bonaccorso S, Meltzer HY, Li Z, Dai J, Alboszta AR, Ichikawa J (2002) SR46349-B, a 5-HT(2A/2C) receptor antagonist, potentiates haloperidol-induced dopamine release in rat medial prefrontal cortex and nucleus accumbens. Neuropsychopharmacology 27(3):430–441. https://doi.org/10.1016/S0893-133X(02)00311-1
166. De Deurwaerdere P, Spampinato U (1999) Role of serotonin(2A) and serotonin(2B/2C) receptor subtypes in the control of accumbal and striatal dopamine release elicited in vivo by dorsal raphe nucleus electrical stimulation. J Neurochem 73(3):1033–1042
167. Liegeois JF, Ichikawa J, Meltzer HY (2002) 5-HT(2A) receptor antagonism potentiates haloperidol-induced dopamine release in rat medial prefrontal cortex and inhibits that in the nucleus accumbens in a dose-dependent manner. Brain Res 947(2):157–165
168. Lucas G, De Deurwaerdere P, Caccia S, Umberto S (2000) The effect of serotonergic agents on haloperidol-induced striatal dopamine release in vivo: opposite role of 5-HT(2A) and 5-HT(2C) receptor subtypes and significance of the haloperidol dose used. Neuropharmacology 39(6):1053–1063
169. Schmidt CJ, Black CK, Taylor VL, Fadayel GM, Humphreys TM, Nieduzak TR et al (1992a) The 5-HT2 receptor antagonist, MDL 28,133A, disrupts the serotonergic-dopaminergic interaction mediating the neurochemical effects of 3,4-methylenedioxymethamphetamine. Eur J Pharmacol 220(2–3):151–159
170. Schmidt CJ, Fadayel GM, Sullivan CK, Taylor VL (1992b) 5-HT2 receptors exert a state-dependent regulation of dopaminergic function: studies with MDL 100,907 and the amphetamine analogue, 3,4-methylenedioxymethamphetamine. Eur J Pharmacol 223(1):65–74
171. Schmidt CJ, Sullivan CK, Fadayel GM (1994) Blockade of striatal 5-hydroxytryptamine2 receptors reduces the increase in extracellular concentrations of dopamine produced by the amphetamine analogue 3,4-methylenedioxymethamphetamine. J Neurochem 62(4): 1382–1389
172. Dewey SL, Smith GS, Logan J, Alexoff D, Ding YS, King P et al (1995) Serotonergic modulation of striatal dopamine measured with positron emission tomography (PET) and in vivo microdialysis. J Neurosci 15(1 Pt 2):821–829
173. Palfreyman MG, Schmidt CJ, Sorensen SM, Dudley MW, Kehne JH, Moser P et al (1993) Electrophysiological, biochemical and behavioral evidence for 5-HT2 and 5-HT3 mediated control of dopaminergic function. Psychopharmacology 112(1 Suppl):S60–S67
174. Schmidt CJ, Fadayel GM (1996) Regional effects of MK-801 on dopamine release: effects of competitive NMDA or 5-HT2A receptor blockade. J Pharmacol Exp Ther 277(3):1541–1549
175. Waldmeier PC, Delini-Stula AA (1979) Serotonin--dopamine interactions in the nigrostriatal system. Eur J Pharmacol 55(4):363–373
176. Nash JF (1990) Ketanserin pretreatment attenuates MDMA-induced dopamine release in the striatum as measured by in vivo microdialysis. Life Sci 47(26):2401–2408
177. Schmidt CJ, Abbate GM, Black CK, Taylor VL (1990) Selective 5-hydroxytryptamine2 receptor antagonists protect against the neurotoxicity of methylenedioxymethamphetamine in rats. J Pharmacol Exp Ther 255(2):478–483
178. Schmidt CJ, Taylor VL, Abbate GM, Nieduzak TR (1991) 5-HT2 antagonists stereoselectively prevent the neurotoxicity of 3,4-methylenedioxymethamphetamine by blocking the acute stimulation of dopamine synthesis: reversal by L-dopa. J Pharmacol Exp Ther 256(1):230–235
179. Auclair A, Blanc G, Glowinski J, Tassin JP (2004) Role of serotonin 2A receptors in the D-amphetamine-induced release of dopamine: comparison with previous data on alpha1b-adrenergic receptors. J Neurochem 91(2):318–326. https://doi.org/10.1111/j.1471-4159.2004.02714.x
180. Egerton A, Ahmad R, Hirani E, Grasby PM (2008) Modulation of striatal dopamine release by 5-HT2A and 5-HT2C receptor antagonists: [11C]raclopride PET studies in the rat. Psychopharmacology 200(4):487–496. https://doi.org/10.1007/s00213-008-1226-4
181. Ichikawa J, Meltzer HY (1992) Amperozide, a novel antipsychotic drug, inhibits the ability of d-amphetamine to increase dopamine release in vivo in rat striatum and nucleus accumbens. J Neurochem 58(6):2285–2291

182. Murnane KS, Andersen ML, Rice KC, Howell LL (2013a) Selective serotonin 2A receptor antagonism attenuates the effects of amphetamine on arousal and dopamine overflow in non-human primates. J Sleep Res 22(5):581–588. https://doi.org/10.1111/jsr.12045
183. Balcioglu A, Wurtman RJ (1998) Dexfenfluramine enhances striatal dopamine release in conscious rats via a serotoninergic mechanism. J Pharmacol Exp Ther 284(3):991–997
184. Murnane KS, Winschel J, Schmidt KT, Stewart LM, Rose SJ, Cheng K et al (2013b) Serotonin 2A receptors differentially contribute to abuse-related effects of cocaine and cocaine-induced nigrostriatal and mesolimbic dopamine overflow in nonhuman primates. J Neurosci 33(33):13367–13374. https://doi.org/10.1523/jneurosci.1437-13.2013
185. Kuroki T, Meltzer HY, Ichikawa J (2003) 5-HT 2A receptor stimulation by DOI, a 5-HT 2A/2C receptor agonist, potentiates amphetamine-induced dopamine release in rat medial prefrontal cortex and nucleus accumbens. Brain Res 972(1–2):216–221
186. Li Z, Ichikawa J, Huang M, Prus AJ, Dai J, Meltzer HY (2005) ACP-103, a 5-HT2A/2C inverse agonist, potentiates haloperidol-induced dopamine release in rat medial prefrontal cortex and nucleus accumbens. Psychopharmacology 183(2):144–153. https://doi.org/10.1007/s00213-005-0170-9
187. Kankaanpaa A, Meririnne E, Lillsunde P, Seppala T (1998) The acute effects of amphetamine derivatives on extracellular serotonin and dopamine levels in rat nucleus accumbens. Pharmacol Biochem Behav 59(4):1003–1009
188. Yamamoto BK, Nash JF, Gudelsky GA (1995) Modulation of methylenedioxymethamphetamine-induced striatal dopamine release by the interaction between serotonin and gamma-aminobutyric acid in the substantia nigra. J Pharmacol Exp Ther 273(3):1063–1070
189. Millan MJ, Brocco M, Gobert A, Joly F, Bervoets K, Rivet J et al (1999) Contrasting mechanisms of action and sensitivity to antipsychotics of phencyclidine versus amphetamine: importance of nucleus accumbens 5-HT2A sites for PCP-induced locomotion in the rat. Eur J Neurosci 11(12):4419–4432
190. Sorensen SM, Humphreys TM, Taylor VL, Schmidt CJ (1992) 5-HT2 receptor antagonists reverse amphetamine-induced slowing of dopaminergic neurons by interfering with stimulated dopamine synthesis. J Pharmacol Exp Ther 260(2):872–878
191. Lucas G, Spampinato U (2000) Role of striatal serotonin2A and serotonin2C receptor subtypes in the control of in vivo dopamine outflow in the rat striatum. J Neurochem 74(2):693–701
192. Nash JF, Brodkin J (1991) Microdialysis studies on 3,4-methylenedioxymethamphetamine-induced dopamine release: effect of dopamine uptake inhibitors. J Pharmacol Exp Ther 259(2):820–825
193. Koch S, Galloway MP (1997) MDMA induced dopamine release in vivo: role of endogenous serotonin. J Neural Transm 104(2–3):135–146
194. De Deurwaerdere P, Bonhomme N, Le Moal M, Spampinato U (1995) D-fenfluramine increases striatal extracellular dopamine in vivo independently of serotonergic terminals or dopamine uptake sites. J Neurochem 65(3):1100–1108
195. Benloucif S, Galloway MP (1991) Facilitation of dopamine release in vivo by serotonin agonists: studies with microdialysis. Eur J Pharmacol 200(1):1–8
196. Benloucif S, Keegan MJ, Galloway MP (1993) Serotonin-facilitated dopamine release in vivo: pharmacological characterization. J Pharmacol Exp Ther 265(1):373–377
197. Bonhomme N, De Deurwaerdere P, Le Moal M, Spampinato U (1995) Evidence for 5-HT4 receptor subtype involvement in the enhancement of striatal dopamine release induced by serotonin: a microdialysis study in the halothane-anesthetized rat. Neuropharmacology 34(3):269–279
198. de Belleroche JS, Bradford HF (1980) Presynaptic control of the synthesis and release of dopamine from striatal synaptosomes: a comparison between the effects of 5-hydroxytryptamine, acetylcholine, and glutamate. J Neurochem 35(5):1227–1234
199. De Deurwaerdere P, L'Hirondel M, Bonhomme N, Lucas G, Cheramy A, Spampinato U (1997) Serotonin stimulation of 5-HT4 receptors indirectly enhances in vivo dopamine release in the rat striatum. J Neurochem 68(1):195–203

200. Jacocks HM III, Cox BM (1992) Serotonin-stimulated release of [3H]dopamine via reversal of the dopamine transporter in rat striatum and nucleus accumbens: a comparison with release elicited by potassium, N-methyl-D-aspartic acid, glutamic acid and D-amphetamine. J Pharmacol Exp Ther 262(1):356–364
201. Yan QS (2000) Activation of 5-HT2A/2C receptors within the nucleus accumbens increases local dopaminergic transmission. Brain Res Bull 51(1):75–81
202. Bowers BJ, Henry MB, Thielen RJ, McBride WJ (2000) Serotonin 5-HT(2) receptor stimulation of dopamine release in the posterior but not anterior nucleus accumbens of the rat. J Neurochem 75(4):1625–1633
203. Ng NK, Lee HS, Wong PT (1999) Regulation of striatal dopamine release through 5-HT1 and 5-HT2 receptors. J Neurosci Res 55(5):600–607
204. Haber SN, Fudge JL, McFarland NR (2000) Striatonigrostriatal pathways in primates form an ascending spiral from the shell to the dorsolateral striatum. J Neurosci 20(6): 2369–2382
205. Bankson MG, Yamamoto BK (2004) Serotonin-GABA interactions modulate MDMA-induced mesolimbic dopamine release. J Neurochem 91(4):852–859. https://doi.org/10.1111/j.1471-4159.2004.02763.x
206. Mijnster MJ, Galis-de Graaf Y, Voorn P (1998) Serotonergic regulation of neuropeptide and glutamic acid decarboxylase mRNA levels in the rat striatum and globus pallidus: studies with fluoxetine and DOI. Brain Res Mol Brain Res 54(1):64–73
207. Hensler JG, Artigas F, Bortolozzi A, Daws LC, De Deurwaerdere P, Milan L et al (2013) Catecholamine/serotonin interactions: systems thinking for brain function and disease. Adv Pharmacol 68:167–197. https://doi.org/10.1016/B978-0-12-411512-5.00009-9
208. Bortolozzi A, Diaz-Mataix L, Scorza MC, Celada P, Artigas F (2005) The activation of 5-HT receptors in prefrontal cortex enhances dopaminergic activity. J Neurochem 95(6):1597–1607. https://doi.org/10.1111/j.1471-4159.2005.03485.x
209. Mocci G, Jimenez-Sanchez L, Adell A, Cortes R, Artigas F (2014) Expression of 5-HT2A receptors in prefrontal cortex pyramidal neurons projecting to nucleus accumbens potential relevance for atypical antipsychotic action. Neuropharmacology 79:49–58. https://doi.org/10.1016/j.neuropharm.2013.10.021
210. Abarca J, Bustos G (1999) Differential regulation of glutamate, aspartate and gamma-amino-butyrate release by N-methyl-D-aspartate receptors in rat striatum after partial and extensive lesions to the nigro-striatal dopamine pathway. Neurochem Int 35(1):19–33
211. De Deurwaerdere P, Di Giovanni G, Millan MJ (2017) Expanding the repertoire of L-DOPA's actions: a comprehensive review of its functional neurochemistry. Prog Neurobiol 151:57. https://doi.org/10.1016/j.pneurobio.2016.07.002
212. Marti M, Guerrini R, Beani L, Bianchi C, Morari M (2002) Nociceptin/orphanin FQ receptors modulate glutamate extracellular levels in the substantia nigra pars reticulata. A microdialysis study in the awake freely moving rat. Neuroscience 112(1):153–160
213. Ferguson MC, Nayyar T, Ansah TA (2014) Reverse microdialysis of a 5-HT2A receptor antagonist alters extracellular glutamate levels in the striatum of the MPTP mouse model of Parkinson's disease. Neurochem Int 71:36–46. https://doi.org/10.1016/j.neuint.2014.03.016
214. Robinson S, Freeman P, Moore C, Touchon JC, Krentz L, Meshul CK (2003) Acute and subchronic MPTP administration differentially affects striatal glutamate synaptic function. Exp Neurol 180(1):74–87
215. Garratt JC, Kidd EJ, Wright IK, Marsden CA (1991) Inhibition of 5-hydroxytryptamine neuronal activity by the 5-HT agonist, DOI. Eur J Pharmacol 199(3):349–355
216. Kidd EJ, Garratt JC, Marsden CA (1991) Effects of repeated treatment with 1-(2,5-dimethoxy-4-iodophenyl)-2-aminopropane (DOI) on the autoregulatory control of dorsal raphe 5-HT neuronal firing and cortical 5-HT release. Eur J Pharmacol 200(1):131–139
217. Wright IK, Garratt JC, Marsden CA (1990) Effects of a selective 5-HT2 agonist, DOI, on 5-HT neuronal firing in the dorsal raphe nucleus and 5-HT release and metabolism in the frontal cortex. Br J Pharmacol 99(2):221–222

218. Martin-Ruiz R, Puig MV, Celada P, Shapiro DA, Roth BL, Mengod G et al (2001) Control of serotonergic function in medial prefrontal cortex by serotonin-2A receptors through a glutamate-dependent mechanism. J Neurosci 21(24):9856–9866
219. Puig MV, Celada P, Diaz-Mataix L, Artigas F (2003) In vivo modulation of the activity of pyramidal neurons in the rat medial prefrontal cortex by 5-HT2A receptors: relationship to thalamocortical afferents. Cereb Cortex 13(8):870–882
220. Walker PD, Riley LA, Hart RP, Jonakait GM (1991) Serotonin regulation of neostriatal tachykinins following neonatal 6-hydroxydopamine lesions. Brain Res 557(1–2):31–36
221. Basura GJ, Walker PD (2001) Serotonin 2A receptor regulation of striatal neuropeptide gene expression is selective for tachykinin, but not enkephalin neurons following dopamine depletion. Brain Res Mol Brain Res 92(1–2):66–77
222. Gresch PJ, Walker PD (1999b) Serotonin-2 receptor stimulation normalizes striatal preprotachykinin messenger RNA in an animal model of Parkinson's disease. Neuroscience 93(3):831–841
223. Walker PD, Capodilupo JG, Wolf WA, Carlock LR (1996) Preprotachykinin and preproenkephalin mRNA expression within striatal subregions in response to altered serotonin transmission. Brain Res 732(1–2):25–35
224. Gresch PJ, Walker PD (1999a) Acute p-chloroamphetamine increases striatal preprotachykinin mRNA: role of the serotonin 2A/2C receptor. Brain Res Mol Brain Res 67(1):190–193
225. Radja F, Descarries L, Dewar KM, Reader TA (1993) Serotonin 5-HT1 and 5-HT2 receptors in adult rat brain after neonatal destruction of nigrostriatal dopamine neurons: a quantitative autoradiographic study. Brain Res 606(2):273–285
226. Laprade N, Radja F, Reader TA, Soghomonian JJ (1996) Dopamine receptor agonists regulate levels of the serotonin 5-HT2A receptor and its mRNA in a subpopulation of rat striatal neurons. J Neurosci 16(11):3727–3736
227. Numan S, Lundgren KH, Wright DE, Herman JP, Seroogy KB (1995) Increased expression of 5HT2 receptor mRNA in rat striatum following 6-OHDA lesions of the adult nigrostriatal pathway. Brain Res Mol Brain Res 29(2):391–396
228. Li L, Qiu G, Ding S, Zhou FM (2013) Serotonin hyperinnervation and upregulated 5-HT2A receptor expression and motor-stimulating function in nigrostriatal dopamine-deficient Pitx3 mutant mice. Brain Res 1491:236–250. https://doi.org/10.1016/j.brainres.2012.11.010
229. Navailles S, De Deurwaerdere P (2012a) Contribution of serotonergic transmission to the motor and cognitive effects of high-frequency stimulation of the subthalamic nucleus or levodopa in Parkinson's disease. Mol Neurobiol 45(1):173–185. https://doi.org/10.1007/s12035-011-8230-0
230. Jorgensen CV, Jacobsen JP, Caron MG, Klein AB, Knudsen GM, Mikkelsen JD (2013) Cerebral 5-HT2A receptor binding, but not mGluR2, is increased in tryptophan hydroxylase 2 decrease-of-function mice. Neurosci Lett 555:118–122. https://doi.org/10.1016/j.neulet.2013.08.073
231. Morin N, Morissette M, Gregoire L, Di Paolo T (2015) Effect of a chronic treatment with an mGlu5 receptor antagonist on brain serotonin markers in parkinsonian monkeys. Prog Neuro-Psychopharmacol Biol Psychiatry 56:27–38. https://doi.org/10.1016/j.pnpbp.2014.07.006
232. Li Y, Huang XF, Deng C, Meyer B, Wu A, Yu Y et al (2010) Alterations in 5-HT2A receptor binding in various brain regions among 6-hydroxydopamine-induced parkinsonian rats. Synapse 64(3):224–230. https://doi.org/10.1002/syn.20722
233. Lopez-Gimenez JF, Villazon M, Brea J, Loza MI, Palacios JM, Mengod G et al (2001b) Multiple conformations of native and recombinant human 5-hydroxytryptamine(2a) receptors are labeled by agonists and discriminated by antagonists. Mol Pharmacol 60(4):690–699
234. Darchen F, Scherman D, Laduron PM, Henry JP (1988) Ketanserin binds to the monoamine transporter of chromaffin granules and of synaptic vesicles. Mol Pharmacol 33(6):672–677
235. Leysen JE, Eens A, Gommeren W, van Gompel P, Wynants J, Janssen PA (1988) Identification of nonserotonergic [3H]ketanserin binding sites associated with nerve terminals in rat brain

and with platelets; relation with release of biogenic amine metabolites induced by ketanserin- and tetrabenazine-like drugs. J Pharmacol Exp Ther 244(1):310–321
236. Roth BL, McLean S, Zhu XZ, Chuang DM (1987) Characterization of two [3H]ketanserin recognition sites in rat striatum. J Neurochem 49(6):1833–1838
237. Burnet PW, Chen CP, McGowan S, Franklin M, Harrison PJ (1996) The effects of clozapine and haloperidol on serotonin-1A, –2A and -2C receptor gene expression and serotonin metabolism in the rat forebrain. Neuroscience 73(2):531–540
238. Melse M, Tan SK, Temel Y, van Kroonenburgh MJ, Leentjens AF (2014) Changes in 5-HT2A receptor expression in untreated, de novo patients with Parkinson's disease. J Parkinsons Dis 4(2):283–287. https://doi.org/10.3233/jpd-130300
239. Navailles S, De Deurwaerdere P (2012b) Imbalanced dopaminergic transmission mediated by serotonergic neurons in L-DOPA-induced dyskinesia. Parkinsons Dis 2012:323686. https://doi.org/10.1155/2012/323686
240. Charron A, Hage CE, Servonnet A, Samaha AN (2015) 5-HT2 receptors modulate the expression of antipsychotic-induced dopamine supersensitivity. Eur Neuropsychopharmacol 25(12):2381–2393. https://doi.org/10.1016/j.euroneuro.2015.10.002
241. Huang XF, Tan YY, Huang X, Wang Q (2007) Effect of chronic treatment with clozapine and haloperidol on 5-HT(2A and 2C) receptor mRNA expression in the rat brain. Neurosci Res 59(3):314–321. https://doi.org/10.1016/j.neures.2007.08.001
242. Huang XF, Han M, Huang X, Zavitsanou K, Deng C (2006) Olanzapine differentially affects 5-HT2Aand2C receptor mRNA expression in the rat brain. Behav Brain Res 171(2):355–362. https://doi.org/10.1016/j.bbr.2006.03.040
243. Carli M, Invernizzi RW (2014) Serotoninergic and dopaminergic modulation of cortico-striatal circuit in executive and attention deficits induced by NMDA receptor hypofunction in the 5-choice serial reaction time task. Front Neural Circuits 8:58. https://doi.org/10.3389/fncir.2014.00058
244. Adams KH, Hansen ES, Pinborg LH, Hasselbalch SG, Svarer C, Holm S et al (2005) Patients with obsessive-compulsive disorder have increased 5-HT2A receptor binding in the caudate nuclei. Int J Neuropsychopharmacol 8(3):391–401. https://doi.org/10.1017/s1461145705005055
245. Perani D, Garibotto V, Gorini A, Moresco RM, Henin M, Panzacchi A et al (2008) In vivo PET study of 5HT(2A) serotonin and D(2) dopamine dysfunction in drug-naive obsessive-compulsive disorder. NeuroImage 42(1):306–314. https://doi.org/10.1016/j.neuroimage.2008.04.233
246. Simpson HB, Slifstein M, Bender J Jr, Xu X, Hackett E, Maher MJ et al (2011) Serotonin 2A receptors in obsessive-compulsive disorder: a positron emission tomography study with [11C] MDL 100907. Biol Psychiatry 70(9):897–904. https://doi.org/10.1016/j.biopsych.2011.06.023
247. Selvaraj S, Arnone D, Cappai A, Howes O (2014) Alterations in the serotonin system in schizophrenia: a systematic review and meta-analysis of postmortem and molecular imaging studies. Neurosci Biobehav Rev 45:233–245. https://doi.org/10.1016/j.neubiorev.2014.06.005
248. Herrick-Davis K (2013) Functional significance of serotonin receptor dimerization. Exp Brain Res 230(4):375–386. https://doi.org/10.1007/s00221-013-3622-1
249. Teitler M, Herrick-Davis K (2014) Determining the oligomer number of native GPCR using florescence correlation spectroscopy and drug-induced inactivation-reactivation. Curr Pharm Biotechnol 15(10):927–937
250. Ciruela F, Vallano A, Arnau JM, Sanchez S, Borroto-Escuela DO, Agnati LF et al (2010) G protein-coupled receptor oligomerization for what? J Recept Signal Transduct Res 30(5):322–330. https://doi.org/10.3109/10799893.2010.508166
251. Borroto-Escuela DO, Romero-Fernandez W, Tarakanov AO, Marcellino D, Ciruela F, Agnati LF et al (2010) Dopamine D2 and 5-hydroxytryptamine 5-HT((2)a) receptors assemble into functionally interacting heteromers. Biochem Biophys Res Commun 401(4):605–610. https://doi.org/10.1016/j.bbrc.2010.09.110
252. Borroto-Escuela DO, Romero-Fernandez W, Narvaez M, Oflijan J, Agnati LF, Fuxe K (2014) Hallucinogenic 5-HT2AR agonists LSD and DOI enhance dopamine D2R protomer recogni-

tion and signaling of D2-5-HT2A heteroreceptor complexes. Biochem Biophys Res Commun 443(1):278–284. https://doi.org/10.1016/j.bbrc.2013.11.104
253. Fuxe K, Borroto-Escuela DO, Tarakanov AO, Romero-Fernandez W, Ferraro L, Tanganelli S et al (2014) Dopamine D2 heteroreceptor complexes and their receptor-receptor interactions in ventral striatum: novel targets for antipsychotic drugs. Prog Brain Res 211:113–139. https://doi.org/10.1016/b978-0-444-63425-2.00005-2
254. Le Moine C, Bloch B (1991) Rat striatal and mesencephalic neurons contain the long isoform of the D2 dopamine receptor mRNA. Brain Res Mol Brain Res 10(4):283–289
255. Le Moine C, Normand E, Bloch B (1991) Phenotypical characterization of the rat striatal neurons expressing the D1 dopamine receptor gene. Proc Natl Acad Sci U S A 88(10):4205–4209
256. Aznar S, Hervig Mel S (2016) The 5-HT2A serotonin receptor in executive function: implications for neuropsychiatric and neurodegenerative diseases. Neurosci Biobehav Rev 64: 63–82. https://doi.org/10.1016/j.neubiorev.2016.02.008
257. Zhang G, Stackman RW Jr (2015) The role of serotonin 5-HT2A receptors in memory and cognition. Front Pharmacol 6:225. https://doi.org/10.3389/fphar.2015.00225
258. Barre A, Berthoux C, De Bundel D, Valjent E, Bockaert J, Marin P et al (2016) Presynaptic serotonin 2A receptors modulate thalamocortical plasticity and associative learning. Proc Natl Acad Sci U S A 113(10):E1382–E1391. https://doi.org/10.1073/pnas.1525586113
259. James TA, Starr MS (1980) Rotational behaviour elicited by 5-HT in the rat: evidence for an inhibitory role of 5-HT in the substantia nigra and corpus striatum. J Pharm Pharmacol 32(3):196–200
260. Filip M, Nowak E, Papla I (2001) On the role of serotonin2A/2C receptors in the sensitization to cocaine. J Physiol Pharmacol 52(3):471–481
261. Kehne JH, Ketteler HJ, McCloskey TC, Sullivan CK, Dudley MW, Schmidt CJ (1996) Effects of the selective 5-HT2A receptor antagonist MDL 100,907 on MDMA-induced locomotor stimulation in rats. Neuropsychopharmacology 15(2):116–124. https://doi.org/10.1016/0893-133X(95)00160-F
262. McMahon LR, Cunningham KA (2001) Antagonism of 5-hydroxytryptamine(2a) receptors attenuates the behavioral effects of cocaine in rats. J Pharmacol Exp Ther 297(1):357–363
263. O'Neill MF, Heron-Maxwell CL, Shaw G (1999) 5-HT2 receptor antagonism reduces hyperactivity induced by amphetamine, cocaine, and MK-801 but not D1 agonist C-APB. Pharmacol Biochem Behav 63(2):237–243
264. Filip M, Bubar MJ, Cunningham KA (2004) Contribution of serotonin (5-hydroxytryptamine, 5-HT) 5-HT2 receptor subtypes to the hyperlocomotor effects of cocaine: acute and chronic pharmacological analyses. J Pharmacol Exp Ther 310(3):1246–1254. https://doi.org/10.1124/jpet.104.068841
265. Howell LL, Cunningham KA (2015) Serotonin 5-HT2 receptor interactions with dopamine function: implications for therapeutics in cocaine use disorder. Pharmacol Rev 67(1): 176–197. https://doi.org/10.1124/pr.114.009514
266. Bishop C, Daut GS, Walker PD (2005) Serotonin 5-HT2A but not 5-HT2C receptor antagonism reduces hyperlocomotor activity induced in dopamine-depleted rats by striatal administration of the D1 agonist SKF 82958. Neuropharmacology 49(3):350–358. https://doi.org/10.1016/j.neuropharm.2005.03.008
267. Bishop C, Tessmer JL, Ullrich T, Rice KC, Walker PD (2004) Serotonin 5-HT2A receptors underlie increased motor behaviors induced in dopamine-depleted rats by intrastriatal 5-HT2A/2C agonism. J Pharmacol Exp Ther 310(2):687–694. https://doi.org/10.1124/jpet.104.066365
268. Bishop C, Walker PD (2003) Combined intrastriatal dopamine D1 and serotonin 5-HT2 receptor stimulation reveals a mechanism for hyperlocomotion in 6-hydroxydopamine-lesioned rats. Neuroscience 121(3):649–657
269. Bastide MF, Meissner WG, Picconi B, Fasano S, Fernagut PO, Feyder M et al (2015) Pathophysiology of L-dopa-induced motor and non-motor complications in Parkinson's disease. Prog Neurobiol 132:96–168. https://doi.org/10.1016/j.pneurobio.2015.07.002
270. Berthet A, Bezard E (2009) Dopamine receptors and L-dopa-induced dyskinesia. Parkinsonism Relat Disord 15(Suppl 4):S8–12

271. Berthet A, Porras G, Doudnikoff E, Stark H, Cador M, Bezard E et al (2009) Pharmacological analysis demonstrates dramatic alteration of D1 dopamine receptor neuronal distribution in the rat analog of L-DOPA-induced dyskinesia. J Neurosci 29(15):4829–4835
272. Visanji NP, Gomez-Ramirez J, Johnston TH, Pires D, Voon V, Brotchie JM et al (2006) Pharmacological characterization of psychosis-like behavior in the MPTP-lesioned non-human primate model of Parkinson's disease. Mov Disord 21(11):1879–1891. https://doi.org/10.1002/mds.21073
273. Baron MS, Dalton WB (2003) Quetiapine as treatment for dopaminergic-induced dyskinesias in Parkinson's disease. Mov Disord 18(10):1208–1209. https://doi.org/10.1002/mds.10551
274. Durif F, Debilly B, Galitzky M, Morand D, Viallet F, Borg M et al (2004) Clozapine improves dyskinesias in Parkinson disease: a double-blind, placebo-controlled study. Neurology 62(3):381–388
275. Reddy S, Factor SA, Molho ES, Feustel PJ (2002) The effect of quetiapine on psychosis and motor function in parkinsonian patients with and without dementia. Mov Disord 17(4): 676–681. https://doi.org/10.1002/mds.10176
276. Wolters EC, Hurwitz TA, Mak E, Teal P, Peppard FR, Remick R et al (1990) Clozapine in the treatment of parkinsonian patients with dopaminomimetic psychosis. Neurology 40(5):832–834
277. Vanover KE, Weiner DM, Makhay M, Veinbergs I, Gardell LR, Lameh J et al (2006) Pharmacological and behavioral profile of N-(4-fluorophenylmethyl)-N-(1-methylpiperidin-4-yl)-N'-(4-(2-methylpropyloxy)phenylmethyl) carbamide (2R,3R)-dihydroxybutanedioate (2,1) (ACP-103), a novel 5-hydroxytryptamine(2A) receptor inverse agonist. J Pharmacol Exp Ther 317(2):910–918. https://doi.org/10.1124/jpet.105.097006
278. Vanover KE, Betz AJ, Weber SM, Bibbiani F, Kielaite A, Weiner DM et al (2008) A 5-HT2A receptor inverse agonist, ACP-103, reduces tremor in a rat model and levodopa-induced dyskinesias in a monkey model. Pharmacol Biochem Behav 90(4):540–544. https://doi.org/10.1016/j.pbb.2008.04.010
279. Cummings J, Isaacson S, Mills R, Williams H, Chi-Burris K, Corbett A et al (2014) Pimavanserin for patients with Parkinson's disease psychosis: a randomised, placebo-controlled phase 3 trial. Lancet 383(9916):533–540. https://doi.org/10.1016/s0140-6736(13)62106-6
280. Taylor JL, Bishop C, Ullrich T, Rice KC, Walker PD (2006) Serotonin 2A receptor antagonist treatment reduces dopamine D1 receptor-mediated rotational behavior but not L-DOPA-induced abnormal involuntary movements in the unilateral dopamine-depleted rat. Neuropharmacology 50(6):761–768. https://doi.org/10.1016/j.neuropharm.2005.12.004
281. Ebadi M, Srinivasan SK (1995) Pathogenesis, prevention, and treatment of neuroleptic-induced movement disorders. Pharmacol Rev 47(4):575–604
282. Seeman P, Chau-Wong M, Tedesco J, Wong K (1975) Brain receptors for antipsychotic drugs and dopamine: direct binding assays. Proc Natl Acad Sci U S A 72(11):4376–4380
283. Seeman P, Lee T, Chau-Wong M, Wong K (1976) Antipsychotic drug doses and neuroleptic/dopamine receptors. Nature 261(5562):717–719
284. Kapur S, Remington G (2001) Atypical antipsychotics: new directions and new challenges in the treatment of schizophrenia. Annu Rev Med 52:503–517. https://doi.org/10.1146/annurev.med.52.1.503
285. Roth BL, Ciaranello RD, Meltzer HY (1992) Binding of typical and atypical antipsychotic agents to transiently expressed 5-HT1C receptors. J Pharmacol Exp Ther 260(3):1361–1365
286. Balsara JJ, Jadhav JH, Chandorkar AG (1979) Effect of drugs influencing central serotonergic mechanisms on haloperidol-induced catalepsy. Psychopharmacology 62(1):67–69
287. Carter CJ, Pycock CJ (1977) Possible importance of 5-hydroxytryptamine in neuroleptic-induced catalepsy in rats [proceedings]. Br J Pharmacol 60(2):267P–268P
288. Costall B, Fortune DH, Naylor RJ, Mardsen CD, Pycock C (1975) Serotonergic involvement with neuroleptic catalepsy. Neuropharmacology 14(11):859–868
289. Bersani G, Grispini A, Marini S, Pasini A, Valducci M, Ciani N (1990) 5-HT2 antagonist ritanserin in neuroleptic-induced parkinsonism: a double-blind comparison with orphenadrine and placebo. Clin Neuropharmacol 13(6):500–506

290. Miller CH, Fleischhacker WW, Ehrmann H, Kane JM (1990) Treatment of neuroleptic induced akathisia with the 5-HT2 antagonist ritanserin. Psychopharmacol Bull 26(3):373–376
291. Lucas G, Bonhomme N, De Deurwaerdere P, Le Moal M, Spampinato U (1997a) 8-OH-DPAT, a 5-HT1A agonist and ritanserin, a 5-HT2A/C antagonist, reverse haloperidol-induced catalepsy in rats independently of striatal dopamine release. Psychopharmacology 131(1):57–63
292. Ashby CR, Wang RY (1996) Pharmacological actions of the atypical antipsychotic drug clozapine: a review. Synapse 24(4):349–394. https://doi.org/10.1002/(SICI)1098-2396(199612)24:4<349::AID-SYN5>3.0.CO;2-D
293. Herrick-Davis K, Grinde E, Teitler M (2000) Inverse agonist activity of atypical antipsychotic drugs at human 5-hydroxytryptamine2C receptors. J Pharmacol Exp Ther 295(1):226–232
294. Navailles S, De Deurwaerdere P, Spampinato U (2006) Clozapine and haloperidol differentially alter the constitutive activity of central serotonin2C receptors in vivo. Biol Psychiatry 59(6):568–575
295. Rauser L, Savage JE, Meltzer HY, Roth BL (2001) Inverse agonist actions of typical and atypical antipsychotic drugs at the human 5-hydroxytryptamine(2C) receptor. J Pharmacol Exp Ther 299(1):83–89
296. Schmidt CJ, Sorensen SM, Kehne JH, Carr AA, Palfreyman MG (1995) The role of 5-HT2A receptors in antipsychotic activity. Life Sci 56(25):2209–2222
297. Gardell LR, Vanover KE, Pounds L, Johnson RW, Barido R, Anderson GT et al (2007) ACP-103, a 5-hydroxytryptamine 2A receptor inverse agonist, improves the antipsychotic efficacy and side-effect profile of haloperidol and risperidone in experimental models. J Pharmacol Exp Ther 322(2):862–870. https://doi.org/10.1124/jpet.107.121715
298. Reavill C, Kettle A, Holland V, Riley G, Blackburn TP (1999) Attenuation of haloperidol-induced catalepsy by a 5-HT2C receptor antagonist. Br J Pharmacol 126(3):572–574. https://doi.org/10.1038/sj.bjp.0702350
299. Gunes A, Dahl ML, Spina E, Scordo MG (2008) Further evidence for the association between 5-HT2C receptor gene polymorphisms and extrapyramidal side effects in male schizophrenic patients. Eur J Clin Pharmacol 64(5):477–482. https://doi.org/10.1007/s00228-007-0450-x
300. Richtand NM, Welge JA, Logue AD, Keck PE, Strakowski SM, McNamara RK (2007) Dopamine and serotonin receptor binding and antipsychotic efficacy. Neuropsychopharmacology 32(8):1715–1726. https://doi.org/10.1038/sj.npp.1301305
301. Richtand NM, Welge JA, Logue AD, Keck PE, Strakowski SM, McNamara RK (2008) Role of serotonin and dopamine receptor binding in antipsychotic efficacy. Prog Brain Res 172:155–175. https://doi.org/10.1016/S0079-6123(08)00908-4
302. Zhang ZJ, Zhang XB, Sha WW, Reynolds GP (2002) Association of a polymorphism in the promoter region of the serotonin 5-HT2C receptor gene with tardive dyskinesia in patients with schizophrenia. Mol Psychiatry 7(7):670–671. https://doi.org/10.1038/sj.mp.4001052
303. Gobert A, Rivet JM, Lejeune F, Newman-Tancredi A, Adhumeau-Auclair A, Nicolas JP, Cistarelli L, Melon C, Millan MJ (2000) Serotonin(2C) receptors tonically suppress the activity of mesocortical dopaminergic and adrenergic, but not serotonergic, pathways: a combined dialysis and electrophysiological analysis in the rat. Synapse 36:205–221

The Involvement of 5-HT$_{2A}$ Receptor in the Regulation of Sleep and Wakefulness, and the Potential Therapeutic Use of Selective 5-HT$_{2A}$ Receptor Antagonists and Inverse Agonists for the Treatment of an Insomnia Disorder

Jaime M. Monti, Seithikurippu R. Pandi Perumal, D. Warren Spence, and Pablo Torterolo

Abstract Several agents have been shown to improve sleep induction and/or maintenance in patients with primary or comorbid insomnia. These include benzodiazepine and non-benzodiazepine receptor allosteric modulators, melatonin and the melatonin receptor agonist ramelteon, low dose doxepin, and suvorexant. However, benzodiazepines induce a further reduction of N3 sleep [slow wave sleep (SWS) or delta sleep] and rapid-eye-movement sleep (REMS), whereas values corresponding to these variables remain decreased during non-benzodiazepine, melatonin, ramelteon or low-dose doxepin administration. By contrast, suvorexant increases REMS. There is evidence indicating that non-selective (ritanserin, ketanserin, sertindole, ICI-170809, ICI-169369, RP-62203, SR-46349B) and selective (volinanserin, pruvanserin, eplivanserin) 5-HT$_{2A}$ receptor antagonists, as well as 5-HT$_{2A}$ receptor inverse agonists (nelotanserin, pimavanserin) increase SWS in laboratory animals and N3 sleep in subjects with normal sleep and/or patients with an insomnia disorder. Thus, the association of a selective 5-HT$_{2A}$ receptor antagonist or a 5-HT$_{2A}$ receptor inverse agonist with a hypnotic drug could be a valid alternative to normalize N3 sleep in patients with an insomnia complaint.

J.M. Monti, M.D. (✉)
Department of Pharmacology and Therapeutics, School of Medicine Clinics Hospital, University of the Republic, Montevideo, Uruguay
e-mail: jmonti@mednet.org.uy

S.R. Pandi Perumal
Somnogen Canada Inc., Toronto, ON, Canada

D. Warren Spence
Sleep and Altertness Clinic, University Health Network, Toronto, ON, Canada

P. Torterolo
Department of Physiology, School of Medicine, University of the Republic, Montevideo, Uruguay

B.P. Guiard, G. Di Giovanni (eds.), *5-HT$_{2A}$ Receptors in the Central Nervous System*, The Receptors 32, https://doi.org/10.1007/978-3-319-70474-6_13

Keywords Sleep • Wakefulness • REM sleep • Slow wave sleep • Serotonin • 5-HT_{2A} receptor antagonist • 5-HT_{2A} receptor inverse agonist • Insomnia disorder

Abbreviations

5-HT	5-Hydroxytryptamine
BFB	Basal forebrain
BZD	Benzodiazepine
CNS	Central nervous system
DRN	Dorsal raphe nucleus
EEG	Electroencephalogram
EMG	Electromyogram
EOG	Electro-oculogram
GABA	γ-Aminobutyric acid
GAD	Generalized anxiety disorder
LC	Locus coeruleus
LDT	Laterodorsal tegmental nucleus
LS	Light sleep
MRN	Median raphe nucleus
NREM	Non-rapid-eye movement
PPT	Pedunculopontine tegmental nucleus
REM	Rapid-eye-movement
SE	Sleep efficiency
SNc	Substantia nigra pars compacta
SOL	Sleep onset latency
SWS	Slow wave sleep
TST	Total sleep time
vPAG	Ventral periaqueductal gray matter
VTA	Ventral tegmental area
W	Wakefulness
WASO	Wake time after sleep onset

Introduction

Sleep has an impact on every facet of daily life. In this respect, disturbed sleep affects not only our health and well-being but also our quality of life. Since the 1950s much progress has been made in scientific efforts to understand the basic need for sleep. The sleep-wakefulness cycle in man can now be characterized by the polysomnographic recording of three basic parameters: the electroencephalogram

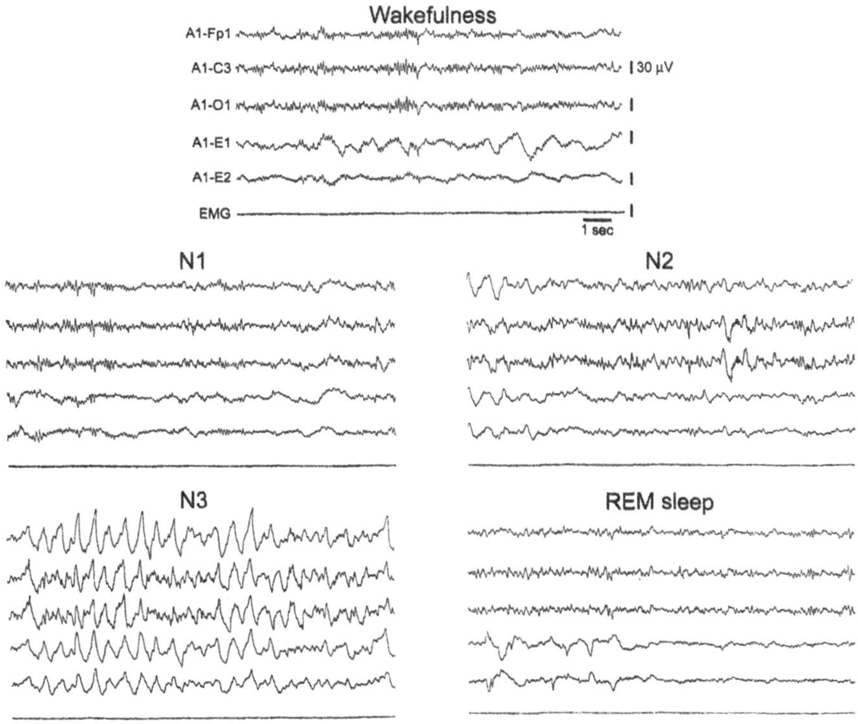

Fig. 1 Patterns of electroencephalogram (EEG) activity during wakefulness, stage N1, N2 and N3 sleep and REM sleep in the young adult. A1-Fp1, frontal cortex; A1-C3, central cortex; A1-O1, occipital cortex; A1-E1, right electro-oculogram; A1-E2, left electro-oculogram; EMG, chin electromyogram. Time in seconds

(EEG), electro-oculogram (EOG), and electromyogram (EMG). Dement and Kleitman [1] and Rechtschaffen and Kales [2] provided a classification system of sleep stages that comprise a waking state, non-rapid-eye-movement sleep (NREMS) and rapid-eye-movement sleep (REMS). Active wakefulness (W) with eyes open is characterized by a low voltage mixed-frequency EEG profile, high tonic EMG activity and voluntary eye movements. In contrast, quiet W with eyes closed is characterized by an EEG with sinusoidal waves (alpha activity), slow or rapid EOG activity and a relatively high tonic EMG (Fig. 1). Concerning NREMS, the Rechtschaffen and Kales [2] scoring manual considers four NREMS stages (NREMS stages 1 to 4), while the more recent American Academy of Sleep Medicine scoring manual [3] distinguishes three stages (N1, N2, N3) mainly on the basis of EEG criteria (Fig. 1). Stage N1 is characterized by the presence of relatively low-voltage waves with a prominence of activity in the theta range, slow and predominantly horizontal eye movements and a decrease of EMG activity; stage N2 is defined by the presence of sleep spindles and K-complexes; and stage N3 is characterized by the occurrence of slow high-amplitude or delta waves. During REMS the subject is flacid and even more unresponsive than during stage N3. Periodically, his eyes move rapidly under closed lids. If the subject is awakened, he might actually say that he was dreaming. The polysomnogram is characterized by the presence of low voltage mixed frequency EEG activity that closely resembles that of stage N1. In this context, theta activity is often observed in conjunction with bursts of REMs. Despite this activity, the muscles are completely relaxed and only periodically interrupted by muscle twitches (Fig. 1).

During stage N3 a restorative process has been proposed to occur in the central nervous system (CNS). In this respect, it has been shown that protein concentration is augmented in neurons and glial tissue, particularly astrocytes, as a result of both an increase of protein synthesis and a reduction in degradation [3]. It has been additionally proposed, that the increase in synaptic strength occurring during W results in higher energy consumption and, as a consequence, the ability to process or learn new information is temporarily restricted, but is restored following the occurrence of sleep [4–6]. Sleep has been associated, in addition, with a 60% increase in the interstitial space, resulting in a striking increase in the convective exchange between cerebrospinal fluid and interstitial fluid [7]. A consideration of this process led the study authors to postulate that the enhanced removal of potentially neurotoxic waste products that accumulate in the brain during W is an important component of the restorative function of sleep. According to Achermann and Borbely [8], a normal homeostatic regulation of sleep strongly depends on the amount of slow wave sleep [(SWS)—N3 sleep] relative to the time awake before. This led the authors to propose that the reduction of N3 sleep that occurs in elderly subjects and in patients with a diagnosis of primary insomnia, might be an important contributor to the cognitive decline and memory deficits that are seen in these patient groups, and hence to the deterioration in their performance of daily tasks. The authors further hypothesized that measures directed to enhance N3 sleep could tentatively normalize these functions in affected patients. Follow-on research has supported proposals for non-pharmacological and pharmacological approaches to enhance N3 sleep. The former includes, among others, acoustic stimulation [5], while the latter involves the administration of compounds that either induce the selective blockade of serotonin 5-HT_{2A} receptor or attenuate its basal constitutive signaling activity.

Electrographic activity of the rat (Fig. 2) and mouse, two species currently used in preclinical studies, has been classified, based on the associated wave-form, into several categories: (1) W that is defined by the presence of low voltage fast waves in frontal cortex, a mixed theta rhythm in occipital cortex and relatively high EMG activity; (2) light sleep (LS) that is characterized by the occurrence of high voltage slow cortical waves interrupted by low voltage fast EEG activity; (3) SWS that is identified by the occurrence of continuous high amplitude slow frontal and occipital waves combined with a reduced EMG; and (4) REMS that is distinguished by the presence of low voltage fast frontal waves, a regular theta rhythm in the occipital cortex and a silent EMG except for occasional myoclonic twitching [9].

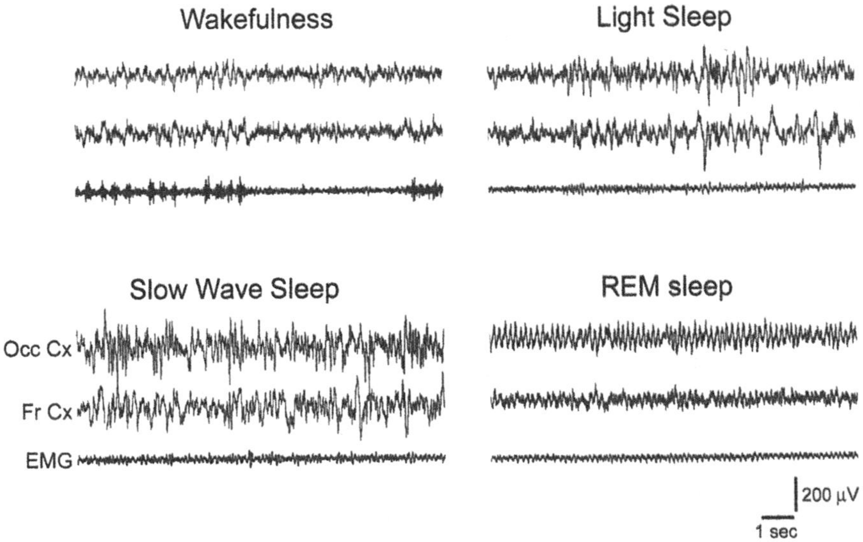

Fig. 2 Patterns of electroencephalogram (EEG) activity during wakefulness, light sleep, slow wave sleep and REM sleep in the rat. Occ Cx, occipital cortex; Fr Cx, frontal cortex; EMG, dorsal neck muscles electromyogram. Time in seconds

Insomnia: Diagnostic Criteria

Insomnia is characterized by one or more of the following: difficulty falling asleep (initial insomnia), insufficient sleep, numerous nocturnal awakenings (intermediate or middle insomnia), early morning awakenings with inability to resume sleep (terminal insomnia), or nonrestorative sleep. Common daytime complaints of insomnia sufferers include somnolence, fatigue, irritability, and difficulty concentrating and performing every day tasks. The duration of insomnia is considered to be an important guide to its evaluation and treatment. In this respect individuals can complain of transient, short-term or chronic insomnia. Chronic insomnia in adult patients has been classified as primary or comorbid [10].

Treatment Options of Chronic Primary Insomnia

Several classes of medications have been prescribed as hypnotics over the years. These have included the benzodiazepine (BZD) receptor allosteric modulators [either BZD or non-BZD (zolpidem, eszopiclone, zaleplon) agents]; melatonin and the melatonin receptor agonist ramelteon, low-dose doxepin (a tricyclic antidepressant), and the dual orexin receptor antagonist suvorexant.

The sleep induced by BZD hypnotics including temazepam, flunitrazepam, quazepam and flurazepam in patients with chronic primary insomnia is characterized by a shortened sleep onset latency (SOL), decreased number of nocturnal awakenings, reduced time spent awake, increases in N2 sleep, consistent reductions in N3 sleep, dose-dependent decrease of REMS, and improvement in the subjective quality of sleep when compared with no treatment (Table 1) [11, 12]. The commonly reported adverse effects of BZD hypnotics are drowsiness, tiredness and anterograde amnesia [12]. In addition, daytime functioning can be negatively affected by long-acting BDZ derivatives [12]. Rebound insomnia has been reported more often with BZDs that have short half-lives, such as triazolam and midazolam. The elderly are more susceptible to the adverse effects of the BZDs, due to age-related alterations in pharmacokinetics as a result of changes in hepatic metabolism and renal excretion [13].

Concerning the non-BZD agents, zolpidem immediate-release is capable of reducing SOL and wake time after sleep onset (WASO) in patients with chronic primary insomnia. The predominant reduction of WASO during the first part of the night, and the absence of an effect on the number of nocturnal awakenings and total sleep time (TST) could be related to zolpidem's short elimination half-life. With respect to sleep architecture, zolpidem immediate-release tends to increase N2 sleep, whereas N3 sleep, REMS latency and REMS in min or as percentage of TST are not significantly modified [14]. On the other hand, zolpidem extended-release is effective for the treatment of insomnia characterized by difficulties with sleep onset and/or sleep maintenance (Table 1) [15, 16]. Adverse events observed during the administration of zolpidem immediate-release and extended-release include drowsiness, headache, dizziness, nausea, diarrhea, myalgia and bizarre behaviors such as sleep eating, sleep walking, sleep conversation, sleep driving and sleep shopping, frequently with amnesia for the event [14, 16, 17]. Rebound insomnia was not evident after discontinuation of the hypnotic agent [14]. In addition, 12 months of nightly zolpidem immediate-release administration did not lead to dose escalation in non-elderly primary insomniacs [18].

The effects of eszopiclone have been assessed also in patients with chronic primary insomnia. Eszopiclone administration significantly reduced SOL and WASO whereas TST and sleep efficiency (SE) showed significant increments. N2 sleep was significantly augmented at the expense of non-significant reductions of N1 and N3 sleep, and REM sleep (Table 1). The most commonly reported complaints or side-effects in patients who received eszopiclone were unpleasant or bitter taste, headache, dyspepsia, pain, diarrhea, dry mouth and dizziness [19]. Rebound insomnia did not occur after eszopiclone withdrawal, nor dit it show addictive potential in individuals without a known history of drug abuse [19].

Polysomnographic data and subjective measures showed that, relative to placebo, zaleplon significantly decreased SOL in adults with chronic primary insomnia. No statistically significant effects on WASO, TST or number of awakenings were seen with the hypnotic drug. Furthermore, zaleplon administration induced inconsistent changes of sleep architecture, including N2 sleep, N3 sleep and REMS (Table 1). Evening administration of zaleplon to patients with an insomnia disorder was associated with minimal next-day residual sedation and impairment [20, 21].

Table 1 The effects of hypnotic drugs on sleep induction and maintenance and sleep architecture in patients with chronic primary insomnia

Variable	Benzodiazepines	Zolpidem	Eszopiclone	Zaleplon	Ramelteon	Doxepin	Suvorexant
Sleep onset latency	Decrease	Decrease	Decrease	Decrease	Decrease	No change or decrease	Decrease
Number of awakenings	Decrease	No change[a]	Decrease	Variable effects	No change	Decrease	No change or decrease
		Decrease[b]					
Wake time after sleep onset	Decrease	Decrease[a]	Decrease	No change	No change	Decrease	Decrease
		Decrease[b]					
Total sleep time	Increase	Inconsistent[a]	Increase	Variable effects	Small increase	Increase	Increase
		Increase[b]					
Stage N2 sleep	Increase	Increase	Increase	Variable effects	Increase	Increase	Increase
Stage N3 sleep	Decrease	No change	No change	Variable effects	No change	No change	No change
REM sleep	Decrease	No change	No change	Variable effects	No change	Decrease	Increase

[a]Immediate release
[b]Extended release

Moreover, rebound insomnia effects were either weak or absent following discontinuation of the medication [21].

The effects of ramelteon have been evaluated also in adult subjects with chronic primary insomnia. Compared with the placebo group, the melatonin receptor agonist reduced polysomnographically recorded SOL. Total sleep time was also increased, but the effect was restricted to the first week of treatment. The number of awakenings, WASO and sleep architecture were not significantly modified (Table 1) [22]. Ramelteon was well tolerated and without significant cognitive, memory or psychomotor effects. Additionally, the drug did not produce rebound insomnia nor dependency in study subjects [22].

In adults with chronic primary insomnia low-dose doxepin significantly improved WASO, TST and SE. There was a significant increase in min of N2 sleep whereas REMS percentage was reduced. Values corresponding to N3 sleep remained unchanged [23]. Overall, the data showed that low-dose doxepin improved sleep, including SE in patients with chronic primary insomnia. Additionally, it exhibited a side-effect profile comparable to placebo as shown by the absence of next-day sedation, memory impairment, anticholinergic effects or rebound insomnia [23].

Presently available evidence tends to indicate that suvorexant induces the reduction of SOL and WASO, while TST and SE are increased in middle-aged and elderly patients with primary insomnia. The increase of TST is related to greater time spent in REMS and, to a lesser extent, N2 sleep (Table 1). Rebound insomnia and withdrawal effects were not detected when the compound was discontinued after long-term administration. Commonly reported adverse effects have included somnolence, headache, dizziness, fatigue, diarrhea and abnormal dreams. Somnolence has usually been shown to resolve with continued use [24].

In conclusion, current therapeutic approaches for the treatment of insomnia focus on improving SOL (zolpidem immediate-release, zaleplon, ramelteon) and/or sleep maintenance (temazepam, quazepam, flurazepam, zolpidem extended-release, eszopiclone, low-dose doxepin, suvorexant) in patients with chronic primary insomnia. However, during the administration of these agents (with the exception of suvorexant) N3 and REMS do not regain normal levels or can be even further reduced.

Brain Regions and Neurotransmitter Systems Involved in the Regulation of Sleep and Wakefulness

The CNS structures involved in the promotion of W include neurons containing serotonin [5-HT: dorsal raphe nucleus (DRN), median raphe nucleus (MRN)]; norepinephrine [locus coeruleus (LC)]); dopamine [ventral tegmental area (VTA), substantia nigra pars compacta (SNpc), ventral periaqueductal gray matter (vPAG)]; acetylcholine [W-on: laterodorsal and pedunculopontine tegmental nuclei (LDT/PPT)]; histamine (tuberomammillary nucleus); and orexin (posterior lateral hypothalamus around the fornix). The serotonergic, noradrenergic, histaminergic and cholinergic (W-on) neurons that participate in the regulation of W give rise to mainly ascending projections to the thalamus and the basal forebrain (BFB) which in turn project to the cerebral cortex and the hippocampus. The dopamine-containing cells of the VTA and SNpc project to the basal ganglia and the prefrontal cortex, while those corresponding to the vPAG project predominantly to the BFB and midline thalamus. Furthermore, orexin-containing neurons carry projections to the entire forebrain and brainstem arousal systems. Under normal conditions, the neuroanatomical structures that promote W inhibit neurons that promote and/or induce NREMS and REMS [25–28].

Neurons of the preoptic area and adjacent BFB constitute the sleep-inducing system. Sleep active neurons of the preoptic area are mainly located in the ventrolateral preoptic area. A majority of these neurons contain γ-aminobutyric acid (GABA) and galanin, and inhibit brainstem and hypothalamic cells involved in the promotion of W [29]. Cholinergic REM-on neurons of the LDT/PPT have been identified as promoting REMS [26]. Moreover, the subcoeruleus nucleus has been proposed as the critical area for REMS generation in the cat [30]. Its equivalent in the rat and mouse is called the sublaterodorsal tegmental nucleus. The REMS generation region includes predominantly glutamatergic and GABAergic neurons [30]. More recently, melanin-concentrating hormone neurons located in the lateral hypothalmus and zona incerta have been proposed also to participate in the regulation of REMS [31].

Role of Serotonin 5-HT_{2A} Receptor in the Regulation of the Behavioral State

Serotonin shares with other neurotransmitters the ability to promote W and to suppress REMS. A number of 5-HT receptors have been characterized in central sites that can be classified into at least seven classes. The 5-HT_2 class consists of three (5-$HT_{2A\text{-}B\text{-}C}$) subtypes and is structurally related to the superfamily of G-protein-coupled receptors. The 5-HT_{2A} receptor is primarily coupled to Gq and its action is mediated by the activation of phospholipase C, with a resulting depolarization of the host cell. It is located within postsynaptic structures, predominantly on proximal and distal dendritic shafts [32]. 5-HT_{2A} receptors are distributed in brain regions involved in the promotion of: (1) the waking state [BFB (nucleus of the diagonal band of Broca, bed nucleus of the stria terminalis), VTA/SNpc, DRN/MRN, LC, tuberomammillary nuclei]; (2) the NREMS state [medial and lateral preoptic area, anterior and lateral hypothalamic areas, thalamus (nonspecific nuclei)]; (3) the REMS state (LDT/PPT, medial pontine reticular formation). Additionally, 5-HT_{2A} receptors have been characterized in the cerebral cortex, limbic system and basal ganglia [33].

Strategies aimed at determining the role of 5-HT$_{2A}$ receptor in the regulation of the behavioral state have included: (1) mutant mice that do not express 5-HT$_{2A}$ receptors; (2) local brain delivery of 5-HT$_{2A}$ receptor ligands; (3) systemic administration of nonselective 5-HT$_{2A}$ receptor agonists and antagonists; (4) systemic administration of selective 5-HT$_{2A}$ receptor antagonists and inverse agonists.

Sleep Patterns in Mutant Mice That Do Not Express 5-HT$_{2A}$ Receptors

Popa et al. [34] characterized the sleep-wake cycle in 5-HT$_{2A}$ receptor knockout mice of the 129Sv/Ev Tac background. Mice lacking 5-HT$_{2A}$ receptor showed a significant increase of W and a reduction of SWS throughout the entire light/dark cycle compared to the wild-type animals. Additionally, the duration of W bouts was augmented during the dark period. Values corresponding the REMS were similar in the mutants and the wild-type rodents. Sleep deprivation during a 6 h period starting at the beginning of the light phase, resulted in a significant increase of the EEG power density in the delta frequency of the wild-type mice only. Since the 5-HT$_{2A}$ receptor is positively coupled to phospholipase C via Gq proteins and mobilizes intracellular Ca^{2+}, a reduction in W and an increase in SWS should have been expected in the mice lacking 5-HT$_{2A}$ receptor. In other words, opposite effects of gene deletion versus acute pharmacological activation of the same protein with a 5-HT$_2$ receptor agonist (to be dealt with in the next section) should have been expected on W and SWS. In support of this contention, systemic administration of the selective 5-HT$_{2A}$ receptor antagonist volinanserin (2.0–5.0 mg/kg, i.p.) significantly augmented SWS and decreased W and REMS in the wild-type mice only [34]. The unexpected increase of W and reduction of SWS in the mutants led Popa et al. [34] and Adrien [35] to propose that the genetic deletion of 5-HT$_{2A}$ receptor would give rise to hyposensitivity of 5-HT$_{2B}$ receptor and hypersensitivity of 5-HT$_{2C}$ receptor. Accordingly, systemic injection of the selective 5-HT$_{2B}$ receptor antagonist SB 215505 (0.5–2.5 mg/kg, i.p.) produced a decrease of SWS and REMS, and an enhancement of W in the wild-type animals only [34]. Moreover, selective blockade of 5-HT$_{2C}$ receptor by compound SB 242084 (0.5–2.5 mg/kg, i.p.) induced the reduction of REMS in the wild-type mice, while SWS and W remained unchanged. Although SB 242084 failed to modify REMS values in the mutants, the dose response-curve of REMS was shifted to the left, which tends to indicate an increased influence of 5-HT$_{2C}$ receptor-mediated regulation of REMS in the 5-HT$_{2A}$ receptor knockout mice.

The increase of W and reduction of SWS in animals lacking 5-HT$_{2A}$ receptor could also be related, either wholly or at least in part, to the increased release of norepinephrine and dopamine at central sites [36, 37]. In this respect, Quesseveur et al. [38] showed that the DOI-induced decrease of DRN 5-HT neurons firing rate in wild-type mice did not occur in animals lacking 5-HT$_{2A}$ receptor. Moreover, the

DOI-induced inhibitory effect on 5-HT neurons activity was attenuated by the loss of noradrenergic neurons provoked by the neurotoxin DSP4. It should be mentioned that 5-HT_{2A} receptors that participate in the control of dopamine and norepinephrine release are expressed by inhibitory GABAergic interneurons [39]. Consequently, the reduction of GABA release at critical sites in the CNS of 5-HT_{2A} receptor knock-out mice would be indirectly responsible for the increased availability of norepinephrine and dopamine at neuronal networks involved in regulating behavioral states.

In conclusion, genetic deletion of 5-HT_{2A} receptor has been shown to produce a phenotype similar to the effect of acute pharmacological activation. In addition, compensatory mechanisms, including hyposensitivity of 5-HT_{2B} receptor and hypersensitivity of 5-HT_{2C} receptor occurred in the mutants. The increased release of norepinephrine and dopamine at central sites could also play a role in the changes of the sleep-wake cycle described in the knockout mice.

Sleep Patterns in Laboratory Animals Administered Non-Selective 5-HT_{2A} Receptor Agonists

To date no studies have been published on the effect of selective 5-HT_{2A} receptor agonists on sleep variables in laboratory animals due to the absence of highly selective 5-HT_{2A} receptor derivatives. Presently available evidence was obtained from studies involving the administration of nonselective 5-$HT_{2A/2C}$ receptor agonists, including the phenylalkylamine derivatives DOM and DOI (Table 2) [40, 41]. Systemic injection of DOM (0.16–2.5 mg/kg, i.p.) at the onset of the light phase produced a significant reduction of SWS and REMS, and an increase of W during the first 4-h period in the rat (Table 3). The reduction of SWS and REMS was due to a decrease in the number of bouts and of their mean duration. In contrast, the increase of W was related to an enhancement in the duration of bouts [42]. Of note, pretreatment with the nonselective 5-$HT_{2A/2C}$ receptor antagonist ritanserin (0.16–2.5 mg/kg, i.p.) dose-dependently prevented the DOM (0.63 mg/kg)-induced SWS deficit and W enhancement. In contrast, REMS remained substantially reduced [42].

Systemic administration of DOI (0.25 mg/kg, i.p.) at the onset of the light period also reduced SWS and REMS and augmented W in the rat [9] (Table 3). Furthermore, systemic (1.0–2.0 mg/kg, s.c.) or intrathalamic (10–50 μg) injection of DOI decreased the neocortical high-voltage spindle activity that occurs during relaxed W in the rat [43]. Ritanserin (0.25 mg/kg, i.p.) antagonized the increase of W and reduction of SWS produced by DOI (0.25 mg/kg). However, DOI-related suppression of REMS was not prevented by the 5-$HT_{2A/2C}$ receptor antagonist [9].

Thus, systemic administration of non-selective 5-$HT_{2A/2C}$ receptor agonists to laboratory animals induced a consistent increase of W and a reduction of SWS and REMS.

Table 2 Affinities to 5-HT_2 receptor subtypes of non-selective 5-HT_{2A} receptor agonists and antagonists and selective 5-HT_{2A} receptor antagonists and inverse agonists

Compound	Function	5-HT_{2A}	5-HT_{2B}	5-HT_{2C}
DOM[1]	Non-selective agonist	8.44[a]	n.a.	7.44[a]
DOI[2]	Non-selective agonist	7.30[a]	7.40[a]	7.80[a]
Ketanserin[1]	Non-selective antagonist	8.90[a]	5.40[a]	7.00[a]
Ritanserin[1]	Non-selective antagonist	8.80[a]	8.80[a]	8.90[a]
ICI 170809[1]	Non-selective antagonist	9.10[a]	n.a.	8.30[a]
RP-62203[2]	Non-selective antagonist	9.60[a]	n.a.	7.60[a]
Sertindole[3]	Non-selective antagonist	9.41[a]	n.a	8.72[a]
Volinanserin[4]	Selective antagonist	0.85[b]	n.a.	88.00[b]
Pruvanserin[5]	Selective antagonist	0.35[c]	n.a.	1334.00[c]
Eplivanserin[6]	Selective antagonist	1.30[c]	n.a.	120.00[c]
Nelotanserin[7]	Inverse agonist	0.35[b]	2000.00[b]	100.00[b]
Pimavanserin[8]	Inverse agonist	9.70[b]	n.a.	8.00[b]

From: [1]Barnes and Sharp [88]; [2]Leysen [89]; [3]Leysen et al. [90]; [4]Kehne et al. [75]; [5]Bartoszyk et al. [76]; [6]Rinaldi-Carmona et al. [81]; [7]Al-Shamma et al. [83]; [8]Vanover et al. [85]
[a]pKi
[b]Ki values are in nanomolar
[c]pIC50; n.a., not available

Table 3 The effect of non-selective 5-HT_{2A} receptor agonists and antagonists and selective 5-HT_{2A} receptor antagonists and inverse agonists on sleep variables in laboratory animals

Compound	Function	W	SWS	REMS	Reference
DOM	Non-selective agonist	+	–	–	Dugovic et al. [42]
DOI	Non-selective agonist	+	–	–	Monti et al. [9]
Ketanserin	Non-selective antagonist	–	+	–	[55])
Ritanserin	Non-selective antagonist	–	+	–	Monti et al. [9]
Ici 170809	Non-selective antagonist	–	+	–	Tortella et al. [59]
RP-62203	Non-selective antagonist	–	+	n.s.	Stutzmann et al. [60]
Sertindole	Non-selective antagonist	–	+	–	Coenen et al. [61]
Volinanserin	Selective antagonist	–	+	–	Popa et al. [34]
Pruvanserin	Selective antagonist	n.s.	+	–	Monti and jantos [79]
Eplivanserin	Selective antagonist	Significant increase of mean number and duration of SWS bouts			Griebel et al. [82]
Nelotanserin	Inverse agonist	–	+	–	Al-shamma et al. [83]

W wakefulness; *SWS* slow wave sleep; *REMS* rapid-eye-movement sleep; + significant increase; – significant decrease; *n.s.* non significant

Sleep Patterns in Laboratory Animals Following Local Microinjection into the DRN or LDT of a Non-Selective 5-HT_{2A} Receptor Agonist

The existence of 5-HT_{2A} receptors in the DRN tends to suggest that they are involved in the regulation of REMS in the rat (for review see [27]). In order to test the hypothesis, DOI (1.4, 2.8, or 5.6 mmol) was infused into the DRN of rats implanted for chronic sleep recordings. Following the microinjection of the 2.8 and 5.6-mmol doses of the 5-$HT_{2A/2C}$ receptor ligand into the DRN, REMS was significantly reduced during the first 4-h of recording. The number of REM periods was decreased also after the 5.6 mmol dose of the 5-HT_2 receptor ligand. Pretreatment with the selective 5-HT_{2A} or 5-HT_{2C} receptor antagonists EMD 281014 (5.6 mmol) or SB-243213 (2.8–5.6 mmol), respectively, effectively prevented the DOI (5.6 mmol)-induced suppression of REMS. These findings indicate that the effect was mediated by the participation of both, 5-HT_{2A} and 5-HT_{2C} receptors located in the DRN [44].

It has been shown that systemic, intra-raphe or microiontophoretic administration of DOI inhibits the firing of serotonergic neurons in the DRN and reduces the extracellular concentration of 5-HT [45, 46]. The inhibition of 5-HT neurons in the DRN of the rat after systemic injection of DOI was reverted by the selective 5-HT_{2A} antagonist volinanserin and by the GABA-A receptor antagonist bicuculline, whereas the 5-HT_{2C} receptor antagonist SB 242084 was less effective in this respect [47]. Thus, within the DRN, DOI inhibits 5-HT neurons firing by activating GABAergic interneurons via 5-HT_{2A} and 5-HT_{2C} receptors [48]. Notwithstanding this, microinjection of DOI into the DRN resulted in the suppression of REMS. In this respect, it should be considered that many GABAergic neurons in the DRN contribute to long projections that reach, among other nuclei, the LDT/PPT [49]. Thus, it can be proposed that DOI activation of projection GABAergic cells inhibits the activity of cholinergic REM-on neurons in the LDT/PPT and suppresses REMS sleep. The recent finding by Vasudeva and Waterhouse [50] that non-serotonergic neurons corresponding to the DRN lateral wings in the rat are immediately rostral to and in line with the cholinergic LDT, tends to support our proposal.

Amici et al. [51] microinjected DOI (1 nmol, 1 μmol, 1 mmol) or ketanserin (1–5 nmol) into the LDT in rats to determine whether 5-$HT_{2A/2C}$ receptors were involved in the regulation of sleep variables. Local administration of DOI into the LDT significantly decreased the number of sequential REMS episodes during the first h of recording, whereas ketanserin induced the opposite effect during the second and third h of recording. The duration of REMS episodes was not modified by the compounds. The reduction in the number of REMS periods after local administration of DOI into the LDT could be also associated with a non-serotonergic mechanism involving GABA. Accordingly, within the LDT/PPT there is a demonstrable role for GABA in regulating REMS. In this respect, GABAergic inhibition arises from both local inhibitory GABAergic interneurons and GABAergic cells located in the brainstem nuclei and hypothalamus that project to the LDT/PPT [52–54]. It is tentatively proposed that DOI activation of 5-$HT_{2A/2C}$ receptors expressed by

GABAergic interneurons decreases the activation of cholinergic REM-on cells in the LDT and reduces REMS. On the other hand, their blockade by ketanserin would induce the opposite effect.

In conclusion, local microinjection of DOI into the DRN or LDT has been found to decrease REMS in the rat. These findings tend to indicate that a 5-HT_2-receptor mechanism participates in DRN and LDT regulation of REMS.

Sleep Patterns in Laboratory Animals Administered Non-Selective 5-HT_{2A} Receptor Antagonists

The effects of the non-selective 5-$HT_{2A/2C}$ receptor antagonists ketanserin, ritanserin, ICI-170809, RP-62203 and sertindole (Table 2) on the sleep-wake cycle have been characterized in the rat. Accordingly, Kirov and Moyanova [55] examined the effect of ketanserin (1.0–6.0 mg/kg, i.p.) administered at the beginning of the light phase on sleep variables in young, middle-aged and old rats. The 5-$HT_{2A/2C}$ receptor antagonist reduced W in the young and middle-aged animals only. On the other hand, SWS was enhanced and REMS was reduced in the three age groups (Table 3). However, the SWS increase was smaller, shorter-lasting, and dose-independent in the old rats, which could be indicative of a loss of 5-HT_2 receptors with aging. In addition, ketanserin increased the amplitude of neocortical high voltage spindles in young and middle-aged rats, and prevented their reduction in animals infused with DOI (10–50 μg) at the level of the thalamus [43, 56].

Systemic injection of ritanserin (0.25–0.63 mg/kg, i.p.) at light onset was found to induce a significant increase of SWS and a decrease of REMS during the first four recording hours in the rat. Concurrently, a decrease of W was observed after administration of the greatest dose (Table 3) [9, 42, 57]. Furthermore, injection of ritanserin (0.3 mg/kg, i.p.) at light onset caused an increase in the EEG power density in the low frequency range (mainly delta activity) and a reduction in the high frequency range that paralleled the changes of sleep and W [58].

Tortella et al. [59] measured the effects of the phenyl-quinoline derivative ICI-170809 on sleep variables in the rat. Oral administration of ICI-170809 (10–20 mg/kg) 5 h after the beginning of the light phase suppressed REMS for 6 h postinjection. The effect was due to a decrease in the total number of REMS periods. At the greatest dose tested the compound significantly increased SWS during the dark phase of the 24 h sleep-wake cycle (Table 3).

Administration of RP-62203, a naphtosulfam derivative at doses ranging from 0.5 to 4.0 mg/kg p.o. induced a significant and dose-dependent increase in SWS and a reduction of W in the rat. At the 4 mg/kg dose, these effects persisted for a period of 4 h. REMS values were not significantly altered following the range of doses tested (Table 3). Coadministration of RP-62203 (0.5 mg/kg, p.o.) with DOI (0.63 mg/kg, i.p) prevented the decrease of SWS induced by the 5-$HT_{2A/2C}$ receptor agonist [60].

Coenen et al. [61] assessed the effects of sertindole (0.08–0.32-1.28 mg/kg, i.p.) on sleep-wake states during the first 4 h of the dark phase, on two successive days in rats. Sertindole 0.32 mg/kg significantly augmented SWS, whereas REMS was diminished after the whole range of doses during the first recording period of 4 h. The amount of W was decreased only after injection of the 1.28 mg/kg dose during the second recording period of 4 h (Table 3). Moreover, the reduction of REMS was still present on the second day. The 5-$HT_{2A/2C}$ receptor ligand had no significant effects on the frequency spectrum of the background EEG.

In conclusion, systemic and oral administration of non-selective 5-$HT_{2A/2C}$ receptor antagonists has been found to increase of SWS in the rat. Waking and REMS were found to be reduced in most studies.

Sleep Patterns in Normal Subjects and Patients with an Insomnia Disorder Administered Non-Selective 2-HT_{2A} Receptor Antagonists

Idzikowski et al. [62] were the first to document the effects of acute administration of ritanserin 10.0 mg in the morning or the evening on nocturnal sleep in healthy subjects (mean age 33.3 years). Administration of the 5-$HT_{2A/2C}$ receptor ligand in the morning (8.00 a.m.) significantly augmented N3 sleep and reduced N2 sleep. REMS was decreased also when the compound was given in the evening (10:30 p.m.) (Table 4). The authors characterized, in addition, the effects on sleep of repeated morning administrations of ritanserin 10.0 in sleep of healthy male volunteers (mean age 28.4 years). After 2 weeks treatment ritanserin-induced changes in sleep variables remained constant. Thus, tolerance did not develop to the increase in N3 sleep. Subjective ratings including sleep quality, morning vigilance and evening alertness were not affected by the compound [63]. The effects of 1.0, 3.0, 10.0 and 30.0 mg ritanserin administered in the morning (0.800 h), were investigated also in healthy subjects (mean age 32.6 years). A clear dose-response relationship was observed for the compound, with greater doses inducing increased duration of N3 sleep. REMS as a percentage of TST was decreased after the 10.0 mg dose. The 30.0 mg dose induced, in addition, a mild hypnogenic effect and an improvement in subjective sleep quality [64]. As shown by Kamali et al. [65], ritanserin 10.0 mg discontinuation following daily administration for 8 weeks in healthy volunteers aged 18–39 years (median 20 years), was not associated with withdrawal symptoms.

The effects of ritanserin 5.0 mg on sleep variables have been compared with those of ketanserin 20.0 mg and 40.0 mg in subjects with normal sleep (mean age 30 years). Both, ritanserin 5.0 mg and ketanserin 40 mg significantly increased N3 sleep and reduced N2 sleep. REMS was reduced following administration of either compound (Table 4) [66].

Table 4 The effect of non-selective and selective 5-HT_{2A} receptor antagonists and inverse agonists on sleep in healthy subjects and patients with primary or comorbid insomnia

Compound	Function	W	N1	N2	N3	REMS	Reference
Ketanserin	Non-selective antagonist						
	– Healthy subjects	n.a.	n.a.	–	+	–	Sharpley et al. [66]
Ritanserin	Non-selective antagonist						
	– Healthy subjects	n.a.	n.a.	–	+	–	Idzikowski et al. [62]
	– Generalized anxiety disorder	n.s.	–	n.s	+	+	da Roza Davis et al. [72]
	– Chronic primary insomnia	n.s.	n.s.	n.s.	+	n.s.	Ruiz-primo et al. [70]
	– Major depressive disorder	n.s.	n.s.	n.s.	+	n.s.	Staner et al. [74]
Eplivanserin	Selective antagonist						
	– Healthy subjects	n.s.	n.s.	–	+	n.s.	Landolt et al. [68]
Nelotanserin	Inverse agonist						
	– Healthy subjects (postnap insomnia)	n.s.	–	n.s.	+	n.s.	Al-shamma et al. [83]
	– Chronic primary insomnia	n.s.	–	–	+	n.s.	Rosenberg et al. [86]
Pimavanserin	Inverse agonist						
	– Healthy subjects	n.s.	n.s.	–	+	n.s.	Ancoli-Israel et al. [87]

W wakefulness; *N1, N2, N3* non-rapid-eye-movement sleep stages; *REMS* rapid-eye-movement sleep; + significant increase; – significant decrease; *n.s.* non-significant; *n.a.* not available

The effects of the 5-$HT_{2A/2C}$ receptor antagonists ICI-169369 (phenyl-quinoline derivative) and SR-46349B (fluorophenyl-propen derivative) have been studied also on the sleep EEG of subjects with normal sleep. Administration of 100.0 mg ICI-169369 to young subjects (mean age 26.8 years) 90 min before retiring, significantly increased N3 sleep. No other significant differences were observed for the drug on any sleep parameter [67].

In a study by Landolt et al. [68], 1 mg SR-46349B was given to male volunteers 3 h before bedtime. The drug enhanced N3 sleep and reduced N2 sleep. There was no significant difference in the duration of REMS.

Analysis of the EEG power spectra showed an increase of delta activity and a reduction of spindle frequency activity. There was no significant difference between placebo and treatment with respect to the subjective quality of sleep. Of concern, central temperature showed a significant decline in the first 6 h of sleep after SR-46349B administration.

Ritanserin was given also to poor sleepers, patients with chronic primary insomnia and psychiatric patients with a generalized anxiety disorder (GAD), dysthymia disorder or major depression.

Ritanserin 5.0 mg taken for 20 days by middle-aged (mean age 58 years) poor sleepers, caused a significant increase of N3 sleep during the early and the late drug period compared with baseline placebo nights. Concomitantly with the increase in N3 sleep there was a reduction of N2 sleep and of the frequency of awakenings. REMS values were not affected by the drug. During the period of ritanserin intake there was a sustained increase in the quality of sleep [69].

The administration of ritanserin 10.0 mg for 5 days in the mornings to a group of chronic primary insomnia patients (mean age 31.6 years) was followed by an increase of N3 sleep, while values corresponding to N2 sleep and REMS remained unchanged. Ritanserin did not induce significant changes in subjective sleep quality, as compared to placebo (Table 4) [70].

The effects of ritanserin on sleep variables have been characterized also in abstinent alcoholic patients (mean age 42.1 years) with comorbid insomnia. The 5-HT$_{2A/2C}$ receptor antagonist was given at a daily dose of 10.0 mg for 28 days. Ritanserin reduced WASO and increased TST. The latter was related to the occurrence of greater amounts of NREMS. N3 sleep and REMS were not significantly modified. The subjective ratings of the perceived ease of getting to sleep and the duration of sleep showed no consistent changes from baseline. Of note, ritanserin did not impair the patients' psychomotor performance on the morning after administration [71].

da Roza Davis et al. [72] studied the acute effects of ritanserin 5.0 mg on sleep variables in patients (mean age 40.6 years) with GAD and matched healthy controls. Ritanserin produced a significant increase of N3 sleep together with a reduction of N1 sleep and WASO. Unexpectedly, the 5-HT$_{2A/2C}$ receptor ligand augmented REMS (Table 3). The derivative also significantly increased sleep efficiency and subjective sleep quality.

Polysomnographic recordings of patients with a dysthymia disorder according to DSM-III (mean age 36.2 years) who received morning administration of ritanserin 10.0 mg for 4 weeks, showed significant increases in N3 sleep and sleep efficiency. No other variables were modified by the drug [73]. Similarly, acute administration of ritanserin 5.0 mg in major depression patients (melancholic type, DSM-III-R) (mean age 39 years), induced a significant increase in N3 without changing N2 or REMS duration (Table 4) [74].

In conclusion, the 5-HT$_{2A/2C}$ antagonists ketanserin, ritanserin, ICI-169369 and SR-46349B consistently increased N3 sleep in subjects with normal sleep. Additionally, ritanserin was shown to augment N3 in poor sleepers and in patients with chronic primary insomnia, GAD or a mood disorder.

Sleep Patterns in Laboratory Animals Administered Selective 5-HT$_{2A}$ Receptor Antagonists or Inverse Agonists

There are two types of drugs that can inhibit the activity of 5-HT$_{2A}$ receptor, namely the silent 5-HT$_{2A}$ receptor antagonists and the inverse agonists. The former block agonist-induced responses, while the latter attenuate the basal constitutive signaling activity of 5-HT$_{2A}$ receptors.

Currently, there are at least five compounds that behave either as selective 5-HT_{2A} antagonists or inverse agonists.

Kehne et al. [75] evaluated the intrinsic *in vitro* activity of the selective 5-HT_{2A} receptor antagonist **volinanserin** (M100907), a chiral phenethyl piperidine. Volinaserin showed a high degree of selectivity for the 5-HT_{2A} receptor (average Ki = 0.85), with a more than 100-fold separation from a great number of receptors including the serotonin 5-HT_{2C}, dopamine D_2, α1-adrenergic and histamine H1 receptors (Table 2). Moreover, the compound was devoid of inverse agonist intrinsic activity. Volinanserin *in vitro* (IC50 = 0.6 nM) antagonized the 5-HT-stimulated inositol triphosphate accumulation in cells transfected with the rat 5-HT_{2A} receptor. *In vivo*, volinanserin reduced 5-methoxy-*N*,*N*-dimethyltryptamine (ED50 = 0.03 mg/kg, i.p.) or DOI (ED50 = 0.01 mg/kg, s.c.)-induced head twitches in mice, and 5-hydroxytryptophan (ED50 = 0.07 mg/kg, i.p.)-induced head twitches in rats [75, 76].

Injection of volinanserin (2.0–5.0 mg/kg, i.p.) to male mice 2–3 months of age 3 h after the beginning of the light phase, produced a significant dose-dependent increase of SWS during the first 3 h after administration. The compound also caused a reduction of W and REMS time during the same period (Table 3) [34].

Administration of volinanserin (0.1–3.0 mg/kg, i.p.) to rats 6 h after the beginning of the dark phase, significantly reduced the latency to sleep onset and W values relative to vehicle. The compound also induced a significant increase of SWS and EEG delta power. REMS time was not affected by the treatment [77].

Although the mechanisms involved in the volinanserin-induced increase of SWS and reduction of W have not been elucidated, there are preclinical data suggesting that they could be partly related to a decrease in the functional activity of the dopaminergic and serotonergic systems. Accordingly, orally administered volinanserin (ED50 = 0.62 mg/kg) antagonized D-amphetamine-stimulated locomotion in mice, which is indicative of a reduction of mesolimbic dopaminergic activity [75]. Moreover, following the administration of volinanserin (10.0 mg/kg, i.p.) in the rat, using the α-[14C]methyl-L-tryptophan autoradiographic method significant reductions in 5-HT synthesis rates were described in the frontal cortex, sensory-motor cortex, cingulate cortex, caudate-putamen, dorsal thalamus, SNpc, and the medulla raphe nuclei (raphe magnus nucleus and raphe pallidum nucleus). On the other hand, there was no effect in the midbrain raphe nuclei (median raphe nucleus and dorsal raphe nucleus) as a whole [78].

Pruvanserin (EMD 281014), a phenethyl piperazine structurally related to volinanserin, has been found to bind with high affinity to human (IC50 = 0.35 nM) and rat (IC50 = 1.0 nM) 5-HT_{2A} receptors. Moreover, IC50 values of ≥ 1000 nM were described for serotonin 5-HT_{1A}, 5-HT_{2B}, 5-HT_{2C} or 5-HT_7, α1 or α2 adrenergic and dopamine D_1 receptors, while those corresponding to serotonin 5-HT_{1B}, 5-HT_3 or 5-HT_6, dopamine D_2 and histamine H1 receptors amounted to ≥ 10.000 nM (Table 2) [76]. Furthermore, pruvanserin prevented 5-HT-stimulated [35S]guanosine 5′-*O*-3-thiotriphosphate accumulation in hamster ovary cells transfected with the human 5-HT_{2A} receptor (IC50 = 9.3 nM), and antagonized an N-ethoxycarbonyl-

2-ethoxy-1,2-dihydroquinoline-induced decrease of [3H]ketanserin binding in rat frontal cortex (ID50 = 0.4 mg/kg p.o), as well as DOI-induced head-twitch behavior in mice (ID50 = 0.01 mg/kg, s.c.) [76].

The effects of pruvanserin on sleep and W have been determined in the rat during both phases of the light-dark cycle. Injection of pruvanserin (2.5–10.0 mg/kg, i.p.) 2 h after the beginning of the light phase, significantly increased SWS during the second 2 h of recording and reduced REMS during the first 2 h period (Table 3) [79]. REMS time was decreased also after the 10 mg/kg dose during the second 2 h period. REMS latency was augmented after the whole range of doses, while the number of REMS periods was diminished during the first and second 2 h following dosing. Administration of pruvanserin 2 h after the beginning of the dark period gave rise to a significant increase in SWS during the second 2 h of recording. Compared with the control vehicle, the 5-HT_{2A} receptor antagonist did not significantly modify the time spent by the rats in W, LS or REMS [79].

Eplivanserin (SR 46349B) is a propenone ether derivative with potent 5-HT_{2A} receptor blocking properties and a relatively long duration of action. It has high affinity (pIC50 = 1.30) for 5-HT_{2A} receptor, moderate affinity for 5-HT_{2C} receptor and low affinity for 5-HT_{2B} receptor (Table 2). Additionally, eplivanserin has low affinity for the serotonin 5-HT_{1A}, 5-HT_{1B} or 5-HT_{1D}, dopamine D_1 or D_2, α1 or α2 adrenergic, and histamine H1 receptors [80]. *In vivo* (3H)- eplivanserin was found to bind predominantly in mouse brain regions containing 5-HT_{2A} receptors, and this binding was displaced by the non-selective 5-HT_{2A} receptor antagonists ritanserin and ketanserin [81].

The effect of eplivanserin (3.0 and 10.0 mg/kg, i.p.) administered 3 h after the beginning of the light phase on the sleep-wake cycle, was studied in male rats prepared for chronic sleep recordings [82]. The compound did not induce significant changes of values corresponding to W or SWS over the 6 h recording period, nor did it alter the latency to sleep onset. In contrast, both doses of eplivanserin increased the mean duration of SWS episodes, while the mean number of SWS episodes was augmented after injection of the 3.0 mg/kg dose only (Table 3). Since EMG recordings were omitted in the study by Griebel et al. [82], it was considered difficult to extract REMS from the data. However, in a previous study Rinaldi-Carmona et al. reported a significant reduction of REMS in rats treated with eplivanserin (1.0 and 10.0 mg/kg, i.p.).

Nelotanserin (APD125), a phenylpyrazole urea, is a potent inverse agonist of the 5-HT_{2A} receptor. Radioligand binding assays using HEK293 cells (American Type Culture Collection, Rockville, MD) stably expressing human 5-HT_{2A}, 5-HT_{2B} and 5-HT_{2C} receptors, have shown that nelotanserin displays high affinity (Ki = 0.35 nM) for the 5-HT_{2A} receptor, while its affinity to 5-HT_{2C} and 5-HT_{2B} receptor is moderate (Ki = 100 nM) and low (Ki = 2000 nM), respectively (Table 1). The Ki of nelotanserin for rat HEK293 cells stably expressing 5-HT_{2A} receptor is 6-fold greater as compared to human HEK293 cells expressing 5-HT_{2A} receptor. In contrast, its Ki for 5-HT_{2C} and 5-HT_{2B} receptor is similar [83]. Furthermore, nelotanserin blocked 5-HT-induced inositol phosphate accumulation with a potency of 4.7 ± 0.4 nM and 170 ± 39 nM for rat 5-HT_{2A} and 5-HT_{2C} receptor, respectively. In *in vivo* experiments nelotanserin (1.0–10.0 mg/kg, p.o.) dose-dependently prevented a DOI-induced (1.0 mg/kg, p.o.) decrease of rearing in rats [83]. The acute effects of nelotanserin on sleep variables have been tested in rats where the compound was orally administered at doses of 1.0, 3.0 and 10.0 mg/kg, 2 h before the beginning of the light phase. Nelotanserin induced a significant increase of SWS and EEG delta power, while W and REMS were reduced (Table 3). Sleep onset latency remained unchanged. In addition, nelotanserin increased SWS consolidation, as judged by the reductions in bout number and increases in bout duration for the 3.0 and 10.0 mg/kg groups. Subchronic administration of nelotanserin (10.0 mg/kg, p.o.) 2 h before lights on for 5 days, also induced an increase in SWS consolidation and of EEG delta power. Neither tolerance with repeated dosing nor rebound after dosing cessation was detected in any of the parameters quantified [83, 84].

Pimavanserin (ACP-103) is a 5-HT_{2A} inverse agonist that utilizes a urea core to bridge hydrophobic and quaternary amine moieties. Receptor selection and amplification technology allowed to determine that pimavanserin is a potent inverse agonist at human 5-HT_{2A} receptors located in NIH-3 T3 cells (pIC50 = 8.73). The inverse agonist potency of the compound is significantly less for the 5-HT_{2C} receptor (pIC50 = 7.04), and absent for the 5-HT_{2B} receptor. Radioligand binding studies in whole human cells have shown that pimavanserin displays high affinity for the 5-HT_{2A} (pKi = 9.7) and 5-HT_{2C} (pKi = 8.0) receptor. On the other hand, no affinity for the human dopamine D_2 receptor could be detected (Table 2) [85]. The *in vivo* activity of pimavanserin has been evaluated, in a number of animal models. The inverse agonist (3 mg/kg, p.o.) attenuated DOI (2.5 mg/kg, i.p.)-induced head twitches in rats. In addition, pimavanserin (1.0–10.0 mg/kg, s.c.) restored DOI (0.5 mg/kg, s.c.)-disrupted prepulse inhibition of the acoustic startle response in rats [85].

Pimavanserin was reported to increase SWS in rats, although no data on dose, route of administration, time of injection nor number and mean duration of SWS bouts were provided by the authors (unpublished observations).

It can be concluded that with the exception of eplivanserin all other compounds reviewed here significantly increased SWS in laboratory animals. Additionally, W was reduced following the administration of volinanserin and nelotanserin, whereas REMS showed a significant decrease after injection of volinanserin, pruvanserin, nelotanserin and eplivanserin. With respect to pimavanserin, values corresponding to W and REMS were not available. Differences in strains of rats, route of drug administration or concentration, and the use of different approaches to analyze data could possibly account for the different effects provoked by the compounds on SWS, W and REMS.

Sleep Patterns in Normal Subjects and Patients with an Insomnia Disorder Following Administration of Selective 5-HT_{2A} Receptor Antagonists or Inverse Agonists

To date four research articles have been published on the effects of selective 5-HT_{2A} receptor antagonists and inverse agonists on sleep variables in healthy volunteers and/or patients with an insomnia disorder. They refer exclusively to eplivanserin, nelotanserin and pimavanserin.

Landolt et al. [68] investigated the effect of **eplivanserin** (1.0 mg, p.o.) on sleep EEG and power spectra in 10 healthy men with a mean age of 22 ± 0.3 years. The study protocol consisted of 2 sessions of 2 consecutive nights. On each session, an adaptation night was followed by an experimental night. Eplivanserin or placebo was administered 3 h prior to the beginning of the recording session, according to a randomized, double-blind, cross-over design. Fifteen minutes after the end of the recording session, subjective sleep quality (estimated sleep latency, perceived number of awakenings, estimated wake duration after sleep onset, sound vs. fragmented sleep, and superficial vs. deep sleep), and self-rated state (recuperated vs. tired, bad mood vs. good mood, lack of energy vs. full of energy, tense vs. relaxed, and unable to concentrate vs. able to concentrate) were assessed by a questionnaire. Eplivanserin induced a significant increase of N3 sleep while N2 sleep was reduced. No significant differences between placebo and the 5-HT_{2A} receptor antagonist were observed for SOL, REMS latency, TST, SE and REMS in min (Table 4). However, when REMS was expressed as a percentage of TST, there was a significant decrease. NREMS power within 0.75–4.5 Hz was augmented, while that corresponding to spindle frequency activity (12.5–15 Hz) was reduced.

Subjective sleep quality was not affected by the compound.

Al-Shamma et al. [83] assessed the effects of **nelotanserin** (10.0, 20.0 and 40.0 mg, p.o.) in a postnap insomnia model in healthy subjects aged 18–45 years. The compound was administered at 10.30 p.m. and polysomnographic recording were carried out for 8 h. A significant increase of N3 sleep and reduction of N1 sleep was observed 2–4 h after administration of the 40.0 mg dose (Table 4). Sleep onset latency and TST were not affected by the treatment. Of note, the three doses of the drug significantly reduced the number of awakenings and of bouts of sleep, whereas the duration of bouts of sleep was augmented. Next morning effects on psychomotor skills and memory were minimal and had no functional consequences.

The effects of nelotanserin have been evaluated also in patients with a diagnosis of chronic primary insomnia according to DSM-IV-TR [86]. The double-blind, placebo-controlled, 3-way crossover study included 147 patients with a mean age of 45.1 ± 11.8 years who received nelotanserin (10.0 and 40.0 mg, p.o.) for 7 days with at least 7 days of washout period between treatments. Polysomnographic recordings were performed at the initial 2 screening nights and at nights 1/2 and 6/7 of each treatment period. Both doses of nelotanserin induced a statistically significant increase of N3 sleep at nights 6/7, whereas values corresponding to N1 and N2 sleep were diminished (Table 4). In addition, TST and SE were significantly increased following administration of nelotanserin 40 mg at nights 1/2. No decrease in SOL could be observed at either nelotanserin dose. On the other hand, the number of arousals, number of awakenings and WASO were reduced by both doses of the inverse agonist at both time points. Reduction in WASO values occurred between hours 3 and 6. The subjective improvements in sleep were generally consistent with those from polysomnographic parameters. Adverse events during nelotanserin administration were mild and predominantly included somnolence and fatigue. According to Rosenberg et al. [86], nelotanserin administration to patients with chronic primary insomnia resulted in a significant improvement in sleep maintenance and sleep consolidation. Moreover, it was not associated with next-morning cognitive nor psychomotor impairment.

A study has been published by Ancoli-Israel et al. [87] in which the effects of **pimavanserin** on sleep were characterized in healthy adult volunteers. This was a randomized, placebo-controlled, double-blind study that included 45 subjects with a mean age of 51.8 ± 6.9 years. Pimavanserin (1, 2.5, 5.0 or 20.0 mg) or placebo was administered once daily in the morning, for 13 consecutive days. The morning administration of the compound was supported by its long Tmax (6 h) and t½ (55 h). Each of the 5 treatment arms included 9 subjects, and polysomnographic recordings were carried out on nights 1 and 13. The 2.5, 5.0 and 20.0 mg doses significantly increased N3 sleep and reduced N2 sleep during day 1 (Table 4). A similar effect was observed for the 5.0 and 20.0 mg doses on day 13. Thus, the effect was not decreased with repeated administration. The number of awakenings was significantly decreased by the whole range of doses on day 1. In contrast, SOL, REMS latency and duration, WASO, TST, number of stage shifts and early morning wake were not affected by pimavanserin. Concerning spectral power density parameters, pimavanserin significantly increased slow delta (0.5–1 Hz), fast delta (1.5–3-5 Hz) and theta (4–7.5 Hz) activities, and reduced spindle (11.5–15 Hz) frequency during NREMS. In addition, beta1 (13–21.5 Hz) activity was diminished during REMS. Daytime functioning, as judged from results obtained with a Continuous Performance Test, was not impaired. The most frequent adverse events included headache and gastrointestinal disorders that were mild to moderate in nature. On the basis of their findings, the authors concluded that additional studies evaluating the effect of pimavanserin on sleep variables in patients with an insomnia disorder are warranted.

In can be concluded that irrespective of their mechanism of action, eplivanserin, nelotanserin and pimavanserin significantly increase N3 sleep in subjects with normal sleep. A similar effect was decribed when nelotanserin was given to patients with chronic primary insomnia.

Concluding Remarks

Compared to healthy subjects, a considerable number of patients with an insomnia disorder show a significant reduction in N3 sleep and REMS. Compounds approved for the treatment of an insomnia disorder are effective in treating the sleep initiation difficulties and/or in supporting sleep maintenance. However, following their administration N3 sleep does not revert to normal levels and indeed can be even further suppressed in a considerable number of patients. This is an important issue, because N3 sleep reduction, particularly in patients with a chronic insomnia disorder or in elderly patients, could contribute to the deterioration of their day-to-day waking performance due to cognitive and memory deficits. During the search for compounds that can increase N3 sleep, our attention was directed to the 5-HT_{2A} receptor selective antagonists and inverse agonists. There are polysomnographic data showing that isolated administration of either eplivanserin, pimavanserin or nelotanserin to subjects with normal sleep significantly increases the duration of N3 sleep. A similar outcome was described when patients with a chronic insomnia disorder were treated with nelotanserin. Of note, there is preclinical evidence indicating that the coadministration of small doses of eplivanserin and zopidem significantly increases SWS in the rat, in addition to other effects [82]. Thus, this finding supports the conclusion that the association of a compound that inhibits the activity of 5-HT_{2A} receptor and a hypnotic drug could be a valid alternative for normalizing N3 sleep values in patients with an insomnia disorder.

References

1. Dement W, Kleitman N (1957) Cyclic variations in EEG during sleep and their relation to eye movements, body motility and dreaming. Electroencephalogr Clin Neurophysiol 9:673–690
2. Rechtschaffen A, Kales A (1968) A manual of standardized terminology, techniques and scoring system for sleep stages in human subjects. National Institute of Health Publ, Washington, DC. Government Printing Office 204
3. Adam K, Oswald I (1983) Protein synthesis, bodily renewal and the sleep-wake cycle. Clin Sci (Lond) 65:561–567
4. Attwell D, Laughlin SB (2001) An energy budget for signaling in the grey matter of the brain. J Cereb Blood Flow Metab 21:1133–1145
5. Bellesi M, Riedner BA, Garcia-Molina GN et al (2014) Enhancement of sleep slow waves: underlying mechanisms and practical consequences. Front Syst Neurosci 8:208
6. Whitlock JR, Heynen AJ, Shuler MG (2006) Learning induces long-term potentiation in the hippocampus. Science 313:1093–1097
7. Xie L, Kang H, Xu Q et al (2013) Sleep drives metabolic clearance from the adult brain. Science 342:373–377
8. Achermann P, Borbely AA (2003) Mathematical models of sleep regulation. Front Biosci 8:s683–s693
9. Monti JM, Orellana C, Boussard M et al (1990) 5-HT receptor agonists 1-(2,5-methoxy-4-iodophenyl)-2-aminopro- pane (DOI) and 8-OH-DPAT increase wakefulness in the rat. Biogen Amines 7:145–151

10. ICSD 2 (2005) International Classification of Sleep Disorders: Diagnostic and Coding Manual, 2nd edn. American Academy of Sleep Medicine, Westchester, IL
11. Kales A, Kales JD (1984) Evaluation and treatment of insomnia. Oxford University Press, New York
12. Monti JM (1981) Sleep laboratory and clinical studies of the effects of triazolam, flunitrazepam and flurazepam in insomniac patients. Methods Find Exp Clin Pharmacol 3:303–326
13. Greenblatt DJ (1991) Benzodiazepine hypnotics: sorting the pharmacokinetic facts. J Clin Psychiatry 52(Suppl):4–10
14. Monti JM, Pandi-Perumal SR, Langer SZ (2008) Zolpidem: its use in the treatment of sleep disorders. In: Pandi-Perumal SR, Verster JC, Monti JM, Langer SZ (eds) Sleep disorders—diagnosis and therapeutics. Informa Healthcare, London, pp 295–323
15. Krystal AD, Erman M, Zammit G et al (2008) Long-term efficacy and safety of zolpidem extended-release 12.5 mg, administered 3 to 7 nights per week for 24 weeks, in patients with chronic primary insomnia: a 6-month, randomized, double-blind, placebo-controlled, parallel-group, multicenter study. Sleep 31:79–90
16. Monti JM, Spence DW, Pandi-Perumal SR et al (2009) Pharmacotherapy of insomnia. Focus on zolpidem extended release. Clin Med Therap 1:123–140
17. Dolder CR, Nelson MH (2008) Hypnosedative-induced complex behaviors: incidence, mechanisms and management. CNS Drugs 22:1021–1036
18. Roehrs TA, Randall S, Harris E et al (2012) Twelve months of nightly zolpidem does not lead to rebound insomnia or withdrawal symptoms: a prospective placebo-controlled study. J Psychopharmacol 26:1088–1095
19. McCall WV, Erman M, Krystal AD et al (2006) A polysomnographic study of eszopiclone in elderly patients with insomnia. Curr Med Res Opin 22:1633–1642
20. Scharf M (1999) A new option for insomnia. Health News 5:4
21. Walsh JK, Vogel GW, Scharf M et al (2000) A five week, polysomnographic assessment of zaleplon 10 mg for the treatment of primary insomnia. Sleep Med 1:41–49
22. Zammit G, Erman M, Wang-Weigand S et al (2007) Evaluation of the efficacy and safety of ramelteon in subjects with chronic insomnia. J Clin Sleep Med 3:495–504
23. Krystal AD, Lankford A, Durrence HH (2011) Efficacy and safety of doxepin 3 mg and 6 mg in a 35-day sleep laboratory trial in adults with chronic primary insomnia. Sleep 34:1433–1434
24. Patel KV, Aspesi AV, Evoy KE (2015) Suvorexant: a dual orexin receptor antagonist for the treatment of sleep onset and sleep maintenance insomnia. Ann Pharmacother 49:477–483
25. Berridge CW, Schmeichel BE, España RA (2012) Noradrenergic modulation of wakefulness/arousal. Sleep Med Rev 16:187–197
26. Brown RE, Basheer R, McKenna JT et al (2012) Control of sleep and wakefulness. Physiol Rev 92:1087–1187
27. Monti JM (2011) Serotonin control of sleep-wake behavior. Sleep Med Rev 15:269–281
28. Thakkar MM (2011) Histamine in the regulation of wakefulness. Sleep Med Rev 15:65–74
29. Szymusiak R, Gvilia I, McGinty D (2007) Hypothalamic control of sleep. Sleep Med 8:291–301
30. Reinoso-Suárez F, de Andrés I, Rodrigo-Angulo ML et al (2001) Brain structures and mechanisms involved in the generation of REM sleep. Sleep Med Rev 5:63–77
31. Monti JM, Torterolo P, Lagos P (2013) Melanin-concentrating hormone control of sleep-wake behavior. Sleep Med Rev 17:293–298
32. Hannon J, Hoyer D (2008) Molecular biology of 5-HT receptors. In: Monti JM, Pandi-Perumal SR, Jacobs BL, Nutt DJ (eds) Serotonin and sleep: molecular, functional and clinical aspects. Birkäuser Verlag, Basel, pp 155–182
33. Cornea-Hébert V, Riad M, Wu C (1999) Cellular and subcellular distribution of the serotonin 5-HT2A receptor in the central nervous system of adult rat. J Comp Neurol 409:187–209
34. Popa D, Lena C, Favre V et al (2005) Contribution of 5-HT2 receptor subtypes to sleep-wakefulness and respiratory control, and functional adaptations in knock-out mice lacking 5-HT2A receptors. J Neurosci 25:11231–11238

35. Adrien J (2008) Sleep and waking in mutant mice that do not express various proteins involved in serotonergic neurotransmission such as the serotonergic transporter, monoamine oxidase A, and the 5-HT1A, 5-HT1B, 5-HT2A, 5-HT2C and 5-HT7 receptors. In: Monti JM, Pandi-Perumal SR, Jacobs BL, Nutt DJ (eds) Serotonin and sleep: molecular, functional and clinical aspects. Birkäuser Verlag, Basel, pp 457–475
36. Done CJ, Sharp T (1994) Biochemical evidence for the regulation of central noradrenergic activity by 5-HT1A and 5-HT2 receptors: microdyalisis studies in the awake and anaesthetized rat. Neuropharmacology 33:411–421
37. Schmidt CJ, Fadayel GM (1995) The selective 5-HT2A receptor antagonist MDL 100,907, increases dopamine efflux in the prefrontal cortex of the rat. Eur J Pharmacol 273:273–279
38. Quesseveur G, Repérant C, David DJ, Gardier AM, Sanchez C, Guiard BP (2013) 5-HT2A receptor inactivation potentiates the acute antidepressant-like activity of escitalopram: involvement of the noradrenergic system. Exp Brain Res 226:285–295
39. Fink KB, Göthert M (2007) 5-HT receptor regulation of neurotransmitter release. Pharmacol Rev 59:360–417
40. Glennon RA, Titeler M, McKenney JD (1984) Evidence for 5-HT2 involvement in the mechanism of action of hallucinogenic agents. Life Sci 35:2505–2511
41. Shannon M, Battaglia G, Glennon RA et al (1984) 5-HT1 and 5-HT2 binding properties of derivatives of the hallucinogen 1-(2,5-dimethoxyphenyl)-2-aminopropane (2,5-DMA). Eur J Pharmacol 102:23
42. Dugovic C, Wauquier A, Leysen JE et al (1989) Functional role of 5-HT2 receptors in the regulation of sleep and wakefulness in the rat. Psychopharmacology 97:436–442
43. Jäkälä P, Sirviö J, Koivisto E et al (1995) Modulation of rat neocortical high-voltage spindle activity by 5-HT1/5-HT2 receptor subtype specific drugs. Eur J Pharmacol 282:39–55
44. Monti JM, Jantos H (2006) Effects of the serotonin 5-HT2A/2C receptor agonist DOI and of the selective 5-HT2A or 5-HT2C receptor antagonists EMD 281014 and SB-243213, respectively, on sleep and waking in the rat. Eur J Pharmacol 553:163–170
45. Boothman LJ, Allers KA, Rasmussen K et al (2003) Evidence that central 5-HT2A and 5-HT2B/C receptors regulate 5-HT cell firing in the dorsal raphe nucleus of the anaesthetised rat. Br J Pharmacol 139:998–1004
46. Garratt JC, Kidd EJ, Wright IK et al (1984) Inhibition of 5-hydroxytryptamine neuronal activity by the 5-HT agonist. Eur J Pharmacol 199:349–355
47. Liu R, Jolas T, Aghajanian G (2000) Serotonin 5-HT2 receptors activate local GABA inhibitory inputs to serotonergic neurons of the dorsal raphe nucleus. Brain Res 873:34–45
48. Boothman LJ, Sharp T (2005) A role for midbrain raphe gamma aminobutyric acid neurons in 5-hydroxytryptamine feedback control. Neuroreport 16:891–896
49. Ford B, Holmes CJ, Mainville L et al (1995) GABAergic neurons in the rat pontomesencephalic tegmentum: codistribution with cholinergic and other tegmental neurons projecting to the posterior lateral hypothalamus. J Comp Neurol 363:177–196
50. Vasudeva RK, Waterhouse BD (2014) Cellular profile of the dorsal raphe lateral wing subregion: relationship to the lateral dorsal tegmental nucleus. J Chem Neuroanat 58:15–23
51. Amici R, Sanford LD, Kearney K et al (2004) A serotonergic (5-HT2) receptor mechanism in the laterodorsal tegmental nucleus participates in regulating the pattern of rapid-eye-movement occurrence in the rat. Brain Res 996:9–18
52. Rodrigo-Angulo ML, Heredero S, Rodríguez-Veiga E et al (2008) GABAergic and non-GABAergic, thalamic, hypothalamic and basal forebrain projections to the oral pontine reticular nucleus: their implication in REM sleep regulation. Brain Res 1210:116–125
53. Vertes RP, Kocsis B (1994) Projections of the dorsal raphe nucleus to the brainstem: PHA-L analysis of the rat. J Comp Neurol 340:11–26
54. Wang HL, Morales M (2009) Pedunculopontine and laterodorsal tegmental nuclei contain distinct populations of cholinergic, glutamatergic and GABAergic neurons in the rat. Eur J Neurosci 29:340–358
55. Kirov R, Moyanova S (1998) Age-dependent effect of ketanserin on the sleep-waking phases in rats. Int J Neurosci 93:257–264

56. Moyanova S, Kortenska L, Kirov R (1998) High-voltage electroencephalogram spindles in rats, aging and 5-HT2 antagonism. Brain Res 786:55–63
57. Silhol S, Glin L, Gottesmann C (1991) Study of the 5-HT2 antagonist ritanserin on sleep-waking cycle in the rat. Physiol Behav 41:241–243
58. Kantor S, Jakus R, Bodizs R et al (2002) Acute and long-term effects of the 5-HT2 receptor antagonist ritanserin on EEG power spectra, motor activity and sleep: changes at the light-dark phase shift. Brain Res 943:105–111
59. Tortella FC, Echevarria E, Pastel RH et al (1989) Suppressant effects of selective 5-HT2 antagonists on rapid eye movement sleep in rats. Brain Res 485:294–300
60. Stutzmann JM, Eon B, Lucas M (1992) RP 62203, a 5-hydroxytryptamine2 antagonist, enhances deep NREM sleep in rats. Sleep 15:119–124
61. Coenen AML, Ates N, Skarsfeldt T et al (1995) Effects of sertindole on sleep-wake states, electroencephalogram, behavioral patterns, and epileptic activity of rats. Pharmacol Biochem Behav 51:353–357
62. Idzikowski C, Mills FJ, Glennard R (1986) 5-Hydroxytryptamine-2-antagonist ritanserin increases human slow wave sleep. Brain Res 378:164–168
63. Idzikowski C, Cowen PJ, Mills FJ (1987) The effect of chronic ritanserin treatment on sleep and on the neuroendocrine response to L-tryptophan. Psychopharmacology 93:416–420
64. Idzikowski C, Mills FJ, James RJ (1991) A dose-response study examining the effects of ritanserin on human slow wave sleep. Br J Pharmacol 31:193–196
65. Kamali F, Stanfield SC, Ashton CH et al (1992) Absence of withdrawal effects of ritanserin following chronic dosing in healthy volunteers. Psychopharmacology 108:213–217
66. Sharpley AL, Elliott JM, Attenburrow MJ (1994) Slow wave sleep in humans: role of 5-HT2A and 5-HT2C receptors. Neuropharmacology 33:467–471
67. Sharpley AL, Solomon RA, Fernando AI (1990) Dose-related effects of selective 5-HT2 receptor antagonists on slow wave sleep in humans. Psychopharmacology 101:568–569
68. Landolt HP, Viola M, Burgess HJ et al (1999) Serotonin-2 receptors and human sleep: effect of a selective antagonist on EEG power spectra. Neuropsychopharmacology 21:455–466
69. Adam K, Oswald I (1989) Effects of repeated ritanserin on middle-aged poor sleepers. Psychopharmacology 99:219–221
70. Ruiz-Primo E, Haro R, Valencia M (1989) Polysomnographic effects of ritanserin in insomniacs. A crossed double-blind controlled study. Sleep Res 18:72
71. Monti JM, Alterwain P, Estévez F et al (1993) The effects of ritanserin on mood and sleep in abstinent alcoholic patients. Sleep 16:647–654
72. da Roza Davis JM, Sharpley AL, Cowen PJ (1992) Slow wave sleep and 5-HT2 receptor sensitivity in generalized anxiety disorder. Psychopharmacology 108:387–389
73. Paiva T, Arriaga F, Wauquier A et al (1988) Effects of ritanserin on sleep disturbances of dysthymic patients. Psychopharmacology 96:395–399
74. Staner L, Kempenaers C, Simonnet MP (1992) 5-HT2 receptor antagonism and slow-wave sleep in major depression. Acta Psychiatr Scand 86:133–137
75. Kehne JH, Baron BM, Carr AA et al (1996) Preclinical characterization of the potential of the putative atypical antipsychotic MDL 100,907 as a potent 5-HT2A antagonist with a favorable CNS safety profile. J Pharmacol Exp Ther 277:968–981
76. Bartoszyk GD, van Amsterdam C, Böttcher H (2003) EMD 281014, a new selective serotonin 5-HT2A receptor antagonist. Eur J Pharmacol 473:229–230
77. Morairty SR, Hedley L, Flores J et al (2008) Selective 5-HT2A and 5-HT6 receptor antagonists promote sleep in rats. Sleep 31:34–44
78. Hasegawa S, Fikre-Merid M, Diksic M (2012) 5-HT2A receptor antagonist M100907 reduces serotonin synthesis: an autoradiographic study. Brain Res Bull 87:44–49
79. Monti JM, Jantos H (2011) Effects of the 5-HT6 receptor antagonists SB-399885 and RO-4368554 and of the 5-HT2A receptor antagonist EMD 281014 on sleep and wakefulness in the rat during both phases of the light-dark cycle. Behav Brain Res 216:381–388

80. Rinaldi-Carmona M, Congy C, Santucci V et al (1992) Biochemical and pharmacological properties of SR 46349B, a new potent and selective 5-hydroxytryptamine2 receptor antagonist. J Pharmacol Exp Ther 262:759–768
81. Rinaldi-Carmona M, Congy C, Pointeau P et al (1994) Identification of binding sites for SR 463449B, a 5-hydroxytryptamine2 receptor antagonist, in rodent brain. Life Sci 54:119–127
82. Griebel G, Beeské S, Jacquet A et al (2013) Further evidence for the sleep-promoting effects of 5-HT2A receptor antagonists and demonstration of synergistic effects with the hypnotic zolpidem in rats. Neuropharmacology 70:19–26
83. Al-Shamma HA, Anderson C, Chuang E et al (2010) Nelotanserin, a novel selective human 5-hydroxytryptamine2A inverse agonist for the treatment of insomnia. J Pharmacol Exp Ther 322:281–290
84. Teegarden BR, Li H, Jayakumar H, Strah-Pleynet S, Dosa PI, Selaya SD et al (2010) Discovery of 1-[3-(4-Bromo-2-methyl-2H-pyrazol-3-yl)-4-methoxyphenyl]-3-(2,4-difluorophenyl) urea (Nelotanserin) and related 5-hydroxytryptamine2A inverse agonists for the treatment of insomnia. J Med Chem 53:1923–1936
85. Vanover KE, Weiner DM, Makhay M et al (2006) Pharmacological and behavioral profile of N-(4-fluorophenylmethyl)-N-(1-methylpiperidin-4-yl)-N'-4-(2-methylpropylo-xymethyl) carbamide (2R,3R)-dihydroxybu- tanedionate (2,1) (ACP-103), a novel 5-hydroxytryotamine(2A) receptor inverse agonist. L Pharmacol Exp Ther 317:910–918
86. Rosenberg R, Seiden DJ, Hull SG et al (2008) APD125, a selective serotonin 5-HT2A receptor inverse agonist, significantly improves sleep maintenance in primary insomnia. Sleep 31:1663–1671
87. Ancoli-Israel S, Vanover KE, Weiner DM et al (2011) Pimavanserin tartrate, a 5-HT2A receptor inverse agonist, increases slow wave sleep as measured by polysomnography in healthy adult volunteers. Sleep Med 12:134–141
88. Barnes NM, Sharp T (1999) A review of central 5-HT receptors and their functions. Neuropharmacology 38:1083–1152
89. Leysen JE (2004) 5-HT2 receptors. Curr Drug Targets CNS Neurol Disord 3:11–26
90. Leysen JE, Gommeren W, Schotte A (1996) Serotonin receptor subtypes: possible roles and implications in antipsychotic drug action. In: Kane JM, Möller H-J, Awouters F (eds) Serotonin and antipsychotic treatment. Marcel Dekker, New York, pp 51–75

5-HT_{2A} Receptors and Pain

Christine Courteix, Amandine Dupuis, Pierre-Yves Martin, and Benoit Sion

Abstract Serotonin (5-hydroxytryptamine, 5-HT) is a key modulator of spinal nociceptive transmission. Among 5-HT receptors, the 5-HT_{2A} subtype plays a critical role in the modulation of nociceptive information. Both pro- and antinociceptive effects of 5-HT_{2A} receptor activation have been reported but converging evidence indicates an excitatory role for peripheral 5-HT_{2A} receptors on pain transmission in acute, sub-chronic and chronic pain conditions. The central effects of 5-HT_{2A} agonists which produce either anti-hyperalgesic or anti-allodynic effect seem to depend on the pathophysiology of pain. Neverthless, some data indicate that 5-HT acting drugs such as selective serotonin reuptake inhibitor (SSRI) antidepressants involve the 5-HT_{2A} receptor to produce analgesia and that restoring 5-HT_{2A} receptor functionality may contribute to enhance the analgesic efficacy of SSRI in metabolic and traumatic neuropathic pain.

Keywords Serotonin • 5-HT_{2A} receptor • Chronic pain • Acute pain • PDZ-proteins • Antidepressants

Introduction

Pain is defined as "an unpleasant sensory and emotional experience associated with actual or potential tissue damage, or described in terms of such damage" [1]. It is usual to distinguish acute pain, a normal sensation that alerts to possible body injury, from chronic (or persistent) pain. Chronic pain is pain that persists or recurs for several months despite adequate medication or treatment, and its pathophysiology is much more complex than that of acute pain. Chronic pain involves pathological alterations that occur in the peripheral and central nervous systems. It is characterized by an increase in the excitatory synaptic transmission from primary afferents to

C. Courteix (✉) • A. Dupuis • P.-Y. Martin • B. Sion
INSERM U1107, NEURO-DOL, Clermont-Ferrand, France

Laboratoire de Physiologie, UFR Pharmacie, Université Clermont Auvergne, Clermont-Ferrand, France
e-mail: christine.courteix@uca.fr

B.P. Guiard, G. Di Giovanni (eds.), *5-HT_{2A} Receptors in the Central Nervous System*, The Receptors 32, https://doi.org/10.1007/978-3-319-70474-6_14

dorsal horn neurons. This increase is mediated by glutamate, AMPA and NMDA receptors and results in persistent plasticity in the dorsal nociceptive network, i.e. central sensitization. In addition, synaptic inhibition of transmission mediated by inhibitory GABAergic and glycinergic interneurons, which is excited both by primary afferent inputs and by inputs from descending neurons (including serotoninergic neurons), is reduced. A series of changes occurring in microglia involving BDNF/TrkB and ATP/P2X4 receptors also plays a role in persistent pain caused by peripheral nerve injury.

Serotonin (5-hydroxytryptamine, 5-HT) is an important neurotransmitter/ neuromodulator involved in numerous physiological functions (appetite regulation, sleep, thermoregulation, mood, learning) and brain diseases (depression, anxiety, obsessive compulsive disorders, schizophrenia, addiction, autism, Alzheimer). It is also a key neuromodulator of pain transmission acting at both peripheral and central levels and is involved in the pathophysiology of pain disorders [2]. Its predominant inhibitory role in persistent pain has been established using mice lacking the Lmx1b gene, which is critical for differentiation of 5-HT neurons [3]. Lmx1b conditional knock-out mice ($Lmx1b^{f/f/p}$), which lack central 5-HT neurons, exhibit enhanced persistent inflammatory pain response to formalin or capsaicin injection, which is attenuated by intrathecal (i.t.) injection of 5-HT [4]. In the central nervous system (CNS) serotonin-containing neurons are distributed in the brainstem to nuclei labeled caudally to rostrally B1–B9, giving rise to a dense and large innervation of nearly all divisions of the CNS, including the spinal cord. The dorsal horn spinal cord receives serotonergic fibers that originate mainly from the raphe magnus nucleus. These descending serotonergic pathways are specifically implicated in the inhibition of pain transmission, which provides a rationale for the use of antidepressants (either tricyclic antidepressants [TCA] or serotonin-noradrenalin reuptake inhibitors [SNRIs]) in the relief of chronic pain. The analgesic effects of the selective serotonin reuptake inhibitor (SSRI) fluoxetine, which blocks the 5-HT transporter and increases the synaptic levels of 5-HT, are abolished in $Lmx1b^{f/f/p}$ mice, indicating that pain relief by SSRI antidepressants is completely dependent on the central 5-HT system [4]. The effect of 5-HT acting drugs on spinal nociceptive transmission involves at least six subtypes of receptors. The 5-HT_{2A} receptor has been identified as one of the key 5-HT receptors contributing to 5-HT-induced modulation of pain transmission in certain pain conditions.

In this chapter, particular attention will be given to spinal 5-HT_{2A} receptors and the serotonergic pathways that modulate nociceptive information at the spinal level in conditions of acute and chronic pain.

Serotonergic Innervation and 5-HT_{2A} Receptors

The caudal group of serotonergic neurons, which is part of the rostroventral medulla, i.e. the raphe pallidus nucleus (B1), the raphe obscurus nucleus (B2), the raphe magnus nucleus and neurons in the lateral medullary reticular formation (B3),

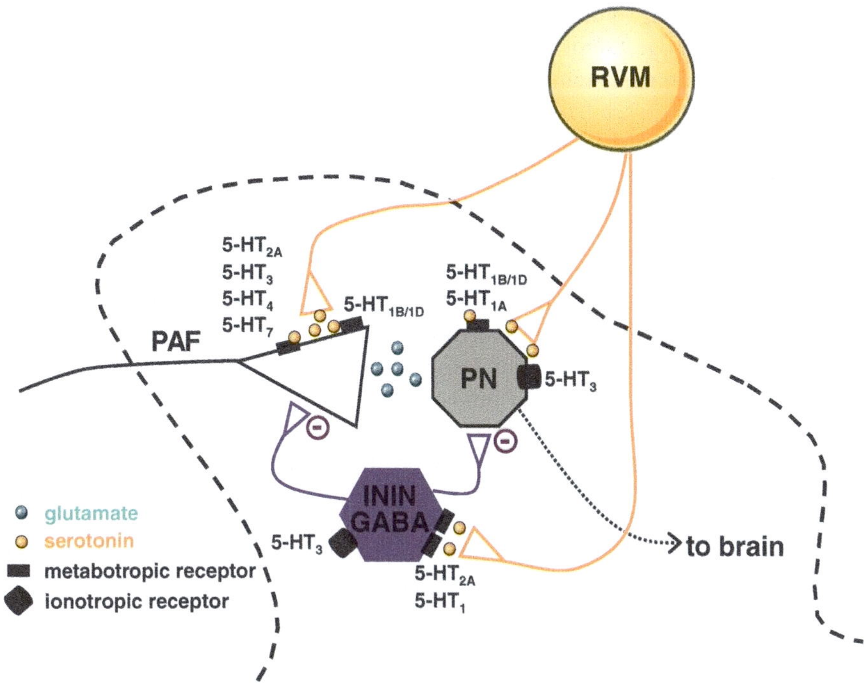

Fig. 1 Opposite effects of serotonin acting on multiple 5-HT receptor subtypes in dorsal horn spinal cord. For example, the activation of peripheral primary afferent fiber (PAF) and inhibitory interneuron (ININ) $5\text{-}HT_{2A}$ receptors may lead to pronociceptive and antinociceptive effects, respectively, whereas the activation of peripheral PAF and ININ $5\text{-}HT_{1A}$ receptors may lead to antinociceptive and pronociceptive effects, respectively. *RVM* rostral ventromedial medulla, *PN*, projecting neurons

provides the majority of 5-HT neurons sending projections into the spinal cord [5]. 5-HT axon terminals project onto all laminae of the gray matter of the spinal cord but particularly onto the superficial laminae (I, II, the substantia gelatinosa) of the dorsal horn. There, 5-HT axon terminals establish contact on neuronal cell bodies (axo-somatic synapses) and dendrites (axo-dendritic synapses) to facilitate or inhibit nociceptive transmission depending on (1) 5-HT receptor subtypes ($5\text{-}HT_1$, $5\text{-}HT_2$, $5\text{-}HT_3$, $5\text{-}HT_4$, $5\text{-}HT_6$ and $5\text{-}HT_7$ receptors) [6–9], (2) their location on terminals of primary afferent fibers (PAF), on local circuit neurons (excitatory or inhibitory interneurons), on nociceptive-thalamic projection neurons (PN), or on descending serotonergic projections (Fig. 1), (3) their signaling pathways and (4) their functionality.

$5\text{-}HT_{2A}$ receptors belong to the superfamily of G-protein-coupled receptors. They have a depolarizing effect on neuronal membranes by increasing intracellular Ca^{2+} (thereby increasing spontaneous excitatory postsynaptic currents) and inositol phosphate concentrations secondary to their coupling to Gq/11 and activation of PLC, PLA_2 and the ERK pathway [10]. The $5\text{-}HT_2$-family receptors have canonical

Type I PDZ-binding motifs (PDZ for postsynaptic density protein [PSD95]), Drosophila disc large tumor suppressor (DlgA), and Zonula occludens-1 protein (zo- 1) at their extreme C-terminus, and interact with multiprotein complexes organized in part around specific PDZ proteins [11–13]. Interaction of 5-HT_2 receptors with accessory proteins affects their signal transduction by influencing their desensitization and coupling to G proteins, and their subcellular compartmentation by influencing localization at the plasma membranes [10]. *In vivo* studies performed in experimental models of chronic neuropathic and inflammatory pain have shown the crucial role of these interactions in pain modulation and pain treatment [14, 15].

Using immunohistochemistry, autoradiography and hybridization techniques, it is possible to reveal distribution patterns of 5-HT_{2A} receptors in the spinal cord, dorsal root ganglia (DRG) and brain [16–18]. Studies have shown that in the dorsal horn spinal cord, 5-HT_{2A} receptor immunoreactivity is greater in the inner part of lamina II (lamina IIi) than in lamina I and in the outer part of lamina II (lamina IIo) [19, 20]. Others observed 5-HT_{2A}-LI neurons in laminae I–III and IV–VI [21]. A large population of 5-HT_{2A} receptors was found in rat lumbar DRG, in small to intermediate-sized neuronal cell bodies (20–35 μm in diameter) [20] and a few in larger cell bodies. This distribution differs from that previously reported by Maeshima [19], who reported intense immunolabeling in large and intermediate-sized neuronal cell bodies. Among nociceptive fibers, those binding isolectin B4 specific for non-peptidergic C-fibers expressed 5-HT_{2A} receptors more frequently than substance P-containing fibers, suggesting that 5-HT_{2A} receptors are also expressed in the non-peptidergic neurons. Interestingly, double immunofluorescent staining of 5-HT_{2A} receptors and GABAergic neurons specifically labeled with glutamate decarboxylase (GAD67) green fluorescent protein (GFP) showed that 6.5–16.5% of GFP positive neurons co-express 5-HT_{2A} receptors in the dorsal horn of mice spinal cord (the highest proportion of double-labeled cells on the whole GFP-positive neurons was found in lamina I) [21], suggesting an effect of 5-HT_{2A} receptors on spinal GABAergic regulation in pain modulation. This role was confirmed later in experimental models of chronic pain (inflammatory and neuropathic) using pharmacological techniques [14, 15].

Pronociceptive and Antinociceptive Roles for 5-HT_{2A} Receptors

Most studies exploring the role of 5-HT_{2A} receptors in pain have looked at the pronociceptive or analgesic effects of 5-HT_{2A} ligands, agonists or antagonists, either peripherally (locally, intraperitoneal (i.p.), per os (p.o.)) or centrally (intracisternally, i.t.) administered to awake or anesthetized rodents (Table 1). Several studies have tested the effects of these agents on central neuronal activity, others on acute pain reaction or chronic pain behavior. This section will review key experiments on the involvement of 5-HT_{2A} receptors in acute and chronic pain processing.

Table 1 An overview of 5-HT_{2A} receptor-mediated effcts in animal models of acute and chronic pain, and proposed mechanisms

Reference(s)	Drug/Route of administration	Pain model	Effect	Proposed mechanisms
Liu et al. [22]	α-methyl-5-HT/ spinal	C-fiber activation	Inhibits WDR response	Excitation of inhibitory interneurons
	Ketanserin/spinal		Inhibits spinal inhibitory effect of 5-HT	
Sasaki et al. [23]	DOI/spinal	Formalin test (paw)	Inhibits pain-related behavior to formalin injection	Spinal release of an inhibitory neurotransmitter, such as GABA, by interneurons
Sasaki et al. [24]	α-methyl-5-HT/ spinal	Formalin test (paw)	Inhibits pain-related behavior to formalin injection	Activation of inhibitory interneurons and ACh release
Okamoto et al. [25]	DOI/trigeminal subnucleus caudalis, spinal	Fiber activation	Inhibits WDR response	Activation of 5HT_{2A}Rs in the Vc/C2 region
	DOI/intracisternal	Formalin injection (masseter muscle)	Inhibits nociceptive behavioral activities	
Kayser et al. [26]	5-$HT_{2A}^{-/-}$ KO mice	Formalin test (paw)	Decrease in nociceptive response of phase 2	Pronociceptive role for peripheral and central 5-HT_{2A}Rs
	M100907/i.p.			Blockade of both peripheral and central 5-HT_{2A}Rs
Sasaki et al. [27]	Sarpogrelate i.p./ locally (paw)	Thermal injury	Decreases primary and scondary hyperalgesia	Blockade of 5-HT_{2A}Rs expressed on peripheral terminals of PAF
Kjorsvik Bertelsen et al. [28]	DOI/spinal	Spinal capsaicin in CARRA-induced inflammatory pain	Facilitates the capsaicin-evoked sP release	Facilitatory action of spinal 5-HT_{2A}Rs on sP release from presynaptic PAF
Kjorsvik et al. [29]	DOI/spinal	Formalin test (paw)	Increases nociceptive response	Release of Glu and sP from PAF at the spinal cord due to the 5-HT_{2A}R activation

(continued)

Table 1 (continued)

Reference(s)	Drug/Route of administration	Pain model	Effect	Proposed mechanisms
Rahman et al. [30]	Ketanserin/spinal	Mechanical and thermal evoked responses	Inhibits WDR response	Spinal 5-HT_{2A}Rs mediate descending facilitation of spinal nociceptive processing
	DOI/spinal	Mechanical and thermal evoked responses	Increases WDR response	
Liu et al. [32]	Ketanserin/spinal	SNL-induced neuropathic pain/ C-fiber activation	Inhibits 5-HT-induced inhibition of WDR response	5-HT_{2A}R activation may cause the release of sP, which in turn causes the release of acetylcholine
Obata et al. [33]	α-methyl-5-HT, DOI/spinal	SNL-induced neuropathic pain	Antiallodynic effect	Activation of 5-HT_{2A}Rs present on central terminals of PAF
	Ketanserin/spinal		Reverses the antillodynic effect of 5-HT_{2A}R agonists	
Sasaki et al. [24]	α-methyl-5-HT/ spinal	CCI-induced neuropathic pain, thermal test	Decreases thermal hyperalgesia	Activation of spinal inhibitory interneurons (ACh, +/−GABA release)
	α-methyl-5-HT/ spinal	CCI-induced neuropathic pain, formalin test (paw)	Decreases phases 1 and 2	
	Ketanserin/spinal		Reverses both effects of the 5-HT_{2A}R agonist	
Song et al. [34]	SCS + α-methyl-5-HT/spinal	PSNL-induced neuropathic pain	Increases antiallodynic and antihyperalgesic effects of SCS	Activation of spinal 5-HT_{2A}Rs and partly via GABAergic interneurons
	SCS + Ketanserin/ spinal	PSNL-induced neuropathic pain	Decreases antiallodynic and antihyperalgesic effects of SCS	
Pichon et al. [14]	α-methyl-5-HT/ spinal	STZ-induced diabetic neuropathic pain	Lack of analgesic effect	Loss of functionality of 5-HT_{2A}Rs
	Fluoxetine + TAT--2A/i.p. + spinal	STZ-induced diabetic neuropathic pain	Enhances SSRI efficacy	Restores spinal 5-HT_{2A}R functionality

Reference(s)	Drug/Route of administration	Pain model	Effect	Proposed mechanisms
Van Steenwinckel et al. [37]	M11,939/spinal	2′,3′-ddC-induced neuropathic pain	Decreases mechanical hypersensitivity	Over-expression and activation of spinal and peripheral $5\text{-}HT_{2A}$Rs
	$5\text{-}HT_{2A}^{-/-}$ KO mice	2′,3′-ddC-induced neuropathic pain	Fails to develop mechanical hypersensitivity	Absence of peripheral and central $5\text{-}HT_{2A}$ Rs
Thibault et al. [38]	M11,939/spinal	VCT-induced neuropathic pain	Decreases mechanical hypersensitivity	Over-expression and activation of central and peripheral $5\text{-}HT_{2A}$Rs
	$5\text{-}HT_{2A}^{-/-}$ KO mice	VCT-induced neuropathic pain	Fails to develop mechanical hypersensitivity	Absence of peripheral and central $5\text{-}HT_{2A}$ Rs
Aira et al. [39]	TCB-2, BW723C86/spinal	SNL-induced neuropathic pain	Increases C-fiber evoked potentials	$5\text{-}HT_{2A}$Rs render silent synapses active. Increase in $5\text{-}HT_{2A}$Rs in ascending neurons. $5\text{-}HT_{2A}$Rs are the mediators of descending facilitatory pathways becoming tonically active after SNL
Aira et al. [40]	M100907/i.p.	SNL-induced neuropathic pain	Decreases C-fiber evoked potentials	$5\text{-}HT_{2A}$Rs mediate late-onset thermal and mechanical allodynia after SNL by facilitating spinal nociceptive transmission. Influence the expression and function of Glu receptor mGluR1
	TCB-2/spinal		Increases C-fiber evoked potentials	
Nitanda et al. [41]	Sarpogrelate, ketanserin/i.p.	CCI-induced neuropathic pain	Decreases mechanical hyperalgesia	Peripheral $5\text{-}HT_{2A}$Rs may have a constitutive activity induced by CCI
Kato et al. [42]	Sarpogrelate/p. os	Lumbar disc herniation-induced pain syndrome	Attenuates pain-related behavior	Blockade of peripheral $5\text{-}HT_{2A}$Rs
Huang et al. [43]	Ketanserin/locally (paw)	CFA-induced inflammatory pain	Inhibits thermal hyperalgesia	Blockade of peripheral $5\text{-}HT_{2A}$Rs

Abbreviations and symbols: *ACh* acetylcholine, *CARRA* carrageenan, *CCI* chronic constriction injury, *CFA* complete Freund's adjuvant, *DOI* 1-[2,5-dimethoxy-4-iodophenyl]-2-aminopropane, *2′,3′-ddC* 2′,3′-dideoxycytidine, *i.p.* intraperitoneal, *KO* knock out, *PAF* primary afferent fibers, *SCS* spinal cord stimulation, *SNL* spinal nerve ligation, *SSRI* selective serotonin re-upatke inhibitor, *sP* substance P, *STZ* steptozocin, *PSNL* partial sciatic nerve ligation, *VCT* vincristine, *WDR* wide dynamic range, *$5\text{-}HT_{2A}R$* $5\text{-}HT_{2A}$ receptor

Acute Pain

Activation of peripheral nociceptors (A-delta or C-fibers) by peripheral stimuli will elicit activity of low-threshold, high-threshold, wide dynamic range (WDR) neurons in the spinal dorsal horn. One study using electrophysiological recording of WDR neurons showed that, under basal conditions, topical application of 5-HT to the spinal cord dose-dependently inhibited the C-fiber responses, which is consistent with the inhibitory effect of 5-HT on spinal nociceptive transmission. Topical administration of a selective 5-HT_{2A} receptor agonist, α-methyl-5-hydroxytryptamine maleate (α-methyl-5-HT), also inhibited C-fiber responses of WDR neurons [22], and the 5-HT_{2A} receptor antagonist ketanserin reversed the inhibitory effect of 5-HT [22]. These observations, which clearly show that spinal 5-HT_{2A} receptors mediate an inhibitory effect of 5-HT on spinal nociceptive transmission, are consistent with previous results obtained in healthy rats subjected to the formalin test, which consists in subcutaneous (s.c.) injection of formalin into the plantar surface of the hindpaw and the quantification of pain-related behavior (i.e. spontaneous flinching movements). In these experiments, the i.t. injection of 5-HT_{2A} receptor agonists (α-methyl-5-HT or 1-[2,5-dimethoxy-4-iodophenyl]-2-aminopropane (DOI)) dose-dependently suppressed the number of flinches in both acute and tonic phases 1 (minutes 1–6) and 2 (minutes 10–60) [23, 24]. Orofacial nociception is also sensitive to descending serotonergic systems via 5-HT_{2A} receptor activation. The biphasic response of deep-nociceptive unit discharges to formalin injection into the masseter muscle is reduced after topical administration of DOI onto the caudal portion of the trigeminal subnucleus caudalis and upper cervical (C1-C2) spinal cord region [25]. Consistent with the lack of tonic activity of the 5-HT system [22], the sensitivity to thermal (radiant heat applied on the plantar surface of the hindpaw, tail immersion and hot-plate) and mechanical (von Frey hair application, tail clip and tail pressure) stimuli remains intact in mutant mice that do not express 5-HT_{2A} receptors compared to wild type mice [26]. In contrast, the nociceptive response to formalin injection (time spent licking or biting the injected paw) is reduced in phase 2 compared to wild-type mice [26]. This result was confirmed by pharmacological experiments with intraperitoneal (i.p.) administration of the selective 5-HT_{2A} receptor antagonist, M100907 in wild-type mice, which suggests a pronociceptive role of 5-HT_{2A} receptors expressed by PAF [26]. This peripheral pronociceptive effect is not surprising considering the membrane depolarizing effect and the intracellular calcium increase induced by 5-HT_{2A} receptor activation, which underlie pain transmission from nociceptors to spinal cord neurons. Hence, the selective peripheral 5-HT_{2A} receptor antagonist, sarpogrelate (which is used in Japan as an antiplatelet agent to treat patients with arteriosclerosis) either i.p.-administered or locally injected into the hindpaw of rats, attenuated primary thermal hyperalgesia induced by mild thermal injury to the hindpaw and mechanical allodynia in sites adjacent to the primary area (secondary allodynia) [27]. More surprising are the facilitatory effects of spinal 5-HT_{2A} receptors on substance P release in animals with and without carrageenan-induced inflammation [28], the pronociceptive effect of i.t.

injection of the 5-HT_{2A} receptor agonist DOI, which enhances painful reactions to formalin injection in rats [29], and the inhibitory effect of spinal ketanserin on evoked responses of WDR dorsal horn neurons to stimulation of the peripheral receptive field [30].

Chronic Pain

Animal models of chronic inflammatory pain and neuropathic pain share a variety of common neuroplastic changes occurring in the peripheral and central nervous system [31]. These changes occur with different time delays depending on the nature of the injury (nerve lesion or tissue inflammation) and even on the type of injury. While trying to shed light on alterations in pain modulation in these animal models seems commonplace, the published data have highlighted the difficulty to assign a clear role to 5-HT_{2A} receptors. In models of neuropathic pain induced by spinal L5 nerve ligation (SNL model) in rats, besides the fact that the content of 5-HT and its metabolite 5-hydroxy-indol acetic acid (5-HIAA) in the dorsal horn of the lumbar spinal cord is decreased and the turnover rate of 5-HT (5-HIAA/5-HT) increased, only the 5-HT_{2A} receptor antagonist blocks 5-HT-induced inhibition to C-fiber responses of dorsal horn WDR neurons, suggesting that 5-HT_{2A} receptors play a major role in mediating the inhibitory effects of 5-HT on the C-responses after SNL [32]. These observations are consistent with previous findings showing that i.t.-administered 5-HT_{2A} receptor agonists α-methyl-5-HT or DOI produce a dose-dependent antiallodynic action in rats with L5 and L6 nerve ligation [33]. Similarly, in the chronic constriction injury (CCI) model, which consists in loose sciatic nerve ligation, i.t.-administered α-methyl-5-HT attenuates thermal hyperalgesia (paw withdrawal response to radiant heat) [24]. In both studies the antiallodynic and antihyperalgesic effects of 5-HT_{2A} receptor agonists result from activation of the 5-HT_{2A} receptors as they are abolished by ketanserin [24, 33]. Furthermore, the pain-relieving effect of spinal cord stimulation on tactile and cold hypersensitivity (assessed by von Frey filament and cold spray, respectively) in rats with partial sciatic nerve ligation is enhanced by α-methyl-5-HT and suppressed by ketanserin [34].

In contrast, we found that α-methyl-5-HT (i.t.) did not exert any antihyperalgesic action against mechanical (paw-pressure test) and thermal (plate preference) noxious stimulations in diabetic hyperalgesic rats. Saturation experiments performed on lumbar spinal cord membranes from healthy and diabetic hyperalgesic rats showed that neither affinity nor receptor density were altered suggesting that the alteration of 5-HT_{2A} receptor-mediated analgesia in the animals did not result from receptor down-regulation but rather from a decrease in receptor responsiveness [14]. The disruption of interactions between the spinal 5-HT_{2A} receptor and associated PDZ proteins by i.t. injection of a cell-penetrating peptidyl mimetic (TAT-2ASCV) of receptor C-terminus significantly inhibited mechanical hyperalgesia (Fig. 2a) and suppressed the spontaneous temperature preference behavior induced

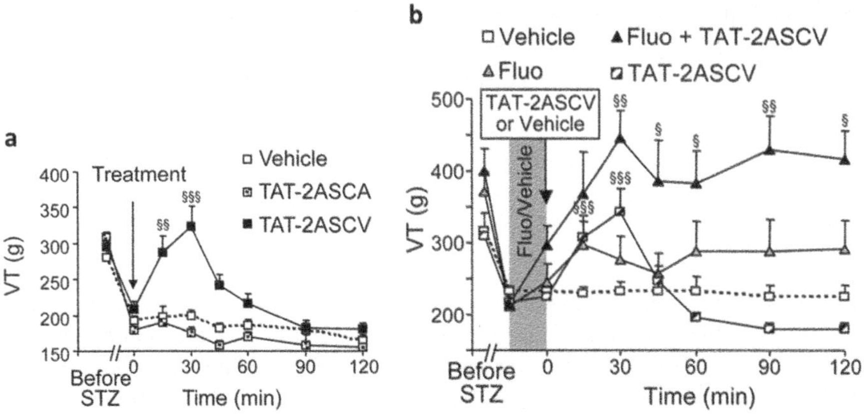

Fig. 2 Disruption of 5-HT_{2A} receptor/PDZ protein interactions with TAT-2ASCV produces antihyperalgesic effects and enhances fluoxetine-induced antihyperalgesia in streptozocin (STZ)-induced diabetic neuropathic rats. (**a**) Intrathecal injection of TAT-2ASCV increased vocalization thresholds (VT) in diabetic hyperalgesic rats, whereas the intrathecal injection of the TAT-2ASCA control peptide did not induce any antihyperalgesic action. (**b**) Administration of fluoxetine (Fluo) combined with an administration of TAT-2ASCV (30 ng/rat, single intrathecal injection) 6 h after the last fluoxetine injection produced a marked antihyperalgesic effect, which was higher and more prolonged than those induced by fluoxetine or the peptide alone. §:$P<0.05$; §§:$P<0.01$; §§§:$P<0.001$ compared with values measured before the treatment [14]

by diabetic neuropathy; disconnection of 5-HT_{2A} receptors from their PDZ partners also strongly enhanced the antihyperalgesic action of fluoxetine in diabetic neuropathic rats (Fig. 2b) [14]. Consistently, thanks to the synthesis of potent non-peptide inhibitors of PSD95-PDZ1 or PDZ2 domains/5-HT_{2A} receptor interaction [35, 36], we confirmed the potential analgesic effect of such agents in traumatic neuropathic pain models (CCI [35, 36] and SNL rats [personal results]).

Using the TAT-2ASCV peptidyl mimetic strategy, we were able to restore 5-HT_{2A} receptor functionality and to reduce mechanical hyperalgesia to paw-pressure in rats with carrageenan induced sub-chronic inflammation and showed that this effect was mediated by activation of the 5-HT_{2A} receptors located on GABAergic interneurons of the spinal cord [15], which is further evidence of the role of GABA neurotransmission in pain modulation.

Conversely, pronociceptive effects resulting from spinal 5-HT_{2A} receptor activation have been described in experimental models of toxic but also traumatic neuropathic pain. In rats treated with 2′,3′-dideoxycytidine (ddC), a nucleoside analogue with reverse transcriptase inhibitory properties used for the treatment of patients infected by HIV, 5-HT_{2A} receptor immunoreactivity is up-regulated in the dorsal lumbar dorsal horn and peripheral nociceptive cells [37]. Since the 5-HT_{2A} receptor antagonist M11,939 spinally injected decreases mechanical hypersensitivity, these receptors play a pronociceptive role in ddC-induced neuropathic pain. Knock-out mice devoid of 5-HT_{2A} receptor gene (5-$HT_{2A}^{-/-}$ mice) failed to develop mechanical hypersensitivity to ddC injection. Similarly, in rats treated with vincristine, an antineoplastic drug, the epidural injection of a $5\text{-}HT_{2A}$ receptor antagonist also decreased mechanical hypersensitivity [38]. Furthermore, as in the ddC model, vincristine-injected $5\text{-}HT_{2A}^{-/-}$ mice failed to develop mechanical hypersensitivity [38]. The authors correlated the over-expression of functional $5\text{-}HT_{2A}$ receptors with hypersensitivity [38] whereas in diabetic rats neither the number nor affinity of $5\text{-}HT_{2A}$ receptors was affected [14].

In the SNL model, spinal hyperexcitation characterized by an increase in C-fiber-evoked potentials after spinal superfusion with a selective $5\text{-}HT_{2A}$ receptor agonist has been reported [39]. Interestingly, in SNL rats, the same authors described the pivotal role of $5\text{-}HT_{2A}$ receptors coupled to NMDA receptor-mediated neurotransmission. They observed that the time course of the $5\text{-}HT_{2A}$ receptor contribution to thermal and mechanical allodynia coincides with that of serotonergic descending facilitatory influences on spinal nociception [40]. In chronic pain, as in acute pain conditions, the proalgesic role of peripheral $5\text{-}HT_{2A}$ receptors is well established. The $5\text{-}HT_{2A}$ receptor contributes to mechanical hyperalgesia consecutive to CCI and the i.p. injection of sarpogrelate and ketanserin reduced pain-related behavior [41]. In a rat model of lumbar disc herniation (LDH), sarpogrelate (p.o.) also attenuated pain-related behavior and suppressed LDH-induced $5\text{-}HT_{2A}$ receptor over- expression in the DRG [42]. Similarly, ketanserin, administered in inflamed paws, inhibited thermal hyperalgesia induced by intraplantar injection of CFA in rats [43].

Conclusion

We report several key findings that emphasize the role of $5\text{-}HT_{2A}$ receptors in acute and chronic pain conditions with specific focus on the spinal cord and peripheral nervous system.

Documented reports suggest an excitatory role for peripheral $5\text{-}HT_{2A}$ receptors on pain transmission in acute, sub-chronic or chronic pain conditions consistent with the molecular mechanisms of second messengers generated that promote PAF depolarization.

The central effects of $5\text{-}HT_{2A}$ agonists producing anti-hyperalgesia or anti-allodynia in animal models of traumatic (CCI, SNL and PSNL) and metabolic (STZ-induced diabetes) neuropathic pain [14, 24, 33, 34] probably mediated by inhibitory interneurons, are not compatible with the analgesic action of M100907 or M11,939 in experimental models of toxic neuropathic pain [37, 38]. It seems difficult to reconcile these differences except if the pathophysiology of ddC- and VCT-induced neuropathic pain involves specific events different from those involved in the other models.

Further studies are needed to understand why $5\text{-}HT_{2A}$ receptors at the level of the spinal cord play controversial roles. The proalgesic and antihyperalgesic effects of $5\text{-}HT_{2A}$ receptor activation resemble the variable effects of $5\text{-}HT_{2A}$ receptors on cognitive processing, with some studies showing no effects, or inhibitory or stimulatory effects on attention and impulsivity [44]. In contrast, evidence indicates that 5-HT acting drugs such as SSRI antidepressants involve different subtypes of receptors including 5-HT_{2A} subtype to produce analgesia [33, 45, 46]. This may potentially open up the field for new drugs that can be beneficial for the treatment of chronic pain in patients.

References

1. Merskey H, Bogduk N, International Association for the Study of Pain. Task Force on Taxonomy (1994) Classification of chronic pain : descriptions of chronic pain syndromes and definitions of pain terms, 2nd edn. IASP Press, Seattle
2. Millan MJ (2002) Descending control of pain. Prog Neurobiol 66(6):355–474
3. Ding YQ, Marklund U, Yuan W, Yin J, Wegman L, Ericson J, Deneris E, Johnson RL, Chen ZF (2003) Lmx1b is essential for the development of serotonergic neurons. Nat Neurosci 6(9):933–938. https://doi.org/10.1038/nn1104nn1104
4. Zhao ZQ, Chiechio S, Sun YG, Zhang KH, Zhao CS, Scott M, Johnson RL, Deneris ES, Renner KJ, Gereau RW, Chen ZF (2007) Mice lacking central serotonergic neurons show enhanced inflammatory pain and an impaired analgesic response to antidepressant drugs. J Neurosci 27(22):6045–6053
5. Muller C, Jacobs BL (2010) Handbook of the behavioral neurobiology of serotonin. Academic, London
6. Castaneda-Corral G, Rocha-Gonzalez HI, Araiza-Saldana CI, Ambriz-Tututi M, Vidal-Cantu GC, Granados-Soto V (2009) Role of peripheral and spinal 5-HT6 receptors according to the rat formalin test. Neuroscience 162(2):444–452. https://doi.org/10.1016/j.neuroscience.2009.04.072
7. Godinez-Chaparro B, Lopez-Santillan FJ, Orduna P, Granados-Soto V (2012) Secondary mechanical allodynia and hyperalgesia depend on descending facilitation mediated by spinal 5-HT(4), 5-HT(6) and 5-HT(7) receptors. Neuroscience 222:379–391. https://doi.org/10.1016/j.neuroscience.2012.07.008
8. Liu J, Reid AR, Sawynok J (2013) Spinal serotonin 5-HT7 and adenosine A1 receptors, as well as peripheral adenosine A1 receptors, are involved in antinociception by systemically administered amitriptyline. Eur J Pharmacol 698(1–3):213–219. https://doi.org/10.1016/j.ejphar.2012.10.042
9. Viguier F, Michot B, Hamon M, Bourgoin S (2013) Multiple roles of serotonin in pain control mechanisms--implications of 5-HT(7) and other 5-HT receptor types. Eur J Pharmacol 716(1–3):8–16. https://doi.org/10.1016/j.ejphar.2013.01.074
10. Bockaert J, Claeysen S, Becamel C, Dumuis A, Marin P (2006) Neuronal 5-HT metabotropic receptors: fine-tuning of their structure, signaling, and roles in synaptic modulation. Cell Tissue Res 326(2):553–572. https://doi.org/10.1007/s00441-006-0286-1
11. Becamel C, Galeotti N, Poncet J, Jouin P, Dumuis A, Bockaert J, Marin P (2002) A proteomic approach based on peptide affinity chromatography, 2-dimensional electrophoresis and mass spectrometry to identify multiprotein complexes interacting with membrane-bound receptors. Biol Proced Online 4:94–104
12. Becamel C, Gavarini S, Chanrion B, Alonso G, Galeotti N, Dumuis A, Bockaert J, Marin P (2004) The serotonin 5-HT2A and 5-HT2C receptors interact with specific sets of PDZ proteins. J Biol Chem 279(19):20257–20266
13. Gavarini S, Becamel C, Chanrion B, Bockaert J, Marin P (2004) Molecular and functional characterization of proteins interacting with the C-terminal domains of 5-HT2 receptors: emergence of 5-HT2 "receptosomes". Biol Cell 96(5):373–381
14. Pichon X, Wattiez AS, Becamel C, Ehrlich I, Bockaert J, Eschalier A, Marin P, Courteix C (2010) Disrupting 5-HT(2A) receptor/PDZ protein interactions reduces hyperalgesia

and enhances SSRI efficacy in neuropathic pain. Mol Ther 18(8):1462–1470. https://doi.org/10.1038/mt.2010.101
15. Wattiez AS, Pichon X, Dupuis A, Hernandez A, Privat AM, Aissouni Y, Chalus M, Pelissier T, Eschalier A, Marin P, Courteix C (2013) Disruption of 5-HT2A receptor-PDZ protein interactions alleviates mechanical hypersensitivity in carrageenan-induced inflammation in rats. PLoS One 8(9):e74661. https://doi.org/10.1371/journal.pone.0074661
16. Doly S, Madeira A, Fischer J, Brisorgueil MJ, Daval G, Bernard R, Verge D, Conrath M (2004) The 5-HT2A receptor is widely distributed in the rat spinal cord and mainly localized at the plasma membrane of postsynaptic neurons. J Comp Neurol 472(4):496–511. https://doi.org/10.1002/cne.20082
17. Fonseca MI, Ni YG, Dunning DD, Miledi R (2001) Distribution of serotonin 2A, 2C and 3 receptor mRNA in spinal cord and medulla oblongata. Brain Res Mol Brain Res 89(1–2):11–19
18. Marlier L, Teilhac JR, Cerruti C, Privat A (1991) Autoradiographic mapping of 5-HT1, 5-HT1A, 5-HT1B and 5-HT2 receptors in the rat spinal cord. Brain Res 550(1):15–23
19. Maeshima T, Ito R, Hamada S, Senzaki K, Hamaguchi-Hamada K, Shutoh F, Okado N (1998) The cellular localization of 5-HT2A receptors in the spinal cord and spinal ganglia of the adult rat. Brain Res 797(1):118–124
20. Van Steenwinckel J, Noghero A, Thibault K, Brisorgueil MJ, Fischer J, Conrath M (2009) The 5-HT2A receptor is mainly expressed in nociceptive sensory neurons in rat lumbar dorsal root ganglia. Neuroscience 161(3):838–846. https://doi.org/10.1016/j.neuroscience.2009.03.087
21. Wang YY, Wei YY, Huang J, Wang W, Tamamaki N, Li YQ, Wu SX (2009) Expression patterns of 5-HT receptor subtypes 1A and 2A on GABAergic neurons within the spinal dorsal horn of GAD67-GFP knock-in mice. J Chem Neuroanat 38(1):75–81. https://doi.org/10.1016/j.jchemneu.2009.04.003
22. Liu FY, Xing GG, Qu XX, Xu IS, Han JS, Wan Y (2007) Roles of 5-hydroxytryptamine (5-HT) receptor subtypes in the inhibitory effects of 5-HT on C-fiber responses of spinal wide dynamic range neurons in rats. J Pharmacol Exp Ther 321(3):1046–1053
23. Sasaki M, Ishizaki K, Obata H, Goto F (2001) Effects of 5-HT2 and 5-HT3 receptors on the modulation of nociceptive transmission in rat spinal cord according to the formalin test. Eur J Pharmacol 424(1):45–52
24. Sasaki M, Obata H, Saito S, Goto F (2003) Antinociception with intrathecal alpha-methyl-5-hydroxytryptamine, a 5-hydroxytryptamine 2A/2C receptor agonist, in two rat models of sustained pain. Anesth Analg 96(4):1072–1078
25. Okamoto K, Imbe H, Kimura A, Donishi T, Tamai Y, Senba E (2007) Activation of central 5HT2A receptors reduces the craniofacial nociception of rats. Neuroscience 147(4):1090–1102
26. Kayser V, Elfassi IE, Aubel B, Melfort M, Julius D, Gingrich JA, Hamon M, Bourgoin S (2007) Mechanical, thermal and formalin-induced nociception is differentially altered in 5-HT1A-/-, 5-HT1B-/-, 5-HT2A-/-, 5-HT3A-/- and 5-HTT-/- knock-out male mice. Pain 130(3):235–248
27. Sasaki M, Obata H, Kawahara K, Saito S, Goto F (2006) Peripheral 5-HT2A receptor antagonism attenuates primary thermal hyperalgesia and secondary mechanical allodynia after thermal injury in rats. Pain 122(1–2):130–136
28. Kjorsvik Bertelsen A, Warsame Afrah A, Gustafsson H, Tjolsen A, Hole K, Stiller CO (2003) Stimulation of spinal 5-HT(2A/2C) receptors potentiates the capsaicin-induced in vivo release of substance P-like immunoreactivity in the rat dorsal horn. Brain Res 987(1):10–16
29. Kjorsvik A, Tjolsen A, Hole K (2001) Activation of spinal serotonin(2A/2C) receptors augments nociceptive responses in the rat. Brain Res 910(1–2):179–181
30. Rahman W, Bannister K, Bee LA, Dickenson AH (2011) A pronociceptive role for the 5-HT2 receptor on spinal nociceptive transmission: an in vivo electrophysiological study in the rat. Brain Res 1382:29–36. https://doi.org/10.1016/j.brainres.2011.01.057
31. Xu Q, Yaksh TL (2011) A brief comparison of the pathophysiology of inflammatory versus neuropathic pain. Curr Opin Anaesthesiol 24(4):400–407. https://doi.org/10.1097/ACO.0b013e32834871df

32. Liu FY, Qu XX, Ding X, Cai J, Jiang H, Wan Y, Han JS, Xing GG (2010) Decrease in the descending inhibitory 5-HT system in rats with spinal nerve ligation. Brain Res 1330:45–60. https://doi.org/10.1016/j.brainres.2010.03.010
33. Obata H, Saito S, Sasaki M, Ishizaki K, Goto F (2001) Antiallodynic effect of intrathecally administered 5-HT(2) agonists in rats with nerve ligation. Pain 90(1–2):173–179
34. Song Z, Meyerson BA, Linderoth B (2011) Spinal 5-HT receptors that contribute to the pain-relieving effects of spinal cord stimulation in a rat model of neuropathy. Pain 152(7):1666–1673. https://doi.org/10.1016/j.pain.2011.03.012
35. Bouzidi N, Deokar H, Vogrig A, Boucherle B, Ripoche I, Abrunhosa-Thomas I, Dorr L, Wattiez AS, Lian LY, Marin P, Courteix C, Ducki S (2013) Identification of PDZ ligands by docking-based virtual screening for the development of novel analgesic agents. Bioorg Med Chem Lett 23(9):2624–2627. https://doi.org/10.1016/j.bmcl.2013.02.100
36. Vogrig A, Dorr L, Bouzidi N, Boucherle B, Wattiez AS, Cassier E, Vallon G, Ripoche I, Abrunhosa-Thomas I, Marin P, Nauton L, Thery V, Courteix C, Lian LY, Ducki S (2013) Structure-based design of PDZ ligands as inhibitors of 5-HT2A receptor/PSD-95 PDZ1 domain interaction possessing anti-hyperalgesic activity. ACS Chem Biol 8(10):2209–2216. https://doi.org/10.1021/cb400308u
37. Van Steenwinckel J, Brisorgueil MJ, Fischer J, Verge D, Gingrich JA, Bourgoin S, Hamon M, Bernard R, Conrath M (2007) Role of spinal serotonin 5-HT2A receptor in 2′,3′-dideoxycytidine-induced neuropathic pain in the rat and the mouse. Pain 137(1):66–80
38. Thibault K, Van Steenwinckel J, Brisorgueil MJ, Fischer J, Hamon M, Calvino B, Conrath M (2008) Serotonin 5-HT2A receptor involvement and Fos expression at the spinal level in vincristine-induced neuropathy in the rat. Pain 140(2):305–322. https://doi.org/10.1016/j.pain.2008.09.006
39. Aira Z, Buesa I, Salgueiro M, Bilbao J, Aguilera L, Zimmermann M, Azkue JJ (2010) Subtype-specific changes in 5-HT receptor-mediated modulation of C fibre-evoked spinal field potentials are triggered by peripheral nerve injury. Neuroscience 168(3):831–841. https://doi.org/10.1016/j.neuroscience.2010.04.032
40. Aira Z, Buesa I, Garcia del Cano G, Salgueiro M, Mendiable N, Mingo J, Aguilera L, Bilbao J, Azkue JJ (2012) Selective impairment of spinal mu-opioid receptor mechanism by plasticity of serotonergic facilitation mediated by 5-HT2A and 5-HT2B receptors. Pain 153(7):1418–1425. https://doi.org/10.1016/j.pain.2012.03.017
41. Nitanda A, Yasunami N, Tokumo K, Fujii H, Hirai T, Nishio H (2005) Contribution of the peripheral 5-HT 2A receptor to mechanical hyperalgesia in a rat model of neuropathic pain. Neurochem Int 47(6):394–400
42. Kato K, Sekiguchi M, Kikuchi S, Konno S (2015) The effect of a 5-HT2A receptor antagonist on pain-related behavior, endogenous 5-hydroxytryptamine production, and the expression 5-HT2A receptors in dorsal root ganglia in a rat lumbar disc herniation model. Spine (Phila Pa 1976) 40(6):357–362. https://doi.org/10.1097/BRS.0000000000000769
43. Huang BQ, Wu B, Hong Y, Hu W (2015) Effects of blockade of 5-HT2A receptors in inflammatory site on complete Freund's adjuvant-induced chronic hyperalgesia and neuropeptide Y expression in the spinal dorsal horn in rats. Sheng Li Xue Bao 67(5):463–469
44. Leiser SC, Li Y, Pehrson AL, Dale E, Smagin G, Sanchez C (2015) Serotonergic regulation of prefrontal cortical circuitries involved in cognitive processing: a review of individual 5-HT receptor mechanisms and concerted effects of 5-HT receptors exemplified by the multimodal antidepressant Vortioxetine. ACS Chem Neurosci 6(7):970–986. https://doi.org/10.1021/cn500340j
45. Anjaneyulu M, Chopra K (2004) Fluoxetine attenuates thermal hyperalgesia through 5-HT1/2 receptors in streptozotocin-induced diabetic mice. Eur J Pharmacol 497(3):285–292
46. Yokogawa F, Kiuchi Y, Ishikawa Y, Otsuka N, Masuda Y, Oguchi K, Hosoyamada A (2002) An investigation of monoamine receptors involved in antinociceptive effects of antidepressants. Anesth Analg 95(1):163–168

5-HT_{2A} Receptors in Eating Disorders

Philip Gorwood, Laurence Lanfumey, Odile Viltart, and Nicolas Ramoz

Abstract Eating disorders consist in *anorexia nervosa* (lack of energy intake and/or excess of caloric consumption) *bulimia nervosa* (episodes of binge eating associated with compensatory behaviors, such as self-induced vomiting, misuse of laxative, diuretics, fasting or excessive exercise) and *binge eating disorder* (recurrent episodes of binge eating without compensatory behavior). The biological mechanisms of these eating disorders have been extensively studied, both in human and animal models, mainly focusing on neuropeptides regulating appetite and on neurotransmitters that may also be involved in mood, appetite and weight, but also impulsivity and rewarding aspects of behavior. Although early preclinical data described a clear role of 5-HT_{2A} receptors in food regulation, the use of specific 5-HT_2 ligands did not confirm these first data. Most of the ligands initially used acted actually through 5-HT_{2C} receptors, and, at least at preclinical level, it is now clearly established that these 5-HT_{2C} receptors are those which regulated food intake. The gene coding for 5-HT_{2A} receptor was the very first gene associated with eating disorders, mainly in anorexia nervosa, raising the scientific interest in the serotonin pathway to explain their genetic vulnerability. The A allele of −1438G/A HTR2A polymorphism was reported as being associated to AN and BN, but with many discrepancies, the association being insufficiently strong to survive the performed meta-analyses. This does not mean that the 5-HT_{2A} receptor is having no role in any eating disorder, but that its contribution, if any, might be too small to be detectable when many types of patients are being gathered.

Keywords Anorexia nervosa • Bulimia nervosa • Binge eating disorder • Appetite • Weight • Food • Genetics

---

P. Gorwood (✉)
CMME (Hôpital Sainte-Anne) and Université Paris-Descartes, Paris, France

Centre de Psychiatrie et Neurosciences (INSERM UMR894), Paris, France
e-mail: p.gorwood@ch-sainte-anne.fr

L. Lanfumey • O. Viltart • N. Ramoz
Centre de Psychiatrie et Neurosciences (INSERM UMR894), Paris, France

B.P. Guiard, G. Di Giovanni (eds.), *5-HT_{2A} Receptors in the Central Nervous System*, The Receptors 32, https://doi.org/10.1007/978-3-319-70474-6_15

Eating Disorders and Their Biological Mechanisms

Clinical Definition and Epidemiology

Three broad categories of eating disorders are usually delineated. The diagnostic and statistical manual of mental disorders fifth edition (DSM-5) mainly distinguishes anorexia nervosa and bulimia nervosa, and describes binge eating disorder under the same umbrella of the latter category (307.51 criteria), even though it is considered as a separate entity, both for the ICD (F50.8) and for many clinicians for easiness of diagnoses. Indeed, patients with *anorexia nervosa* are underweight (because of lack of energy intake and/or excess of caloric consumption), patients with *bulimia nervosa* usually have a normal weight (as episodes of binge eating are associated with compensatory behaviors, such as self-induced vomiting, misuse of laxative, diuretics, fasting or excessive exercise) and the diagnosis of *binge eating disorder* is based on recurrent episodes of binge eating but without compensatory behaviour, therefore associated with overweight.

Those three disorders are characterized by aberrant patterns of feeding behavior and weight regulation, and disturbances in attitudes toward weight and shape and the perception of body shape. In AN, there is an irrational fear of weight gain and obsession with fatness even when patients are extremely thin. In fact this simple statement has been recently challenged [1] as it could indirectly reflect what is really being involved, i.e. a rewarding effect of thinness.

The lifetime prevalence of eating disorders in adults is about 0.6% for anorexia nervosa, 1% for bulimia nervosa, and 3% for binge eating disorder. Women are 2–3 times more frequently affected than are men for all these eating disorders, and are more common in young ones, especially between 10 and 25 years old [2].

Risk Factors and Biological Mechanisms

Because eating disorders are complex diseases, including psychological, metabolism and developmental aspects, many risk factors were tested, and a large set of mechanisms were considered as important in the onset of eating disorders. Because it gives sense to many characteristics of eating disorders [3] serotonin is one of the most frequently studied monoamine in these disorders. Many scientific evidences explain the focus on serotonin while studying the biological mechanisms of eating disorders such as the (1) role of stress as a triggering event, the (2) observed excess of depressive comorbidity [4], the (3) importance of serotonin in appetite/satiety regulation, (4) the use as a treatment strategy of serotonin reuptake inhibitors (SRI) in patients with all eating disorders, and (5) the role of this neurotransmitter in associated traits such as high level of 'harm avoidance' or 'behavioral inhibition' [5], obsessionality [6], anxiety and fear [7].

Facilitated by the use of consensual criteria, many research strategies were used to decipher the biological mechanisms explaining why some subjects are more at risk than others. A large familial study demonstrated that the relative risk of anorexia nervosa was 11 in female relatives of anorexic probands, and 12 in female relatives of bulimic probands [8]. The heritability of the three eating disorders, using aggregation studies but also twin studies, was in fact estimated to range between 50% and 83% [2, 9], stressing the potential benefit of using genetics in order to further understand eating disorders. Genetics also provides an interesting opportunity to analyze the role of genes coding for specific receptors, such as the serotonin receptors and transporters, without being confused by the impact of malnourishment and/or obesity.

The Potential Role of Serotonin in Eating Disorders

Serotonin (5-HT) is synthesized from its precursor tryptophan, an essential amino acid that must be obtained through food (please see other chapters for more details). Treatments that increase synaptic 5-HT, or activate 5-HT receptors, reduce food consumption, whereas interventions that dampen 5-HT neurotransmission or block 5-HT receptors increase food consumption and promote weight gain [10, 11].

When underweight, individuals with AN have a significant reduction in basal concentrations of the 5-HT metabolite (5-HIAA) in the cerebral spinal fluid (CSF) compared to healthy controls [12]. Using a dynamic test, a blunted plasma prolactin response to drugs with 5-HT activity is observed, with reduced 3H-imipramine binding [13–16]. In long-term recovered AN with normal weight, individual CSF concentrations of 5-HIAA are elevated [17]. Serotonin abnormalities are therefore observed in anorexia nervosa according to several lines of evidence, even though it is difficult to prove that it actively participates in the development of the disorder, and does not only represent a long standing associated marker.

In BN, a blunted prolactin response is also observed to 5-HT receptor agonists such as m-chlorophenylpiperazine (mCPP), 5-hydroxytrytophan, and DL-fenfluramine, [14–16, 18]. Dietary depletion of tryptophan has also been associated to increased food intake in individuals with BN compared to healthy controls [19]. Like in AN, women with long-term recovery from BN have been shown to have elevated concentrations of 5-HIAA in the CSF, [12, 15, 17, 20–22] and reduced platelet [3H-] paroxetine binding, which is thought to be a marker of 5-HTT activity [23]. Furthermore, in 4 studies out of 5 testing the efficacy of SSRI in bulimia nervosa, a reduction of binge/purging frequency was more frequently observed in the group treated by SSRIs compared to the group treated by placebo [24].

Serotonin (5HT) might also have a crucial role in binge eating behavior [25]. In both animals and humans, manipulations that decrease 5HT activity precipitate compulsive or binge eating [26]. More precisely, obese women with binge-eating disorder have lowered 5-HT transporter binding in the midbrain, and thereby a reduction in the 5-HT re-uptake sites, which may contribute to concurrent or sequential periods of binge eating [27]. Even more convincing, the 5-HT transporter binding of the symptomatically recovered patients increased significantly (24%) after treatment [28]. Patients with BED have a high lifetime prevalence of major depression diagnosis and often present depressive symptoms coincident with ED [29].

RCTs against placebo have thus been tested in BED patients with citalopram [30], escitalopram [31], fluoxetine [32], fluvoxamine [33] and sertraline [34]. A significant reduction in binge-eating frequencies (except escitalopram) and BMI were observed, with an overall clinical improvement. A review of eight randomized, placebo-controlled trials [35] concluded on the positive action of SSRI in binge eating disorder.

Serotonin might therefore be affected, at least in part, in all types of eating disorder, but its role in the regulation of appetite relies on a vast number of scientific publications.

Biological Analyses of the 5-HT_{2A} Receptors in Feeding Behaviors

Brain monoamine function in eating disorders has been studied in the acute state (which can be confounded by illness effects) and after recovery with specific ligands and positron emission tomography. These findings for anorexia nervosa have been synthesized into an explanatory model, where 5-HT_{2A} receptors are reduced and 5-HT_{1A} receptors are increased in both the acute and recovered state, and dopamine receptors (DA2) within the striatum are increased after recover [36, 37]. $5HT_{2A}$ could therefore be a trait-related disturbance, when dopamine abnormalities could be state-dependent. The 5-HT_{2A} receptor was indeed one of the most studied receptor in eating disorders.

Roles of 5-HT in Feeding Behavior

Serotonin (5-HT) is an old phylogenetically neurotransmitter which recovers various functions from invertebrates to vertebrates. Both peripheral and central 5-HT systems contribute to energy homeostasis, and serotonin is certainly the neurotransmitter the most studied in the field of feeding. Its involvement in the control of food intake is well recognized in many species [38] and most of the molecules used in the pharmacological treatment of eating disorders and weight status are known to have a serotonergic component. As soon as in 1977, Blundell reviewed the experimental evidences showing the involvement of serotonin in feeding behavior [39]. He proposed that increasing the availability of 5-HT in the synaptic cleft or activating the 5-HT receptors result in a reduction of food consumption whereas a reduced availability of the neurotransmitter or a blockade of its receptors could induce feeding. This was further demonstrated using pharmacological agents that increased 5-HT neurotransmission either by selectively inhibiting its reuptake, such as selective serotonin reuptake inhibitors (SSRIs) like fluoxetine [40], or by enhancing 5-HT release and inhibiting its reuptake, such as d-fenfluramine [41]. Interestingly, the majority of compounds reviewed in the guidelines for the pharmacological treatment of eating disorders are acting, at least partially, through the blockade of the serotonin transporter; either as tricyclics (clomipramine, imipramine, amitryptiline, desipramine), SSRI (citalopram, fluoxetine, fluvoxamine, sertraline,) or SNRI (venlafaxine) with few exceptions (mirtazapine, moclobemide, reboxetine, bupropion, mianserin) [42]. All these compounds reduced food intake in both humans and experimental animals [43, 44] clearly evidencing that 5-HT was playing a key role in the brain mediation of satiety and feeding.

The serotoninergic neurons are located in the midbrain and brainstem raphe nuclei [45]. Their projections are diffused and concerned about all the brain areas including the hippocampus, the hypothalamus, the cerebral cortex (prefrontal, parietal, occipital and cingulated), the basal ganglia and more caudally, the cerebellum and the spinal cord. Among these projections, those originated from the anterior raphe nuclei and projecting to the hypothalamus give rise to interactions between the amine and neuroendocrine systems linked to food regulation. Some of these interactions implicate neuropeptide Y (NPY) neurones located in the hypothalamic arcuate nucleus. Indeed, NPY neurons are ones of the first target of feeding peripheral signals, thereby NPY is considered as the most potent brain stimulator of food intake through a complex hypothalamic network projections [46, 47]. This orexigenic neuronal population may indirectly contribute to central anorectic effects of serotonin since mRNA expression of NPY is decreased following a chronic systemic administration of dl-fenfluramine in rats and application of 5-HT_{1B} agonists hyperpolarize these neurons through activation of Gi protein [48, 49].

Finally, some data underlined a positive correlation between body mass index and 5-HT_{2A} receptor binding in the cerebral cortex [50]. However, peripheral 5-HT might have opposite functions to central 5-HT in the regulation of energy homeostasis and weight regulation. Indeed, it has been proposed that central 5-HT could act as an anorexigenic neurotransmitter by activating the 5-HT_{2C} receptors in the brain, while in periphery, 5-HT could inhibit thermogenesis through 5-HT_3 receptors in brown adipose tissue (BAT) and increased lipogenesis through 5-HT_{2A} receptors in white adipose tissue (WAT) [51].

Eating Disorders and 5-HT

The development of antagonists and agonists of the 5-HT receptors has helped to better understand the dysregulated mechanisms of food intake control in the case of obesity or eating disorders like anorexia nervosa (AN) as proposed 30 years ago [26]. As recently reviewed by Kumar and Mann [52], the multiplicity of 5-HT receptors and subtypes in the brain supported the diversity of functions in which 5-HT is involved. Among the 15 receptors engaged in 5-HT signaling, several of them, namely the 5-HT_{1B}, 5-HT_{2A}, 5-HT_{2C}, 5-HT_4 and 5-HT_6 subtypes, are implicated in the pathophysiology of eating disorders. The very first ones to be supposed to play a key role in feeding regulation were the 5-HT_1 and 5-$HT_{2A/2C}$ receptors, which were described to modulate the interactions between 5-HT and the hypothalamic neuroendocrine system [53]. The involvement of 5-HT in the control of feeding have been showed to occur through the modulation of neurons located in the arcuate nucleus where 5-HT_{1B} and 5-HT_{2C} receptors are mainly expressed (see review of [54]). The 5-HT role in the reward control of food intake has also been described to involve 5-HT_4 and 5HT_{2A} receptors located in the nucleus accumbens, ventral pallidum, left hippocampal region and orbitofrontal cortex [55, 56].

In anorexia nervosa, the 5-HT system is altered. In fact, 5-HT contents as well as plasma tryptophan levels are lower than in controls [57, 58] and brain imaging studies described changes in the 5-HT_{1A} and 5-HT_{2A} receptor binding in cortical and limbic areas such as amygdala or hypothalamus in ill patients [59–61]. Indeed, the activity of 5HT_{1A} receptors appears to increase, while 5HT_{2A} activity decreases, more particularly in the prefrontal cortex, in both ill anorexic patients and recovered patients. These altered serotoninergic pathways might be linked to alteration in the reward process of food intake and in the level of anxiety, behavioral inhibition or body image distortion, other behavioral modifications usually observed in patients beside the refusal to eat, hyperactivity or excessive exercise, or impaired impulse control. As mentioned by Kaye [62], numerous evidences exist illustrating the interaction between 5-HT_{1A} and 5-HT_{2A} receptors in the brain. As an example, Krebs-Thomson and Geyer [63] described how the interaction between 5-HT_{1A} and 5-HT_{2A} receptors regulates the inhibition of exploration of novel environments produced by either 5-HT_{1A} or 5-HT_{2A} receptor agonists in rats. Such a behavioral inhibition is also described in ill anorexia nervosa patients as harm avoidance, a multifaceted temperament trait that contains elements of anxiety, inhibition, and inflexibility [64].

Interestingly, a special focus on 5-HT_{2A} receptors originated from genetic data obtained from AN patients where 5-HT_{2A} receptor and 5-HT transporter gene polymorphisms were described to increase the risk of AN [65, 66]. However, the role of this receptor in appetite control remains unclear. The hypophagic effect of agonists acting at 5-HT_{2A} receptors such as mCPP (1-(3-chlorophenyl)piperazine di hydrochloride) or DOI (1-(2,5-dimethoxy-4-iodophenyl)-2-aminopropane) has been abundantly reported in the literature at the end of the twentieth century. mCPP had been shown to increase latency to feed, reduced the size of the first meal and feeding rate during that meal, effects intake that were as profound on water as those on food intake [67, 68]. DOI also produced dose-related decreases in 1-h food intake in the food-deprived paradigm [69]. However, the use of selective 5-HT_2 receptor antagonist compounds evidenced afterwards that most of those effects could be attributed to the 5-HT_{2C} -and not 5-HT_{2A}- properties of these drugs.

In their review, Kumar and Mann reported the role of the 5HT2A receptor in psychiatric disorders like schizophrenia, post-traumatic stress disorder or major depression that has rendered this receptor a main target for different generations of antipsychotic drugs [52]. $5\text{-}HT_{2A}$ receptors have been implicated in the genesis of, as well as the treatment of, psychosis, negative symptoms, mood disturbance, and extrapyramidal symptoms. The antipsychotic effect of several antipsychotics such as clozapine has been attributed, at least in part, to its ability to block excessive $5\text{-}HT_{2A}$ receptor stimulation without excessive blockade of D_2 receptors [70]. Interestingly, several atypical antipsychotics which antagonize 5-HT receptor subtype are known to affect feeding behavior. In particular, it has been reported that a single injection of risperidone or aripiprazole, which have binding affinities to $5\text{-}HT_{1A,\ 2A,\ 2B}$ and $_{2C}$ receptors, decreased food intake in mice [71]. In this study, it was hypothesized that the effect of 5-HT was indirect, through the inhibition of histamine release via $5\text{-}HT_2$ receptors and that antipsychotics enhanced hypothalamic histamine release by blockade of $5\text{-}HT_2$ receptors resulting in a H_1 receptor–related anorexia. However, most of the studies reported an increase in food consumption under antipsychotics, which could lead to obesity. It is the case after both typical antipsychotics such as chlorpromazine and haloperidol, and atypical antipsychotics such as clozapine, olanzapine, risperidone or sulpiride which all induced hyperphagia. The general consensus is that clozapine and olanzapine are associated with the greatest effects on weight gain and decreased insulin sensitivity, followed by risperidone and quetiapine [70]. The mechanism of these effects is not completely understood but it is believed to result from a complex interaction between several pharmacologic actions of these drugs. Recent evidence suggests a role of the α_1 adrenoceptor and $5\text{-}HT_{2A}$ receptor in the metabolic effects of atypical antipsychotics. This is particularly true for clozapine which hyperphagia-like effect was blocked by quipazine, a 5-HT receptor antagonist that bind to $5\text{-}HT_{2A}$ receptors [72].

Rodent Models for Exploring Feeding Behaviour and Eating Disorders

The use of appropriate animal models is necessary to better understand the precise role of 5-HT receptors, and in particular the $5\text{-}HT_{2A}$ subtype, both in food control or/and in the reward process in conjunction with the dopaminergic system. Several models of eating disorders have been extensively described in recent reviews [73, 74], but in the case of anorexia nervosa, the development of appropriate animal models is rendered difficult due to the complex etiology of the disease. In rodents, genetic and environmental models were developed with varying degrees of success.

Environmental Models

In environmental models of eating disorders and particularly of anorexia nervosa, various models were described (see recent reviews of [73, 74]). In the dietary restriction models, different protocols can be achieved by manipulation of the severity of food restriction, of the diet composition (low- or free-fat, low carbohydrate, or amino-acids deficiency diets), of the schedule of feeding with either a fixed percentage of the food that animals normally consume during 24-h or a fixed amount of time to eat food delivered *ad libitum*. The latter model was called the Separation Based Anorexia model (SBA) [75, 76]. The stress-induced appetite loss models apply acute or chronic stressors to induce reduction of food intake. Amongst the procedures classically used, acute stressors are for example tail pinch, cold exposure, social stress, whereas chronic stressors include chronic mild stress procedure [77], or separation stress (also used in the SBA model). Beside the food restriction and stress, another feature usually observed in anorexia nervosa patients is the physical activity. The "activity based anorexia" model (ABA) and its variants include this last parameter and is currently one of the best characterized model of "anorexia nervosa". It combined a limited access to food (1–2 h according the species) with free access to a wheel. This rodent model showed rapidly a decrease in food intake, self-starvation in parallel with an increase of physical activity, and increase of activity just before food delivering, called food-anticipatory behavior, also observed in AN patients [78], and considered an equivalent to the search of food. In a variation of this model, mice receive a fixed percentage of the amount of food (50%) eaten by mice fed *ad libitum* and have unlimited access to running wheels [79]. Despite the disappearance of self-starvation, this protocol allows the follow-up of all the metabolic (like osteopenia or osteoporosis) and brain alterations on the long-term (up to 10 weeks of protocol). In this environmental model ABA, the decreased food intake is associated in particular with increased 5-HT levels in hypothalamus [80]. Moreover, the levels of 5-HT in the nucleus accumbens remained low along the day compared to fed rats and the initiation of food anticipatory behavior failed to increase accumbal 5-HT release [81]. D-fenfluramine, a 5-HT_{2C} receptor agonist leading to suppression of appetite by its action on the hypothalamic arcuate pro-opiomelanocortin neurons, has no effect in this model on food intake and activity except hypodypsia [82]. Since rats often combine food and water intake, the reduction of water intake in ABA rats observed in this study may be exaggerated as compared with ad libitum fed rats which eat more. Once again, the literature poorly documented the precise role of 5-HT_{2A} receptor in environmental models of eating disorders. However, it should be noted that chronic hyperphagia increased hypothalamic 5-HT_{2A}, 5-HT_{2C} and 5-HT_{1B} gene expression, whereas inactivation of 5-HT_{2A} receptors inhibited overfeeding and obesity in A(y) mice expressing the ectopic agouti protein [83, 84].

Genetic Models

Among genetic models, two categories are commonly used: genetically engineered models, which consist in the constitutive or inducible deletion or overexpression of candidate genes and spontaneous mutations like the mutant *anx/anx* mice.

There are currently very few genetically modified animal models focused on 5HT receptors in relation to feeding behavior. Some models manipulating the genetic expression of one of these 5-HT receptors gave evidence about their involvement in the control of food intake and particularly on the motivation to eat. Mice knock-out for $5\text{-}HT_4$ receptors displayed a decreased "anorexia-like" behavior in response to restrained stress as well as an attenuation of novelty-induced exploratory activity [85]. More specifically, stimulation of $5\text{-}HT_4$ receptor in the nucleus accumbens, striatal structure involved in the reward pathway, increased satiety in fed mice and reduced the physiological drive to eat after food deprivation [86]. Similarly, the use of specific antagonist or the inactivation of $5\text{-}HT_4$ accumbal receptors by $si5\text{-}HT_4R$ increased the food intake in fed mice but not in motivated mice (food-deprived). These data clearly demonstrate the involvement of these receptors in the physiological drive to eat. Moreover, mice lacking $5\text{-}HT_{1A}$ receptor or wild type mice chronically treated subcutaneously with a $5\text{-}HT_{1A}/5\text{-}HT_7$ receptor agonist decreased their food intake [87]. Likewise, food restricted (20%, 3 days) mice lacking $5\text{-}HT_{1B}$ receptor self-imposed food restriction compared to wild type mice when standard food ration is given after the restriction period and showed locomotor hyperactivity [88]. Regarding $5\text{-}HT_{2A}$ receptors, the consequence of their deletion in relation to food has not been deeply studied because $5\text{-}HT_{2A}$ knock-out mice did not develop any modification in food intake or body weight [89]. However, in the case of the $5\text{-}HT_{2C}$ subtype, mice with deletion of this receptor display an overweight phenotype due to appetite disruptions, thereby establishing a clear role for this receptor in the serotonergic control of appetite [90]. Its function in feeding regulation was further confirmed by the study of mice displaying the full edited form of $5\text{-}HT_{2C}$ receptor, which were characterized by a massive $5\text{-}HT_{2C}$ membrane expression, and increased energy expenditure and decreased body fat/cholesterol levels [91] .

Data obtained in these experimental conditions open new issues to better understand the role of these receptors to predispose to anorexia-like symptoms, but are often difficult to interpret due to the large number of receptors distributed throughout the brain and the various effects they have depending on their location at synaptic level (pre- or post-synaptic). Taking into account these comments, the best way to elucidate the role of these receptors, in particular the $5\text{-}HT_{2A}$, in the feeding behavior and the drive to eat might be to use conditioned deletion or the cre-lox technology to avoid large effects that might be more the result of compensatory mechanisms than a real action of the neurotransmitter.

In the mutant *anx/anx* mice, which present a spontaneous mutation on the chromosome 2, and are characterized by an emaciated appearance, a spontaneous reduction of food intake from the 5th postnatal day and a death by starvation around

3–5 weeks after birth [92], several studies reported a 5-HT hyper-innervation in various parts of the brain including frontal cortex, hippocampus or cerebellum that might contribute to the phenotype observed (for review see [93]). However, to our knowledge, no data described the evolution of the 5-HT receptors in this interesting model.

Conclusions for Animal Models in Eating Disorders

Although early preclinical data described a clear role of 5-HT_{2A} receptors in food regulation, the use of specific 5-HT_2 ligands did not confirm these first data. Most of the ligands initially used acted actually through 5-HT_{2C} receptors, and, at least at preclinical level, it is now clearly established that these 5-HT_{2C} receptors are those which regulated food intake. However, given the various studies showing a clear link between changes in 5-HT_{2A} receptors binding or polymorphisms in both anorexia and overweight individuals, it is legitimate to further perform experimental investigations to address the regulation of this receptor in different situations of feeding (overeating or scheduled food restriction) to better understand both its precise role in the homeostatic and non-homeostatic control of food intake.

Genetic of Eating Disorders and 5-HT_{2A} Gene

Among the neurobiological pathways involving monoamines that could be involved in the eating disorders, the serotonin pathway could be altered, especially in anorexia nervosa (AN). One of the candidate in this pathway is the serotonin receptor 2A (5-HT_{2A}) that is encoded by the *HTR2A* gene located on the 13q14-q21 chromosomal band. A linkage was found between the "drive for thinness" phenotype and this chromosomal region [94]. The *HTR2A* gene encompass less than 70 kilobases (kb) and more than 2200 polymorphisms have been identified to date. The genomic structure of the gene is composed of two haplotype blocks, one is covering the majority of the gene and its promoter region and the other is encompassing 15 kb of the last intron and the last exon. The most investigated variant is the single nucleotide polymorphism (SNP) rs6311, usually named −1438G/A, a promoter variant located 1438 nucleotides before the transcription start of the gene where the common G allele is changed to an A allele.

Case Control Study

A first association between *HTR2A* gene and anorexia nervosa was found with the promoter polymorphism −1438G/A in a population of 81 AN patients from UK compared to 226 normal white controls including 88 females [95]. The frequency of the A allele (or the AA genotype) was significantly higher in patients compared to all controls or only female controls (respectively 0.51 & 0.31 in AN versus 0.41 & 0.15 in controls, and 0.42 & 0.125 in female controls only). Association study was also reported in an Italian sample of 77 AN patients, including 43 RAN subtype and 34 ANBP subtype, compared to 107 female controls [96]. A significantly higher frequency of the A allele and AA genotype was reported in patients (0.57 and 0.299) compared to controls (0.36 and 0.09). Interestingly, the association was driven by the subset of RAN patients where the frequencies of the A allele and AA genotype were the highest, respectively 0.66 and 0.42. Another work also observed an association between AN and the 5HT2A promoter variant in two independent cohorts, respectively of 68 AN patients versus 69 controls from USA, and 20 AN patients from Italy [97].

In contrast, no evidence for association was observed in a population of 100 AN patients from Germany [98]. In fact, a similar distribution of the frequency of the A allele was reported in the AN German patients (0.40) compared to 101 underweight non-AN subjects (0.43), 254 obese patients (0.42) or the UK female controls from Collier et al. [95]. A lack of association between AN and rs6311 was also observed in an independent UK cohort of 152 AN patients compared to 150 female controls [99]. In AN patients, the frequencies of the A allele (0.48) and the AA genotype (0.25) were not significantly different from those of controls, respectively 0.42 and 0.20. A work on an independent cohort of 78 AN patients compared to 170 controls was also not able to detect any association between any allele or genotype of rs6311 and AN [100]. This study included a meta-analysis of the six performed association studies, and failed to show an association either for a genotype nor for an allele [100]. Then, 99 bulimia nervosa (BN) patients were also compared to 170 controls and, no evidence for association was observed [100]. Absence of association between AN and −1438G/A *5HT2A* promoter polymorphism was also reported in a Japanese population by comparing 62 patients to 374 controls [101]. However, a significant association was observed for eating disorders when 62 AN and 110 BN patients were combined and compared to 374 controls. Furthermore, no evidence for association was observed for rs6311 and AN, nor in 75 AN Japanese patients, including 37 RAN and 38 BPAN cases, when compared to 127 controls [102]. This work did not find an association with the "restrictor" phenotype in the 44 AN patients [102]. No association was reported in a French sample of 145 AN patients compared to 98 controls [103]. However, the authors reported a significant excess of the A allele in AN patients with an older age at onset, suggesting that the A allele of rs6311 could be a "modifying factor", potentially explaining variations of frequency the A allele across the different published cohorts [103].

Sorbi's group increased its Italian sample and confirmed an association while comparing 148 AN patients to 115 controls [104]. The AA genotype and the A allele were significantly more frequent in AN patients (25.7% and 52%) than in controls (10.4% and 36%). The association was driven by the RAN subset of 74 patients where AA genotype and A allele were the highest, respectively 35.1% and 60.1%. Association was also reported with an excess of the A allele for 86 BN binge/purging patients and 54 binge eating disorder (BED) patients compared to 115 controls, but no difference was observed when comparing patients with 132 obese subjects [104]. However, other groups still failed to find an association between −1438G/A *5HT2A* promoter polymorphism with AN, RAN subset or BN, in samples from different origins [105–107]. As AN have unstable subtypes throughout lifetime (the majority of patients switch from one form to the other), the definition of subtypes might also have an important impact, especially as some studies used a minimal length to conclude for a stable subtype, while others relied only on the present form during the inclusion process of the protocol.

In conclusion, although an increased frequency was found for the A allele of the 5-HT_{2A} gene (−1438G/A polymorphism, in the promoter region), in many studies, and in different types of eating disorders, there are even more or larger negative studies, and the meta-analyses are usually concluding in favor of an *absence* of significant role for this allele.

Family-Based Study

To address the controversy results of the case control studies of −1438G/A *5HT2A* promoter polymorphism in AN and to reduce both the risk of false positives and the impact of a lack of statistical power due to small samples or population stratification, a large cohort of 316 AN families from six European centers was analyzed, using transmission disequilibrium test (TDT) for each cohort and for the entire group of families [108]. No excess of transmission of the A allele was observed with such TDT approach. No evidence for association was observed using the alternative haplotype relative risk method. Finally, no heterogeneity of the A allele frequency between the six samples was found, either according to minimal lifetime BMI, or according to age at onset [108].

Other variants among the *HTR2A* gene were investigated, including amino acid substitutions Thr25Asn and His452Tyr (rs1805055 and rs6314), and case control studies have showed no evidence for association with AN [98, 109].

Genome-Wide Association Study

Recently, international consortiums have emerged to contribute to the creation of large cohorts of patients and controls to perform high throughput genetic study at the pan-genomic level, the genome-wide association study (GWAS). Thus, 5151 SNPs of 182 candidate genes, including 55 in the HTR2A gene, were screened in 1085 AN patients (including 421 RAN) compared to 677 controls [110]. None of the *HTR2A* SNPs were in the top 25 associated SNPs, with a minimum trend P-value of 1.9×10^{-4}. In a first GWAS investigating more than 598,000 SNPs in 1033 AN patients (including 394 RAN) compared to 3733 pediatric controls, no evidence for association was found for *HTR2A* SNPs [111]. Recently, another GWAS was performed, including patients from all around the world, for which 1,185,559 SNPs were genotyped in 2907 AN patients (1606 RAN and 1445 BPAN cases) and 14,860 controls. 287 SNPs showed evidence for association in the discovery stage (i.e., with a p-value below $<10^{-4}$), but none were located among the *HTR2A* gene [112] (Table 1).

The A allele of −1438G/A HTR2A polymorphism was thus reported as being associated to AN and BN with discrepancies. This variant may play a role as a modifying factor potentially explaining variations of frequency across cohorts [103]. This A allele of rs6311 was also associated with a reduction of energy and fat intake, according to a work performed in 370 children and adolescent of the Stanislas family study [113]. The AA homozygote subjects, either males or females, had a reduced food intake of energy, with less fat, especially monounsaturated and saturated fat. Thus, additional studies investigating more specific phenotypes of food intake, and perhaps also energetic balance and metabolism of the analyzed subjects, are needed to conclude on the role of HTR2A in the pathophysiology of eating disorders.

Conclusions

Serotonin has a major role in appetite and energy regulation, the impact of stress and might be a core neurotransmitter in many biological aspects of the three main eating disorders (anorexia nervosa, bulimia nervosa and binge eating). Different serotonin receptors have been studied, with a major excitement for one of the polymorphism of the gene coding for the 5-HT_{2A} receptor in anorexia nervosa. But even if more than four positive association studies were published [108], the initial excitement vanished away because of a large set of non-replications [108], negative meta-analyses [9], the absence of an excess of transmitted alleles from heterozygous parents to affected probands [108] and no signals from the largest GWAS performed up to now on anorexia nervosa [112]. This does not mean that the 5-HT_{2A} receptor is having no role in any eating disorder, but that its contribution, if any, might be too small to be detectable when many types of patients are being gathered. One example of such a limited role could be a *modifying* rather than a *vulnerability* role. We showed for example [103] that anorexia nervosa patients with the A allele of the 5-HT_{2A} gene had a significantly later age at onset of the disease ($p = 0.032$) and was also transmitted with an older age at onset ($p = 0.023$) using a quantitative-trait TDT approach. The A allele may thus act as a modifying factor (delaying onset), potentially explaining variations of allele frequency across samples, in which differences in average age at onset are not only possible, but also expected.

Table 1 Genetic studies of SNP rs6311 (−1438A/G) of *HTR2A* gene in AN and subtypes

						−1438/A				−1438/G				p value and OR [95%CI]					
Refs.	Country	Controls	AN	RAN	BPAN	Controls	AN	RAN	BPAN	Controls	AN	RAN	BPAN	Controls vs AN					
Cases controls studies																			
[95]	United Kingdom	221	81	–	–	184 (0.41)	83 (0.51)			266 (0.59)	79 (0.49)			**0.023**	**1.52 [1.06–2.18]**				
[98]	Germany	101[a]	100			86 (0.43)	79 (0.40)			116 (0.57)	121 (0.60)			0.531	–				
[96]	Italy	107	77	43	34	76 (0.36)	88 (0.57)	57 (0.66)	30 (0.44)	138 (0.64)	66 (0.43)	29 (0.34)	38 (0.56)	**<0.001**	**2.42 [1.58–3.70]**	**<0.001**	**3.57 [2.11–6.05]**	0.202	–
[104]	Italy	115[b]	148	74	74	83 (0.36)	154 (0.52)	89 (0.60)	65 (0.44)	147 (0.64)	142 (0.48)	59 (0.40)	48 (0.56)	**<0.001**	**1.92 [1.35–2.73]**	**<0.001**	**2.67 [1.75–4.09]**	**<0.001**	**2.40 [1.51–3.80]**
[97]	USA	69	68			50 (0.36)	69 (0.51)			88 (0.64)	67 (0.49)			**0.015**	**1.81 [1.18–2.94]**				
[97]	Italy	69	20			50 (0.36)	26 (0.65)			88 (0.64)	14 (0.35)			**0.001**	**3.27 [1.56–6.83]**				
[99]	United Kingdom	160	152			127 (0.42)	146 (0.48)			173 (0.58)	158 (0.52)			0.160	–				
[100]	Germany	170	78			102 (0.30)	46 (0.29)			238 (0.70)	110 (0.71)			0.908	–				
[101]	Japan	374	62			403 (0.54)	57 (0.46)			345 (0.46)	67 (0.54)			0.102	–				
[102]	Japan	127	75	37	38	126 (0.50)	82 (0.55)	39 (0.53)	42 (0.55)	128 (0.50)	68 (0.45)	35 (0.47)	34 (0.45)	0.325	–	0.639	–	0.387	–

						−1438/A				−1438/G				p value and OR [95%CI]					
[103]	France	98	145			70 (0.36)	103 (0.36)			126 (0.64)	187 (0.64)			0.965	–				
[107]	Poland	89	131	96	38	101 (0.57)	170 (0.65)	127 (0.66)	43 (0.61)	77 (0.43)	92 (0.35)	65 (0.34)	27 (0.39)	0.085	–	0.063	–	0.501	–
[106]	Czech Republic		75			55 (0.42)	71 (0.47)			75 (0.58)	79 (0.53)			0.399	–				
Family studies																			
[108]	France	87	101			73 (0.42)	84 (0.36)			101 (0.58)	120 (0.59)			0.878	–				
[108]	Italy	27	29			24 (0.44)	31 (0.53)			30 (0.56)	27 (0.47)			0.341	–				
[108]	Italy & Spain	16	17			18 (0.56)	13 (0.38)			14 (0.44)	21 (0.62)			0.143	–				
[108]	United Kingdom	74	78			60 (0.41)	63 (0.40)			88 (0.59)	93 (0.60)			0.978	–				
[108]	Germany	82	90			69 (0.42)	62 (0.34)			95 (0.58)	118 (0.66)			0.146	–				
Pan-genomic studies																			
[110]	USA	677	1085	415	664	na[d]	na			na	na			ns[e]		ns		ns	
[111]	USA	3733[c]	1033	394	373	na	na			na	na			ns		ns		ns	
[112]	Worldwide	21,080	5551	1606	1445	na	na			na	na			ns		ns		ns	

[a]Underweight without AN
[b]Increased sample from [96]
[c]Pediatric controls 12.75 ± 4.2 years
[d]Not available
[e]Not significant

References

1. Clarke J, Ramoz N, Fladung AK, Gorwood P (2016) Higher reward value of starvation imagery in anorexia nervosa and association with the Val66Met BDNF polymorphism. Transl. Psychiatry 6(6):e829
2. Treasure J, Claudino AM, Zucker N (2010) Eating disorders. Lancet 375(9714):583–593. https://doi.org/10.1016/S0140-6736(09)61748-7
3. Gorwood P (2004) Eating disorders, serotonin transporter polymorphisms and potential treatment response. Am J Pharmacogenomics 4(1):9–17
4. Grahame-Smith DG (1992) Serotonin in affective disorders. Int Clin Psychopharmacol. 6(Suppl 4):5–13
5. Núñez-Navarro A, Agüera Z et al (2012) Do men with eating disorders differ from women in clinics, psychopathology and personality? Eur Eat Disord Rev 20(1):23–31
6. Barr LC, Goodman WK, Price LH, McDougle CJ, Charney DS (1992) The serotonin hypothesis of obsessive compulsive disorder: implications of pharmacologic challenge studies. J Clin Psychiatry. 53(Suppl):17–28
7. Charney DS, Deutch A (1996) A functional neuroanatomy of anxiety and fear: implications for the pathophysiology and treatment of anxiety disorders. Crit Rev Neurobiol. 10(3-4):419–446
8. Strober M, Freeman R, Lampert C, Diamond J, Kaye W (2000) Controlled family study of anorexia nervosa and bulimia nervosa: evidence of shared liability and transmission of partial syndromes. Am J Psychiatry. 157(3):393–401
9. Gorwood P, Kipman A, Foulon C (2003) The human genetics of anorexia nervosa. Eur J Pharmacol 480(1–3):163–170
10. Blundell JE (1984) Serotonin and appetite. Neuropharmacology 23(12B):1537–1551
11. Leibowitz SF, Shor-Posner G (1986) Brain serotonin and eating behavior. Appetite (7 Suppl):1–14
12. Kaye WH, Ebert MH, Gwirtsman HE et al (1984) Differences in brain serotonergic metabolism between nonbulimic and bulimic patients with anorexia nervosa. Am J Psychiatry 141(12):1598–1601
13. Brewerton TD (1995) Toward a unified theory of serotonin dysregulation in eating and related disorders. Psychoneuroendocrinology 20(6):561–590
14. Jimerson DC, Lesem MD, Kaye WH et al (1990) Eating disorders and depression: is there a serotonin connection? Biol Psychiatry 28(5):443–454
15. Kaye W, Gendall K, Strober M (1998) Serotonin neuronal function and selective serotonin reuptake inhibitor treatment in anorexia and bulimia nervosa. Biol Psychiatry 44(9):825–838
16. Kaye WH, Weltzin TE (1991) Serotonin activity in anorexia and bulimia nervosa: relationship to the modulation of feeding and mood. J Clin Psychiatry (52 Suppl):41–48
17. Kaye WH, Gwirtsman HE, George DT et al (1991) Altered serotonin activity in anorexia nervosa after long-term weight restoration. Does elevated cerebrospinal fluid 5-hydroxyindoleacetic acid level correlate with rigid and obsessive behavior? Arch Gen Psychiatry 48(6):556–562
18. Steiger H, Young SN, Kin NM (2001) Implications of impulsive and affective symptoms for serotonin function in bulimia nervosa. Psychol Med 31(1):85–95
19. Smith KA, Fairburn CG, Cowen PJ (1999) Symptomatic relapse in bulimia nervosa following acute tryptophan depletion. Arch Gen Psychiatry 56(2):171–176
20. Jimerson DC, Lesem MD, Kaye W et al (1992) Low serotonin and dopamine metabolite concentrations in cerebrospinal fluid from bulimic patients with frequent binge episodes. Arch Gen Psychiatry 49(2):132–138
21. Kaye WH, Gwirtsman HE, George DT et al (1988) CSF 5-HIAA concentrations in anorexia nervosa: reduced values in underweight subjects normalize after weight gain. Biol Psychiatry 23(1):102–105

22. Kaye WH, Ballenger JC, Lydiard RB et al (1990) CSF monoamine levels in normal-weight bulimia: evidence for abnormal noradrenergic activity. Am J Psychiatry 147(2):225–229
23. Steiger H, Richardson J, Israel M et al (2005) Reduced density of platelet-binding sites for [3H]paroxetine in remitted bulimic women. Neuropsychopharmacology 30(5):1028–1032
24. Flament MF, Bissada H, Spettigue W (2012) Evidence-based pharmacotherapy of eating disorders. Int J Neuropsychopharmacol 15(2):189–207
25. Monteleone P, Tortorella A, Castaldo E, Maj M (2006) Association of a functional serotonin transporter gene polymorphism with binge eating disorder. Am J Med Genet B Neuropsychiatr Genet. 141B(1):7–9
26. Blundell JE (1986) Serotonin manipulations and the structure of feeding behaviour. Appetite (7 Supplement):39–56
27. Kuikka JT, Tammela L, Karhunen L et al (2001) Reduced serotonin transporter binding in binge eating women. Psychopharmacology 155:310–314
28. Tammela LI, Rissanen A, Kuikka JT et al (2003) Treatment improves serotonin transporter binding and reduces binge eating. Psychopharmacology 170(1):89–93
29. Keski-Rahkonen A, Mustelin L (2016) Epidemiology of eating disorders in Europe: prevalence, incidence, comorbidity, course, consequences, and risk factors. Curr Opin Psychiatry 29(6):340–345
30. McElroy SL, Hudson JI, Malhotra S et al (2003) Citalopram in the treatment of binge-eating disorder: a placebo-controlled trial. J Clin Psychiatry 64:807–813
31. Guerdjikova AI, McElroy SL, Welge JA et al (2009) Lamotrigine in the treatment of binge-eating disorder with obesity: a randomized, placebo-controlled monotherapy trial. Int Clin Psychopharmacol 24:150–158
32. Arnold LM, McElroy SL, Hudson JI et al (2002) A placebo-controlled, randomized trial of fluoxetine in the treatment of binge-eating disorder. J Clin Psychiatry 63:1028–1033
33. Hudson JI, McElroy SL, Raymond NC et al (1998) Fluvoxamine in the treatment of binge-eating disorder: a multicenter placebo-controlled, double-blind trial. Am J Psychiatr 155:1756–1762
34. McElroy SL, Casuto LS, Nelson EB et al (2000) Placebo-controlled trial of sertraline in the treatment of binge eating disorder. Am J Psychiatr 157:1004–1006
35. McElroy SL, Guerdjikova AI, Mori N et al (2012) Pharmacological management of binge eating disorder: current and emerging treatment options. Ther Clin Risk Manag 8:219–241
36. Frank GK, Kaye WH, Meltzer CC et al (2002) Reduced 5-HT2A receptor binding after recovery from anorexia nervosa. Biol Psychiatry 52(9):896–906
37. Frank GK, Bailer UF, Henry SE et al (2005) Increased dopamine D2/D3 receptor binding after recovery from anorexia nervosa measured by positron emission tomography and [(11) C]raclopride. Biol Psychiatry 58:908–912
38. Hen R (1992) Of mice and flies: commonalities among 5-HT receptors. Trends Pharmacol Sci 13:160–165
39. Blundell JE (1977) Is there a role for serotonin (5-hydroxytryptamine) in feeding? Int J Obes. 1(1):15–42
40. Tao R, Fray A, Aspley S (2002) Effects on serotonin in rat hypothalamus of D-fenfluramine, aminorex, phentermine and fluoxetine. Eur J Pharmacol 445:69–81
41. Gibson EL, Kennedy AJ, Curzon G (1993) d-Fenfluramine- and d-norfenfluramine-induced hypophagia: differential mechanisms and involvement of postsynaptic 5-HT receptors. Eur J Pharmacol 242:83–90
42. Aigner M, Treasure J, Kaye W et al (2011) WFSBP Task Force On Eating Disorders. World Federation of Societies of Biological Psychiatry (WFSBP) guidelines for the pharmacological treatment of eating disorders. World J Biol Psychiatry 12(6):400–443
43. Heisler LK, Kanarek RB, Homoleski B (1999) Reduction of fat and protein intakes but not carbohydrate intake following acute and chronic fluoxetine in female rats. Pharmacol Biochem Behav 63:377–385
44. Halford JCG, Harrold JA, Boyland EJ et al (2007) Serotonergic drugs: effects on appetite expression and use for the treatment of obesity. Drugs 67:27–55

45. Hornung JP (2010) The neuronatomy of the serotonergic system. In: Muller CP, Jacobs BL (eds) Handbook of behavioral neuroscience, Handbook of the behavioral neurobiology of serotonin. Elsevier, Oxford, pp 51–64
46. Clark JT, Kalra PS, Crowley WR et al (1984) Neuropeptide Y and human pancreatic polypeptide stimulate feeding behavior in rats. Endocrinology 115:427–429
47. Williams G, Bing C, Cai XJ et al (2001) The hypothalamus and the control of energy homeostasis: different circuits, different purposes. Physiol Behav 74(4–5):683–701
48. Choi S, Blake V, Cole S, Fernstrom JD (2006) Effects of chronic fenfluramine administration on hypothalamic neuropeptide mRNA expression. Brain Res 1087:83–86
49. Heisler LK, Jobst EE, Sutton GM et al (2006) Serotonin reciprocally regulates melanocortin neurons to modulate food intake. Neuron 51:239–249
50. Erritzoe D, Frokjaer VG, Haugbol S (2009) Brain serotonin 2A receptor binding: relations to body mass index, tobacco and alcohol use. NeuroImage 46:23–30
51. Oh CM, Namkung J, Go Y (2015) Regulation of systemic energy homeostasis by serotonin in adipose tissues. Nat Commun 6:6794
52. Kumar JD, Mann JJ (2014) PET tracers for serotonin receptors and their applications. Cent Nerv Syst Agents Med Chem 14:96
53. Kennett GA, Curzon G (1988) Evidence that mCPP may have behavioural effects mediated by central 5-HT1C receptors. Br J Pharmacol 94:137–147
54. Voigt JP, Fink H (2015) Serotonin controlling feeding and satiety. Behav Brain Res 277:14–31
55. Erritzoe D, Frokjaer VG, Haahr MT et al (2010) Cerebral serotonin transporter binding is inversely related to body mass index. NeuroImage 52:284–289
56. Haahr ME, Rasmussen PM, Madsen K (2012) Obesity is associated with high serotonin 4 receptor availability in the brain reward circuitry. NeuroImage 61:884–888
57. Gauthier C, Hassler C, Mattar L (2014) Symptoms of depression and anxiety in anorexia nervosa: links with plasma tryptophan and serotonin metabolism. Psychoneuroendocrinology 39:170–178
58. Gauthier C, Hassler C, Mattar L, Launay JM, Callebert J, Steiger H, Melchior JC, Falissard B, Berthoz S, Mourier-Soleillant V, Lang F, Delorme M, Pommereau X, Gerardin P, Bioulac S, Bouvard M, Godart N, EVHAN Group (2014) Symptoms of depression and anxiety in anorexia nervosa: links with plasma tryptophan and serotonin metabolism. Psychoneuroendocrinology. 39:170–178
59. Bailer UF, Frank GK, Henry SE et al (2007) Exaggerated 5-HT1A but normal 5-HT2A receptor activity in individuals ill with anorexia nervosa. Biol Psychiatry 61:1090–1099
60. Galusca B, Costes N, Zito NG (2008) Organic background of restrictive-type anorexia nervosa suggested by increased Serotonin1A receptor binding in right Frontotemporal cortex of both lean and recovered patients: [18F]MPPF PET scan study. Biol Psychiatry 64:1009–1013
61. Kaye WH, Bailer UF, Frank GK et al (2006) Persistent alterations of serotonin and dopamine activity after recovery from anorexia and bulimia nervosa. Proceedings of the 18th World Congress on Psychosomatic Medicine, held in Kobe, Japan, 1287:45–48
62. Kaye W (2008) Neurobiology of anorexia and bulimia nervosa. Physiol Behav 94(1):121–135. https://doi.org/10.1016/j.physbeh.2007.11.037
63. Krebs-Thomson K, Geyer MA (1998) Evidence for a functional interaction between 5-HT1A and 5-HT2A receptors in rats. Psychopharmacology 140:69–74
64. Cloninger CR, Przybeck TR, Svrakic DM et al (1994) The temperament and character inventory (TCI): a guide to its development and use. Center for Psychobiology of Personality, Washington University, St. Louis, MO, pp 19–28
65. Clarke TK, Weiss ARD, Berrettini WH (2012) The genetics of anorexia nervosa. Clin Pharmacol Ther 91:181–188
66. Hinney A, Scherag S, Hebebrand J (2010) Genetic findings in anorexia and bulimia nervosa. Prog Mol Biol Transl Sci 94:241–270
67. Clifton PG, Barnfield AM, Curzon G (1993) Effects of food deprivation and mCPP treatment on the microstructure of ingestive behaviour of male and female rats. J Psychopharmacol 7(3):257–264

68. Maurel S, De Vry J, De Beun R et al (1999) 5-HT2A and 5-HT2C/5-HT1B receptors are differentially involved in alcohol preference and consummatory behavior in cAA rats. Pharmacol Biochem Behav 62(1):89–96
69. Aulakh CS, Hill JL, Yoney HT et al (1992) Evidence for involvement of 5-HT1C and 5-HT2 receptors in the food intake suppressant effects of 1-(2,5-dimethoxy-4-iodophenyl)-2-amino-propane (DOI). Psychopharmacology 109(4):444–448
70. Meltzer HY (2002) Action of atypical antipsychotics. Am J Psychiatry 159(1):153–154
71. Murotani T, Ishizuka T, Isogawa Y et al (2011) Possible involvement of serotonin 5-HT2 receptor in the regulation of feeding behavior through the histaminergic system. Neuropharmacology 61:228–233
72. Kaur G, Kulkarni SK (2002) Studies on modulation of feeding behavior by atypical antipsychotics in female mice. Prog Neuro-Psychopharmacol Biol Psychiatry 26:277–285
73. Méquinion M, Chauveau C, Viltart O (2015a) The use of animal models to decipher physiological and neurobiological alterations of anorexia nervosa patients. Front Endocrinol (Lausanne) 6:68
74. van Gestel MA, Kostrzewa E, Adan RAH (2014) Pharmacological manipulations in animal models of anorexia and binge eating in relation to humans: pharmacological interventions for eating disorders. Br J Pharmacol 171:4767–4784
75. Siegfried Z, Berry EM, Hao S (2003) Animal models in the investigation of anorexia. Physiol Behav 79:39–45
76. Zgheib S, Méquinion M, Lucas S (2014) Long-term physiological alterations and recovery in a mouse model of separation associated with time-restricted feeding: a tool to study anorexia nervosa related consequences. PLoS One 9:e103775
77. Willner P (2005) Chronic mild stress (CMS) revisited: consistency and behavioural-neurobiological concordance in the effects of CMS. Neuropsychobiology 52(2):90–110
78. Scheurink AJW, Boersma GJ, Nergårdh R et al (2010) Neurobiology of hyperactivity and reward: agreeable restlessness in anorexia nervosa. Physiol Behav 100:490–495
79. Méquinion M, Caron E, Zgheib S (2015b) Physical activity: benefit or weakness in metabolic adaptations in a mouse model of chronic food restriction? Am J Physiol Endocrinol Metab 308:E241–E255
80. Avraham Y, Hao S, Mendelson S et al (2001) Tyrosine improves appetite, cognition, and exercise tolerance in activity anorexia. Med Sci Sports Exerc 33(12):2104–2110
81. Verhagen LAW, Luijendijk MCM, Korte-Bouws GAH (2009) Dopamine and serotonin release in the nucleus accumbens during starvation-induced hyperactivity. Eur Neuropsychopharmacol 19:309–316
82. Hillebrand JJG, Heinsbroek ACM, Kas MJH (2006) The appetite suppressant d-fenfluramine reduces water intake, but not food intake, in activity-based anorexia. J Mol Endocrinol 36:153–162
83. Nonogaki K, Nozue K, Oka Y (2006a) Hyperphagia alters expression of hypothalamic 5-HT2C and 5-HT1B receptor genes and plasma des-acyl ghrelin levels in Ay mice. Endocrinology 147:5893–5900
84. Nonogaki K, Nozue K, Oka Y (2006b) Increased hypothalamic 5-HT2A receptor gene expression and effects of pharmacologic 5-HT2A receptor inactivation in obese Ay mice. Biochem Biophys Res Commun 351:1078–1082
85. Compan V, Zhou M, Grailhe R (2004) Attenuated response to stress and novelty and hypersensitivity to seizures in 5-HT4 receptor knock-out mice. J Neurosci 24:412–419
86. Jean A, Conductier G, Manrique C (2007) Anorexia induced by activation of serotonin 5-HT4 receptors is mediated by increases in CART in the nucleus accumbens. Proc Natl Acad Sci U S A 104:16335–16340
87. Butt I, Hong A, Di J et al (2014) The effects of serotonin1A receptor on female mice body weight and food intake are associated with the differential expression of hypothalamic neuropeptides and the GABAA receptor. Neuropeptides 48:313–318
88. Jean A, Laurent L, Bockaert J (2012) The nucleus accumbens 5-HTR_4-CART pathway ties anorexia to hyperactivity. Transl Psychiatry 2:e203

89. Weisstaub NV, Zhou M, Lira A (2006) Cortical 5-HT2A receptor signaling modulates anxiety-like behaviors in mice. Science 313:536–540
90. Tecott LH, Sun LM, Akana SF (1995) Eating disorder and epilepsy in mice lacking 5-HT2c serotonin receptors. Nature 374:542–546
91. Martin CB, Martin VS, Trigo JM et al (2014) 5-HT2C receptor desensitization moderates anxiety in 5-HTT deficient mice: from behavioral to cellular evidence. Int J Neuropsychopharmacol 31:18(3)
92. Maltais LJ, Lane PW, Beamer WG (1984) Anorexia, a recessive mutation causing starvation in preweanling mice. J Hered 75:468–472
93. Nilsson IAK, Lindfors C, Schalling M et al (2013) Anorexia and Hypothalamic Degeneration. In: Vitamins & Hormones. Elsevier, pp. 27–60
94. Ramoz N, Versini A, Gorwood P (2007) Eating disorders: an overview of treatment responses and the potential impact of vulnerability genes and endophenotypes. Expert Opin Pharmacother 8:2029–2044
95. Collier DA, Arranz MJ, Li T et al (1997) Association between 5-HT2A gene promoter polymorphism and anorexia nervosa. Lancet 350:412
96. Sorbi S, Nacmias B, Tedde A et al (1998) 5-HT2A promoter polymorphism in anorexia nervosa. Lancet 351:1785
97. Enoch MA, Kaye WH, Rotondo A et al (1998) 5-HT2A promoter polymorphism -1438G/A, anorexia nervosa, and obsessive-compulsive disorder. Lancet 351:1785–1786
98. Hinney A, Ziegler A, Nöthen MM et al (1997) 5-HT2A receptor gene polymorphisms, anorexia nervosa, and obesity. Lancet 350:1324–1325
99. Campbell DA, Sundaramurthy D, Markham AF et al (1998) Lack of association between 5-HT2A gene promoter polymorphism and susceptibility to anorexia nervosa. Lancet 351:499
100. Ziegler A, Hebebrand J, Görg T et al (1999) Further lack of association between the 5-HT2A gene promoter polymorphism and susceptibility to eating disorders and a meta-analysis pertaining to anorexia nervosa. Mol Psychiatry 4:410–412
101. Nishiguchi N, Matsushita S, Suzuki K et al (2001) Association between 5HT2A receptor gene promoter region polymorphism and eating disorders in Japanese patients. Biol Psychiatry 50:123–128
102. Ando T, Komaki G, Karibe M et al (2001) 5-HT2A promoter polymorphism is not associated with anorexia nervosa in Japanese patients. Psychiatr Genet 11:157–1160
103. Kipman A, Bruins-Slot L, Boni C et al (2002) 5-HT(2A) gene promoter polymorphism as a modifying rather than a vulnerability factor in anorexia nervosa. Eur Psychiatry 17:227–229
104. Ricca V, Nacmias B, Cellini E et al (2002) 5-HT2A receptor gene polymorphism and eating disorders. Neurosci Lett 323:105–108
105. Fuentes JA, Lauzurica N, Hurtado A et al (2004) Analysis of the -1438 G/A polymorphism of the 5-HT2A serotonin receptor gene in bulimia nervosa patients with or without a history of anorexia nervosa. Psychiatr Genet 14:107–109
106. Martásková D, Slachtová L, Kemlink D et al (2009) Polymorphisms in serotonin-related genes in anorexia nervosa. The first study in Czech population and metaanalyses with previously performed studies. Folia Biol (Praha) 55:192–197
107. Rybakowski F, Slopien A, Dmitrzak-Weglarz M et al (2006) The 5-HT2A -1438 A/G and 5-HTTLPR polymorphisms and personality dimensions in adolescent anorexia nervosa: association study. Neuropsychobiology 53:33–39
108. Gorwood P, Adès J, Bellodi L et al (2002) The 5-HT(2A) -1438G/A polymorphism in anorexia nervosa: a combined analysis of 316 trios from six European centres. Mol Psychiatry 7:90–94
109. Nacmias B, Ricca V, Tedde A et al (1999) 5-HT2A receptor gene polymorphisms in anorexia nervosa and bulimia nervosa. Neurosci Lett 277:134–136
110. Pinheiro AP, Bulik CM, Thornton LM et al (2010) Association study of 182 candidate genes in anorexia nervosa. Am J Med Genet B Neuropsychiatr Genet 153B:1070–1080

111. Wang K, Zhang H, Bloss CS et al (2011) A genome-wide association study on common SNPs and rare CNVs in anorexia nervosa. Mol Psychiatry 16:949–959
112. Boraska V, Franklin CS, Floyd JA et al (2014) A genome-wide association study of anorexia nervosa. Mol Psychiatry 19(10):1085–1094
113. Herbeth B, Aubry E, Fumeron F et al (2005) Polymorphism of the 5-HT2A receptor gene and food intakes in children and adolescents: the Stanislas Family Study. Am J Clin Nutr 82:467–470

Role of Serotonin2A (5-HT_{2A}) Receptors in Epilepsy

Vincenzo Crunelli, Marcello Venzi, Philippe De Deurwaerdère, and Giuseppe Di Giovanni

Abstract 5-Hydroxytryptamine 2A receptors (5-HT_{2A}Rs), have been implicated in various psychiatric and neurological disorders, including epilepsy. Interestingly, epileptic patients commonly present comorbid psychiatric symptoms, and a bidirectional link between depression and epilepsy has been suggested. Therefore, the alteration of 5-HT_{2A} signalling might represent a common anatomical and neurobiological substrate of both pathologies.

After a brief presentation of the role of 5-HT in epilepsy, this chapter illustrates how 5-HT_{2A} receptors may directly or indirectly control neuronal excitability in networks involved in different types of epilepsy. It also synthetizes the preclinical and clinical evidence, demonstrating the role of these receptors in antiepileptic responses.

Keywords 5-HT • 5-HT_{2A} receptor • Antidepressants • Antipsychotics • Depression • Epilepsy

Abbreviations

5-HT	5-hydroxytryptamine or serotonin
5-HT_{2A}-Rs	Serotonin 2A receptors
AD	After discharge
DA	Dopamine

V. Crunelli (✉) • G. Di Giovanni (✉)
Neuroscience Division, School of Bioscience, Cardiff University, Cardiff, UK

Department of Physiology and Biochemistry, University of Malta, Msida, Malta
e-mail: crunelli@cardiff.ac.uk; giuseppe.digiovanni@um.edu.mt

M. Venzi
AstraZeneca Translational Science Centre at Karolinska Institutet, Stockholm, Sweden

P. De Deurwaerdère
Université de Bordeaux, Bordeaux, France

Centre National de la Recherche Scientifique, UMR 5287, Bordeaux, France
e-mail: deurwaer@u-bordeaux.fr

B.P. Guiard, G. Di Giovanni (eds.), *5-HT_{2A} Receptors in the Central Nervous System*, The Receptors 32, https://doi.org/10.1007/978-3-319-70474-6_16

DG	Dentate gyrus
DOI	2,5-Dimethoxy-4-iodoamphetamine
DRN	Dorsal raphe nucleus
eGABA	Extrasynaptic $GABA_A$
GAERS	Genetic absence epilepsy in rats from Strasbourg
GPCRs	G protein coupled receptors
LC	Locus coeruleus
MDA	Maximal dentate activation
mPFC	Medial prefrontal cortex
MRN	Medial raphe nucleus
NE	Norepinephrine
NRT	Nucleus reticulari thalami
PAG	Periaqueductal grey
SERT	Serotonin transporter
SSRI	Selective serotonin reuptake inhibitor
SUDEP	Sudden unexpected death in epilepsy
SWDs	Spike and wave discharges
VB	Ventrobasal thalamus
VTA	Ventral tegmental area

Introduction

Serotonin (5-HT) is an important neurotransmitter in the brain, as it is involved in many psychiatric and neurological diseases, including epilepsy. By activating its fourteen receptor subtypes, 5-HT may directly or indirectly depolarize or hyperpolarize neurons by modulating various g-protein coupled channels, controlling the release of other neurotransmitters and the activation of intracellular pathways [1, 2]. Therefore, it is not surprising that 5-HT is involved in the sequence of events that can turn a normal neuronal network into a hyperexcitable network [3–5]. Apart from epileptogenesis, 5-HT likely plays a role in the initiation, propagation and sustainment of seizure activity. Here, we will focus on the evidence of a 5-HT_{2A}R control of epilepsy. 5-HT_{2A}Rs are G protein-coupled receptors (GPCRs), which are members of the metabotropic 7 transmembrane-spanning receptors superfamily. In particular, 5-HT_{2A}Rs, along with 5-HT_{2B} and 5-$HT_{2C,}$ belong to the 5-HT_2 receptor subfamily that consists of three Gq/G11-coupled receptors. 5-HT_{2A}Rs in general mediate excitatory effects of 5-HT on CNS neurons [6, 7].

Conventionally, epilepsy syndromes are classified into two distinct categories, focal and generalized, according to seizure onset (arising from a specific brain area or from both hemispheres), electroencephalographic and behavioural characteristics and the brain circuitry that sustain the paroxysms [8]. Focal and generalized epilepsy may also differ in terms of the pathological, neurochemical imbalance observed in the brain areas between glutamate and γ-aminobutyric acid (GABA) function. This leads to a different therapeutic approach; for instance, drugs that increase extracellular GABA levels and/or GABA transmission are first choice in focal/generalized convulsive epilepsy, whereas they exacerbate generalized non-convulsive seizures. Indeed, gabapentin, a structural GABA analogue, which increases GABA synthesis, can exacerbate absence seizures and is not indicated in non-convulsive epilepsies [9]. Consistently, we have shown that an increase of tonic GABA$_A$ inhibition is a *conditio sine qua non* for the generation of absence seizure in rodents [10–12].

The majority of the focal and generalized seizures are convulsive (60–70%), and temporal lobe epilepsy (TLE) is one of the most common forms of epilepsy with this type of seizures. TLE is traditionally associated to many disorders localized in the cortex (neocortex and entorhinal cortex), the hippocampal formation, or both. Moreover, histological reports of TLE patients and animal models of epilepsy have consistently demonstrated that pathology is not limited to these areas but is also found in the thalamus; therefore, the epileptogenic network in TLE is broad [13].

The remaining seizures are generalized non-convulsive (e.g., absence seizures) that because are mostly not associated with obvious cell death or other tissue pathology are classified are idiopathic and typically associated with genetic abnormalities [14].

Typical absence seizures of idiopathic generalized epilepsies consist of sudden, brief periods of loss of consciousness that are accompanied by synchronous, generalized spike and wave discharges (SWDs) in the EEG [14]. SWDs originate from abnormal firing in thalamic and cortical networks, and GABA$_A$ inhibition is integral to their appearance [10, 14].

The involvement of the serotonergic system in epilepsy was suggested in the late 1950s [15]. All the brain regions involved in epilepsy receive 5-HT innervation and express different 5-HTRs. Using *in situ* hybridization, western blot, and immunohistochemical analyses in rodents, 5-HT$_{2A}$R mRNA or protein have been identified in various brain regions involved in focal epilepsy, such as the amygdala, the hippocampus [16, 17], entorhinal and temporal cortex, the frontal and somatosensory cortices, and the brain circuitry of absence epilepsy, such as various nuclei of the thalamus, i.e. ventrobasal (VB) thalamus and nucleus reticulari thalami (NRT) [18]. 5-HT$_{2A}$Rs have also been detected in the majority of monoaminergic nuclei; i.e., the median and dorsal raphe nucleus, the locus coeruleus, and the ventral tegmental area (VTA) [19–21], which strongly suggests their indirect role in regulating cell normo- and hyperexcitability via monoaminergic systems. Moreover, 5-HT$_{2A}$Rs located somatodendritically or on nerve terminals of glutamatergic and GABAergic neurons might indirectly control network excitability [22].

5-HT$_{2A}$ Receptor Modulation of Focal (Limbic) and Generalized (Convulsive) Seizures

In general, drugs that elevate extracellular 5-HT levels, such as 5-hydroxytryptophan and 5-HT reuptake blockers, inhibit seizures and increase seizure threshold [23, 24]. Conversely, depletion of brain 5-HT lowers the threshold for audiogenically-, chemically-, and electrically-evoked convulsions [25]. More recently, an increased threshold to kainic acid-induced seizures was observed in mice with genetically 5-HT hyperinnervation [26]. These findings were corroborated by data showing that mice lacking the 5-HT_{1A} [27, 28], 5-HT_{2C} [29], 5-HT_4 [30], and 5-HT_7Rs [31] were more prone to develop chemical and electrical-induced seizures. On the other hand, rats which had 5-HT_{2A}R knock-down by antisense oligonucleotide treatment [32] had significantly reduced convulsions and body tremors induced by tryptamine. Nevertheless, since only 5-HT_{2C}R KO mice are prone to develop spontaneous convulsive seizures [33], and seizures have not been reported with pharmacological blockade of different 5-HTRs, adaptive changes involving different mechanisms may play a role in the low seizure thresholds observed in different 5-HTR KO mice. Therefore, it seems that serotonergic neurotransmission by activating different 5-HTRs suppresses neuronal network hyperexcitability and seizure activity [3] although opposite effects have also been reported, especially for 5-$HT_{3/4/6/7}$Rs [34].

The role of pharmacological activation of 5-HT_{2A}Rs in epilepsy modulation is far from being well established (Table 1). However, it might be an important and potential target in light of recent evidence suggesting that their activation might not only be anticonvulsant but also capable of reducing seizure-related mortality due to sudden, unexpected death in epilepsy (SUDEP) [35], which is the leading cause of death in patients with refractory epilepsy [46]. In addition, we recently showed that mCPP and lorcaserin, two 5-$HT_{2A/2C}$R agonists with different pharmacological profiles [47, 48], stop the elongation of maximal dentate activation (MDA) afterdischarge (AD) induced by repetitive perforant path electrical stimulation recorded in granular cell layer of the hippocampal dentate gyrus (DG) of urethane-anesthetized rats. The elongation of the MDA has been considered an electroencephalographic representation of epileptogenic phenomena that occurs after the first electrical insult [49], [50]. mCPP and lorcaserin effects on MDA were not blocked by SB 242084 [38], a selective 5-HT_{2C}R antagonist [51, 52], suggesting the involvement of other 5-HTRs. Interestingly, preliminary results from our laboratory seem to indicate that mCPP and lorcaserin effects on MDA elongation might be due to the activation of 5-HT_{2A}Rs rather than 5-HT_{2C}Rs since TCB-2, a potent 5-HT_{2A} agonist [53], mimicked mCPP and lorcaserin effects (unpublished observations). Conversely, evidence from other groups showed that the mixed 5-$HT_{2A/2C}$ agonist 2,5-Dimethoxy-4-iodoamphetamine (DOI) strongly facilitated kindling development and reduced the number of stimulations needed to produce generalized seizures in the amygdala kindled rats [42], while it was ineffective in all the parameters of hippocampal partial seizures generated by low-frequency electrical stimulation of the hippocampus in rats [54]. Similarly, Wada and colleagues showed that in feline hippocampal-kindled seizures, DOI had no effect, displaying only a tendency to be anti-epileptic [39]. In the same model, the selective 5-HT_{2A}R antagonist, M100907 [55], had no effect on seizure thresholds and duration and latency of secondary ADs [37]. However, the 1 mg/kg dose of MDL 100907 significantly increased the primary AD duration, suggesting that at this dose, MDL 100907 increased seizure severity in this model although the high AD control levels might have invalidated the 5-HT_{2A}R antagonist effect [37]. The 5-$HT_{2A/2C}$R antagonist, ritanserin, and the preferential 5-HT_{2A}R antagonist, ketanserin, decreased the threshold in the maximal electroshock seizures

Table 1 Antiepileptic and pro-epileptic role of the 5-HT_{2A} receptors activation in temporal lobe epilepsy

		Model	Effect	Refs
Antiepileptic role of the 5-HT_{2A} receptors activation in temporal lobe epilepsy				
Antagonists	MDL 11,939 (5-HT_{2A})	MEST test in *Lmx1b$^{f/f/p}$* mice	Blocked DOI-TCB-2 effect in preventing seizure-induced respiratory arrest and death	[35]
	Ketanserin (5-HT_{2A})	MEST test in mice	Decreases the threshold for seizures	[36]
	Ritanserin (5-$HT_{2A/2B/2C}$)	MEST test in mice	Decreases the threshold for seizures	
	MDL 100907 (5-HT_{2A})	Electroshock-induced hippocampal partial seizures in rats	Increases primary AD duration	[37]
Agonists	DOI (5-$HT_{2A/2C}$) TCB-2 (5-HT_{2A})	MEST test in *Lmx1b$^{f/f/p}$* mice	Prevented seizure-induced respiratory arrest and death	[35]
	mCPP (5-$HT_{2A/2B/2C}$)	MDA in rats	Stop MDA elongation (not blocked by SB242084)	[38]
	Lorcaserin (5-$HT_{2B/2C}$)	MDA in rats	Stop MDA elongation (not blocked by SB242084)	
	DOI (5-$HT_{2A/2C}$)	Hippocampal kindled seizures in rats	Reduces AD duration	[39]
Pro-epileptic role of the 5-HT_{2A} receptors activation in temporal lobe epilepsy				
Antagonists	Antisense oligonucleotide designed to inhibit 5-HT_{2A} expression	Tryptamine-induced serotonergic syndrome-associated convulsions	Inhibited tryptamine-induced bilateral convulsions and body tremors	[32]
	MDL100907 (5-HT_{2A})	Feline hippocampal-kindled seizures	No effect on seizure thresholds, secondary AD duration or latency of secondary AD	[37]
	Ritanserin (5-$HT_{2A/2B/2C}$)	Kainic acid-induced seizures in rats	Has no effect	[40]
	Ketanserin (5-HT_{2A})	Cocaine-induced convulsions in mice	Dose-dependently inhibits seizures	[41]
		Hippocampal kindled seizures in cats	Increases latency to generalized convulsions	[39]
		Amygdala kindling in rats	Delays the development of kindling	[42]
		Picrotoxin-induced seizures in stressed and unstressed mice	Has no effect on seizure thresholds	[43]
		Ethanol-withdrawal seizures in mice	Has no effect on seizure severity	[44]

(continued)

Table 1 (continued)

Antiepileptic role of the 5-HT$_{2A}$ receptors activation in temporal lobe epilepsy				
		Model	Effect	Refs
	Cinanserin (5-HT$_{2A/2C}$)	Cocaine-induced convulsions in mice	Dose-dependently inhibits seizures	[41]
	Pirenperone (5-HT$_{2A/2C}$)	Cocaine-induced convulsions in mice	Dose-dependently inhibits seizures	
	Dotarizine (5-HT$_{2A/2C}$)	Electroshock-induced seizures in rats	Increases the threshold for seizures	[45]
		PTZ-induced seizures in rats	Has no effect on seizure thresholds	
Agonists	DOI (5-HT$_{2A/2C}$)	Hippocampal kindled seizures in cats	Decreases latency to generalized convulsions	[39]
		Amygdala kindling in rats	Facilitates kindling and reduces the number of stimulations needed to elicit generalized convulsions	[42]
		Picrotoxin-induced seizures in stressed and unstressed mice	Has no effect on seizure thresholds	[43]

MEST maximal electroshock threshold, *PTZ* pentylenetetrazole, *SE* status epilepticus

threshold (MEST) test in mice [36]. In other experimental models, 5-HT_{2A}R antagonists failed to be effective in seizure control. Ritanserin was ineffective on kainic acid-induced seizures [40], and ketanserin did not affect the seizure threshold for picrotoxin [43] or ethanol withdrawal seizures in mice [44]; instead, it antagonized cocaine-induced convulsions in a dose-dependent manner [41]. Moreover, the 5-$HT_{2A/2C}$R and calcium antagonist, dotarizine [56], inhibited electroconvulsive shock-induced seizures but had no effect on pentylenetetrazol (PTZ)-induced convulsions in rats [45].

5-HT_{2A} Receptor Modulation of Generalized (Non-convulsive) Seizures

As far as the 5-HT control of idiopathic generalized absence seizures is concerned, most of the limited available evidence has been obtained in a well-established polygenic rat model of absence seizures, the WAG/Rij rats [57], with 5-HT_{1A}, 5-HT_{2C}, and 5-HT_7Rs appearing as the most critical for the expression of this form of epilepsy [58, 59]. Briefly, activation or inhibition of 5-HT_{1A} and 5-HT_7Rs, respectively increases or decreases SWDs. While 5-HT_{2C}R agonists inhibit epileptiform activity, 5-HT_{2C}R antagonism does not produce any effects on absence seizures [59, 60]. SSRIs such as fluoxetine and citalopram caused a moderate (non-significant) decrease in SWDs, which was inhibited or potentiated by pre-treatment with the 5-HT_{2C}R antagonist SB 242084 and the 5-HT_{1A}R antagonist, WAY-100635, respectively [59]. This suggests that 5-HT binding both 5-HT_{1A} and 5-HT_{2C}Rs does not have an important role in absence epilepsy due to the counterbalancing activation of the pro-conversant 5-HT_{1A}R and anti-convulsant 5-HT_{2C}R. On the other hand, the role of 5-HT_{2A}Rs has not been investigated in WAG/Rij rats yet. In another genetic animal model of absence epilepsy, the Groggy (GRY) rats [61], increasing 5-HT levels by treatment with the 5-HT reuptake inhibitors, fluoxetine and clomipramine, inhibited SWD generation, an effect mimicked by DOI and blocked by ritanserin pre-treatment [62].

There are two types of absence epilepsy observed both clinically in children and experimentally in rodents: typical and atypical [63]. Atypical absence seizures involve thalamo-hippocampal circuitry, while the epileptiform activity of the typical absence seizures is confined to thalamocortical neuronal pathways, sparing limbic circuitry [64]. The pharmacological profiles of the two absence types are the same because anti-absence drugs act upon thalamic circuitry which is common to both typical and atypical absence seizures. Indeed, anticonvulsant drugs that act through GABAergic mechanisms exacerbate typical and atypical absence seizures both clinically and experimentally [65]. In the atypical absence seizures induced by AY-9944, DOI reduced the total duration and number of SWDs, where ketanserin exacerbated the number of SWDs. On the other hand, in contrast to the evidence obtained in WAG/Rij rats, 5-HT_{2C}R activation by mCPP had no effect on total duration or number of SWD in a model of atypical absence epilepsy [66].

In contrast to these findings, earlier evidence had shown that serotonergic neurotransmission and 5-HT_{2A}Rs did not appear to be involved in the pathogenesis or control of absence seizures in another well-established genetic rat model, the Genetic Absence Epilepsy in Rats from Strasbourg (GAERS) [67, 68]. Although this discrepancy could be simply due to differences between the two experimental models, it is more likely explained by the lack of selectivity of the serotoninergic drugs that were used in the earlier study in GAERS. The role of 5-HT and especially the different areas in which the modulation of absence seizures might occur have not been examined thoroughly and are currently the subjects of investigation in our laboratories. We recently re-evaluated the effects of pharmacological manipulation of 5-HT_2Rs in the expression of absence seizures in GAERS, using selective 5-HT_{2A} drugs [69].

The potent 5-HT_{2A} agonist TCB-2 produced a drastic decrease in the time spent in seizures and reduced the occurrence and length of SWDs in GAERS, which, at the higher dose, persisted through the 2-hour observation window. The block of absence seizures was accompanied by typical 5-HT_{2A}-R mediated behaviours in rodents (e.g., wet dog shakes, head twitches) [70, 71]. Of importance, the effect of TCB-2 on both absence seizures and animal behavior was blocked by pretreatment with the 5-HT_{2A} antagonist, MDL11,939, confirming that the effects were indeed driven by the activation of 5-HT_{2A} receptors. Importantly, the 5-HT_{2A} antagonists, MDL11,939 and M100,907, induced a small, albeit significant increase in the time spent in seizure, driven by an increase in seizure length. The possibility that either suppression or aggravation of seizures through 5-HT_{2A}R activation or inhibition, respectively, supports the idea that these receptors tonically and phasically affect the occurrence of seizures in GAERS.

It is difficult to pinpoint the brain target(s) of 5-HT_{2A}R agonists that form the basis of the modulation of absence seizures. However, since these receptors are diffusely expressed in the thalamus and the cortex, direct changes in excitability in these areas could be hypothesized to be pivotal for the development and sustainment of absence seizures [14].

Although a recent study showed no immunohistochemical staining for the 5-HT_{2A}Rs in thalamocortical (TC) neurons in the mouse dorsal lateral geniculate nucleus (dLGN) [72], 5-HT_{2A}R mRNA was detected in GABAergic interneurons of the dLGN and the receptors were shown to control phasic $GABA_A$R inhibition [73]. These authors also showed that the intracellular pathways that couple the 5-HT_2Rs to the Ca^{2+}-influx mechanism seemed to depend on the phospholipase C (PLC) system without involving Ca^{2+} release or voltage-gated Ca^{2+} channels in the plasma membrane, and the effects of 5-HT_{2A}R activation were critically dependent on the transient receptor potential (TRP) protein, TRPC4 [73]. We also showed that α-methyl-5-hydroxytryptamine (a 5-$HT_{2A/2C}$ receptor agonist) increased extrasynaptic $GABA_A$R (e$GABA_A$R) tonic and phasic inhibition in dLGN neurons in rats, an effect that seems to be mediated by 5-HT_{2A}Rs since it was blocked by pretreatment with ketanserin and not by SB 242084 [74]. The increased synaptic GABAergic modulation by 5-HT_{2A}Rs seems to be limited to the dLGN because it is present in somatosensory VB thalamus, where TCB-2 was incapable of modifying phasic inhibition but was capable of increasing tonic inhibition [75], probably because interneurons are almost absent in rodent VB thalamus [76]. Again, this control is phasic because MDL11,939 did not have any effects on its own but blocked TCB-2 effect. Nevertheless, MDL11,939 decreased the aberrant $GABA_A$ tonic current in GAERS [75], normalizing the gain-of-function of this current that is considered a necessary condition for the development of absence seizures [10].

Interestingly, it appears that different monoaminergic neurotransmitters, including dopamine [77], can modulate $GABA_AR$ function in normal and pathological conditions of the CNS (see for a recent review [78]), likely acting on the phosphorylation state of $GABA_ARs$ [79, 80].

The strong converging corticothalamic input, during SWDs in NRT neurons, produces bursts of excitatory postsynaptic potentials that trigger T-type Ca^{2+}-channel-mediated low-threshold spikes (LTS) and bursts of action potentials [81], leading to enhanced phasic inhibition and a tonic hyperpolarization in TC neurons [82]. Activation of the 5-$HT_{2A}Rs$ in the NRT would cause depolarization and a shift in voltage dependency of the hyperpolarization activated cation current, I_h, associated a shift from bursts to single-spike activity [83]. During ictal activity, TC neurons are generally silent [82, 84], as a result of much stronger corticothalamic excitatory inputs into NRT neurons compared to TC neurons. It would be thus be expected that during SWDs, the reduced NRT GABAergic input into TC neurons by 5-$HT_{2A}R$ activation would produce a corresponding reduction in GABA release in the VB and thus in $GABA_A$ tonic current. However, we found that 5-$HT_{2A}R$ activation increased tonic current in the VB [75], an effect that should aggravate absence seizures. In contrast, the *in vivo* effect of systemic 5-$HT_{2A}R$ activation is a clear anti-absence effect. The effects of 5-$HT_{2A}R$ activation on CNS neurons is complex and they may mediate depolarizing effects of both principal excitatory and inhibitory interneurons. It is thus impossible at this stage to pinpoint the precise site of action within the thalamocortical network that mediate the predominant anti-absence effect of systemically injected 5-$HT_{2A}R$ agonist. Moreover, 5-$HT_{2A}Rs$ are highly expressed in different areas of the CNS that are known to indirectly affect the expression of absence seizures. For instance, 5-$HT_{2A}Rs$ are found in the substantia nigra pars reticulata (SNr), basal ganglia, and ventral striatum, all areas which are known to regulate the expression and maintenance of absence seizures although they are not involved in their generation [85]. Moreover, 5-$HT_{2A}Rs$ can modulate dopamine, 5-HT, and noradrenaline neuronal activity, changing the arousal state and thus reverting thalamic and cortical pathological oscillations [22, 86, 87].

Another common action of monoamine neurotransmitters in the thalamus is a depolarization of the membrane potential of TC neurons, causing rhythmic bursts to cease and tonic activity to commence [88]. The membrane depolarization caused by 5-HT is to a large extent mediated by the inhibition of a leak of K^+ conductance (IKL) [89] and by modulation of the hyperpolarization-activated nonselective cation current, I_h [90, 91]. 5-HT and α-methyl-5-hydroxytryptamine produced comparable membrane depolarization depending on Gq-coupled intracellular signalling cascades [72]. In support of this evidence, the local application of 5-HT in the NRT inhibited the burst firing of NRT neurons (associated with slow-wave sleep, inattentiveness, drowsiness), resulting in the occurrence of the single-spike activity that is a characteristic feature of aroused and attentive states [83, 92]. The administration of the

5-HT_{2A}R antagonist, ketanserin, blocked the switch from bursts to tonic NRT activity. Together, these findings suggest a serotonergic involvement in sleep-waking behaviour via modulation of pacemaking GABAergic neurons of the NRT [83, 92]. In fact, 5-HT shares with other neurotransmitters the ability to promote waking and to suppress rapid-eye-movement (REM) sleep. 5-HT_{2A}R KO mice showed a significant increase of waking and a reduction of non-rapid eye movement sleep (NREMS). Paradoxically, systemic administration of the selective 5-HT_{2A} antagonists induced a decrease of waking and an increase in NREMS during the first 3 h after injection in mice [93]. The similar pattern of effects induced by the non-selective 5-HT_2 antagonist S32212, opposite to those induced by selective 5-HT_{2C} antagonists, was attributed to its 5-HT_{2A} antagonist component [94]. A summary of the evidence of 5-HT_{2A}R modulation of absence epilepsy is shown in Table 2.

Antiepileptic Drugs and 5-HT_{2A}Rs

Lamotrigine is the only antiepileptic drug (AED) that also has a clear benefit for bipolar disorder and is approved by the FDA for maintenance treatment [95]. Interestingly, chronic lamotrigine treatment in Long-Evans rats with spontaneous SWDs suppressed seizures and ameliorated comorbid anxiety and depression, indicating that patients with absence epilepsy can benefit from this treatment [96]. There is evidence that indicates that 5-HT_{2A}R activation potentiates the inhibitory effect of lamotrigine, a widely used antiepileptic agent for both absence [97] and focal epilepsy [98] that targets voltage-gated sodium channels [99]. Therefore, the polytherapy of a non-hallucinogenic 5-HT_{2A}R agonists in combination with lamotregine may be a rational strategy in the treatment of many patients with refractory epilepy. An experimental validation of this co-administration in GAERS or WAG/rij rats might produce interesting results.

Furthermore, some ligand-binding studies in animals have shown that valproate, the first line treatment for both generalized tonic-clonic seizures and absence seizures [100], produces an increase in 5-HT_{2A}R expression [101, 102] although an *in vivo* imaging study did not confirm this in acute mania [103]. However, this study cannot exclude the possibility that valproate improves mood symptoms by altering second messenger signalling cascades linked to 5-HT_{2A}-Rs. Indeed, brain 5-HT_{2A}Rs are coupled to the phosphoinositol pathway via G-proteins, and there is a growing body of evidence that suggests that valproate and lithium have multiple effects on this pathway [104].

Role of 5-HT_{2A}Rs in the Comorbidity of Epilepsy and Depression

It is estimated that between 15 and 30% of people with epilepsy develop several psychiatric disorders, such as anxiety, depression, and different levels of cognitive impairments [105–107]. The patients with partial complex epilepsy, such as TLE, or who have poorly-controlled epilepsy have the highest frequency rate of comorbid

Table 2 Role of the 5-HT_{2A} receptors in absence epilepsy

Role of the 5-HT_{2A} receptors in absence epilepsy					
Anticonvulsant role of the 5-HT_{2A} receptors			Model	Effect	References
Typical absence epilepsy	Agonists	DOI (5-$HT_{2A/2C}$)	GRY rats	Inhibits SWDs	[62]
		m-CPP (5-$HT_{2A/2B/2C}$)	WAG/Rij rats	Decreases the duration and frequency of SWDs	[59]
		TCB-2	GAERS	Decreases the duration and frequency of SWDs	[69]
	Antagonists	Ritanserin (5-$HT_{2A/2B/2C}$)	GRY	Increases SWDs	[62]
		MDL11,939 (5-HT_{2A})	GAERS	Increases SWDs	[69]
		M100,907	GAERS	Increases SWDs	
		Ritanserin (5-$HT_{2A/2B/2C}$)	GAERS	Has no effect	[124]
		Ketanserin (5-HT_{2A})	GAERS	Has no effect	
Atypical absence epilepsy	Agonists	DOI (5-$HT_{2A/2C}$)	AY-9944 rats	Reduces the frequency and duration of slow SWDs	[66]
	Antagonists	m-CPP (5-$HT_{2A/2B/2C}$)	AY-9944 rats	Has no effect	[66]
		Ketanserin (5-HT_{2A})	AY-9944 rats	Increases the frequency and duration of slow SWDs	[66]

GRY groggy, *WAG/Rij* Wistar Albino Glaxo rats from Rijswijk, *SWD* spike-wave discharge, *GAERS* genetic absence epilepsy in rats from Strasbourg, AY-9944 trans-N,N-bis[2-Chlorophenylmethyl]-1,4-cyclohexanedimethanamine dihydrochloride. Modified from [34]

affective disorders [108]. In addition, depression-like behaviour has also been found in generalized epilepsy, such as childhood absence epilepsy [109]. This clear link between epilepsy, comorbid psychiatric disorders, and monoaminergic (specifically serotoninergic) dysfunction has also been observed in humans [110] and in different animal models of epilepsy [111, 112]. Moreover, animal and human evidence has revealed that the relationship between depression and epilepsy is, in reality, bidirectional. Indeed patients with depression, especially suicide attempters, have an increased seizure risk compared to the normal population [113]. Thus, the fact that epilepsy and depression may share common pathogenic mechanism(s) and dysfunction(s) of the serotonergic system is an obvious explanation for this bidirectional comorbidity, considering defects in the serotonergic system are linked to both conditions [111, 114]. In addition, we have shown evidence of the involvement of both serotonergic and dopaminergic systems in the pathogenesis of epilepsy [38, 49, 69, 75, 77–80] and in depression and its pharmacological treatments [87, 115]. Compelling data on the involvement of 5-HT_{1A} and 5-HT_7Rs in epilepsy and depression has been described. Therefore, it is possible to infer that agonists at these receptors might have both antiepileptic and antidepressant activity which may ameliorate cognitive deficits [49]. On the other hand, the role of the 5-HT_{2A}Rs has been less investigated, and this field is still in its infancy and has many issues that still need to be addressed. Regarding the 5-HT_{2A}R as a drug target for treating depression and epilepsy, it was recently shown in WAG/Rij rats that sub-chronic treatment with aripiprazole, a new antipsychotic with antagonism at 5-HT_{2A}/5-HT_6Rs and also partial agonism at D_2 DA and 5-HT_{1A} and 5-HT_7Rs, has an anti-absence seizure effect and positive, modulatory actions on depression, anxiety, and memory, which might also be beneficial in other epileptic syndromes [116]. Nevertheless, this study did not identify which receptor subtype was underlying these promising aripiprazole therapeutic properties. Of note, clozapine, the first atypical antipsychotic to be developed with some 5-HT_{2A}R antagonist effects, increases seizure risk even at therapeutic serum levels [117], and it is the only psychotropic drug to have received an FDA "black box" warning regarding seizures.

Improved seizure control was also observed in epileptic patients treated for psychiatric disorders with antidepressants which elevated extracellular 5-HT levels in the epileptic foci, leading to an anticonvulsant effect [118]. However, which 5-HTR subtype(s) contribute to this effect has not yet been revealed.

As far as cognitive impairments are concerned, preclinical studies have shown that the 5-HT_{2A}R inhibition also has some therapeutic benefits. For instance, ketanserin inhibited the impairment of short-term memory, which is observed after seizures in spontaneous alternation behaviour during the Y-maze task [119]. In addition, ketanserin inhibited electroconvulsive shock-induced retrograde amnesia in the step-down passive avoidance task, suggesting that 5-HT_{2A}Rs impede consolidation and/or retrieval of memory after seizures [120]. A summary of the evidence of 5-HT_{2A}R comorbidity between epilepsy and depression is shown in Table 3.

Table 3 5-HT_{2A} receptors in comorbidity between epilepsy and depression

5-HT_{2A} receptors in comorbidity between epilepsy and depression			
	Model	Effect	References
Lamotrigine	Chronic pain states in rats	+ m-CPP (5-$HT_{2A/2B/2C}$) increased the reflex inhibitory action of lamotrigine	[99]
	Chronic pain states in rats	Decreased the reflex inhibitory action of + Ketanserin (5-HT_{2A})lamotrigine	
	Humans	Bipolar disorders	[95]
	WAG/Rij rats	Suppression of AS and amelioration of comorbid anxiety and depression	[96]
Aripiprazole (5-HT_{2A}/5-HT_6 antagonist)	WAG/Rij rats	Suppression of AS amelioration of comorbid anxiety depression and memory impairment	[116]
Valproate	Humans	Increases 5-HT_{2A}-R expression	[101, 102]
	ECS	Inhibited impairment of spontaneous Alternation behavior	[119]
SSRIs (5-HT-R?)	Different models	Anticonvulsant	[118]
Ketanserin (5-HT_{2A} antagonist)	ECS	Inhibited the impairment of short-term memory	[119]
	ECS	Inhibited electroconvulsive shock-induced retrograde amnesia	[120]

WAG/Rij Wistar Albino Glaxo rats from Rijswijk, *ECS* electroconvulsive shock

Conclusion

The findings reviewed in this chapter highlight an important role for 5-HT_{2A}Rs in both pathologic neuronal excitability in epilepsy and comorbid affective disorders. The available literature suggests that antagonism at 5-HT_{2A}Rs might have beneficial effects on psychiatric disorders of TLE patients, while their activation seems to be anti-epileptic. Nevertheless, 5-HT_{2A}R role in this type of epilepsy is far from being fully clarified. Indeed, both 5-HT_{2A}R agonists and antagonists appear to be useful in focal/convulsive epilepsy treatment. On the other hand, 5-HT_{2A}R activation shows a clear anti-absence effect. These paradoxical anticonvulsant efficacy of 5-HT_{2A} antagonists and agonists can be reconciled taking into consideration that (a) the two types of epilepsy have a different network substrate, (b) both agonism and antagonism induce 5-HT_{2A}R desensitization or downregulation [121]. And/or (c) different populations of 5-HT_{2A}Rs with different signal transduction mechanisms. Moreover, the anti- versus pro-epileptic effects of the 5-HT_{2A}R activation might depend on the dose of the ligands used, with the pro-convulsive effects when the receptors are excessively activated.

The main hindrance for the development for example of 5-HT_{2A}R agonists as AEDs would be their potential hallucinogenic effects [122]. As a result, new 5-HT_{2A}R compounds with higher selectivity that lack these aversive side effects are needed.

More research is needed to clarify the role of 5-HT_{2A}Rs in seizures. Thus, increasing our understanding of the role of 5-HT_{2A}Rs and their modulation of other neurotransmitter systems, such as GABA and glutamate, might reveal novel mechanism(s) of potential translational significance.

Acknowledgments Our work in this area was supported by the ERUK (grant P1202 to VC and GDG), the Malta Council of Science and Technology (grant R&I-2013-14 to GDG and VC) and EU COST Action CM1103 (GDG and PDD).

References

1. Barnes NM, Sharp T (1999) A review of central 5-HT receptors and their function. Neuropharmacology 38:1083–1152
2. D'Adamo MC, Servettini I, Guglielmi L, Di Matteo V, Di Maio R, Di Giovanni G et al (2013) 5-HT2 receptors-mediated modulation of voltage-gated K+ channels and neurophysiopathological correlates. Exp Brain Res 230:453–462
3. Bagdy G, Kecskemeti V, Riba P, Jakus R (2007) Serotonin and epilepsy. J Neurochem 100:857–873
4. Ghanbari R, El Mansari M, Blier P (2012) Electrophysiological impact of trazodone on the dopamine and norepinephrine systems in the rat brain. Eur Neuropsychopharmacol 22:518–526
5. Jakus R, Bagdy G (2011a) The role of 5-HT2C receptor in epilepsy. In: Di Giovanni G et al (eds) 5-HT2C receptors in the pathophysiology of CNS disease, vol 22. Humana Press, Totowa, pp 429–444
6. Di Giovanni G, Di Matteo V, Pierucci M, Benigno A, Esposito E (2006) Central serotonin2C receptor: from physiology to pathology. Curr Top Med Chem 6:1909–1925
7. Millan MJ, Marin P, Bockaert J, Mannoury la Cour C (2008) Signaling at G-protein-coupled serotonin receptors: recent advances and future research directions. Trends Pharmacol Sci 29:454–464
8. Berg AT, Berkovic SF, Brodie MJ, Buchhalter J, Cross JH, van Emde Boas W et al (2010) Revised terminology and concepts for organization of seizures and epilepsies: report of the ILAE Commission on Classification and Terminology, 2005-2009. Epilepsia 51:676–685
9. Manning JP, Richards DA, Bowery NG (2003) Pharmacology of absence epilepsy. Trends Pharmacol Sci 24:542–549
10. Cope DW, Di Giovanni G, Fyson SJ, Orban G, Errington AC, Lorincz ML et al (2009) Enhanced tonic GABAA inhibition in typical absence epilepsy. Nat Med 15:1392–1398
11. Errington AC, Di Giovanni G, Crunelli V (eds) (2014) Extrasynapitic GABAA receptors. Springer, New York
12. Errington AC, Gibson KM, Crunelli V, Cope DW (2011) Aberrant GABA(A) receptor-mediated inhibition in cortico-thalamic networks of succinic semialdehyde dehydrogenase deficient mice. PLoS One 6:e19021
13. Bernhardt BC, Hong S, Bernasconi A, Bernasconi N (2013) Imaging structural and functional brain networks in temporal lobe epilepsy. Front Hum Neurosci 7:624
14. Crunelli V, Leresche N (2002) Childhood absence epilepsy: genes, channels, neurons and networks. Nat Rev Neurosci 3:371–382
15. Bonnycastle DD, Giarman NJ, Paasonen MK (1957) Anticonvulsant compounds and 5-hydroxytryptamine in rat brain. Br J Pharmacol Chemother 12:228–231
16. Bombardi C (2012) Neuronal localization of 5-HT2A receptor immunoreactivity in the rat hippocampal region. Brain Res Bull 87:259–273

17. Bombardi C, Di Giovanni G (2013) Functional anatomy of 5-HT2A receptors in the amygdala and hippocampal complex: relevance to memory functions. Exp Brain Res 230:427–439
18. Li QH, Nakadate K, Tanaka-Nakadate S, Nakatsuka D, Cui YL, Watanabe Y (2004) Unique expression patterns of 5-HT2A and 5-HT2C receptors in the rat brain during postnatal development: western blot and immunohistochemical analyses. J Comp Neurol 469:128–140
19. Cornea-Hebert V, Riad M, Wu C, Singh SK, Descarries L (1999) Cellular and subcellular distribution of the serotonin 5-HT2A receptor in the central nervous system of adult rat. J Comp Neurol 409:187–209
20. Doherty MD, Pickel VM (2000) Ultrastructural localization of the serotonin 2A receptor in dopaminergic neurons in the ventral tegmental area. Brain Res 864:176–185
21. Nocjar C, Roth BL, Pehek EA (2002) Localization of 5-HT(2A) receptors on dopamine cells in subnuclei of the midbrain A10 cell group. Neuroscience 111:163–176
22. Di Giovanni G (2013) Serotonin in the pathophysiology and treatment of CNS disorders. Exp Brain Res 230:371–373
23. Prendiville S, Gale K (1993) Anticonvulsant effect of fluoxetine on focally evoked limbic motor seizures in rats. Epilepsia 34:381–384
24. Yan QS, Jobe PC, Dailey JW (1994) Evidence that a serotonergic mechanism is involved in the anticonvulsant effect of fluoxetine in genetically epilepsy-prone rats. Eur J Pharmacol 252:105–112
25. Statnick MA, Maring-Smith ML, Clough RW, Wang C, Dailey JW, Jobe PC et al (1996) Effect of 5,7-dihydroxytryptamine on audiogenic seizures in genetically epilepsy-prone rats. Life Sci 59:1763–1771
26. Tripathi PP, Di Giovannantonio LG, Viegi A, Wurst W, Simeone A, Bozzi Y (2008) Serotonin hyperinnervation abolishes seizure susceptibility in Otx2 conditional mutant mice. J Neurosci 28:9271–9276
27. Parsons LH, Kerr TM, Tecott LH (2001) 5-HT(1A) receptor mutant mice exhibit enhanced tonic, stress-induced and fluoxetine-induced serotonergic neurotransmission. J Neurochem 77:607–617
28. Sarnyai Z, Sibille EL, Pavlides C, Fenster RJ, McEwen BS, Toth M (2000) Impaired hippocampal-dependent learning and functional abnormalities in the hippocampus in mice lacking serotonin(1A) receptors. Proc Natl Acad Sci U S A 97:14731–14736
29. Applegate CD, Tecott LH (1998) Global increases in seizure susceptibility in mice lacking 5-HT2C receptors: a behavioral analysis. Exp Neurol 154:522–530
30. Compan V, Zhou M, Grailhe R, Gazzara RA, Martin R, Gingrich J et al (2004) Attenuated response to stress and novelty and hypersensitivity to seizures in 5-HT4 receptor knock-out mice. J Neurosci 24:412–419
31. Witkin JM, Baez M, Yu J, Barton ME, Shannon HE (2007) Constitutive deletion of the serotonin-7 (5-HT(7)) receptor decreases electrical and chemical seizure thresholds. Epilepsy Res 75:39–45
32. Van Oekelen D, Megens A, Meert T, Luyten WH, Leysen JE (2003) Functional study of rat 5-HT2A receptors using antisense oligonucleotides. J Neurochem 85:1087–1100
33. Tecott LH, Sun LM, Akana SF, Strack AM, Lowenstein DH, Dallman MF et al (1995) Eating disorder and epilepsy in mice lacking 5-HT2c serotonin receptors. Nature 374:542–546
34. Gharedaghi MH, Seyedabadi M, Ghia JE, Dehpour AR, Rahimian R (2014) The role of different serotonin receptor subtypes in seizure susceptibility. Exp Brain Res 232:347–367
35. Buchanan GF, Murray NM, Hajek MA, Richerson GB (2014) Serotonin neurones have anticonvulsant effects and reduce seizure-induced mortality. J Physiol 592:4395–4410
36. Przegalinski E, Baran L, Siwanowicz J (1994) Role of 5-hydroxytryptamine receptor subtypes in the 1-[3- (trifluoromethyl)phenyl] piperazine-induced increase in threshold for maximal electroconvulsions in mice. Epilepsia 35:889–894
37. Watanabe K, Ashby CR Jr, Katsumori H, Minabe Y (2000) The effect of the acute administration of various selective 5-HT receptor antagonists on focal hippocampal seizures in freely-moving rats. Eur J Pharmacol 398:239–246

38. Orban G, Bombardi C, Marino Gammazza A, Colangeli R, Pierucci M, Pomara C et al (2014) Role(s) of the 5-HT2C receptor in the development of maximal dentate activation in the hippocampus of anesthetized rats. CNS Neurosci Ther 20:651–661
39. Wada Y, Nakamura M, Hasegawa H, Yamaguchi N (1992) Role of serotonin receptor subtype in seizures kindled from the feline hippocampus. Neurosci Lett 141:21–24
40. Velisek L, Bohacenkova L, Capkova M, Mares P (1994) Clonidine, but not ritanserin, suppresses kainic acid-induced automatisms in developing rats. Physiol Behav 55:879–884
41. Ritz MC, George FR (1997) Cocaine-induced convulsions: pharmacological antagonism at serotonergic, muscarinic and sigma receptors. Psychopharmacology 129:299–310
42. Wada Y, Shiraishi J, Nakamura M, Koshino Y (1997) Role of serotonin receptor subtypes in the development of amygdaloid kindling in rats. Brain Res 747:338–342
43. Pericic D, Lazic J, Jazvinscak Jembrek M, Svob Strac D (2005) Stimulation of 5-HT 1A receptors increases the seizure threshold for picrotoxin in mice. Eur J Pharmacol 527:105–110
44. Grant KA, Hellevuo K, Tabakoff B (1994) The 5-HT3 antagonist MDL-72222 exacerbates ethanol withdrawal seizures in mice. Alcohol Clin Exp Res 18:410–414
45. Lazarova M, Petkova B, Petkov VD (1995) Effect of dotarizine on electroconvulsive shock or pentylenetetrazol-induced amnesia and on seizure reactivity in rats. Methods Find Exp Clin Pharmacol 17:53–58
46. Shorvon S, Tomson T (2011) Sudden unexpected death in epilepsy. Lancet 378:2028–2038
47. Fletcher A, Higgins GA (2011) Serotonin and reward-related behaviour: focus on 5-HT2C receptors. In: Di Giovanni G et al (eds) 5-HT2C receptors in the pathophysiology of CNS disease. Springer, New York, pp 293–324
48. Higgins GA, Silenieks LB, Lau W, de Lannoy IA, Lee DK, Izhakova J et al (2013) Evaluation of chemically diverse 5-HT(2)c receptor agonists on behaviours motivated by food and nicotine and on side effect profiles. Psychopharmacology 226:475–490
49. Orban G, Pierucci M, Benigno A, Pessia M, Galati S, Valentino M et al (2013) High dose of 8-OH-DPAT decreases maximal dentate gyrus activation and facilitates granular cell plasticity in vivo. Exp Brain Res 230:441–451
50. Stringer JL, Williamson JM, Lothman EW (1989) Induction of paroxysmal discharges in the dentate gyrus: frequency dependence and relationship to afterdischarge production. J Neurophysiol 62:126–135
51. Di Matteo V, Di Giovanni G, Esposito E (2000) SB 242084: a selective 5-HT2C receptor antagonist. CNS Drug Rev 6:195–205
52. Kennett GA, Wood MD, Bright F, Trail B, Riley G, Holland V et al (1997) SB 242084, a selective and brain penetrant 5-HT2C receptor antagonist. Neuropharmacology 36:609–620
53. McLean TH, Parrish JC, Braden MR, Marona-Lewicka D, Gallardo-Godoy A, Nichols DE (2006) 1-Aminomethylbenzocycloalkanes: conformationally restricted hallucinogenic phenethylamine analogues as functionally selective 5-HT2A receptor agonists. J Med Chem 49:5794–5803
54. Watanabe K, Minabe Y, Ashby CR Jr, Katsumori H (1998) Effect of acute administration of various 5-HT receptor agonists on focal hippocampal seizures in freely moving rats. Eur J Pharmacol 350:181–188
55. Sorensen SM, Kehne JH, Fadayel GM, Humphreys TM, Ketteler HJ, Sullivan CK et al (1993) Characterization of the 5-Ht(2) receptor antagonist Mdl 100907 as a putative atypical antipsychotic—behavioral, electrophysiological and neurochemical studies. J Pharmacol Exp Ther 266:684–691
56. Montiel C, Herrero CJ, Garcia-Palomero E, Renart J, Garcia AG, Lomax RB (1997) Serotonergic effects of dotarizine in coronary artery and in oocytes expressing 5-HT2 receptors. Eur J Pharmacol 332:183–193
57. Coenen AM, Drinkenburg WH, Inoue M, van Luijtelaar EL (1992) Genetic models of absence epilepsy, with emphasis on the WAG/Rij strain of rats. Epilepsy Res 12:75–86
58. Graf M, Jakus R, Kantor S, Levay G, Bagdy G (2004) Selective 5-HT1A and 5-HT7 antagonists decrease epileptic activity in the WAG/Rij rat model of absence epilepsy. Neurosci Lett 359:45–48

59. Jakus R, Graf M, Juhasz G, Gerber K, Levay G, Halasz P et al (2003) 5-HT2C receptors inhibit and 5-HT1A receptors activate the generation of spike-wave discharges in a genetic rat model of absence epilepsy. Exp Neurol 184:964–972
60. Jakus R, Bagdy G (2011b) The role of 5-HT2C receptor in epilepsy. In: Di Giovanni G et al (eds) 5-HT2C receptors in the pathophysiology of CNS disease. Springer-Verlag, Wien, pp 429–444
61. Tokuda S, Kuramoto T, Tanaka K, Kaneko S, Takeuchi IK, Sasa M et al (2007) The ataxic groggy rat has a missense mutation in the P/Q-type voltage-gated Ca2+ channel alpha1A subunit gene and exhibits absence seizures. Brain Res 1133:168–177
62. Ohno Y, Sofue N, Imaoku T, Morishita E, Kumafuji K, Sasa M et al (2010) Serotonergic modulation of absence-like seizures in groggy rats: a novel rat model of absence epilepsy. J Pharmacol Sci 114:99–105
63. Cortez MA, McKerlie C, Snead OC 3rd (2001) A model of atypical absence seizures: EEG, pharmacology, and developmental characterization. Neurology 56:341–349
64. Velazquez JL, Huo JZ, Dominguez LG, Leshchenko Y, Snead OC 3rd (2007) Typical versus atypical absence seizures: network mechanisms of the spread of paroxysms. Epilepsia 48:1585–1593
65. Cortez MA, Perez Velazquez JL, Snead OC 3rd (2006) Animal models of epilepsy and progressive effects of seizures. Adv Neurol 97:293–304
66. Bercovici E, Cortez MA, Snead OC 3rd (2007) 5-HT2 modulation of AY-9944 induced atypical absence seizures. Neurosci Lett 418:13–17
67. Danober L, Deransart C, Depaulis A, Vergnes M, Marescaux C (1998) Pathophysiological mechanisms of genetic absence epilepsy in the rat. Prog Neurobiol 55:27–57
68. Depaulis A, David O, Charpier S (2015) The genetic absence epilepsy rat from Strasbourg as a model to decipher the neuronal and network mechanisms of generalized idiopathic epilepsies. J Neurosci Methods
69. Venzi M, David F, Bellet J, Bombardi C, Cavaccini A, Di Giovanni G (2016) Role of serotonin2A (5-HT2A) and 2C (5-HT2C) receptors in experimental absence seizures: an electrophysiological and immunohistochemical study in GAERS and NEC rats. Neuropharmacology 108:292–304
70. Bedard P, Pycock CJ (1977) "Wet-dog" shake behaviour in the rat: a possible quantitative model of central 5-hydroxytryptamine activity. Neuropharmacology 16:663–670
71. Corne SJ, Pickering RW (1967) A possible correlation between drug-induced hallucinations in man and a behavioural response in mice. Psychopharmacologia 11:65–78
72. Coulon P, Kanyshkova T, Broicher T, Munsch T, Wettschureck N, Seidenbecher T et al (2010) Activity modes in thalamocortical relay neurons are modulated by G(q)/G(11) family G-proteins—serotonergic and glutamatergic signaling. Front Cell Neurosci 4:132
73. Munsch T, Freichel M, Flockerzi V, Pape HC (2003) Contribution of transient receptor potential channels to the control of GABA release from dendrites. Proc Natl Acad Sci U S A 100:16065–16070
74. Di Giovanni G, Cope DW, Crunelli V (2008a) Cholinergic and monoaminergic modulation of tonic GABAA inhibition in the rat dorsal lateral geniculate nucleus. In: Annual meeting of Neuroscience Society, San Diego, USA, p 531.532/D533
75. Cavaccini A, Yagüe JG, Errington AC, Crunelli V, Di Giovanni G (2012) Opposite effects of thalamic 5-HT2A and 5-HT2C receptor activation on tonic GABA-A inhibition: implications for absence epilepsy. In: Annual meeting of Neuroscience Society, New Orleans, USA, p 138.103/B157
76. Barbaresi P, Spreafico R, Frassoni C, Rustioni A (1986) GABAergic neurons are present in the dorsal column nuclei but not in the ventroposterior complex of rats. Brain Res 382:305–326
77. Yague JG, Cavaccini A, Errington AC, Crunelli V, Di Giovanni G (2013) Dopaminergic modulation of tonic but not phasic GABA(A)-receptor-mediated current in the ventrobasal thalamus of Wistar and GAERS rats. Exp Neurol 247:1–7
78. Crunelli V, Di Giovanni G (2014) Monoamine modulation of tonic GABAA inhibition. Rev Neurosci 25(2):1–12

79. Connelly WM, Errington AC, Di Giovanni G, Crunelli V (2013) Metabotropic regulation of extrasynaptic GABA(A) receptors. Front Neural Circuits 7:171
80. Connelly WM, Errington AC, Yague JG, Cavaccini A, Crunelli V, Di Giovanni G (2014) GPCR modulation of extrasynapitic GABAA receptors. In: Errington AC et al (eds) Extrasynaptic GABAA receptors, vol 27. Springer, New York, pp 125–153
81. Steriade M (2005) Sleep, epilepsy and thalamic reticular inhibitory neurons. Trends Neurosci 28:317–324
82. Pinault D, Leresche N, Charpier S, Deniau JM, Marescaux C, Vergnes M et al (1998) Intracellular recordings in thalamic neurones during spontaneous spike and wave discharges in rats with absence epilepsy. J Physiol 509(Pt 2):449–456
83. McCormick DA, Wang Z (1991) Serotonin and noradrenaline excite GABAergic neurones of the guinea-pig and cat nucleus reticularis thalami. J Physiol 442:235–255
84. Polack PO, Guillemain I, Hu E, Deransart C, Depaulis A, Charpier S (2007) Deep layer somatosensory cortical neurons initiate spike-and-wave discharges in a genetic model of absence seizures. J Neurosci 27:6590–6599
85. Warter JM, Vergnes M, Depaulis A, Tranchant C, Rumbach L, Micheletti G et al (1988) Effects of drugs affecting dopaminergic neurotransmission in rats with spontaneous petit mal-like seizures. Neuropharmacology 27:269–274
86. Di Giovanni G, Di Matteo V, Esposito E (eds) (2008b) Serotonin–dopamine interaction: experimental evidence and therapeutic relevance. Elsevier, Amsterdam
87. Di Giovanni G, Esposito E, Di Matteo V (2010) Role of serotonin in central dopamine dysfunction. CNS Neurosci Ther 16:179–194
88. McCormick DA (1992) Neurotransmitter actions in the thalamus and cerebral cortex and their role in neuromodulation of thalamocortical activity. Prog Neurobiol 39:337–388
89. Meuth SG, Aller MI, Munsch T, Schuhmacher T, Seidenbecher T, Meuth P et al (2006) The contribution of TWIK-related acid-sensitive K+–containing channels to the function of dorsal lateral geniculate thalamocortical relay neurons. Mol Pharmacol 69:1468–1476
90. Chapin EM, Andrade R (2001) A 5-HT(7) receptor-mediated depolarization in the anterodorsal thalamus. II. Involvement of the hyperpolarization-activated current I(h). J Pharmacol Exp Ther 297:403–409
91. Pape HC, McCormick DA (1989) Noradrenaline and serotonin selectively modulate thalamic burst firing by enhancing a hyperpolarization-activated cation current. Nature 340:715–718
92. McCormick DA, Pape HC (1990) Properties of a hyperpolarization-activated cation current and its role in rhythmic oscillation in thalamic relay neurones. J Physiol 431:291–318
93. Popa D, Lena C, Fabre V, Prenat C, Gingrich J, Escourrou P et al (2005) Contribution of 5-HT2 receptor subtypes to sleep-wakefulness and respiratory control, and functional adaptations in knock-out mice lacking 5-HT2A receptors. J Neurosci 25:11231–11238
94. Dekeyne A, Brocco M, Loiseau F, Gobert A, Rivet JM, Di Cara B et al (2012) S32212, a novel serotonin type 2C receptor inverse agonist/alpha2-adrenoceptor antagonist and potential antidepressant: II. A behavioral, neurochemical, and electrophysiological characterization. J Pharmacol Exp Ther 340:765–780
95. Bowden CL, Calabrese JR, Sachs G, Yatham LN, Asghar SA, Hompland M, Montgomery P, Earl N, Smoot TM, Deveaugh-Geiss J, Lamictal 606 Study, G (2003) A placebo-controlled 18-month trial of lamotrigine and lithium maintenance treatment in recently manic or hypomanic patients with bipolar I disorder. Arch Gen Psychiatry 60:392–400
96. Huang HY, Lee HW, Chen SD, Shaw FZ (2012) Lamotrigine ameliorates seizures and psychiatric comorbidity in a rat model of spontaneous absence epilepsy. Epilepsia 53:2005–2014
97. Glauser TA, Cnaan A, Shinnar S, Hirtz DG, Dlugos D, Masur D et al (2013) Ethosuximide, valproic acid, and lamotrigine in childhood absence epilepsy: initial monotherapy outcomes at 12 months. Epilepsia 54:141–155
98. Iyer A, Marson A (2014) Pharmacotherapy of focal epilepsy. Expert Opin Pharmacother 15:1543–1551
99. Than M, Kocsis P, Tihanyi K, Fodor L, Farkas B, Kovacs G, Kis-Varga A, Szombathelyi Z, Tarnawa I (2007) Concerted action of antiepileptic and antidepressant agents to

depress spinal neurotransmission: Possible use in the therapy of spasticity and chronic pain. Neurochem Int 50:642–652
100. Loscher W (2002) Basic pharmacology of valproate: a review after 35 years of clinical use for the treatment of epilepsy. CNS Drugs 16:669–694
101. Green AR, Johnson P, Mountford JA, Nimgaonkar VL (1985) Some anticonvulsant drugs alter monoamine-mediated behaviour in mice in ways similar to electroconvulsive shock; implications for antidepressant therapy. Br J Pharmacol 84:337–346
102. Sullivan NR, Burke T, Siafaka-Kapadai A, Javors M, Hensler JG (2004) Effect of valproic acid on serotonin-2A receptor signaling in C6 glioma cells. J Neurochem 90:1269–1275
103. Yatham LN, Liddle PF, Lam RW, Adam MJ, Solomons K, Chinnapalli M et al (2005) A positron emission tomography study of the effects of treatment with valproate on brain 5-HT2A receptors in acute mania. Bipolar Disord 7(Suppl 5):53–57
104. Brown KM, Tracy DK (2013) Lithium: the pharmacodynamic actions of the amazing ion. Ther Adv Psychopharmacol 3:163–176
105. Kanner AM (2003) Depression in epilepsy: prevalence, clinical semiology, pathogenic mechanisms, and treatment. Biol Psychiatry 54:388–398
106. Kanner AM, Balabanov A (2002) Depression and epilepsy: how closely related are they? Neurology 58:S27–S39
107. Stafford-Clark D (1954) Epilepsy and depression: implications of empirical therapy. Guys Hosp Rep 103:306–316
108. Kanner AM, Schachter SC, Barry JJ, Hersdorffer DC, Mula M, Trimble M et al (2012) Depression and epilepsy: epidemiologic and neurobiologic perspectives that may explain their high comorbid occurrence. Epilepsy Behav 24:156–168
109. Vega C, Guo J, Killory B, Danielson N, Vestal M, Berman R et al (2011) Symptoms of anxiety and depression in childhood absence epilepsy. Epilepsia 52:e70–e74
110. Harden CL (2002) The co-morbidity of depression and epilepsy: epidemiology, etiology, and treatment. Neurology 59:S48–S55
111. Epps SA, Tabb KD, Lin SJ, Kahn AB, Javors MA, Boss-Williams KA et al (2012) Seizure susceptibility and epileptogenesis in a rat model of epilepsy and depression co-morbidity. Neuropsychopharmacology 37:2756–2763
112. Sarkisova K, van Luijtelaar G (2012) The WAG/Rij strain: a genetic animal model of absence epilepsy with comorbidity of depression [corrected]. Prog Neuro-Psychopharmacol Biol Psychiatry 35:854–876
113. Hesdorffer DC, Allen Hauser W, Olafsson E, Ludvigsson P, Kjartansson O (2006) Depression and suicide attempt as risk factors for incident unprovoked seizures. Ann Neurol 59:35–41
114. Epps SA, Weinshenker D (2013) Rhythm and blues: animal models of epilepsy and depression comorbidity. Biochem Pharmacol 85:135–146
115. Esposito E, Di Matteo V, Di Giovanni G (2008) Serotonin-dopamine interaction: an overview. Prog Brain Res 172:3–6
116. Russo E, Citraro R, Davoli A, Gallelli L, Donato Di Paola E, De Sarro G (2013) Ameliorating effects of aripiprazole on cognitive functions and depressive-like behavior in a genetic rat model of absence epilepsy and mild-depression comorbidity. Neuropharmacology 64:371–379
117. Hedges D, Jeppson K, Whitehead P (2003) Antipsychotic medication and seizures: a review. Drugs Today (Barc) 39:551–557
118. Specchio LM, Iudice A, Specchio N, La Neve A, Spinelli A, Galli R et al (2004) Citalopram as treatment of depression in patients with epilepsy. Clin Neuropharmacol 27:133–136
119. Hidaka N, Suemaru K, Araki H (2010) Serotonin-dopamine antagonism ameliorates impairments of spontaneous alternation and locomotor hyperactivity induced by repeated electroconvulsive seizures in rats. Epilepsy Res 90:221–227
120. Genkova-Papazova M, Lazarova-Bakarova M, Petkov VD (1994) The 5-HT2 receptor antagonist ketanserine prevents electroconvulsive shock- and clonidine-induced amnesia. Pharmacol Biochem Behav 49:849–852

121. Graybiel AM (2004) Network-level neuroplasticity in cortico-basal ganglia pathways. Parkinsonism Relat Disord 10:293–296
122. Krebs-Thomson K, Paulus MP, Geyer MA (1998) Effects of hallucinogens on locomotor and investigatory activity and patterns: influence of 5-HT2A and 5-HT2C receptors. Neuropsychopharmacology 18:339–351
123. Deransart C, Riban V, Le B, Marescaux C, Depaulis A (2000) Dopamine in the striatum modulates seizures in a genetic model of absence epilepsy in the rat. Neuroscience 100:335–344
124. Marescaux C, Vergnes M, Depaulis A (1992) Genetic absence epilepsy in rats from Strasbourg - A review. J Neural Transm 35:37–69

5-HT_{2A} Receptors and BDNF Regulation: Implications for Psychopathology

Minal Jaggar and Vidita A. Vaidya

Abstract Serotonin$_{2A}$ (5-HT_{2A}) receptors are implicated in the pathophysiology of mood disorders and schizophrenia, as well as in mediating the effects of hallucinogens. 5-HT_{2A} receptors also serve as drug targets for specific classes of antidepressants and atypical antipsychotics. Preclinical and clinical studies have identified an important role for brain derived neurotrophic factor (BDNF) in the pathogenesis of depression and schizophrenia, and in the therapeutic actions of antidepressants and antipsychotics. 5-HT_{2A} receptors have been reported to regulate BDNF expression within key limbic neurocircuits, including the prefrontal cortex and hippocampus. Further, alterations in BDNF directly impact 5-HT_{2A} receptor expression, signaling and function. In this book chapter, we have extensively reviewed the current understanding of the regulation of BDNF by 5-HT_{2A} receptors at multiple levels spanning from transcriptional regulation to modulation of BDNF signaling. We have also discussed the impact of perturbations in BDNF on 5-HT_{2A} receptors, primarily focusing on studies from BDNF mouse mutant models. These studies highlight a reciprocal relationship between 5-HT_{2A} receptors and BDNF, and suggest that such a crosstalk may play an important role in the actions of stress, antidepressant and atypical antipsychotic treatments, and in mediating hallucinogenic responses. We also highlight specific open questions hitherto unexplored in understanding the nature of interaction between 5-HT_{2A} receptors and BDNF, and the implications of such a relationship to psychopathology.

Keywords Serotonin • Brain derived neurotrophic factor • Tropomyosin related kinase B • Neurotrophin • Hippocampus • Neocortex • Stress • DOI • Ketanserin

M. Jaggar • V.A. Vaidya (✉)
Department of Biological Sciences, Tata Institute of Fundamental Research, Mumbai, Maharashtra, India
e-mail: vvaidya@tifr.res.in

B.P. Guiard, G. Di Giovanni (eds.), *5-HT_{2A} Receptors in the Central Nervous System*, The Receptors 32, https://doi.org/10.1007/978-3-319-70474-6_17

Introduction

Serotonin$_{2A}$ (5-HT$_{2A}$) receptors and brain derived neurotrophic factor (BDNF) have both been implicated in the pathophysiology and treatment of depression, anxiety [1–5] and schizophrenia [6, 7]. 5-HT$_{2A}$ receptors and BDNF exert powerful effects at the molecular and cellular level, regulating synaptic plasticity [8–11], neuronal survival and progenitor turnover [12–16], as well as playing a key role at the organismal level by influencing cognitive [17–19], anxiety, depressive [4, 5, 20] and psychosis-related behaviors [21–25]. A reciprocal relationship between 5-HT$_{2A}$ receptors and BDNF has been highlighted across several studies, primarily in rodent models [26–33]. The goal of this chapter is to comprehensively review the 5-HT$_{2A}$ receptor-BDNF interactions and to highlight the importance of this reciprocity in the context of psychopathology.

Serotonin$_{2A}$ (5-HT$_{2A}$) Receptors

Serotonin (5-Hydroxytryptamine or 5-HT) is amongst the most phylogenetically ancient neurotransmitters, and is also synthesized in several organisms that lack nervous systems, wherein it is thought to exhibit a trophic factor-like function regulating growth, cellular proliferation, maturation and movement [34]. During evolution, serotonin was likely co-opted to take on a neurotransmitter function where it exerts pleiotropic effects on diverse brain functions, including but not restricted to regulation of emotion, cognition, feeding, body temperature and sleep [35]. These diverse functions of serotonin are mediated via 14 distinct serotonin receptors belonging to seven 5-HT receptor subfamilies of which 5-HT$_1$, 5-HT$_2$, 5-HT$_{4–7}$ are G-protein coupled receptors (GPCRs) and the 5-HT$_3$ receptor is a ligand-gated ion channel [36]. Amongst these different serotonergic receptors that exhibit distinct expression profiles, 5-HT$_{1A}$ and 5-HT$_{2A/2C}$ receptors have been strongly implicated in the regulation of emotionality.

Serotonin$_{2A}$ (5-HT$_{2A}$) receptors are metabotropic seven-transmembrane domain receptors that belong to the 5-HT$_2$ receptor subfamily [36]. The 5-HT$_{2A}$ receptors can signal via multiple intracellular signaling cascades, through the recruitment of Gq/G11-dependent mechanisms [37, 38] or through β-arrestin dependent signaling [39, 40] (Fig. 1a). Further, the ability of 5-HT$_{2A}$ receptors to form heteromeric complexes with glutamatergic mGluR2 [41] and dopaminergic D2R [42] receptors could also play an important role in influencing the ligand-dependent, differential recruitment of signaling pathways by diverse 5-HT$_{2A}$ receptor ligands [43, 44] (Fig. 1b). Different 5-HT$_{2A}$ receptor ligands, including the endogenous ligand serotonin, hallucinogenic agonists such as 4-iodo-2,5-dimethoxyphenylisopropylamine (DOI) and atypical antipsychotics such as clozapine are reported to bias the nature of downstream signaling pathways recruited by the 5-HT$_{2A}$ receptor-mGluR2 [44, 45] and 5-HT$_{2A}$ receptor-D2R heterocomplexes [42, 46], shifting the balance of

Gq-Gi mediated downstream signaling (Fig. 1b). In addition, the 5-HT_{2A} receptor ligands serotonin, DOI, the antagonist ketanserin and the atypical antipsychotic clozapine also exhibit characteristic dynamics for 5-HT_{2A} receptor internalization and recycling as well as distinct endocytosed 5-HT_{2A} receptor-mediated signaling [47, 48] (Fig. 1c, d). The 5-HT_{2A} receptors serve as the major excitatory receptor subtype amongst the metabotropic 5-HT receptors, and regulate excitability through the reduction of an outward potassium current [49–51], and enhanced phosphoinositide hydrolysis and calcium signaling [52, 53] (Fig. 1a). Several distinct classes of drugs including therapeutic agents such as atypical antipsychotics, antidepressants and anxiolytics, and drugs of abuse such as the hallucinogens, lysergic acid diethylamide (LSD) and mescaline target the 5-HT_{2A} receptor [23, 54–56] (Fig. 1c). Both preclinical and clinical studies have linked 5-HT_{2A} receptors to the modulation of anxiety and depressive behavior [3–5, 55, 57], mechanism of action of hallucinogenic and antipsychotic drugs [24, 41, 44], and in the regulation of learning and memory [17, 23, 58–60].

Ligand binding analysis, *in situ* hybridization, and immunohistochemical localization studies indicate that the 5-HT_{2A} receptor is widely distributed within the central nervous system in all species studied [36, 61] (Fig. 1e). 5-HT_{2A} receptors are enriched across multiple cortical regions, including the neocortex, entorhinal and piriform cortex, and the claustrum [62, 63]. Brain regions such as the hippocampus, caudate, nucleus accumbens, amygdala, thalamus, olfactory tubercle, periaqueductal grey, dorsal raphe nucleus and locus coeruleus of several of the species (mouse, rat, sheep, monkey, human) studied also express 5-HT_{2A} receptors ([36, 64–67]) (Fig. 1e). In the rat neocortex, in particular the frontal cortex, 5-HT_{2A} receptors that are densely expressed within the apical dendrites of Layer V pyramidal neurons evoke excitatory postsynaptic potentials (EPSPs) [49, 68]. In contrast, within rat paleocortical brain regions such as the piriform cortex and hippocampus, 5-HT_{2A} receptors primarily induce inhibitory postsynaptic potentials (IPSPs) [69–73]. This is mediated through the facilitation of GABAergic neurotransmission via the activation of 5-HT_{2A} receptor expressing GABAergic interneurons within the hippocampal [73] and piriform cortex microcircuits [74]. Within these distinct cortical circuits, the differential pattern of expression of 5-HT_{2A} receptors on excitatory pyramidal neurons, as well as GABAergic interneurons, plays an important role in driving the eventual functional impact on the local microcircuit [72–74] (Fig. 1e). In addition to neuronal expression of the 5-HT_{2A} receptors, rat astrocytes [75, 76] as well as microglia [77] exhibit expression of the 5-HT_{2A} receptor. 5-HT_{2A} receptors have been implicated in the regulation of calcium signaling in astrocytes [78, 79] and exosome secretion in microglia [80]. The pattern of expression of 5-HT_{2A} receptors is also differentially regulated during development [81–83] and exhibits a sexual dimorphism in specific neurocircuits of rats, such as the ventromedial hypothalamus, where females exhibited lower levels of expression [84].

Several studies indicate a role for serotonergic signaling in the regulation of the neurotrophin, BDNF [85], and the 5-HT_{2A} receptor in particular has been strongly implicated in the regulation of *Bdnf* expression within the neocortex and hippocampus [32, 33, 86]. Given the role of BDNF in regulating synaptic plasticity, anxiety

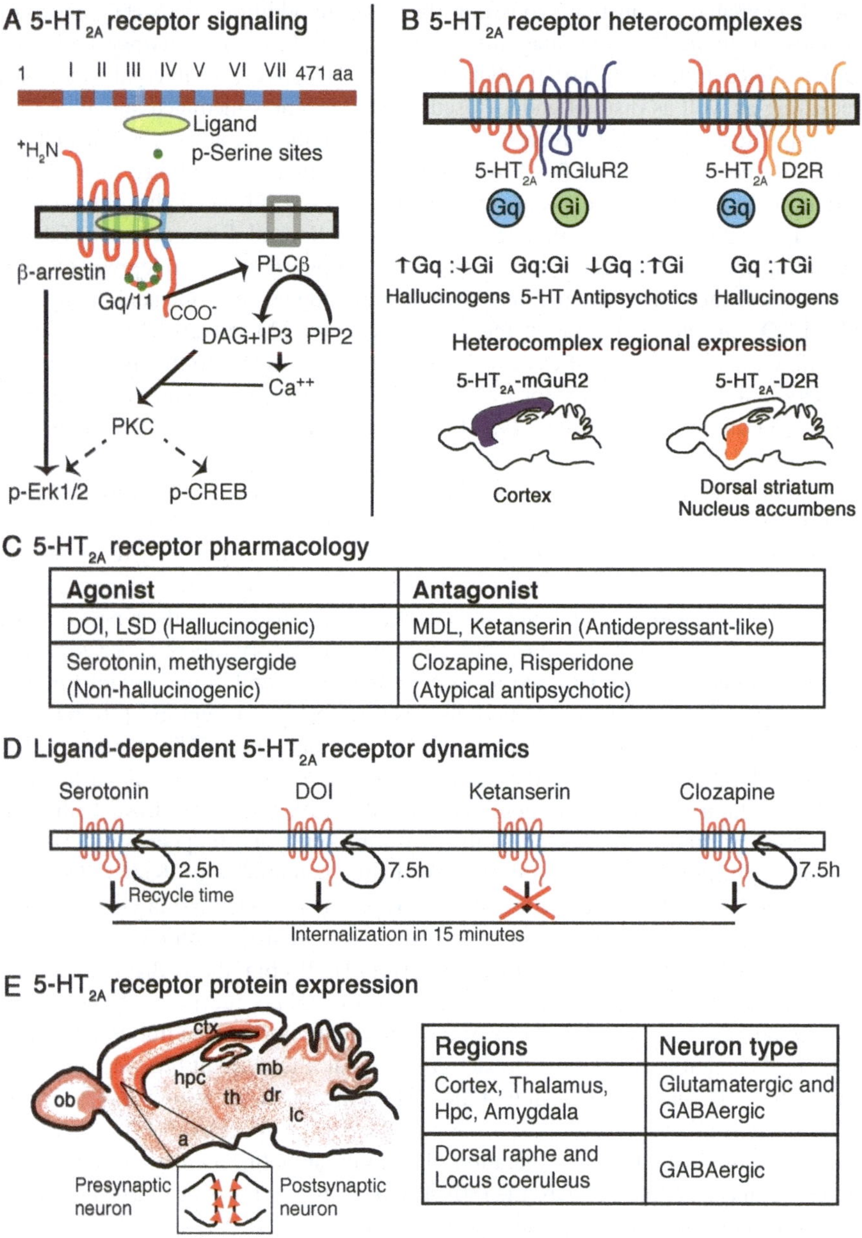

Agonist	Antagonist
DOI, LSD (Hallucinogenic)	MDL, Ketanserin (Antidepressant-like)
Serotonin, methysergide (Non-hallucinogenic)	Clozapine, Risperidone (Atypical antipsychotic)

Regions	Neuron type
Cortex, Thalamus, Hpc, Amygdala	Glutamatergic and GABAergic
Dorsal raphe and Locus coeruleus	GABAergic

Fig. 1 Schematics representing 5-HT$_{2A}$ receptor signaling (**a**), 5-HT$_{2A}$ receptor- heterocomplexes (**b**), 5-HT$_{2A}$ receptor pharmacology (**c**), Ligand-dependent 5-HT$_{2A}$ receptor dynamics (**d**), and 5-HT$_{2A}$ receptor protein expression (**e**). (**a**) Shown is a schematic highlighting transmembrane domains (blue), ligand (yellow) and serine phosphorylation sites (green) on the 471 amino acid long 5-HT$_{2A}$ receptor (red). Illustrations depict 5-HT$_{2A}$ receptor signaling through either G protein regulated cascades (Gq/G11) leading to recruitment of phospholipase C beta (PLCβ), PIP2 hydrolysis, IP3, DAG, calcium signaling, protein kinase C (PKC), extracellular signal regulated kinase

and depressive behavior, as well as learning and memory, it has been hypothesized that BDNF may contribute to specific cellular and behavioral effects of 5-HT_{2A} receptor signaling.

Brain Derived Neurotrophic Factor (BDNF)

Brain Derived Neurotrophic Factor (BDNF), the most widely expressed member of the neurotrophic factor family [87–90], is known to regulate neurogenesis [91–93], neuronal survival [94] and differentiation [95], neurite outgrowth [96], as well as structural and functional modulation of synaptic plasticity [97, 98]. Mature BDNF is generated through the proteolytic cleavage of the precursor form proBDNF [99, 100] (Fig. 2a), which is secreted by neurons [101] and astrocytes [102] in both a constitutive and regulated fashion. BDNF, which is a 14 kDa protein, regulates intracellular signaling through binding to its tyrosine kinase coupled receptor, Tropomyosin receptor kinase type B (TrkB), thus leading to the activation of signaling cascades, in particular the MAP Kinase (MAPK – mitogen activated protein kinase), PI3 kinase and PLCγ signaling pathways [103] (Fig. 2a). In addition, both the precursor form proBDNF and mature BDNF also bind to the low-affinity p75 neurotrophin receptor (p75NTR) [103, 104], thus regulating intracellular signaling via the NF-κB and JNK pathways and influencing cell death [105, 106] (Fig. 2a). Differential processing of BDNF and the recruitment of distinct intracellular signaling cascades can evoke diverse biological outcomes spanning from the recruitment of apoptotic pathways to promoting cell survival [94, 101, 107, 108]. BDNF is trafficked and released in both an anterograde and retrograde fashion [109], and the BDNF receptors, TrkB and p75NTR are expressed both pre- [110] and post-synaptically [111]. Further, the expression of truncated forms of the TrkB receptor (TrkB. T1 and TrkB.T2) lacking the intracellular catalytic domain, provides another level for the regulation of BDNF signaling [112, 113] (Fig. 2a).

Bdnf is expressed at high levels in the brain, with highest expression observed within the neocortex, hippocampus and hypothalamus [114, 115] (Fig. 2b).

Fig. 1 (continued) (Erk1/2) and cAMP response element binding protein (CREB) or β-arrestin-mediated activation of Erk1/2. (**b**) 5-HT_{2A} receptors (red) form heterocomplexes with mGluR2 (purple) or D2R (orange) receptors with ligand-dependent alterations in Gq-Gi signaling and distinct expression patterns. (**c**) Listed are classes of 5-HT_{2A} receptor agonists and antagonists. (**d**) Distinct 5-HT_{2A} receptor ligands including the endogenous agonist serotonin, the hallucinogenic agonist DOI, the antagonist ketanserin and the atypical antipsychotic clozapine exhibit differential timelines for receptor recycling. (E) Shown is the 5-HT_{2A} receptor protein expression within a schematic of a saggital section of the rodent brain, highlighting expression within the olfactory bulb (ob), cortex (ctx) especially in layers II/III and V/VI, hippocampus (hpc), amygdala (a), thalamus (th), midbrain (mb), dorsal raphe nucleus (dr), and locus coeruleus (lc). In the neocortex, 5-HT_{2A} receptors are reported to be present both pre- and post-synaptically (boxed). Further, highlighted is the presence of 5-HT_{2A} receptors in glutamatergic and GABAergic neuronal cell types within specific brain regions

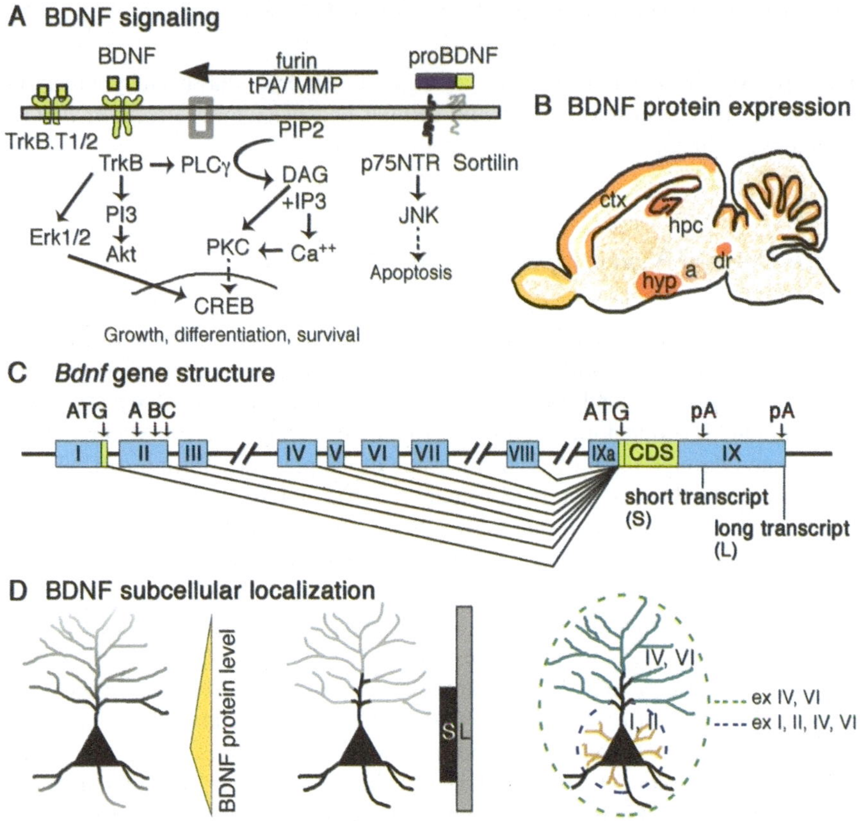

E *Bdnf* exon-specific transcript distribution in mouse brain

Brain region	I	II	III	IV	V	VI	VII	VIII	IXa
Cortex	+++	+	+++	++	++	+++	+	++	+++
Hippocampus	+++	+++	++	++	++	+++	+	++	+++
Thalamus	+++	++	+++	++	+	+++	+	++	++
Midbrain	+++	++	+++	++	+	+++	+	++	+++
Brain stem	++	+	+++	+	+	+++	+	+	++

F *Bdnf* exon-specific transcript distribution in rat brain

Brain region	I	II	III	IV	V	VI	VII	VIII	IXa
Cortex	+++	+	+++	++	++	+++	-	+	+
Hippocampus	+++	++	+++	++	++	++	+++	++	+++
Thalamus	++	+	+++	++	-	+++	-	++	-
Midbrain	++	+	+++	++	+	+++	++	+	-
Brain stem	++	+	++	++	+	+++	+	++	+

Fig. 2 Schematics representing BDNF signaling (**a**), BDNF protein expression (**b**), *Bdnf* gene structure (**c**), BDNF subcellular localization (**d**), and *Bdnf* exon-specific transcript distribution in the mouse (**e**) and rat (**f**) brain. (**a**) Represented is the BDNF signaling cascade in which proBDNF

Expression of *Bdnf* is developmentally regulated [116, 117], activity-dependent [118], and reaches a peak during postnatal development [119, 120]. The *Bdnf* gene locus is complex, with the generation of at least 18 distinct *Bdnf* transcript variants driven through distinct *Bdnf* exon-specific promoters [119, 121] (Fig. 2c). In the rat, the *Bdnf* gene consists of nine exons, with the generation of multiple transcript variants through the alternate splicing of eight non-coding 5′ exons each with its own unique promoter to the common 3′ coding exon that contains the sequence information for the generation of mature BDNF (Fig. 2c). Further, these *Bdnf* transcript variants can also exist in either a short or long 3′-UTR form through the use of two distinct polyadenylation sites, which is reported to result in distinct subcellular localization [119, 122, 123] (Fig. 2c, d). Exon-specific *Bdnf* transcript variants exhibit a distinct localization pattern in the mouse and rat brain, suggesting a differential recruitment of individual exon-specific *Bdnf* promoters in a brain region dependent fashion [119] (Fig. 2e, f). While a single BDNF protein is generated by all of these *Bdnf* transcript variants, it is likely that such a diversity of transcript variants, with unique 5′-UTRs and either a short or long 3′-UTR, contributes to differential localization, variant-specific regulation in response to diverse cues, trafficking and mRNA stability [119, 124–128] (Fig. 2c–f).

Bdnf expression is regulated by diverse environmental cues including stress [129, 130], enriched environment [131, 132] and learning [133, 134], as well as by pharmacological agents such as antidepressants [135], anxiolytics [136, 137], antipsychotics [138, 139], and drugs of abuse [140, 141] including hallucinogens [142, 143]. Regulation of BDNF can occur at multiple levels, including alteration of transcription of distinct *Bdnf* mRNA variants [124, 127], BDNF protein synthesis [144, 145], BDNF proteolytic processing [100, 146], trafficking [147, 148] and release [149, 150], regulation of BDNF receptors [151], including the truncated TrkB

Fig. 2 (continued) (purple-yellow) is proteolytically cleaved to generate the 14 kDa mature BDNF (yellow) protein. Mature BDNF can signal through the trKB receptors recruiting phospholipase C gamma (PLCγ), IP3, DAG, calcium signaling, protein kinase C (PKC), phosphoinositide 3-kinase (PI3K), protein kinase B (Akt), extracellular signal regulated kinase (Erk1/2) and cAMP response element binding protein (CREB). Mature BDNF promotes growth, differentiation and survival. Truncated TrkB.T1/T2 receptors can also sequester mature BDNF. Mature and ProBDNF can also signal through the p75NTR-sortilin complex, a pathway that recruits Jun amino-terminal kinases (JNK) and apoptosis. (**b**) Shown is BDNF protein expression within a schematic of a saggital section of the rodent brain, highlighting expression within the cortex (ctx), hippocampus (hpc), amygdala (a), hypothalamus (hyp), and dorsal raphe nucleus (dr). (**c**) Represented is the *Bdnf* gene structure consisting of *Bdnf* exons I to IX (blue), exon-specific promoters and splice junctions connected to the common coding exon IX (yellow). The 3'UTR in exon IX consists of two polyA (pA) transcription termination sites, generating a short and long form of *Bdnf* exon-specific transcripts. (**d**) Shown is BDNF protein subcellular localization with highest levels in the soma of neuronal cells. The subcellular distribution of the short and long form of *Bdnf* exon-specific transcripts and of *Bdnf* exon-specific transcript variants is also shown (**d**). Shown is the region specific expression pattern of *Bdnf* exon-specific transcript variants (I to IX) in the mouse (**d**) and Rat (**e**) brain. '+' indicates extent of expression, '–' indicates undetectable level of *Bdnf* transcript variant expression

variants [152–154], and the modulation of distinct intracellular BDNF signaling cascades [104, 105]. However, most studies examining the effects of serotonergic pathways, including the 5-HT_{2A} receptors, on the regulation of BDNF have primarily focused at the level of *Bdnf* transcription [32, 86, 135, 138, 143, 155, 156]. Our chapter will review the present understanding of the regulation of BDNF by 5-HT_{2A} receptors, and also highlight the open questions that remain unexplored in understanding the effects of 5-HT_{2A} receptors on *Bdnf* transcription, BDNF protein synthesis, trafficking, release and signaling.

5-HT_{2A} Receptor Mediated Regulation of BDNF

Serotonergic signaling has been shown to alter *Bdnf* mRNA expression within key limbic brain regions such as the frontal cortex and hippocampus, with distinct effects evoked in response to acute or chronic elevation of 5-HT levels [27, 157–159]. An acute increase in 5-HT levels in general evokes an induction of neocortical *Bdnf* mRNA whilst inducing a decline in hippocampal *Bdnf* expression [155, 158–160]. In contrast, a sustained elevation of 5-HT through the administration of selective serotonin reuptake inhibitors (SSRIs), is associated with increased *Bdnf* expression in both the neocortex and hippocampus [135, 161, 162]. Specific effects of 5-HT on *Bdnf* mRNA expression are mediated via 5-HT_{2A} receptors, and this has been explored particularly with respect to the effects of acute 5-HT elevation on *Bdnf* expression [32, 163]. We will review the effects of 5-HT_{2A} receptors on *Bdnf* expression, BDNF protein levels and signaling, as well as discuss possible molecular mechanisms that contribute to the effects of 5-HT_{2A} receptors on BDNF, and the implications of this regulation to psychopathology.

5-HT_{2A} Receptor Mediated Regulation of Bdnf *mRNA Expression*

In vivo studies in rat models indicate that acute treatment with the 5-HT_{2A} receptor agonist DOI evokes a dose- and time-dependent upregulation of *Bdnf* mRNA levels within the neocortex, including the frontal and parietal cortex, and a decline of *Bdnf* expression in the dentate gyrus (DG) subfield of the hippocampus [32]. Within the neocortex, the 5-HT_{2A} receptor-mediated upregulation of *Bdnf* expression exhibits a layer-specific pattern, noted within layers II/III and V/VI. This differential regulation of *Bdnf* expression by DOI is mediated via the 5-HT_{2A} receptor and can be completely prevented by pretreatment with the 5-HT_{2A} receptor antagonist MDL100,907. Interestingly, acute treatment with the 5-$HT_{2A/C}$ receptor antagonist, ketanserin or the 5-HT_{2A} receptor antagonist MDL100,907 does not alter baseline expression of *Bdnf* mRNA within the neocortex or hippocampus [32]. This suggests that while 5-HT_{2A} receptor stimulation does influence *Bdnf* expression, basal *Bdnf* mRNA levels may not be under the control of 5-HT_{2A} receptors. The regulation of *Bdnf* expression in the neocortex and hippocampus is not observed following chronic treatment with DOI for 7 days, suggesting a downregulation of 5-HT_{2A} receptors and signaling pathways upstream of *Bdnf* transcript regulation [32]. Further, while sub-chronic (7 day) treatment with the $5\text{-HT}_{2A/C}$ receptor antagonist ketanserin does not alter hippocampal *Bdnf* expression [164], longer duration treatments for 19 days with the $5\text{-HT}_{2A/C}$ receptor antagonist ritanserin upregulated *Bdnf* expression in the CA1 hippocampal subfield [139] and for 21 days with mianserin enhanced total hippocampal *Bdnf* expression [135].

Studies wherein 5-HT levels were elevated through the combined administration of tranylcypromine and L-tryptophan, or through acute treatment with the 5-HT releasing agent parachloroamphetamine or the SSRI paroxetine, evoke an upregulation of neocortical *Bdnf* mRNA and a decline in hippocampal *Bdnf* expression, mimicking the nature of *Bdnf* regulation induced by a 5-HT_{2A} receptor agonist [32, 160]. Recent studies indicate that the effects of an acute elevation in 5-HT levels on hippocampal downregulation of *Bdnf* expression involve a role for both 5-HT_4 and 5-HT_2 receptors, as pretreatment with 5-HT_4 or 5-HT_2 receptor antagonists can prevent the effects of acute 5-HT elevation on the hippocampal and DG regulation of *Bdnf* expression respectively [163]. *In vitro* studies in C6 glioma cell lines indicate a two-fold induction in *Bdnf* mRNA levels following exposure to 5-HT (100 μM), which is mediated through the 5-HT_{2A} receptor, and involves a role for calcium and protein-kinase dependent signaling [86]. The current understanding of the 5-HT_{2A} receptor-dependent regulation of distinct *Bdnf* transcript variants is limited. A single report using *in vitro* studies and luciferase reporter assays in cortical neurons indicates that DOI enhances transcription through *Bdnf* exon I and exon II, but not exon IV, promoters, and is associated with an enhancement in cAMP response element (CRE) driven transcriptional activation [165]. *Bdnf* exon I and II transcript variants are thought to show a similar pattern of regulation in response to diverse cues, and exhibit a protein-synthesis dependent transcriptional regulation unlike other specific *Bdnf* exon mRNAs (for eg, IV and XI) which are regulated in an immediate-early gene like fashion [166]. It is interesting that acute and chronic immobilization stress biphasically regulate *Bdnf* exon I and II containing transcripts in the rat brain, with a decline observed following acute stress and an increase following chronic stress [167]. Physical activity and antidepressants also upregulate the expression of *Bdnf* exon I and II containing transcripts [168, 169]. Diverse stressors, antidepressants and neuronal activity also influence the expression of the *Bdnf* exon IV containing transcripts, which exhibit an activity-dependent pattern of regulation [170, 171]. The transcription factors recruited downstream of 5-HT_{2A} receptor signaling remain unclear, with a possible role for CREB suggested based on *in vitro* findings [165].

The 5-HT_{2A} receptor-mediated regulation of *Bdnf* transcript expression has been shown in specific cases to be modulated by the hormonal milieu of the organism. In ovariectomized rats, estradiol pretreatment attenuates the cortical upregulation of *Bdnf* expression following acute DOI treatment but does not influence the hippocampal regulation of *Bdnf* [172]. These effects suggest that estradiol and 5-HT_{2A} receptors may regulate *Bdnf* transcript levels through discrete and potentially divergent pathways. In contrast, adrenalectomy, goitrogen administration that lowers circulating thyroid hormone levels or exogenous thyroid hormone administration does not influence the hippocampal regulation of *Bdnf* expression evoked by 5-HT_{2A} receptor stimulation [32, 173]. Studies in C6 glioma cells indicate that the upregulation of *Bdnf* expression evoked by 5-HT is enhanced by progesterone treatment [174], although the role of 5-HT_{2A} receptors in these effects of 5-HT is unexplored. Studies suggest an interaction between estradiol and 5-HT_{2A} receptors in the regulation of *Bdnf* expression [172], however, the interaction of other hormonal pathways with 5-HT_{2A} receptors in the regulation of *Bdnf* expression is still poorly understood. Future studies are required to gain a mechanistic insight into the role of intracellular signaling pathways and transcription factors that control the regulation of *Bdnf* in these contexts.

Potential Mechanisms for the 5-HT_{2A} Receptor-Mediated Regulation of **Bdnf** *Expression*

The molecular and cellular mechanisms that mediate the regulation of neocortical and hippocampal *Bdnf* expression in response to 5-HT_{2A} receptor stimulation are currently poorly elucidated. At the cellular level, it has been hypothesized that 5-HT_{2A} receptor-mediated upregulation of neocortical *Bdnf* expression may arise through activity-dependent mechanisms involving either (1) a 5-HT_{2A} receptor-mediated induction of EPSPs within pyramidal excitatory neurons or (2) increased glutamate release through presynaptic 5-HT_{2A} heteroreceptors on thalamocortical afferents (Fig. 3a, b) [32, 49, 51, 175]. Within hippocampal subfields, 5-HT_{2A} receptors evoke IPSPs in the granule cell layer via an increase in GABAergic neurotransmission through enhanced firing of 5-HT_{2A} receptor-expressing GABAergic neurons within the hilus, and an associated decline in *Bdnf* expression in the DG hippocampal subfield (Fig. 3a, c) [32, 73, 176, 177]. These hypotheses for the 5-HT_{2A} receptor-mediated regulation of *Bdnf* expression remain to be experimentally validated.

The recruitment of specific intracellular signaling cascades downstream of the 5-HT_{2A} receptor and the transcription factors recruited to regulate *Bdnf* expression are poorly delineated. Further, it is unclear how individual *Bdnf* transcript variants and exon-specific *Bdnf* promoters are targeted by 5-HT_{2A} receptor signaling. The *Bdnf* locus has several AP-1 binding sites [178] which may play a critical role in case of activity-dependent mechanisms being recruited downstream of 5-HT_{2A} receptor signaling. In addition, CREB-mediated transcription is reported to be enhanced in cortical cultures following 5-HT_{2A} receptor stimulation, and CREB plays a key role in *Bdnf* expression driven from *Bdnf* exon I promoter, which is targeted by DOI treatment [165, 179]. One can speculate a role for CREB-mediated

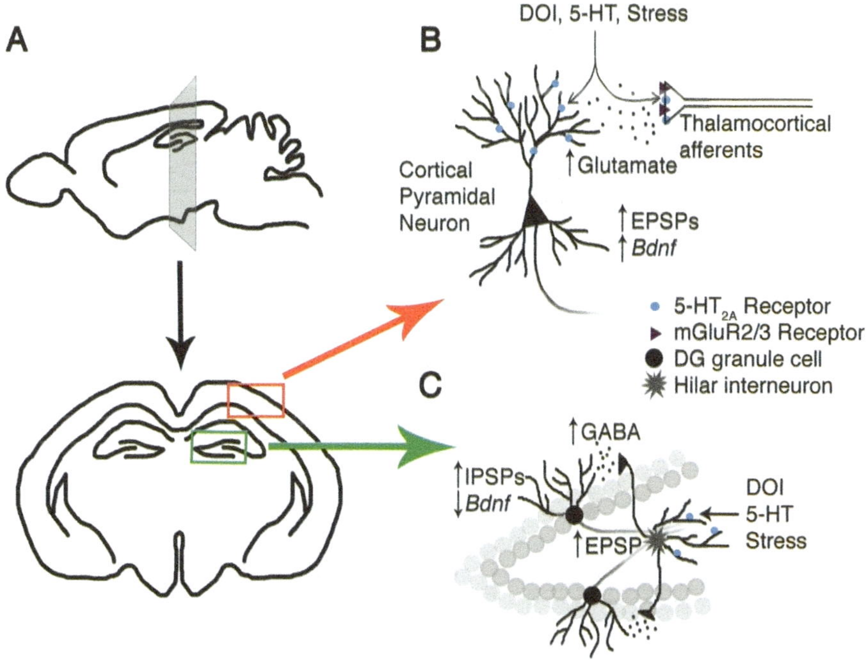

Fig. 3 Cellular model for the 5-HT$_{2A}$ receptor-mediated regulation of *Bdnf* expression in the neocortex and hippocampus. (**a**) Coronal section of the rodent brain highlighting the neocortex (red box) and hippocampus (green box). (**b**) In the neocortex, DOI, 5-HT and stress enhance *Bdnf* expression. This is likely to be mediated through 5-HT$_{2A}$ receptor activation evoked increases in EPSPs in layer V cortical pyramidal neurons and an activity-dependent increase in *Bdnf* expression. Two possible mechanisms for the 5-HT$_{2A}$ receptor mediated increase in cortical pyramidal neuron EPSPs could be through (1) the activation of postsynaptic 5-HT$_{2A}$ receptors on cortical pyramidal neurons thus enhancing EPSPs and *Bdnf* expression or (2) the activation of presynaptic 5-HT$_{2A}$ receptors present on thalamocortical afferents increasing glutamate release and thus enhancing activity-dependent *Bdnf* expression in cortical pyramidal neurons. mGlu2/3 receptor agonists serve to presynaptically inhibit glutamate release and thus can attenuate the effects of stress and DOI on cortical *Bdnf* expression. (**c**). In the hippocampus, DOI, 5-HT and stress evoke a decline in *Bdnf* expression. This is likely to be mediated through 5-HT$_{2A}$ receptor-mediated activation of hilar GABAergic interneurons thus leading to enhanced inhibition of dentate gyrus (DG) granule cells, and hence a reduction in *Bdnf* expression in the DG

transcriptional regulation of *Bdnf* expression in response to 5-HT$_{2A}$ receptor agonist treatment [165] (Fig. 4a). Our unpublished results indicate that CREB plays a critical role in mediating the cortical upregulation of *Bdnf* in response to 5-HT$_{2A}$ receptor stimulation. However, extensive experimental studies are required to provide mechanistic understanding of the nature of 5-HT$_{2A}$ receptor-mediated regulation of *Bdnf* mRNA levels, in particular the role of transcription factors such as CREB and AP-1.

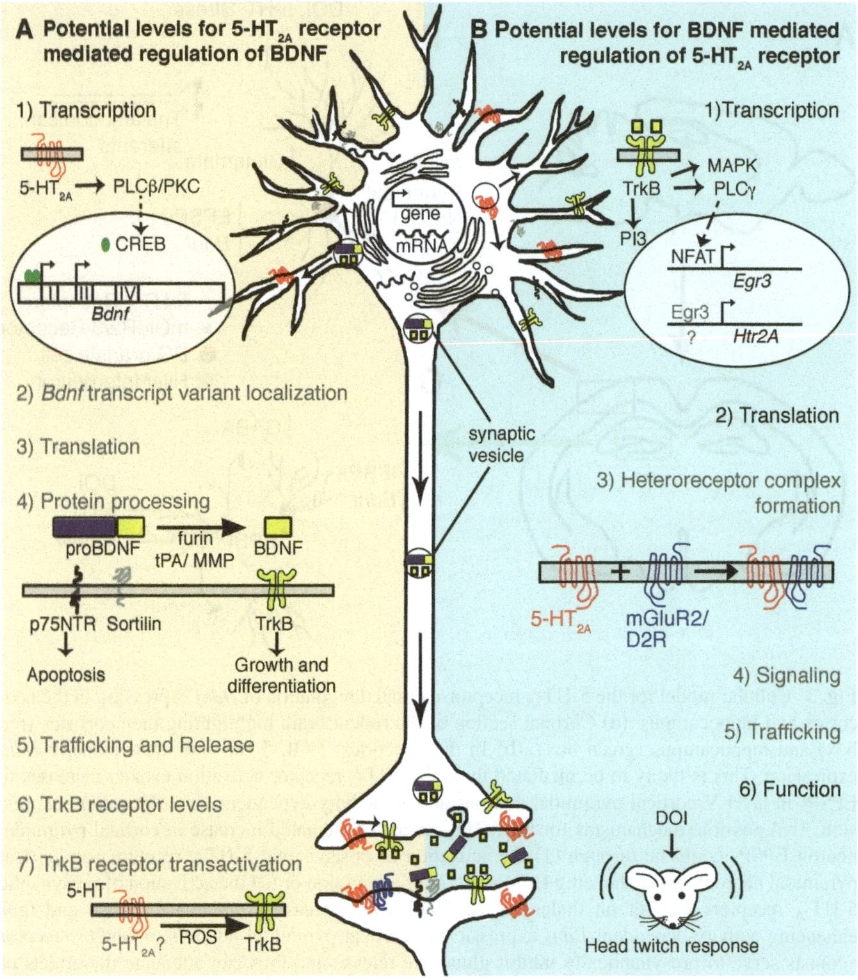

Fig. 4. Potential levels for 5-HT$_{2A}$ receptor and BDNF crosstalk, highlighting the multiple levels for the 5-HT$_{2A}$ receptor-mediated regulation of BDNF (**a**) and the BDNF-mediated regulation of 5-HT$_{2A}$ receptor (**b**). Shown is a schematic of a multipolar neuron. (**a**) 5-HT$_{2A}$ receptor-mediated regulation of BDNF can occur at multiple possible levels spanning from transcriptional regulation of *Bdnf* exon-specific transcript variants, *Bdnf* transcript variant localization, BDNF translation, protein processing, trafficking and release. 5-HT$_{2A}$ receptors may also influence TrkB receptor levels and transactivation. (**b**) BDNF-mediated regulation of 5-HT$_{2A}$ receptors can occur at multiple possible levels from transcriptional and translational regulation, 5-HT$_{2A}$ receptor heterocomplex formation, signaling and trafficking, and 5-HT$_{2A}$ receptor function, inclusive of electrophysiological and behavioral responses (head twitch behavior). The specific levels of regulation for which there is experimental evidence are indicated in black and grey refers to those levels at which the evidence is limited or lacking

$5\text{-}HT_{2A}$ Receptor-Mediated Regulation of BDNF Protein and BDNF Signaling

Regulation of BDNF by $5\text{-}HT_{2A}$ receptors, in addition to modulation at the level of *Bdnf* transcription [32, 86, 135, 139, 155, 164], can also involve effects at the level of BDNF synthesis [145], proteolytic processing of proBDNF into the mature BDNF form [146], regulation of BDNF secretion [150] and trafficking [148], influences on the BDNF receptors, p75NTR [180] and the truncated and full-length TrkB isoforms [153] (Fig. 4a). Further, $5\text{-}HT_{2A}$ receptor-mediated modulation of signaling pathways could also impact BDNF signaling [181–183]. Relatively few studies have addressed the influence of $5\text{-}HT_{2A}$ receptors on these aspects of BDNF regulation. The influence of treatment with the $5\text{-}HT_{2A}$ receptor agonist DOI on BDNF protein levels in either *in vitro* or *in vivo* studies has not been explored, despite several reports of robust effects on cortical and hippocampal *Bdnf* transcript levels [32, 163, 175]. Subchronic (7 day) treatment with the $5\text{-}HT_{2A/C}$ receptor antagonist ketanserin does not alter hippocampal BDNF protein levels, either when administered alone or in conjunction with fluoxetine [164]. There is a paucity of studies focused on examining the influence of $5\text{-}HT_{2A}$ receptors on pathways that impinge on BDNF proteolytic processing [146], trafficking [148] and secretion [150]. Given the roles for intracellular chaperones like sortilin [149] in trafficking, and the involvement of the plasmin-tissue plasminogen activator (tPA) system [184] and matrix metalloproteinases (MMPs) [146] in BDNF proteolytic processing it would be interesting to address whether these processes are targeted by $5\text{-}HT_{2A}$ receptors. Indeed, within cardiac fibroblasts as well as in uterine smooth muscle cells [185], the 5-HT-mediated induction of MMPs is regulated via $5\text{-}HT_{2A}$ receptors [186]. It will be important to identify the effects of $5\text{-}HT_{2A}$ receptors on the conversion of proBDNF to mature BDNF, as the nature of the BDNF ligand can evoke functionally distinct biological outcomes [106].

$5\text{-}HT_{2A}$ receptors could also influence BDNF signaling through modulation of the transcription [32], alternate splicing [121, 125] and translation of the TrkB [135] and p75NTRs [180], and by influencing their intracellular signaling cascades [182, 183]. Few studies have addressed the influence of $5\text{-}HT_{2A}$ receptors on BDNF receptor regulation, with evidence so far indicating no change in cortical and hippocampal *TrkB* expression following chronic treatment with the $5\text{-}HT_{2A}$ receptor antagonist mianserin [135]. TrkB receptors are expressed either as the full-length catalytic variant or as the truncated forms which lack the intracellular catalytic domain and can exhibit dominant negative action [113, 152, 153]. Truncated TrkB variants can also recruit inositol-1,4,5-trisphosphate-dependent calcium release and interact with the p75NTR, thus suggesting that a modulation of the nature of BDNF receptors present could result in the recruitment of distinct signaling cascades and evoke diverse physiological outcomes [106, 187–190]. $5\text{-}HT_{2A}$ receptors may also influence signaling pathways downstream of the BDNF receptors. Interestingly, 5-HT and N-acetylserotonin have both been shown to result in transactivation of the TrkB receptor [191, 192]. 5-HT mediated transactivation occurs via a reactive oxygen species mediated pathway, however the role of 5-HT_{2A} receptors in such TrkB transactivation remains unknown [192] (Fig. 4a). Such interplay between G-protein coupled receptors and receptor tyrosine kinases has been shown to involve (1) a role for metalloproteinase activation which can modulate ligand processing [193] or (2) the recruitment of ligand-independent signaling via intermediate steps involving non-receptor tyrosine kinases or reactive oxygen species [192, 194, 195]. Other sites of action for an influence of 5-HT_{2A} receptors on BDNF signaling include effects on the MAP kinase cascade, which is downstream of both 5-HT_{2A} receptor and BDNF-evoked signaling, with a rapid induction in MAP kinase activity evoked in response to either a 5-HT_{2A} receptor agonist [196, 197] or to BDNF-mediated TrkB activation [198, 199] (Fig. 4a). 5-HT_{2A} receptors could also influence BDNF signaling by modulation of the Rac guanine nucleotide exchange factor (RacGEF) kalirin-7, which is known to regulate the phosphorylation of p21-activated kinase (PAK) which also lies downstream of BDNF signaling [200, 201], thus serving as a potential target for the effects of both 5-HT_{2A} receptors and BDNF on dendrite morphology and actin cytoskeletal dynamics. However, the direct influences of 5-HT_{2A} receptors on BDNF protein synthesis, proteolytic processing, trafficking, secretion and BDNF receptor mediated signaling remain largely unaddressed and require further investigation (Fig. 4a).

5-HT_{2A} Receptor Mediated Regulation of Bdnf *Expression: Relevance to the Actions of Stress, Antidepressants, Antipsychotics and Hallucinogens*

5-HT_{2A} receptors may contribute to the regulation of *Bdnf* expression by stress [33], antidepressant [164, 202] and antipsychotic treatments ([203]; L. [204]), and hallucinogenic drugs [205, 206]. Adult-onset acute and chronic stress, using a range of diverse stressors, have predominantly been shown to evoke a reduction in hippocampal *Bdnf* expression [2, 6, 129]. The acute immobilization stress evoked reduction in *Bdnf* mRNA levels is attenuated following pretreatment with either the 5-$HT_{2A/C}$ receptor antagonist ketanserin [32] or the selective 5-HT_{2A} receptor antagonist, MDL100,907 [33]. In contrast within the frontal cortex, acute stressors have been reported to evoke an induction in *Bdnf* mRNA levels [207, 208], mimicking the effects of 5-HT_{2A} receptor agonist administration [32]. These studies are intriguing as they indicate that immobilization stress, DOI and acute elevation of 5-HT levels similarly upregulate prefrontal *Bdnf* mRNA expression and downregulate hippocampal *Bdnf* mRNA levels. This raises the possibility that the effects of stress are mediated via 5-HT_{2A} receptors, and indeed this has been demonstrated with respect to the stress-mediated decline in hippocampal *Bdnf* expression [33]. Interestingly, within the prefrontal cortex the mGlu2/3 receptor agonist LY354740 prevents the stress-evoked induction of prefrontal *Bdnf* expression [208]. mGlu2/3 receptor agonists prevent 5-HT_{2A} receptor-induced EPSPs in the prefrontal cortex through effects via autoreceptors at glutamatergic terminals resulting in reduced glutamate release [209–211]. This suggests the possibility that the effects of stress on cortical *Bdnf* expression involve a role for $5\text{-}HT_{2A}$ and mGlu2/3 receptors and an activity-dependent transcriptional regulation of *Bdnf* [212, 213]. Indeed, the mGlu2/3 receptor agonist LY354740 can also prevent the DOI-induced upregulation of prefrontal *Bdnf* expression, and the mGlu2/3 antagonist LY341495 can potentiate the DOI-evoked increase in prefrontal *Bdnf* mRNA levels [175]. This suggests that both the stress and DOI-evoked increases in prefrontal *Bdnf* expression can be attenuated by mGlu2/3 agonists, likely via an attenuation of the increased neuronal activity evoked by both stress and DOI through $5\text{-}HT_{2A}$ receptors. It remains possible that this effect could involve $5\text{-}HT_{2A}$ receptors directly present on excitatory pyramidal neurons where they modulate EPSPs, or could be via the $5\text{-}HT_{2A}$ receptor-mediated regulation of glutamate release from thalamocortical afferents that also express mGlu2/3 autoreceptors [68] (Fig. 3). It is intriguing to note that early stress and maternal immune activation (MIA) models also enhance cortical $5\text{-}HT_{2A}$ receptor function, as well as perturbing BDNF levels although the relationship between these specific changes remains as yet unexplored [214–219].

Chronic treatment for 21 days with pharmacological antidepressants belonging to distinct classes, including SSRIs, is associated with an upregulation of both cortical and hippocampal *Bdnf* expression [135]. Strikingly, sub-chronic (7 days) combined treatment with the $5\text{-}HT_{2A/C}$ receptor antagonist, ketanserin and the SSRI, fluoxetine, resulted in an induction of *Bdnf* mRNA within the hippocampal DG and CA3 subfields, accelerating the neurotrophic effects of fluoxetine administration [164]. It is also interesting to note that chronic treatment with the antidepressants mianserin and mirtazapine, which exhibit $5\text{-}HT_{2A}$ receptor antagonism, also show a robust increase in cortical and hippocampal *Bdnf* expression [57, 135, 220]. In addition to pharmacological antidepressants, studies with physical exercise in rodent models have suggested that ketanserin pretreatment prevents the exercise-evoked induction in *Bdnf* mRNA levels within the CA4 hippocampal subfields [156]. This finding differs from other reports wherein $5\text{-}HT_{2A}$ receptor antagonism, either standalone or in combination with pharmacological antidepressants predominantly enhances *Bdnf* expression within the hippocampus [135, 164, 220]. However, it is important to note that in the study with physical exercise the ability of ketanserin to attenuate exercise-evoked *Bdnf* upregulation is restricted to only the CA4 region [156]. While studies do indicate a role for $5\text{-}HT_{2A}$ receptor antagonism in potentiating the antidepressant effects of SSRIs [221] and in preventing the SSRI-evoked decline in dorsal raphe neuronal firing [222], thus far no studies have directly addressed whether $5\text{-}HT_{2A}$ receptors are required to mediate the effects of antidepressants on *Bdnf* regulation.

Clinical studies implicate $5\text{-}HT_{2A}$ receptors and BDNF in both the pathogenesis and treatment of schizophrenia ([6, 223–227]b; [204]). Preclinical studies on *Bdnf* regulation by atypical antipsychotic drugs, which exhibit $5\text{-}HT_{2A}$ receptor antagonism, remain controversial with reports of either an increase, no change or a decline in cortical and hippocampal *Bdnf* expression [138, 139, 203, 228, 229]. Hallucinogenic agents such as LSD have also been suggested to enhance cortical

Bdnf expression, in particular within the prefrontal cortex [143]. While distinct "transcriptome fingerprints" within the neocortex are associated with hallucinogenic and non-hallucinogenic 5-HT_{2A} receptor agonists [230], it is at present unclear whether these two classes of 5-HT_{2A} receptor ligands evoke differential effects on *Bdnf* expression. Future studies are required to elucidate the contribution of 5-HT_{2A} receptors to the effects of antipsychotics and hallucinogens on BDNF.

Taken together, these studies highlight a role for 5-HT_{2A} receptor-mediated regulation of BDNF in both the pathophysiological and treatment of depression and schizophrenia, and motivate future studies to gain a deeper mechanistic understanding of the interplay between 5-HT_{2A} receptors and BDNF.

BDNF Regulation of 5-HT_{2A} Receptors

BDNF regulates the development, survival, differentiation, growth and function of serotonergic neurons [231–235]. BDNF autocrine signaling via the recruitment of TrkB receptors expressed by serotonergic raphe neurons can influence the expression of serotonergic markers such as tryptophan hydroxylase [236, 237]. Further, BDNF enhances 5-HT uptake and increases the firing rate of raphe 5-HT neurons [238, 239]. Several reviews have focused on the relationship between 5-HT and BDNF [27, 29, 30] and we will restrict our discussion to specific interactions between BDNF and 5-HT_{2A} receptors. BDNF may exert an influence on 5-HT_{2A} receptors at multiple levels including modulation of 5-HT_{2A} receptor transcription, synthesis, trafficking, heterocomplex formation, signaling and function [3, 27].

BDNF Regulation of 5-HT_{2A} Receptor Expression

Intracerebroventricular BDNF infusion evokes an upregulation of 5-HT_{2A} receptor mRNA expression in the hippocampus, and a decline or no change in the frontal cortex, with no regulation observed in midbrain regions, in a mouse-strain dependent fashion [240, 241]. While antidepressant sensitive cataleptic (ASC) mice exhibit chronic (17–19 days) BDNF-evoked increase in hippocampal, and decline in frontal, 5-HT_{2A} receptor expression, this was absent in the parental CBA mouse strain [240, 241]. *In vitro* studies in both primary hippocampal cultures and in hippocampal organotypic slices indicate that chronic (7 day), but not acute (1 day) or subchronic (3 or 5 days), exposure to BDNF evokes a dose-dependent decline in 5-HT_{2A} receptor protein expression [31]. At present, a clear cut understanding of the region-specific, dose-, time- and age-dependent effects of BDNF on 5-HT_{2A} receptor expression is lacking, making it difficult to resolve the discrepancies present in the literature. Extensive studies that clearly elucidate the nature of influence of BDNF on 5-HT_{2A} receptor expression, with attention paid to the influence of specific BDNF doses, treatment duration, time-points of analysis, region-specificity and nature of model system are critically required.

Studies performed in BDNF mutant mice indicate region-specific effects on 5-HT_{2A} receptor mRNA and protein expression. Constitutive BDNF heterozygote mice ($BDNF^{+/-}$) that exhibit a 50% reduction in BDNF expression, also show an upregulation of 5-HT_{2A} receptor mRNA in the frontal cortex and hypothalamus, with no change observed in hippocampal 5-HT_{2A} receptor mRNA expression [242]. However, western blotting analysis indicates a two-fold upregulation of 5-HT_{2A} receptor protein levels within the hippocampus of adult $BDNF^{+/-}$ mice [31]. It is important to note that these BDNF heterozygote mice exhibit a decline in BDNF expression through embryonic development and also exhibit a decline in 5-HT_{1A} autoreceptors and an age-dependent reduction in serotonergic innervation and neurotransmission [243, 244]. Studies with conditional BDNF mutant mice ($BDNF^{2L/2LCk\text{-}Cre}$), which exhibit a central depletion of BDNF levels commencing postnatally, show a decline in 5-HT_{2A} receptor protein expression in the frontal cortex and dorsal raphe nucleus when examined in adulthood [245]. Autoradiographic binding studies reveal a reduction in 5-HT_{2A} receptor binding in the frontal cortex and an increase in hippocampal 5-HT_{2A} receptor binding in $BDNF^{2L/2LCk\text{-}Cre}$ mice [28]. Further, 5-HT_{2A} receptor mRNA levels exhibit a significant reduction within the frontal cortex in keeping with the decline noted in protein expression and receptor binding, with no change observed in hippocampal 5-HT_{2A} receptor expression [28, 245]. $BDNF^{2L/2LCk\text{-}Cre}$ mice also exhibit a trend towards a decline in 5-HT_{2A} receptor mRNA levels within the basolateral amygdala, accompanied by a significant reduction in 5-HT_{2C} receptor mRNA expression [26]. 5-HT_{2A} receptor expression is differentially regulated within these two BDNF mutant mouse lines, with enhanced frontal cortex 5-HT_{2A} receptor mRNA levels noted in the $BDNF^{+/-}$ mouse line [242] and a decline in frontal cortex 5-HT_{2A} receptor mRNA, protein and binding in the $BDNF^{2L/2LCk\text{-}Cre}$ mouse line [28, 245]. A single study has compared the effects of prenatal versus postnatal loss of BDNF on 5-HT_{2A} receptor expression using the $BDNF^{2L/1LNes\text{-}cre}$ and the $BDNF^{2L/2LCk\text{-}Cre}$ which deplete BDNF from prenatal and postnatal time-points respectively [246]. While prenatal and postnatal onset of BDNF loss both evoke a decline in frontal cortex 5-HT_{2A} receptor protein levels, the decline in dorsal raphe 5-HT_{2A} receptor protein levels is only observed with the $BDNF^{2L/2LCk\text{-}Cre}$ mouse model [246]. Mutant mice (BDNF-KIV) that lack BDNF expression driven through the activity-responsive BDNF exon IV exhibit a significant decrease in frontal cortex 5-HT_{2A} receptor mRNA expression, with no change observed in the hippocampus [247]. However, a recent study using mutant mouse lines with *Bdnf* promoter-specific deletion of transcripts (exon I, II, IV and VI) show enhanced prefrontal cortex 5-HT_{2A} receptor gene expression only in the *Bdnf* exon I deletion model [248]. In addition, *Bdnf* exon I deletion leads to an enhanced expression of *Bdnf* exon 2c and exon 4 with no change in BDNF protein levels in prefrontal cortex [248]. It is important to note here that removal of only exon-specific *Bdnf* transcripts may not necessarily influence total BDNF levels due to compensation through other *Bdnf* exon promoters. The consistent pattern of 5-HT_{2A} receptor regulation, across several brain regions, that emerges from studies with three distinct

Table 1 Summary showing alterations in 5-HT$_{2A}$ receptor expression and function across distinct BDNF mutant mouse lines

BDNF mutant mouse line	5-HT$_{2A}$ mRNA expression	5-HT$_{2A}$ protein expression	5-HT$_{2A}$ function	References
BDNF (+/−) *Constitutive 50% reduction*	▲ LFC ■ Hpc ▲ Hyp	▲ Hpc	ND	[31, 242]
BDNF$^{2L/2LCk\text{-}Cre}$ *P1-P14 onset ~90% reduction ~70% reduction in hpc*	▼ FC ■ Hpc ▼ Amy	▼ FC ▲ Hpc ▼DRN	▼ EPSPs in mPFC ▼ sPSCs in DRN ▼ IPSCs in BLA ▲ EPSCs in subset of BLA neurons ⊣ DOI evoked ESR but not HTR	[26, 28, 245, 249]
BDNF$^{2L/1LNes\text{-}cre}$ *E9.5-E15.5 onset ~90% reduction throughout the brain*	ND	▼ mPFC ■ DRN	ND	[246]
BDNF-KIV *Constitutive absence of BDNF promoter IV driven transcripts*	▼ FC ■ Hpc	ND	ND	[247, 250]
Bdnf-e1 −/− *~50% reduction in hyp but not in pfc and hpc*	▲ PFC	ND	ND	[248]
Bdnf-e2 −/− *No change in BDNF protein levels in pfc, hyp and hpc*	■ PFC	ND	ND	[248]
Bdnf-e4 −/− *~50% reduction in hyp, and ~25% reduction in pfc and hpc*	■ PFC	ND	ND	[248]
Bdnf-e6 −/− *No change in BDNF protein levels in pfc, hyp and hpc*	■ PFC	ND	ND	[248]

Amy amygdala, *BLA* basolateral amygdala, *DRN* dorsal raphe nuclei, *Hyp* hypothalamus, *FC* frontal cortex, *LFC* lateral frontal cortex, *Hpc* hippocampus, *ESR* ear scratch response, *HTR* head twitch response, *EPSCs* excitatory postsynaptic currents, *EPSPs* excitatory postsynaptic potentials, *IPSCs* inhibitory postsynaptic currents, *sPSCs* spontaneous postsynaptic currents, *ND* not determined

BDNF mouse mutant models, namely the BDNF$^{2L/1LNes\text{-}cre}$ [246], the BDNF$^{2L/2LCk\text{-}Cre}$ [28, 245], and the BDNF-KIV [247] mouse mutant lines which differentially influence BDNF protein levels, is a significant decline in frontal cortex 5-HT$_{2A}$ receptor expression (Table 1). The exception to this is the 5-HT$_{2A}$ receptor upregulation noted in the frontal cortex of BDNF heterozygote (BDNF$^{+/-}$) mutant mice [242] and Bdnf-e1−/− mutant mice (Table 1). The differences noted across the distinct BDNF mutant mouse lines could arise due to variation in the extent and onset of BDNF depletion, as well as potentially confounding effects on serotonergic neurotransmission and innervation which could further influence $5\text{-}HT_{2A}$ receptor expression (Table 1).

While the mechanisms that mediate the influence of BDNF on $5\text{-}HT_{2A}$ receptor expression are currently not understood, recent reports suggest a potential link through the recruitment of the transcription factor, early growth response gene 3 (*egr3*). BDNF induced enhancement of *Egr3* transcriptionally regulates NMDA receptor levels and GABAA receptor alpha 4 subunits levels in cortical and hippocampal neurons respectively ([251, 252]). Strikingly, Egr3 knockout mice (*Egr3*$^{-/-}$) showed a ~70% reduction in ketanserin binding sites in the prefrontal cortex and lowered DOI-induced head twitch responses [253]. This suggests the speculative possibility that BDNF via TrkB signaling may recruit nuclear factor of activated T-cells (NFAT) to transcriptionally upregulate *Egr3* expression, and thus eventually influence $5\text{-}HT_{2A}$ receptor expression through an Egr3-mediated mechanism (Fig. 4b). The $5\text{-}HT_{2A}$ receptor promoter has two putative Egr3 binding sites [254], although direct mechanistic evidence that links Egr3 to the BDNF regulation of $5\text{-}HT_{2A}$ receptors is currently lacking. Studies of the influence of altered BDNF signaling on $5\text{-}HT_{2A}$ receptors have also assessed effects in individuals bearing the BDNF Val66met polymorphism [255]. This polymorphism is associated with alterations in activity-dependent, but not constitutive, BDNF secretion [255]. Previous studies link the BDNF Val66met polymorphism with altered serotonin transporter binding [256]. However, studies revealed no influence of the BDNF Val66met polymorphism on neocortical $5\text{-}HT_{2A}$ receptor binding [257]. Collectively the findings suggest that BDNF is capable of regulating $5\text{-}HT_{2A}$ receptors, at the level of mRNA and protein expression, within diverse neuronal circuits including the frontal cortex, hippocampus and amygdala.

BDNF Regulation of $5\text{-}HT_{2A}$ Receptor Function

BDNF infusion into the lateral ventricle potentiates DOI-evoked head twitch responses in a mouse strain-dependent manner, with effects noted in depressive ASC mice but not observed in the parental CBA mice [240, 241] (Fig. 4b). This suggests that BDNF can enhance $5\text{-}HT_{2A}$ receptor-mediated functional responses. This idea is further supported by studies in BDNF mutant mice, which show region-specific changes in both $5\text{-}HT_{2A}$ receptor expression, as well as exhibiting alterations in $5\text{-}HT_{2A}$ receptor-mediated electrophysiological and behavioral responses [28, 242, 245–247]. Studies in conditional BDNF mutants, namely postnatal-onset loss of function $BDNF^{2L/2LCk\text{-}Cre}$ mice, indicate that the DOI-evoked ear scratch response (ESR), but not the head twitch response (HTR), is lost in the $BDNF^{2L/2LCk\text{-}Cre}$ mice [28]. It is surprising that effects on DOI-evoked behaviors are restricted to a decline in ESR, with no change in HTR responses, in conditional BDNF mutant mice, given that both behaviors are thought to be mediated through cortical 5-HT_{2A} receptors [41, 258]. It has been suggested these differences in DOI-evoked behaviors in the conditional BDNF mutant mice may be a consequence of differential sensitivity of the ESR and HTR behaviors to reductions in 5-HT_{2A} receptor expression which is lowered by about 20% within the frontal cortex of BDNF conditional mutant mice [28]. Taken together, these studies show that while BDNF infusion potentiates 5-HT_{2A} receptor-mediated behavioral responses [240, 241], conditional loss of BDNF results in a loss of specific 5-HT_{2A} receptor-mediated behaviors [28].

$BDNF^{2L/2LCk\text{-}Cre}$ mice exhibit a steep decline in cortical 5-HT_{2A} receptor mRNA and protein expression [28, 245]. This downregulation in frontal cortex 5-HT_{2A} receptor expression is accompanied by a failure of 5-HT to elicit spontaneous EPSCs (sEPSCs) in the prefrontal cortex of conditional BDNF mutant mice. These 5-HT-elicited sEPSCs are mediated via the 5-HT_{2A} receptor and can be blocked by MDL100,907 and are observed in wildtype but not in $BDNF^{2L/2LCk\text{-}Cre}$ mice [245]. 5-HT_{2A} receptor-mediated electrophysiological responses are severely reduced within the prefrontal cortex of $BDNF^{2L/2LCk\text{-}Cre}$ mice. It is important to note however that cortical layer V pyramidal neurons within these mice exhibited normal baseline electrophysiological properties, including no perturbation in glutamate release or glutamatergic receptor responses. This suggests a specific deficit in 5-HT_{2A} receptor-evoked electrophysiological responses in the prefrontal cortex of conditional BDNF mutant mice [245]. Further, the 5-HT_{2A} receptor-mediated increases in both glutamatergic and GABAergic spontaneous postsynaptic currents (sPSCs) within the dorsal raphe nucleus are significantly attenuated in conditional BDNF mutant mice [245]. Attenuated 5-HT_{2A} receptor-mediated activation of both glutamatergic and GABAergic neurotransmission within the prefrontal cortex and dorsal raphe nucleus of conditional BDNF mutant mice may mechanistically contribute to the decline in specific 5-HT_{2A} receptor-mediated behaviors observed in BDNF conditional mutant mice.

In addition to reports of perturbed 5-HT_{2A} receptor-mediated synaptic transmission in the prefrontal cortex and dorsal raphe nucleus of conditional BDNF mutant mice [28, 245], a report also indicates that 5-HT_{2A} receptor-mediated modulation of GABAergic and glutamatergic neurotransmission within the basolateral amygdala (BLA) is disrupted in $BDNF^{2L/2LCk\text{-}Cre}$ mice [26]. The frequency of serotonin-evoked IPSCs is decreased in pyramidal neurons of the BLA pyramidal neurons in $BDNF^{2L/2LCk\text{-}Cre}$ mice, which may arise as a consequence of reduced 5-HT_{2A} receptor density on GABAergic neurons in the BLA thus resulting in a reduced inhibitory tone on BLA pyramidal neurons. Further, the frequency of 5-HT-elicited EPSCs in a subset of BLA pyramidal neurons is enhanced in the $BDNF^{2L/2LCk\text{-}Cre}$ mice, likely through a 5-HT_{2A} receptor-mediated modulation of glutamate release onto BLA pyramidal neurons [26]. This suggests that 5-HT_{2A} receptors contribute to the maintenance of an excitatory-inhibitory balance within the BLA, and such a balance may be tipped by perturbations of BDNF within the BLA circuitry. Indeed it is intriguing that conditional BDNF mutant mice exhibit elevated anxiety associated with enhanced 5-HT-evoked excitation within the BLA, likely through a 5-HT_{2A} receptor-mediated mechanism [26]. While most studies assessing 5-HT_{2A} receptor-mediated functional responses in conditional BDNF mutant mice have focused on the $BDNF^{2L/2LCk\text{-}Cre}$ mouse line, further investigation of the distinct BDNF mutant mouse models, including the adult-onset BDNF loss of function and BDNF overexpression mouse models, would greatly aid in gaining a deeper mechanistic understanding of the effects of BDNF on 5-HT_{2A} receptor function. Similarly studies in normal mice to directly assess the effects of BDNF on 5-HT_{2A} receptor-mediated electrophysiological responses in distinct neurocircuits are required, as studies in BDNF mutant mice come with the caveat of possible confounding variables, including developmental consequences, compensatory adaptations, and effects on 5-HT neurotransmission, as well as other 5HT receptors.

To the best of our knowledge no studies till date have directly assessed an influence of BDNF on 5-HT_{2A} receptor-mediated signaling pathways. However, it is interesting to note that BDNF can stimulate intracellular calcium increase and enhance calcium oscillations through the recruitment of IP3-gated calcium stores [259, 260], activate the MAP kinase, PLC and PI3/Akt signaling cascades [106], thus influencing signaling events that also lie downstream of the 5-HT_{2A} receptors. Further investigation is required to systematically address the influence of BDNF on 5-HT_{2A} receptor function, including studies to assess effects of BDNF on 5-HT_{2A} receptor heterocomplex formation, trafficking, electrophysiological responses to both hallucinogenic and non-hallucinogenic ligands, as well as 5-HT_{2A} receptor driven behaviors (Fig. 4b).

Implication of 5-HT_{2A} Receptor and BDNF Cross-talk in Psychopathology

Several reviews have focused on the interaction between 5-HT and BDNF, in the context of their role in neuronal plasticity and psychopathology [20, 27, 30]. For the purpose of this chapter we will restrict ourselves to discussing the interaction between 5-HT_{2A} receptors and BDNF, and the relevance of such a relationship to the effects of stress, antidepressants, antipsychotics and hallucinogenic drug action.

Stress, 5-HT_{2A} Receptors and BDNF

Both 5-HT_{2A} receptors [3–5, 55, 261] as well as BDNF [6, 262, 263] have been implicated in the pathophysiology of depression and anxiety. Several studies suggest that the specific physiological sequelae of stress exposure, both early [205, 214, 264] and adult-onset [33, 265], are regulated by 5-HT_{2A} receptors, and that stress experience may enhance 5-HT_{2A} receptor function. The general theme emerging across studies suggests that 5-HT_{2A} receptor blockade attenuates diverse downstream consequences of stress, such as conditioned defeat in social defeat models [266], anxiety behavior in both early [267] and predator stress models [268], hyperthermic responses in psychosocial [269] and restraint stress models [270], mesocortical dopamine release in handling stress [271] and ACTH responses evoked by restraint or ether stress [261]. While this is not an exhaustive list of the role of 5-HT_{2A} receptors in the effects of stress, it supports the notion that stress recruits 5-HT_{2A} receptors to mediate specific stress-evoked physiological consequences. Further, a large body of literature that has been extensively reviewed also implicates stress-evoked alterations in BDNF signaling in regulating specific outcomes of stressor experience [2, 272–275]. However, few studies have directly examined the importance of a relationship between 5-HT_{2A} receptors and BDNF in contributing to the effects of stress [33, 205].

Studies using exposure to acute and chronic stress have demonstrated a robust decline in hippocampal *Bdnf* expression [2, 33, 129, 272, 276, 277], as well as a reduction in adult hippocampal neurogenesis [278–281]. The stress-induced *Bdnf* downregulation is mediated, at least in part, through the 5-HT_{2A} receptor [33]. This raises the possibility that a 5-HT_{2A} receptor-mediated decline in BDNF expression may contribute to the effects of chronic stress on long-term potentiation [282, 283], dendritic atrophy [284–287] and neurogenic decline within the hippocampus [12, 279, 288]. Interestingly, studies in BDNF mutant mice suggest that reduced BDNF levels may further evoke enhanced 5-HT_{2A} receptor expression [28, 31], and could act to worsen the effects of stress on *Bdnf* expression, raising the possibility of a positive feedback loop. The crosstalk between 5-HT_{2A} receptors and BDNF in the context of stress is also highly relevant within the amygdala [289]. In the BLA of stressed animals, 5-HT_{2A} receptor-mediated facilitation of GABAergic inhibition is significantly reduced along with a reduction in 5-HT_{2A} receptor expression [289]. One can envisage that this consequence of stress exposure could result in hyperexcitability and a lowered inhibitory tone within the BLA, which may mechanistically contribute to the anxiety associated with stress experience [26, 289]. Within the prefrontal cortex, 5-HT_{2A} receptor activation, as well as acute stress, both enhance glutamate release [49, 51, 286, 290, 291] and *Bdnf* expression [32, 292]. Treatment with a mGluR2/3 agonist can attenuate both the DOI and stress-evoked increase in prefrontal *Bdnf* expression and also dampen the glutamate release within the PFC evoked by 5-HT_{2A} receptor stimulation [175, 208, 211]. Further, enhanced *Arc* mRNA levels observed following either 5-HT_{2A} receptor stimulation with DOI or immobilization stress are substantially diminished in BDNF knockout mice with an adult-onset, forebrain specific BDNF loss [205]. These studies suggests the possibility that stress recruits 5-HT_{2A} receptors to regulate prefrontal immediate early gene expression, as well as contributing to the enhanced glutamate release and altered prefrontal excitability in response to stress, effects that may involve an important role for BDNF. Further studies are required to directly address whether stress-evoked apical dendritic atrophy and perturbations of prefrontal cortical excitability involve a key role for reciprocal interactions between 5-HT_{2A} receptors and BDNF.

Antidepressants, 5-HT$_{2A}$ Receptors and BDNF

5-HT_{2A} receptor antagonism, or knockdown, has been shown to produce antidepressant-like behavioral effects in preclinical studies [164, 293–295]. In clinical studies, single nucleotide polymorphisms (SNP) in the 5-HT_{2A} receptor gene have been linked to major depression [4, 296, 297], as well as to treatment responsivity to SSRIs [298–301]. Further, drugs like mirtazapine and mianserin that exhibit 5-HT_{2A} receptor antagonism are clinically used as antidepressants [302–304]. Enhanced 5-HT_{2A} receptor binding has been observed in both the frontal cortex and in platelets derived from depressed patients or in postmortem samples [305–308]. However, a deeper understanding of the precise functional role of 5-HT_{2A} receptors in the pathogenesis and treatment of depression still remains elusive. Preclinical studies have also implicated BDNF in mediating specific effects of antidepressant treatments on hippocampal neurogenesis [15, 309], neuronal plasticity [310, 311], as well as regulating antidepressant-evoked behavioral outcomes [312–319]. This has lead to the neurotrophic hypothesis of depression which posits that antidepressant enhance BDNF levels, and such an increase contributes mechanistically to the effects of antidepressants on neuronal plasticity, dendritic morphology, hippocampal neurogenesis and behavioral outcomes [20, 320–323]. Further, the BDNFVal66Met SNP that functionally impacts BDNF mRNA trafficking and secretion [255, 324] has been shown to contribute to the genetic risk for major depression [325–327], and in preclinical studies modifies responses to stress and antidepressant treatments [328–331]. The role of BDNF as a target of antidepressants as well as contributing to the pathogenesis of depression has been extensively reviewed [310, 320, 332]. However, other than correlative studies there is a limited understanding of the relevance of a direct relationship between 5-HT_{2A} receptors and BDNF in mechanistically contributing to antidepressant action.

Co-administration of a selective 5-HT_{2A} receptor antagonist has been suggested to accelerate the antidepressant effects of SSRIs [164, 221, 293]. In this regard, it is noteworthy that simultaneous administration of a 5-HT_{2A} receptor antagonist and SSRIs in a subchronic paradigm is sufficient to evoke an upregulation of *Bdnf* expression within the hippocampus [164]. Further, short duration (7 day) treatment with a 5-HT_{2A} receptor antagonist is also sufficient to enhance hippocampal neurogenesis [13, 164]. It possible that enhanced neurotrophic signaling contributes to the hippocampal neurogenic changes evoked by 5-HT_{2A} receptor antagonism. This presents the intriguing possibility that slow-onset hippocampal neurotrophic and neurogenic changes that ensue following chronic antidepressant treatment and contribute to the behavioral effects could be hastened through 5-HT_{2A} receptor antagonism. Temporal delays in the neurotrophic and neurogenic effects of antidepressants may arise in part due to the time required to downregulate the 5-HT_{2A} receptor, a common effect of several classes of antidepressants [333, 334]. It is tempting to speculate that simultaneous 5-HT_{2A} receptor antagonism along with antidepressant treatment could speed up effects on both hippocampal *Bdnf* expression and hippocampal neurogenesis [164], thus providing for the possibility of a hastening of the behavioral outcomes of antidepressant treatments. However, this idea has yet to be experimentally validated and several lines of investigation are required to explore the importance of a cross-talk between 5-HT_{2A} receptors and BDNF in mediating antidepressant-evoked molecular, cellular and behavioral changes.

Atypical Antipsychotic and Hallucinogenic Drug Action, 5-HT_{2A} Receptors and BDNF

Atypical antipsychotics have been distinguished from conventional or first-generation antipsychotics based not only on their improved side-effect profile and enhanced therapeutic efficacy, but also on their ability to block both 5-HT_{2A} and D2 receptors. Atypical antipsychotics are further sub-classified based on their high selectivity for 5-HT_{2A} and D2 receptors and referred to as serotonin-dopamine antagonists (risperidone, ziprasidone, lurasidone) or based on their affinity for 5-HT_{2A} and D2 receptors as well as other cholinergic and histaminergic receptors as multi-acting receptor targeted antipsychotics (clozapine, olanzapine, quetiapine, asenapine) [226]. 5-HT_{2A} receptor antagonism exhibited by atypical antipsychotic drugs has been suggested to contribute to their neuroprotective, neurotrophic and neurogenic actions [24, 226, 335, 336]. Further, *Bdnf* expression is enhanced by the atypical antipsychotics olanzapine [337] and clozapine [161], and this effect has also been implicated in the atypical antipsychotic-mediated effects on cortical neuroprotection and neurogenesis [224, 338, 339]. However, what is still not understood mechanistically is whether the reciprocal interactions between 5-HT_{2A} receptors and BDNF are essential to the survival-promoting and neurogenic actions of atypical antipsychotics.

BDNF-mediated TrkB signaling is attenuated in the prefrontal cortex of schizophrenic patients [340], which may contribute to impaired neuronal survival and plasticity, in particular impacting cortical GABAergic networks [339, 341, 342]. A reduction in the number of parvalbumin-expressing GABAergic interneurons has been reported within the prefrontal cortex of both schizophrenic patients [343–345], as well as in animal models of schizophrenia [346–348]. Schizophrenic patients also exhibit altered gene expression within parvalbumin-expressing fast-spiking GABAergic interneurons, an effect that is thought to be regulated via reduced BDNF-TrkB signaling [250, 341, 349]. The 5-HT_{2A} receptor is expressed both by prefrontal pyramidal neurons, as well as by parvalbumin-expressing fast-spiking GABAergic interneurons, and both of these cell types are excited by serotonin via 5-HT_{2A} receptors [350]. The role of a 5-HT_{2A} receptor-mediated BDNF regulation in mechanistically contributing to altered parvalbumin expressing GABAergic interneuron gene expression and neuronal survival remains unknown, and is of substantial interest given the key role of this cortical microcircuit in the pathophysiology and treatment of schizophrenia. Studies have previously shown that acute 5-HT_{2A} receptor stimulation results in enhanced activity-dependent prefrontal *Bdnf* expression [32, 175], though it is unclear which specific prefrontal neuronal cell types exhibit this increase in *Bdnf* expression. Further, chronic 21 days treatment with the atypical antipsychotics, clozapine, olanzapine and aripiprazole that exhibit 5-HT_{2A} receptor antagonism, also evokes an induction of prefrontal *Bdnf* expression [161, 351]. It is indeed interesting that both acute 5-HT_{2A} receptor stimulation and sustained chronic blockade by the atypical antipsychotics olanzapine and aripiprazole can evoke an increase in prefrontal *Bdnf* expression, however the time scales of treatment required for these effects are quite distinct and point to recruitment of distinct downstream signaling cascades. In this regard it is interesting that the atypical antipsychotics, clozapine and risperidone, have been reported to evoke biased signaling from 5-HT_{2A}-mGluR2 heterocomplexes shifting the balance to enhanced Gi and reduced Gq-driven signaling [44, 45]. The signaling mechanisms that contribute to atypical antipsychotic mediated BDNF regulation remain poorly elucidated.

In addition to studies within the prefrontal cortex, reports indicate an increase in *Bdnf* expression in the DG and the CA fields of the hippocampus following chronic atypical antipsychotic treatment [139, 203]. However, it is important to note that studies with diverse atypical antipsychotics differ in the nature and magnitude of *Bdnf* regulation within specific brain regions based on differences in treatment regime, dose, and duration [138, 139, 161, 203, 351–353]. Predominantly, most reports indicate that atypical antipsychotics, evoke an induction in *Bdnf* expression in hippocampal subfields, an effect that parallels the increase noted in hippocampal neurogenesis suggesting a possible role for BDNF in contributing to the neurogenic consequences of atypical antipsychotics. Given that 5-HT_{2A} receptor stimulation is associated with a decline in hippocampal *Bdnf* expression [32], it is tempting to speculate that the induction noted in *Bdnf* levels following atypical antipsychotic administration is mediated via 5-HT_{2A} receptor antagonism. Further studies are required to address whether the reciprocal relationship between 5-HT_{2A} receptors and BDNF contributes mechanistically to the neurogenic actions of atypical antipsychotics within the hippocampus [354]. While studies do implicate BDNF signaling in contributing to the pathophysiology and treatment of schizophrenia, the role of 5-HT_{2A} receptor-driven BDNF regulation in this context remains relatively poorly explored.

Insights into the role of 5-HT_{2A} receptors in the psychotic symptoms of schizophrenia also come in part due to observations of the psychedelic properties of serotonergic hallucinogens and drugs such as lysergic acid diethylamide (LSD). The underlying mechanism of action implicated in the hallucinogenic properties of these drugs is activation of the 5-HT_{2A} receptor [355, 356]. Interestingly, hallucinogens such as the $5\text{-HT}_{2A/C}$ receptor agonist DOI [32, 205] and LSD [143] induce cortical *Bdnf* expression. It is unclear whether non-hallucinogenic ligands of the 5-HT_2 receptor also regulate cortical *Bdnf* expression in a similar fashion. The role of specific signaling cascades recruited downstream of 5-HT_{2A} receptors by hallucinogenic and non-hallucinogenic ligands is of particular interest and will also clarify whether this contributes to the differential transcriptome fingerprints of these distinct classes of 5-HT_{2A} receptor ligands [230]. Hallucinogenic ligands (DOI) of the 5-HT_{2A} receptor, shift 5-HT_{2A}-mGluR2 heterocomplex signaling towards enhanced Gq-mediated drive and a reduction in Gi-mediated signaling, in comparison to non-hallucinogenic (methysergide) and endogenous (5-HT) ligands which appear to favour higher Gi-mediated signaling downstream of the 5-HT_{2A}-mGluR2 heterocomplex [44]. Similarly, only hallucinogenic 5-HT_{2A} receptor ligands differentially influence downstream signaling from the 5-HT_{2A}-D2R heterocomplex, allosterically enhancing D2R protomer signaling [46]. Such tipping of the Gi-Gq balance downstream of the 5-HT_{2A}-mGluR2 and the 5-HT_{2A}-D2R heterocomplex could drive distinct transcriptional outcomes and differentially influence the regulation of BDNF expression by diverse 5-HT_{2A} receptor ligands. Given hallucinogenic 5-HT_{2A} receptor ligands robustly enhance cortical BDNF expression, BDNF-mediated effects on cytoarchitecture, synaptic plasticity and long term potentiation could also contribute to the persistent effects of hallucinogens that result in the recurrence of symptoms long after cessation of the drug [357]. Future experiments are required to address whether hallucinogenic compounds mediate such long-lasting effects through a 5-HT_{2A} receptor-mediated recruitment of BDNF signaling.

Conclusion

In this book chapter, we have reviewed the regulation of the neurotrophin BDNF by 5-HT_{2A} receptors, the influence of BDNF on 5-HT_{2A} receptor expression and function, and the relevance of such a reciprocal relationship in the pathophysiology and treatment of disorders such as depression and schizophrenia. While a series of studies point to an important crosstalk between 5-HT_{2A} receptors and BDNF within the frontal cortex, hippocampus and amygdala, there are multiple aspects of this interaction that remain to be experimentally investigated. While the focus of studies so far has largely been to gain an understanding of the transcriptional regulation of *Bdnf* by 5-HT_{2A} receptors, the functional consequences of changes in BDNF expression, as well as the influence of 5-HT_{2A} receptors on BDNF synthesis, transport, release, signaling and BDNF-driven cellular and behavioral consequences remain to be investigated (Fig. 4a). Most studies addressing the influence of BDNF on 5-HT_{2A} receptors have focused on using BDNF mouse mutant models to examine changes in 5-HT_{2A} receptor expression and electrophysiological responses (Table 1), with limited understanding of the impact of BDNF on 5-HT_{2A} receptor heterodimerization, internalization and receptor recycling, signaling, network activity, and 5-HT_{2A} receptor-driven cellular and behavioral outcomes (Fig. 4b). Further, despite an understanding of the key role that 5-HT [358–360] and BDNF [361–363] play during critical period plasticity and the activity-dependent refinement of cortical circuits, there is a relative paucity of information on the role of a 5-HT_{2A} receptor-BDNF crosstalk during these early temporal windows. Converging lines of evidence indicate that diverse early life adverse events disrupt 5-HT_{2A} receptor function [214, 216, 264], and also alter gene expression of *Bdnf* [167, 364]. These studies motivate future investigation into the relationship between 5-HT_{2A} receptors and BDNF during development, in particular with a view to understanding the establishment of vulnerability for psychopathology. The vast majority of studies that have examined the crosstalk between 5-HT_{2A} receptors and BDNF are based on preclinical observations in animal models, clinical studies examining this relationship are currently limited. Indeed, it will be important to address the reciprocal regulation of 5-HT_{2A} receptors and BDNF in clinical samples derived from patients with genetic polymorphisms of BDNF and the 5-HT_{2A} receptor that influence risk and treatment responses for diverse psychiatric disorders. The BDNF val66met [25, 325, 365, 366] and specific 5-HT_{2A} receptor SNPs [297, 367, 368] have been strongly linked to both enhanced vulnerability for psychopathology and to altered treatment response to antidepressants and antipsychotics. In this regard the potential use of fibroblast-derived human induced pluripotent stem cells to mechanistically examine the association between 5-HT_{2A} receptors and BDNF would open up important avenues for future studies that examine the relevance of 5-HT_{2A} receptor-BDNF cross-talk in a clinical context. A deeper understanding of the relationship between 5-HT_{2A} receptors and BDNF is of importance to elucidating the mechanisms underlying the selective vulnerability of specific neuronal populations implicated in the pathogenesis of anxiety and depression, and in the cortical microcircuit dysfunction associated with schizophrenia. Such studies bear promise for the identification of novel therapeutic targets for the treatment of psychiatric disorders such as anxiety, depression and schizophrenia.

References

1. Artigas F (2013) Future directions for serotonin and antidepressants. ACS Chem Neurosci 4:5–8
2. Duman RS, Monteggia LM (2006) A Neurotrophic model for stress-related mood disorders. Biol Psychiatry 59(12):1116–1127
3. Guiard BP, Di Giovanni G (2015) Central serotonin-2A (5-HT2A) receptor dysfunction in depression and epilepsy: the missing link? Front Pharmacol 6:46
4. Petit AC, Quesseveur G, Gressier F et al (2014) Converging translational evidence for the involvement of the serotonin 2A receptor gene in major depressive disorder. Prog Neuro-Psychopharmacology Biol Psychiatry 54:76–82
5. Weisstaub NV, Zhou M, Lira A et al (2006) Cortical 5-HT2A receptor signaling modulates anxiety-like behaviors in mice. Science 313:536–540
6. Autry AE, Monteggia LM (2012) Brain-derived neurotrophic factor and neuropsychiatric disorders. Pharmacol Rev 64:238–258
7. Ebdrup BH, Rasmussen H, Arnt J, Glenthøj B (2011) Serotonin 2A receptor antagonists for treatment of schizophrenia. Expert Opin Investig Drugs 20:1211–1223
8. Bramham CR, Messaoudi E (2005) BDNF function in adult synaptic plasticity: the synaptic consolidation hypothesis. Prog Neurobiol 76:99–125
9. Chen A, Hough CJ, Li H (2003) Serotonin type II receptor activation facilitates synaptic plasticity via n-methyl-d-aspartate-mediated mechanism in the rat basolateral amygdala. Neuroscience 119:53–63
10. Edelmann E, Lessmann V, Brigadski T (2014) Pre- and postsynaptic twists in BDNF secretion and action in synaptic plasticity. Neuropharmacology 76:610–627
11. Lesch K-P, Waider J (2012) Serotonin in the modulation of neural plasticity and networks: implications for neurodevelopmental disorders. Neuron 76:175–191
12. Banasr M, Hery M, Printemps R, Daszuta A (2004) Serotonin-induced increases in adult cell proliferation and neurogenesis are mediated through different and common 5-HT receptor subtypes in the dentate gyrus and the subventricular zone. Neuropsychopharmacology 29:450–460

13. Jha S, Rajendran R, Fernandes KA, Vaidya VA (2008) 5-HT2A/2C receptor blockade regulates progenitor cell proliferation in the adult rat hippocampus. Neurosci Lett 441:210–214
14. Klempin F, Babu H, De Pietri Tonelli D et al (2010) Oppositional effects of serotonin receptors 5-HT1a, 2, and 2c in the regulation of adult hippocampal neurogenesis. Front Mol Neurosci 3
15. Sairanen M, Lucas G, Ernfors P et al (2005) Brain-derived neurotrophic factor and antidepressant drugs have different but coordinated effects on neuronal turnover, proliferation, and survival in the adult dentate gyrus. J Neurosci 25:1089–1094
16. Waterhouse EG, An JJ, Orefice LL et al (2012) BDNF promotes differentiation and maturation of adult-born neurons through GABAergic transmission. J Neurosci 32:14318–14330
17. Bekinschtein P, Renner MC, Gonzalez MC, Weisstaub N (2013) Role of medial prefrontal cortex serotonin 2A receptors in the control of retrieval of recognition memory in rats. J Neurosci 33:15716–15725
18. Lu B, Nagappan G, Lu Y (2014) BDNF and synaptic plasticity, cognitive function, and dysfunction. Handb Exp Pharmacol 220:223–250
19. Zhang G, Ásgeirsdóttir HN, Cohen SJ et al (2013) Stimulation of serotonin 2A receptors facilitates consolidation and extinction of fear memory in C57BL/6J mice. Neuropharmacology 64:403–413
20. Martinowich K, Manji H, Lu B (2007) New insights into BDNF function in depression and anxiety. Nat Neurosci 10:1089–1093
21. Geyer MA, Vollenweider FX (2008) Serotonin research: contributions to understanding psychoses. Trends Pharmacol Sci 29:445–453
22. Lu B, Martinowich K (2008) Cell biology of BDNF and its relevance to schizophrenia. Novartis Found Symp 289:119–129
23. Meltzer HY, Massey BW, Horiguchi M (2012) Serotonin receptors as targets for drugs useful to treat psychosis and cognitive impairment in schizophrenia. Curr Pharm Biotechnol 13:1572–1586
24. Mestre TA, Zurowski M, Fox SH (2013) 5-Hydroxytryptamine 2A receptor antagonists as potential treatment for psychiatric disorders. Expert Opin Investig Drugs 22:411–421
25. Notaras M, Hill R, van den Buuse M (2015a) A role for the BDNF gene Val66Met polymorphism in schizophrenia? A comprehensive review. Neurosci Biobehav Rev 51:15–30
26. Daftary SS, Calderon G, Rios M (2012) Essential role of brain-derived neurotrophic factor in the regulation of serotonin transmission in the basolateral amygdala. Neuroscience 224:125–134
27. Homberg JR, Molteni R, Calabrese F, M a R (2014) The serotonin-BDNF duo: developmental implications for the vulnerability to psychopathology. Neurosci Biobehav Rev 43:35–47
28. Klein AB, Santini MA, Aznar S et al (2010b) Changes in 5-HT2A-mediated behavior and 5-HT2A- and 5-HT1A receptor binding and expression in conditional brain-derived neurotrophic factor knock-out mice. Neuroscience 169:1007–1016
29. Martinowich K, Lu B (2008) Interaction between BDNF and serotonin: role in mood disorders. Neuropsychopharmacology 33:73–83
30. Mattson MP, Maudsley S, Martin B (2004) BDNF and 5-HT: a dynamic duo in age-related neuronal plasticity and neurodegenerative disorders. Trends Neurosci 27:589–594
31. Trajkovska V, Santini M a., Marcussen a. B, et al (2009) BDNF downregulates 5-HT2A receptor protein levels in hippocampal cultures. Neurochem Int 55:697–702
32. Vaidya VA, Marek GJ, Aghajanian GK, Duman RS (1997) 5-HT2A receptor-mediated regulation of brain-derived neurotrophic factor mRNA in the hippocampus and the neocortex. J Neurosci 17:2785–2795
33. Vaidya VA, Terwilliger R, Duman R (1999) Role of 5-HT 2A receptors in the stress-induced down-regulation of brain-derived neurotrophic factor expression in rat hippocampus. Neurosci Lett 262:1–4
34. Turlejski K (1996) Evolutionary ancient roles of serotonin: long-lasting regulation of activity and development. Acta Neurobiol Exp (Wars) 56:619–636

35. Weiger WA (1997) Serotonergic modulation of behaviour: a phylogenetic overview. Biol Rev Camb Philos Soc 72:61–95
36. Barnes NM, Sharp T (1999) A review of central 5-HT receptors and their function. Neuropharmacology 38:1083–1152
37. Adlersberg M, Arango V, Hsiung S et al (2000) In vitro autoradiography of serotonin 5-HT2A/2C receptor-activated G protein: Guanosine-5′-(g-[35S])thiotriphosphate binding in rat brain. J Neurosci Res 61:674–685
38. Millan MJ, Marin P, Bockaert J, Mannoury la Cour C (2008) Signaling at G-protein-coupled serotonin receptors: recent advances and future research directions. Trends Pharmacol Sci 29:454–464
39. Bohn LM, Schmid CL (2010) Serotonin receptor signaling and regulation via β-arrestins. Crit Rev Biochem Mol Biol 45:555–566
40. Schmid CL, Bohn LM (2010) Serotonin, but not N-Methyltryptamines, activates the serotonin 2A receptor via a β-Arrestin2 / Src / Akt Signaling complex in vivo. Drugs 30:13513–13524
41. González-Maeso J, Weisstaub NV, Zhou M et al (2007) Hallucinogens recruit specific cortical 5-HT(2A) receptor-mediated signaling pathways to affect behavior. Neuron 53:439–452
42. Albizu L, Holloway T, González-Maeso J, Sealfon SC (2011) Functional crosstalk and heteromerization of serotonin 5-HT2A and dopamine D2 receptors. Neuropharmacology 61:770–777
43. Delille HK, Becker JM, Burkhardt S et al (2012) Heterocomplex formation of 5-HT2A-mGlu2 and its relevance for cellular signaling cascades. Neuropharmacology 62:2184–2191
44. Fribourg M, Moreno JL, Holloway T et al (2011) Decoding the signaling of a GPCR heteromeric complex reveals a unifying mechanism of action of antipsychotic drugs. Cell 147:1011–1023
45. Moreno JL, Miranda-Azpiazu P, García-Bea A et al (2016) Allosteric signaling through an mGlu2 and 5-HT2A heteromeric receptor complex and its potential contribution to schizophrenia. Sci Signal 9:ra5. https://doi.org/10.1126/scisignal.aab0467
46. Borroto-Escuela DO, Romero-Fernandez W, Narvaez M et al (2014) Hallucinogenic 5-HT2AR agonists LSD and DOI enhance dopamine D2R protomer recognition and signaling of D2-5-HT2A heteroreceptor complexes. Biochem Biophys Res Commun 443:278–284
47. Bhattacharyya S, Puri S, Miledi R, Panicker MM (2002) Internalization and recycling of 5-HT2A receptors activated by serotonin and protein kinase C-mediated mechanisms. Proc Natl Acad Sci U S A 99:14470–14475
48. Raote I, Bhattacharyya S, Panicker MM (2013) Functional selectivity in serotonin receptor 2A (5-HT2A) endocytosis, recycling, and phosphorylation. Mol Pharmacol 83:42–50
49. Aghajanian GK, Marek GJ (1999) Serotonin, via 5-HT2A receptors, increases EPSCs in layer V pyramidal cells of prefrontal cortex by an asynchronous mode of glutamate release. Brain Res 825:161–171
50. Andrade R (2011) Serotonergic regulation of neuronal excitability in the prefrontal cortex. Neuropharmacology 61:382–386
51. Martín-Ruiz R, Puig MV, Celada P et al (2001) Control of serotonergic function in medial prefrontal cortex by serotonin-2A receptors through a glutamate-dependent mechanism. J Neurosci 21:9856–9866
52. Day M, Olson PA, Platzer J et al (2002) Stimulation of 5-HT(2) receptors in prefrontal pyramidal neurons inhibits Ca(v)1.2 L type Ca(2+) currents via a PLCbeta/IP3/calcineurin signaling cascade. J Neurophysiol 87:2490–2504
53. Villalobos C, Foehring RC, Lee JC, Andrade R (2011) Essential role for phosphatidylinositol 4,5-Bisphosphate in the expression, regulation, and gating of the slow Afterhyperpolarization current in the cerebral cortex. J Neurosci 31:18303–18312
54. Gresch PJ, Barrett RJ, Sanders-Bush E, Smith RL (2007) 5-Hydroxytryptamine (serotonin)2A receptors in rat anterior cingulate cortex mediate the discriminative stimulus properties of d-lysergic acid diethylamide. J Pharmacol Exp Ther 320:662–669
55. Quesseveur G, Nguyen HT, Gardier AM, Guiard BP (2012) 5-HT_{2A} ligands in the treatment of anxiety and depression. Expert Opin Investig Drugs 21:1701–1725

56. Yadav PN, Kroeze WK, Farrell MS, Roth BL (2011) Antagonist functional selectivity: 5-HT2A serotonin receptor antagonists differentially regulate 5-HT2A receptor protein level in vivo. J Pharmacol Exp Ther 339:99–105
57. Watanabe N, Omori IM, Nakagawa A, et al (2011) Mirtazapine versus other antidepressive agents for depression. Cochrane database Syst Rev. CD006528
58. Gray JA, Roth BL (2007) Molecular targets for treating cognitive dysfunction in schizophrenia. Schizophr Bull 33:1100–1119
59. Ul Haq R, Anderson ML, Hollnagel J-O et al (2015) Serotonin dependent masking of hippocampal sharp wave ripples. Neuropharmacology 101:188–203
60. van Wel JHP, Kuypers KPC, Theunissen EL et al (2011) Blockade of 5-HT2 receptor selectively prevents MDMA-induced verbal memory impairment. Neuropsychopharmacology 36:1932–1939
61. Mengod G, Vilaró MT, Raurich A et al (1996) 5-HT receptors in mammalian brain: receptor autoradiography and in situ hybridization studies of new ligands and newly identified receptors. Histochem J 28:747–758
62. Mengod G, Palacios JM, Cortes R (2015) Cartography of 5-HT1A and 5-HT2A receptor subtypes in prefrontal cortex and its projections. ACS Chem Neurosci 6:1089–1098
63. Willins DL, Deutch AY, Roth BL (1997) Serotonin 5-HT2A receptors are expressed on pyramidal cells and interneurons in the rat cortex. Synapse 27:79–82
64. Ettrup A, da Cunha-Bang S, McMahon B et al (2014) Serotonin 2A receptor agonist binding in the human brain with [11C]Cimbi-36. J Cereb Blood Flow Metab 34:1188–1196
65. Tanaka KF, Samuels BA, Hen R (2012) Serotonin receptor expression along the dorsal-ventral axis of mouse hippocampus. Philos Trans R Soc B Biol Sci 367:2395–2401
66. Szabo ST, Blier P (2001) Functional and pharmacological characterization of the modulatory role of serotonin on the firing activity of locus coeruleus norepinephrine neurons. Brain Res 922:9–20
67. Boothman LJ, Sharp T (2005) A role for midbrain raphe gamma aminobutyric acid neurons in 5-hydroxytryptamine feedback control. Neuroreport 16:891–896
68. Puig MV, Celada P, Díaz-Mataix L, Artigas F (2003) In vivo modulation of the activity of pyramidal neurons in the rat medial prefrontal cortex by 5-HT2A receptors: relationship to thalamocortical afferents. Cereb Cortex 13:870–882
69. Marek GJ, Aghajanian GK (1996) LSD and the phenethylamine hallucinogen DOI are potent partial agonists at 5-HT2A receptors on interneurons in rat piriform cortex. J Pharmacol Exp Ther 278:1373–1382
70. Marek GJ, Aghajanian GK (1994) Excitation of interneurons in piriform cortex by 5-hydroxytryptamine: blockade by MDL 100,907, a highly selective 5-HT2A receptor antagonist. Eur J Pharmacol 259:137–141
71. Narla C, Dunn HA, Ferguson SSG, Poulter MO (2015) Suppression of piriform cortex activity in rat by corticotropin-releasing factor 1 and serotonin 2A/C receptors. Front Cell Neurosci 9:200
72. Sheldon PW, Aghajanian GK (1990) Serotonin (5-HT) induces IPSPs in pyramidal layer cells of rat piriform cortex: evidence for the involvement of a 5-HT2-activated interneuron. Brain Res 506:62–69
73. Shen RY, Andrade R (1998) 5-Hydroxytryptamine2 receptor facilitates GABAergic neurotransmission in rat hippocampus. J Pharmacol Exp Ther 285:805–812
74. Sheldon PW, Aghajanian GK (1991) Excitatory responses to serotonin (5-HT) in neurons of the rat piriform cortex: evidence for mediation by 5-HT1C receptors in pyramidal cells and 5-HT2 receptors in interneurons. Synapse 9:208–218
75. Hamada S, Senzaki K, Hamaguchi-Hamada K et al (1998) Localization of 5-HT2A receptor in rat cerebral cortex and olfactory system revealed by immunohistochemistry using two antibodies raised in rabbit and chicken. Mol Brain Res 54:199–211
76. Hirst WD, Price GW, Rattray M, Wilkin GP (1998) Serotonin transporters in adult rat brain astrocytes revealed by [3H]5-HT uptake into glial plasmalemmal vesicles. Neurochem Int 33:11–22

77. Meller R, Harrison PJ, Elliott JM, Sharp T (2002b) In vitro evidence that 5-hydroxytryptamine increases efflux of glial glutamate via 5-HT(2A) receptor activation. J Neurosci Res 67:399–405
78. Hagberg GB, Blomstrand F, Nilsson M et al (1998) Stimulation of 5-HT2A receptors on astrocytes in primary culture opens voltage-independent Ca2+ channels. Neurochem Int 32:153–162
79. Jalonen TO, Margraf RR, Wielt DB et al (1997) Serotonin induces inward potassium and calcium currents in rat cortical astrocytes. Brain Res 758:69–82
80. Glebov K, Löchner M, Jabs R et al (2015) Serotonin stimulates secretion of exosomes from microglia cells. Glia 63:626–634
81. Béïque J-C, Imad M, Mladenovic L et al (2007) Mechanism of the 5-hydroxytryptamine 2A receptor-mediated facilitation of synaptic activity in prefrontal cortex. Proc Natl Acad Sci U S A 104:9870–9875
82. Marinova Z, Monoranu C-M, Fetz S et al (2015) Region-specific regulation of the serotonin 2A receptor expression in development and ageing in post mortem human brain. Neuropathol Appl Neurobiol 41:520–532
83. Zhang Z-W (2003) Serotonin induces tonic firing in layer V pyramidal neurons of rat prefrontal cortex during postnatal development. J Neurosci 23:3373–3384
84. Zhang L, Ma W, Barker J, Rubinow D (1999) Sex differences in expression of serotonin receptors (subtypes 1A and 2A) in rat brain: a possible role of testosterone. Neuroscience 94:251–259
85. Chen M, Russo-Neustadt A (2013) Kinetics of norepinephrine- and serotonin-induced BDNF release in cultured embryonic hippocampal neurons. Neurosci & Med 04:194–207
86. Meller R, Babity J, Grahame-Smith D (2002a) 5-HT 2A receptor activation leads to increased BDNF mRNA expression in C6 glioma cells. NeuroMolecular Med 1:197–205
87. Conner JM, Lauterborn JC, Yan Q et al (1997) Distribution of brain-derived neurotrophic factor (BDNF) protein and mRNA in the normal adult rat CNS: evidence for anterograde axonal transport. J Neurosci 17:2295–2313
88. Kawamoto Y, Nakamura S, Nakano S et al (1996) Immunohistochemical localization of brain-derived neurotrophic factor in adult rat brain. Neuroscience 74:1209–1226
89. Lewin GR, Barde Y-A (1996) Physiology of the Neurotrophins. Annu Rev Neurosci 19:289–317
90. Yan Q, Rosenfeld R, Matheson C et al (1997) Expression of brain-derived neurotrophic factor protein in the adult rat central nervous system. Neuroscience 78:431–448
91. Benraiss A, Chmielnicki E, Lerner K et al (2001) Adenoviral brain-derived neurotrophic factor induces both neostriatal and olfactory neuronal recruitment from endogenous progenitor cells in the adult forebrain. J Neurosci 21:6718–6731
92. Pencea V, Bingaman KD, Wiegand SJ, Luskin MB (2001) Infusion of brain-derived neurotrophic factor into the lateral ventricle of the adult rat leads to new neurons in the parenchyma of the striatum, septum, thalamus, and hypothalamus. J Neurosci 21:6706–6717
93. Scharfman H, Goodman J, Macleod A et al (2005) Increased neurogenesis and the ectopic granule cells after intrahippocampal BDNF infusion in adult rats. Exp Neurol 192:348–356
94. Lipsky RH, Marini AM (2007) Brain-derived neurotrophic factor in neuronal survival and behavior-related plasticity. Ann N Y Acad Sci 1122:130–143
95. Barnabé-Heider F, Miller FD (2003) Endogenously produced neurotrophins regulate survival and differentiation of cortical progenitors via distinct signaling pathways. J Neurosci 23:5149–5160
96. Barde YA, Edgar D, Thoenen H (1982) Purification of a new neurotrophic factor from mammalian brain. EMBO J 1:549–553
97. Bergami M, Rimondini R, Santi S et al (2008a) Deletion of TrkB in adult progenitors alters newborn neuron integration into hippocampal circuits and increases anxiety-like behavior. Proc Natl Acad Sci U S A 105:15570–15575
98. Causing CG, Gloster A, Aloyz R et al (1997) Synaptic innervation density is regulated by neuron-derived BDNF. Neuron 18:257–267
99. Lu B (2003) Pro-region of neurotrophins: role in synaptic modulation. Neuron 39:735–738

100. Seidah NG, Benjannet S, Pareek S et al (1996) Cellular processing of the neurotrophin precursors of NT3 and BDNF by the mammalian proprotein convertases. FEBS Lett 379:247–250
101. Yang F, Je H-S, Ji Y et al (2009) Pro-BDNF-induced synaptic depression and retraction at developing neuromuscular synapses. J Cell Biol 185:727–741
102. Bergami M, Santi S, Formaggio E et al (2008b) Uptake and recycling of pro-BDNF for transmitter-induced secretion by cortical astrocytes. J Cell Biol 183:213–221
103. Chao MV (2003) Neurotrophins and their receptors: a convergence point for many signalling pathways. Nat Rev Neurosci 4:299–309
104. Fayard B, Loeffler S, Weis J et al (2005) The secreted brain-derived neurotrophic factor precursor pro-BDNF binds to TrkB and p75NTR but not to TrkA or TrkC. J Neurosci Res 80:18–28
105. Koshimizu H, Hazama S, Hara T et al (2010) Distinct signaling pathways of precursor BDNF and mature BDNF in cultured cerebellar granule neurons. Neurosci Lett 473:229–232
106. Reichardt LF (2006) Neurotrophin-regulated signalling pathways. Philos Trans R Soc Lond Ser B Biol Sci 361:1545–1564
107. Taylor AR, Gifondorwa DJ, Robinson MB et al (2012) Motoneuron programmed cell death in response to proBDNF. Dev Neurobiol 72:699–712
108. Teng HK, Teng KK, Lee R et al (2005) ProBDNF induces neuronal apoptosis via activation of a receptor complex of p75NTR and sortilin. J Neurosci 25:5455–5463
109. Adachi N, Kohara K, Tsumoto T (2005) Difference in trafficking of brain-derived neurotrophic factor between axons and dendrites of cortical neurons, revealed by live-cell imaging. BMC Neurosci 6:42
110. Jovanovic JN, Czernik AJ, Fienberg AA et al (2000) Synapsins as mediators of BDNF-enhanced neurotransmitter release. Nat Neurosci 3:323–329
111. Kovalchuk Y, Hanse E, Kafitz KW, Konnerth A (2002) Postsynaptic induction of BDNF-mediated long-term potentiation. Science 295:1729–1734
112. Fryer RH, Kaplan DR, Kromer LF (1997) Truncated trkB receptors on nonneuronal cells inhibit BDNF-induced neurite outgrowth in vitro. Exp Neurol 148:616–627
113. Strohmaier C, Carter BD, Urfer R et al (1996) A splice variant of the neurotrophin receptor trkB with increased specificity for brain-derived neurotrophic factor. EMBO J 15:3332–3337
114. Friedman WJ, Black IB, Kaplan DR (1998) Distribution of the neurotrophins brain-derived neurotrophic factor, neurotrophin-3, and neurotrophin-4/5 in the postnatal rat brain: an immunocytochemical study. Neuroscience 84:101–114
115. Maisonpierre PC, Belluscio L, Friedman B et al (1990) NT-3, BDNF, and NGF in the developing rat nervous system: parallel as well as reciprocal patterns of expression. Neuron 5:501–509
116. Kim J, Yang M, Kim J et al (2014) Developmental and degenerative modulation of brain-derived neurotrophic factor transcript variants in the mouse hippocampus. Int J Dev Neurosci 38:68–73
117. Menshanov PN, Lanshakov DA, Dygalo NN (2015) proBDNF is a major product of bdnf gene expressed in the perinatal rat cortex. Physiol Res 64(6):925–934
118. Zafra F, Hengerer B, Leibrock J et al (1990) Activity dependent regulation of BDNF and NGF mRNAs in the rat hippocampus is mediated by non-NMDA glutamate receptors. EMBO J 9:3545–3550
119. Aid T, Kazantseva A, Piirsoo M et al (2007) Mouse and ratBDNF gene structure and expression revisited. J Neurosci Res 85:525–535
120. Sathanoori M, Dias BG, Nair AR et al (2004) Differential regulation of multiple brain-derived neurotrophic factor transcripts in the postnatal and adult rat hippocampus during development, and in response to kainate administration. Mol Brain Res 130:170–177
121. Liu Q-R, Walther D, Drgon T et al (2005) Human brain derived neurotrophic factor (BDNF) genes, splicing patterns, and assessments of associations with substance abuse and Parkinson's disease. Am J Med Genet B Neuropsychiatr Genet 134B:93–103

122. An JJ, Gharami K, Liao G-Y et al (2008) Distinct role of long 3′ UTR BDNF mRNA in spine morphology and synaptic plasticity in hippocampal neurons. Cell 134:175–187
123. Baj G, Leone E, Chao MV, Tongiorgi E (2011) Spatial segregation of BDNF transcripts enables BDNF to differentially shape distinct dendritic compartments. Proc Natl Acad Sci U S A 108:16813–16818
124. Metsis M, Timmusk T, Arenas E, Persson H (1993) Differential usage of multiple brain-derived neurotrophic factor promoters in the rat brain following neuronal activation. Proc Natl Acad Sci U S A 90:8802–8806
125. Pruunsild P, Kazantseva A, Aid T et al (2007) Dissecting the human BDNF locus: bidirectional transcription, complex splicing, and multiple promoters. Genomics 90:397–406
126. Timmusk T, Belluardo N, Persson H, Metsis M (1994a) Developmental regulation of brain-derived neurotrophic factor messenger RNAs transcribed from different promoters in the rat brain. Neuroscience 60:287–291
127. Timmusk T, Lendahl U, Funakoshi H et al (1995) Identification of brain-derived neurotrophic factor promoter regions mediating tissue-specific, axotomy-, and neuronal activity-induced expression in transgenic mice. J Cell Biol 128:185–199
128. Timmusk T, Persson H, Metsis M (1994b) Analysis of transcriptional initiation and translatability of brain-derived neurotrophic factor mRNAs in the rat brain. Neurosci Lett 177:27–31
129. Smith MA, Makino S, Kvetnansky R, Post RM (1995) Stress and glucocorticoids affect the expression of brain-derived neurotrophic factor and neurotrophin-3 mRNAs in the hippocampus. J Neurosci 15:1768–1777
130. Tsankova NM, Berton O, Renthal W et al (2006) Sustained hippocampal chromatin regulation in a mouse model of depression and antidepressant action. Nat Neurosci 9:519–525
131. Bekinschtein P, Oomen CA, Saksida LM, Bussey TJ (2011) Effects of environmental enrichment and voluntary exercise on neurogenesis, learning and memory, and pattern separation: BDNF as a critical variable? Semin Cell Dev Biol 22:536–542
132. Falkenberg T, Mohammed AK, Henriksson B et al (1992) Increased expression of brain-derived neurotrophic factor mRNA in rat hippocampus is associated with improved spatial memory and enriched environment. Neurosci Lett 138:153–156
133. Hall J, Thomas KL, Everitt BJ (2000) Rapid and selective induction of BDNF expression in the hippocampus during contextual learning. Nat Neurosci 3:533–535
134. Radecki DT, Brown LM, Martinez J, Teyler TJ (2005) BDNF protects against stress-induced impairments in spatial learning and memory and LTP. Hippocampus 15:246–253
135. Nibuya M, Morinobu S, Duman RS (1995) Regulation of BDNF and trkB mRNA in rat brain by chronic electroconvulsive seizure and antidepressant drug treatments. J Neurosci 15:7539–7547
136. Casarotto PC, dos Santos PC, Lucas GA et al (2015) BDNF-TRKB signaling system of the dorsal periaqueductal gray matter is implicated in the panicolytic-like effect of antidepressant drugs. Eur Neuropsychopharmacol 25:913–922
137. Licata SC, Shinday NM, Huizenga MN et al (2013) Alterations in brain-derived neurotrophic factor in the mouse hippocampus following acute but not repeated benzodiazepine treatment. PLoS One 8:e84806
138. Bai O, Chlan-Fourney J, Bowen R et al (2003) Expression of brain-derived neurotrophic factor mRNA in rat hippocampus after treatment with antipsychotic drugs. J Neurosci Res 71:127–131
139. Chlan-Fourney J, Ashe P, Nylen K et al (2002) Differential regulation of hippocampal BDNF mRNA by typical and atypical antipsychotic administration. Brain Res 954:11–20
140. Braun AA, Herring NR, Schaefer TL et al (2011) Neurotoxic (+)-methamphetamine treatment in rats increases brain-derived neurotrophic factor and tropomyosin receptor kinase B expression in multiple brain regions. Neuroscience 184:164–171
141. Schmidt HD, Sangrey GR, Darnell SB et al (2012) Increased brain-derived neurotrophic factor (BDNF) expression in the ventral tegmental area during cocaine abstinence is associated with increased histone acetylation at BDNF exon I-containing promoters. J Neurochem 120:202–209

142. Butovsky E, Juknat A, Goncharov I et al (2005) In vivo up-regulation of brain-derived neurotrophic factor in specific brain areas by chronic exposure to Delta-tetrahydrocannabinol. J Neurochem 93:802–811
143. Martin DA, Marona-Lewicka D, Nichols DE, Nichols CD (2014) Chronic LSD alters gene expression profiles in the mPFC relevant to schizophrenia. Neuropharmacology 83:1–8
144. Caputo V, Sinibaldi L, Fiorentino A et al (2011) Brain derived neurotrophic factor (BDNF) expression is regulated by microRNAs miR-26a and miR-26b allele-specific binding. PLoS One 6:e28656
145. Fukuchi M, Tsuda M (2010) Involvement of the 3′-untranslated region of the brain-derived neurotrophic factor gene in activity-dependent mRNA stabilization. J Neurochem 115:1222–1233
146. Hwang JJ, Park M-H, Choi S-Y, Koh J-Y (2005) Activation of the Trk signaling pathway by extracellular zinc. Role of metalloproteinases. J Biol Chem 280:11995–12001
147. Oe S, Yoneda Y (2010) Cytoplasmic polyadenylation element-like sequences are involved in dendritic targeting of BDNF mRNA in hippocampal neurons. FEBS Lett 584:3424–3430
148. Wu YC, Williamson R, Li Z et al (2011) Dendritic trafficking of brain-derived neurotrophic factor mRNA: regulation by translin-dependent and -independent mechanisms. J Neurochem 116:1112–1121
149. Evans SF, Irmady K, Ostrow K et al (2011) Neuronal brain-derived neurotrophic factor is synthesized in excess, with levels regulated by sortilin-mediated trafficking and lysosomal degradation. J Biol Chem 286:29556–29567
150. Wong Y-H, Lee C-M, Xie W et al (2015) Activity-dependent BDNF release via endocytic pathways is regulated by synaptotagmin-6 and complexin. Proc Natl Acad Sci U S A 112:E4475–E4484
151. Rohe M, Hartl D, Fjorback AN et al (2013) SORLA-mediated trafficking of TrkB enhances the response of neurons to BDNF. PLoS One 8:e72164
152. Eide FF, Vining ER, Eide BL et al (1996) Naturally occurring truncated trkB receptors have dominant inhibitory effects on brain-derived neurotrophic factor signaling. J Neurosci 16:3123–3129
153. Fenner BM (2012) Truncated TrkB: beyond a dominant negative receptor. Cytokine Growth Factor Rev 23:15–24
154. Fryer RH, Kaplan DR, Feinstein SC et al (1996) Developmental and mature expression of full-length and truncated TrkB receptors in the rat forebrain. J Comp Neurol 374:21–40
155. Coppell A, Pei Q, Zetterström TS (2003) Bi-phasic change in BDNF gene expression following antidepressant drug treatment. Neuropharmacology 44:903–910
156. Ivy a S, Rodriguez FG, Garcia C et al (2003) Noradrenergic and serotonergic blockade inhibits BDNF mRNA activation following exercise and antidepressant. Pharmacol Biochem Behav 75:81–88
157. De Foubert G, Carney SL, Robinson CS et al (2004) Fluoxetine-induced change in rat brain expression of brain-derived neurotrophic factor varies depending on length of treatment. Neuroscience 128:597–604
158. Donnici L, Tiraboschi E, Tardito D et al (2008) Time-dependent biphasic modulation of human BDNF by antidepressants in neuroblastoma cells. BMC Neurosci 9:61
159. Khundakar AA, Zetterström TSC (2006) Biphasic change in BDNF gene expression following antidepressant drug treatment explained by differential transcript regulation. Brain Res 1106:12–20
160. Zetterström TSC, Pei Q, Madhav TR et al (1999) Manipulations of brain 5-HT levels affect gene expression for BDNF in rat brain. Neuropharmacology 38:1063–1073
161. Balu DT, Hoshaw BA, Malberg JE et al (2008) Differential regulation of central BDNF protein levels by antidepressant and non-antidepressant drug treatments. Brain Res 1211:37–43
162. Mannari C, Origlia N, Scatena A et al (2008) BDNF level in the rat prefrontal cortex increases following chronic but not acute treatment with duloxetine, a dual acting inhibitor of noradrenaline and serotonin re-uptake. Cell Mol Neurobiol 28:457–468

163. Zetterström TSC, A a C, A a K (2014) The role of 5-hydroxytryptamine receptor subtypes in the regulation of brain-derived neurotrophic factor gene expression. J Pharm Pharmacol 66:53–61
164. Pilar-Cuéllar F, Vidal R, Pazos a (2012) Subchronic treatment with fluoxetine and ketanserin increases hippocampal brain-derived neurotrophic factor, β-catenin and antidepressant-like effects. Br J Pharmacol 165:1046–1057
165. Musazzi L, Rimland JM, Ieraci a, et al (2014) Pharmacological characterization of BDNF promoters I, II and IV reveals that serotonin and norepinephrine input is sufficient for transcription activation. Int J Neuropsychopharmacol 17:779–791
166. Lauterborn JC, Rivera S, Stinis CT et al (1996) Differential effects of protein synthesis inhibition on the activity-dependent expression of BDNF transcripts: evidence for immediate-early gene responses from specific promoters. J Neurosci 16:7428–7436
167. Nair A, Vadodaria KC, Banerjee SB et al (2007) Stressor-specific regulation of distinct brain-derived neurotrophic factor transcripts and cyclic AMP response element-binding protein expression in the postnatal and adult rat hippocampus. Neuropsychopharmacology 32:1504–1519
168. Dias BG, Banerjee SB, Duman RS, V a V (2003) Differential regulation of brain derived neurotrophic factor transcripts by antidepressant treatments in the adult rat brain. Neuropharmacology 45:553–563
169. Russo-Neustadt A, Beard R, Huang Y, Cotman C (2000) Physical activity and antidepressant treatment potentiate the expression of specific brain-derived neurotrophic factor transcripts in the rat hippocampus. Neuroscience 101:305–312
170. Lubin FD, Roth TL, Sweatt JD (2008) Epigenetic regulation of BDNF gene transcription in the consolidation of fear memory. J Neurosci 28:10576–10586
171. Zheng F, Zhou X, Moon C, Wang H (2012) Regulation of brain-derived neurotrophic factor expression in neurons. Int J Physiol Pathophysiol Pharmacol 4:188–200
172. Cavus I, Duman RS (2003) Influence of estradiol, stress, and 5-HT2A agonist treatment on brain-derived neurotrophic factor expression in female rats. Biol Psychiatry 54:59–69
173. Vaidya VA, Castro ME, Pei Q et al (2001) Influence of thyroid hormone on 5-HT1A and 5-HT2A receptor-mediated regulation of hippocampal BDNF mRNA expression. Neuropharmacology 40:48–56
174. Morita K, Her S (2008) Progesterone Pretreatment enhances serotonin-stimulated BDNF gene expression in rat C6 Glioma cells through production of 5α-reduced Neurosteroids. J Mol Neurosci 34:193–200
175. Gewirtz JC, Chen AC, Terwilliger R et al (2002) Modulation of DOI-induced increases in cortical BDNF expression by group II mGlu receptors. Pharmacol Biochem Behav 73:317–326
176. Bombardi C (2012) Neuronal localization of 5-HT2A receptor immunoreactivity in the rat hippocampal region. Brain Res Bull 87:259–273
177. Piguet P, Galvan M (1994) Transient and long-lasting actions of 5-HT on rat dentate gyrus neurones in vitro. J Physiol 481(Pt 3):629–639
178. Hayes VY, Towner MD, Isackson PJ (1997) Organization, sequence and functional analysis of a mouse BDNF promoter. Brain Res Mol Brain Res 45:189–198
179. Tabuchi A, Sakaya H, Kisukeda T et al (2002) Involvement of an upstream stimulatory factor as well as cAMP-responsive element-binding protein in the activation of brain-derived neurotrophic factor gene promoter I. J Biol Chem 277:35920–35931
180. Gao X, Daugherty RL, Tourtellotte WG (2007) Regulation of low affinity neurotrophin receptor (p75(NTR)) by early growth response (Egr) transcriptional regulators. Mol Cell Neurosci 36:501–514
181. Chen A, Xiong L-J, Tong Y, Mao M (2013) Neuroprotective effect of brain-derived neurotrophic factor mediated by autophagy through the PI3K/Akt/mTOR pathway. Mol Med Rep 8:1011–1016
182. Lee FS, Rajagopal R, Chao MV (2002) Distinctive features of Trk neurotrophin receptor transactivation by G protein-coupled receptors. Cytokine Growth Factor Rev 13:11–17

183. Rajagopal R, Chen Z-Y, Lee FS, Chao MV (2004) Transactivation of Trk neurotrophin receptors by G-protein-coupled receptor ligands occurs on intracellular membranes. J Neurosci 24:6650–6658
184. Lee R, Kermani P, Teng KK, Hempstead BL (2001) Regulation of cell survival by secreted proneurotrophins. Science 294:1945–1948
185. Shum JKS, Melendez JA, Jeffrey JJ (2002) Serotonin-induced MMP-13 production is mediated via phospholipase C, protein kinase C, and ERK1/2 in rat uterine smooth muscle cells. J Biol Chem 277:42830–42840
186. Yabanoglu S, Akkiki M, Seguelas M-H et al (2009) Platelet derived serotonin drives the activation of rat cardiac fibroblasts by 5-HT2A receptors. J Mol Cell Cardiol 46:518–525
187. Baxter GT, Radeke MJ, Kuo RC et al (1997) Signal transduction mediated by the truncated trkB receptor isoforms, trkB.T1 and trkB.T2. J Neurosci 17:2683–2690
188. Michaelsen K, Zagrebelsky M, Berndt-Huch J et al (2010) Neurotrophin receptors TrkB.T1 and p75NTR cooperate in modulating both functional and structural plasticity in mature hippocampal neurons. Eur J Neurosci 32:1854–1865
189. Rose CR, Blum R, Pichler B et al (2003) Truncated TrkB-T1 mediates neurotrophin-evoked calcium signalling in glia cells. Nature 426:74–78
190. Sherrard RM, Dixon KJ, Bakouche J et al (2009) Differential expression of TrkB isoforms switches climbing fiber-Purkinje cell synaptogenesis to selective synapse elimination. Dev Neurobiol 69:647–662
191. Jang S-W, Liu X, Pradoldej S et al (2010) N-acetylserotonin activates TrkB receptor in a circadian rhythm. Proc Natl Acad Sci U S A 107:3876–3881
192. Kruk JS, Vasefi MS, Heikkila JJ, Beazely MA (2013) Reactive oxygen species are required for 5-HT-induced transactivation of neuronal platelet-derived growth factor and TrkB receptors, but not for ERK1/2 activation. PLoS One 8:e77027
193. Overland AC, Insel PA (2015) Heterotrimeric G proteins directly regulate MMP14/membrane type-1 matrix metalloprotease: a novel mechanism for GPCR-EGFR transactivation. J Biol Chem 290:9941–9947
194. Cattaneo F, Guerra G, Parisi M et al (2014) Cell-surface receptors transactivation mediated by g protein-coupled receptors. Int J Mol Sci 15:19700–19728
195. Kamato D, Rostam MA, Bernard R et al (2015) The expansion of GPCR transactivation-dependent signalling to include serine/threonine kinase receptors represents a new cell signalling frontier. Cell Mol Life Sci 72:799–808
196. Marinova Z, Walitza S, Grünblatt E (2013) 5-HT2A serotonin receptor agonist DOI alleviates cytotoxicity in neuroblastoma cells: role of the ERK pathway. Prog Neuro-Psychopharmacology Biol Psychiatry 44:64–72
197. Seitz PK, Bremer NM, McGinnis AG et al (2012) Quantitative changes in intracellular calcium and extracellular-regulated kinase activation measured in parallel in CHO cells stably expressing serotonin (5-HT) 5-HT2A or 5-HT2C receptors. BMC Neurosci 13:25
198. Cavanaugh JE, Ham J, Hetman M et al (2001) Differential regulation of mitogen-activated protein kinases ERK1/2 and ERK5 by neurotrophins, neuronal activity, and cAMP in neurons. J Neurosci 21:434–443
199. Ying S-W, Futter M, Rosenblum K et al (2002) Brain-derived neurotrophic factor induces long-term potentiation in intact adult hippocampus: requirement for ERK activation coupled to CREB and upregulation of arc synthesis. J Neurosci 22:1532–1540
200. Jones KA, Srivastava DP, Allen JA et al (2009) Rapid modulation of spine morphology by the 5-HT2A serotonin receptor through kalirin-7 signaling. Proc Natl Acad Sci U S A 106:19575–19580
201. Ohira K, Homma KJ, Hirai H et al (2006) TrkB-T1 regulates the RhoA signaling and actin cytoskeleton in glioma cells. Biochem Biophys Res Commun 342:867–874
202. Yoshimura R, Ikenouchi-Sugita A, Hori H et al (2010) Adding a low dose atypical antipsychotic drug to an antidepressant induced a rapid increase of plasma brain-derived neurotrophic factor levels in patients with treatment-resistant depression. Prog Neuro-Psychopharmacology Biol Psychiatry 34:308–312

203. Pillai A, Terry AV, Mahadik SP (2006) Differential effects of long-term treatment with typical and atypical antipsychotics on NGF and BDNF levels in rat striatum and hippocampus. Schizophr Res 82:95–106
204. L. Huang T (2013) Effects of antipsychotics on the BDNF in schizophrenia. Curr Med Chem 20:345–350
205. Benekareddy M, Nair AR, Dias BG et al (2013) Induction of the plasticity-associated immediate early gene arc by stress and hallucinogens: role of brain-derived neurotrophic factor. Int J Neuropsychopharmacol 16:405–415
206. Vollenweider FX, Kometer M (2010) The neurobiology of psychedelic drugs: implications for the treatment of mood disorders. Nat Rev Neurosci 11:642–651
207. Bland ST, Tamlyn JP, Barrientos RM et al (2007) Expression of fibroblast growth factor-2 and brain-derived neurotrophic factor mRNA in the medial prefrontal cortex and hippocampus after uncontrollable or controllable stress. Neuroscience 144:1219–1228
208. Lee Y, Duman RS, Marek GJ (2006) The mGlu2/3 receptor agonist LY354740 suppresses immobilization stress-induced increase in rat prefrontal cortical BDNF mRNA expression. Neurosci Lett 398:328–332
209. Gewirtz J (2000) Behavioral evidence for interactions between a hallucinogenic drug and group II metabotropic glutamate receptors. Neuropsychopharmacology 23:569–576
210. Kłodzinska A, Bijak M, Tokarski K, Pilc A (2002) Group II mGlu receptor agonists inhibit behavioural and electrophysiological effects of DOI in mice. Pharmacol Biochem Behav 73:327–332
211. Zhang C, Marek GJ (2007) Group III metabotropic glutamate receptor agonists selectively suppress excitatory synaptic currents in the rat prefrontal cortex induced by 5-Hydroxytryptamine2A receptor activation. J Pharmacol Exp Ther 320:437–447
212. Menezes MM, Santini MA, Benvenga MJ et al (2013) The mGlu2/3 receptor agonists LY354740 and LY379268 differentially regulate restraint-stress-induced expression of c-Fos in rat cerebral cortex. Neurosci J 2013:1–8
213. Zhai Y, George CA, Zhai J et al (2003) Group II metabotropic glutamate receptor modulation of DOI-induced c-fos mRNA and excitatory responses in the cerebral cortex. Neuropsychopharmacology 28:45–52
214. Benekareddy M, Goodfellow NM, Lambe EK, V a V (2010) Enhanced function of prefrontal serotonin 5-HT(2) receptors in a rat model of psychiatric vulnerability. J Neurosci 30:12138–12150
215. Hemmerle AM, Ahlbrand R, Bronson SL et al (2015) Modulation of schizophrenia-related genes in the forebrain of adolescent and adult rats exposed to maternal immune activation. Schizophr Res 168:411–420
216. Malkova NV, Gallagher JJ, Yu CZ et al (2014) Manganese-enhanced magnetic resonance imaging reveals increased DOI-induced brain activity in a mouse model of schizophrenia. Proc Natl Acad Sci U S A 111:E2,492–E2,500
217. Moreno JL, Kurita M, Holloway T et al (2011) Maternal influenza viral infection causes schizophrenia-like alterations of $5\text{-}HT_{2A}$ and $mGlu_2$ receptors in the adult offspring. J Neurosci 31:1863–1872
218. Suri D, Bhattacharya A, Vaidya VA (2014) Early stress evokes temporally distinct consequences on the hippocampal transcriptome, anxiety and cognitive behaviour. Int J Neuropsychopharmacol 17:289–301
219. Wang Q, Shao F, Wang W (2015) Maternal separation produces alterations of forebrain brain-derived neurotrophic factor expression in differently aged rats. Front Mol Neurosci 8:49
220. Rogóz Z, Skuza G, Legutko B (2005) Repeated treatment with mirtazepine induces brain-derived neurotrophic factor gene expression in rats. J Physiol Pharmacol 56:661–671
221. Marek GJ, Martin-Ruiz R, Abo A, Artigas F (2005) The selective 5-HT2A receptor antagonist M100907 enhances antidepressant-like behavioral effects of the SSRI fluoxetine. Neuropsychopharmacology 30:2205–2215
222. Liu R, Jolas T, Aghajanian G (2000) Serotonin 5-HT2 receptors activate local GABA inhibitory inputs to serotonergic neurons of the dorsal raphe nucleus. Brain Res 873:34–45

223. Abdolmaleky HM, Yaqubi S, Papageorgis P et al (2011) Epigenetic dysregulation of HTR2A in the brain of patients with schizophrenia and bipolar disorder. Schizophr Res 129:183–190
224. Favalli G, Li J, Belmonte-de-Abreu P et al (2012) The role of BDNF in the pathophysiology and treatment of schizophrenia. J Psychiatr Res 46:1–11
225. Golimbet VE, Lavrushina OM, Kaleda VG et al (2007) Supportive evidence for the association between the T102C 5-HTR2A gene polymorphism and schizophrenia: a large-scale case-control and family-based study. Eur Psychiatry 22:167–170
226. Horacek J, Bubenikova-Valesova V, Kopecek M et al (2006) Mechanism of action of atypical antipsychotic drugs and the neurobiology of schizophrenia. CNS Drugs 20:389–409
227. Lin P-Y (2012) Increase in brain-derived Neurotrophic factor in patients with schizophrenia treated with olanzapine: a systemic review and meta-analysis. J Exp Clin Med 4:119–124
228. Angelucci F (2000) Mathé A a., Aloe L. Brain-derived neurotrophic factor and tyrosine kinase receptor TrkB in rat brain are significantly altered after haloperidol and risperidone administration J Neurosci Res 60:783–794
229. Park DI, Kim HG, Jung WR et al (2011a) Mecamylamine attenuates dexamethasone-induced anxiety-like behavior in association with brain derived neurotrophic factor upregulation in rat brains. Neuropharmacology 61:276–282
230. González-Maeso J, Yuen T, Ebersole BJ et al (2003) Transcriptome fingerprints distinguish hallucinogenic and nonhallucinogenic 5-hydroxytryptamine 2A receptor agonist effects in mouse somatosensory cortex. J Neurosci 23:8836–8843
231. Djalali S, Höltje M, Grosse G et al (2005) Effects of brain-derived neurotrophic factor (BDNF) on glial cells and serotonergic neurones during development. J Neurochem 92:616–627
232. Galter D, Unsicker K (1999) Regulation of the transmitter phenotype of rostral and caudal groups of cultured serotonergic raphe neurons. Neuroscience 88:549–559
233. Galter D, Unsicker K (2000a) Brain-derived neurotrophic factor and trkB are essential for cAMP-mediated induction of the serotonergic neuronal phenotype. J Neurosci Res 61:295–301
234. Mamounas LA, Altar CA, Blue ME et al (2000) BDNF promotes the regenerative sprouting, but not survival, of injured serotonergic axons in the adult rat brain. J Neurosci 20:771–782
235. Mamounas LA, Blue ME, Siuciak JA, Altar CA (1995) Brain-derived neurotrophic factor promotes the survival and sprouting of serotonergic axons in rat brain. J Neurosci 15:7929–7939
236. Galter D, Unsicker K (2000b) Sequential activation of the 5-HT1(a) serotonin receptor and TrkB induces the serotonergic neuronal phenotype. Mol Cell Neurosci 15:446–455
237. Siuciak JA, Clark MS, Rind HB et al (1998) BDNF induction of tryptophan hydroxylase mRNA levels in the rat brain. J Neurosci Res 52:149–158
238. Celada P, Siuciak JA, Tran TM et al (1996) Local infusion of brain-derived neurotrophic factor modifies the firing pattern of dorsal raphé serotonergic neurons. Brain Res 712:293–298
239. Deltheil T, Guiard BP, Cerdan J et al (2008) Behavioral and serotonergic consequences of decreasing or increasing hippocampus brain-derived neurotrophic factor protein levels in mice. Neuropharmacology 55:1006–1014
240. Naumenko VS, Kondaurova EM, Bazovkina DV et al (2012) Effect of brain-derived neurotrophic factor on behavior and key members of the brain serotonin system in genetically predisposed to behavioral disorders mouse strains. Neuroscience 214:59–67
241. Tsybko AS, Il'chibaeva TV, Kondaurova EM et al (2014) Effect of central administration of the neurotrophic factors BDNF and GDNF on the functional activity and expression of 5-HT2A serotonin receptors in mice genetically predisposed to depressive-like behavior. Mol Biol 48:864–869
242. Lyons WE, L a M, G a R et al (1999) Brain-derived neurotrophic factor-deficient mice develop aggressiveness and hyperphagia in conjunction with brain serotonergic abnormalities. Proc Natl Acad Sci U S A 96:15239–15244
243. Hensler JG, Ladenheim EE, Lyons WE (2003) Ethanol consumption and serotonin-1A (5-HT1A) receptor function in heterozygous BDNF (+/−) mice. J Neurochem 85:1139–1147
244. Luellen BA, Bianco LE, Schneider LM, Andrews AM (2007) Reduced brain-derived neurotrophic factor is associated with a loss of serotonergic innervation in the hippocampus of aging mice. Genes Brain Behav 6:482–490

245. Rios M, Lambe EK, Liu R et al (2006) Severe deficits in 5-HT2A -mediated neurotransmission in BDNF conditional mutant mice. J Neurobiol 66:408–420
246. Chan JP, Unger TJ, Byrnes J, Rios M (2006) Examination of behavioral deficits triggered by targeting Bdnf in fetal or postnatal brains of mice. Neuroscience 142:49–58
247. Sakata K, Duke SM (2014) Lack of BDNF expression through promoter IV disturbs expression of monoamine genes in the frontal cortex and hippocampus. Neuroscience 260:265–275
248. Maynard KR, Hill JL, Calcaterra NE et al (2015) Functional role of BDNF production from unique promoters in aggression and serotonin Signaling. Neuropsychopharmacology. https://doi.org/10.1038/npp.2015.349
249. Rios M, Fan G, Fekete C, Kelly J, Bates B, Kuehn R, Lechan RM, Jaenisch R (2001) Conditional Deletion Of Brain-Derived Neurotrophic Factor in the Postnatal Brain Leads to Obesity and Hyperactivity. Mol Endocrinol 15:1748–1757
250. Sakata K, Woo NH, Martinowich K et al (2009) Critical role of promoter IV-driven BDNF transcription in GABAergic transmission and synaptic plasticity in the prefrontal cortex. Proc Natl Acad Sci U S A 106:5942–5947
251. Kim JH, Roberts DS, Hu Y, Lau GC, Brooks-Kayal AR, Farb DH, Russek SJ (2012) Brain-derived neurotrophic factor uses CREB and Egr3 to regulate NMDA receptor levels in cortical neurons. J Neurochem 120:210–219
252. Roberts DS, Hu Y, Lund IV, Brooks-Kayal AR, Russek SJ (2006) Brain-derived Neurotrophic Factor (BDNF)-induced Synthesis of Early Growth Response Factor 3 (Egr3) Controls the Levels of Type A GABA Receptorα4 Subunits in Hippocampal Neurons. J Biol Chem 281:29431–29435
253. Williams AA, Ingram WM, Levine S et al (2012) Reduced levels of serotonin 2A receptors underlie resistance of Egr3-deficient mice to locomotor suppression by clozapine. Neuropsychopharmacology 37:2285–2298
254. Maple AM, Zhao X, Elizalde DI et al (2015) Htr2a expression responds rapidly to environmental stimuli in an Egr3-dependent manner. ACS Chem Neurosci 6:1137–1142
255. Egan MF, Kojima M, Callicott JH et al (2003) The BDNF val66met polymorphism affects activity-dependent secretion of BDNF and human memory and hippocampal function. Cell 112:257–269
256. Henningsson S, Borg J, Lundberg J et al (2009) Genetic variation in brain-derived neurotrophic factor is associated with serotonin transporter but not serotonin-1A receptor availability in men. Biol Psychiatry 66:477–485
257. Klein AB, Trajkovska V, Erritzoe D et al (2010a) Cerebral 5-HT2A receptor and serotonin transporter binding in humans are not affected by the val66met BDNF polymorphism status or blood BDNF levels. J Cereb Blood Flow Metab 30:e1–e7
258. Willins DL, Meltzer HY (1997) Direct injection of 5-HT2A receptor agonists into the medial prefrontal cortex produces a head-twitch response in rats. J Pharmacol Exp Ther 282:699–706
259. Kuribara M, Eijsink VD, Roubos EW et al (2010) BDNF stimulates Ca2+ oscillation frequency in melanotrope cells of Xenopus Laevis: contribution of IP3-receptor-mediated release of intracellular Ca2+ to gene expression. Gen Comp Endocrinol 169:123–129
260. Li YX, Zhang Y, Lester HA et al (1998) Enhancement of neurotransmitter release induced by brain-derived neurotrophic factor in cultured hippocampal neurons. J Neurosci 18:10231–10240
261. Jørgensen H, Knigge U, Kjaer A et al (1998) Serotonergic involvement in stress-induced ACTH release. Brain Res 811:10–20
262. Bath KG, Schilit A, Lee FS (2013) Stress effects on BDNF expression: effects of age, sex, and form of stress. Neuroscience 239:149–156
263. Castrén E, Võikar V, Rantamäki T (2007) Role of neurotrophic factors in depression. Curr Opin Pharmacol 7:18–21
264. Holloway T, Moreno JL, Umali A et al (2013) Prenatal stress induces schizophrenia-like alterations of serotonin 2A and metabotropic glutamate 2 receptors in the adult offspring: role of maternal immune system. J Neurosci 33:1088–1098

265. Matuszewich L, Yamamoto BK (2003) Long-lasting effects of chronic stress on DOI-induced hyperthermia in male rats. Psychopharmacology 169:169–175
266. Harvey ML, Swallows CL, Cooper MA (2012) A double dissociation in the effects of 5-HT2A and 5-HT2C receptors on the acquisition and expression of conditioned defeat in Syrian hamsters. Behav Neurosci 126:530–537
267. Benekareddy M, Vadodaria KC, Nair AR, Vaidya VA (2011) Postnatal serotonin type 2 receptor blockade prevents the emergence of anxiety behavior, dysregulated stress-induced immediate early gene responses, and specific transcriptional changes that arise following early life stress. Biol Psychiatry 70:1024–1032
268. Adamec R, Creamer K, Bartoszyk GD, Burton P (2004) Prophylactic and therapeutic effects of acute systemic injections of EMD 281014, a selective serotonin 2A receptor antagonist on anxiety induced by predator stress in rats. Eur J Pharmacol 504:79–96
269. Beig MI, Baumert M, Walker FR et al (2009) Blockade of 5-HT2A receptors suppresses hyperthermic but not cardiovascular responses to psychosocial stress in rats. Neuroscience 159:1185–1191
270. Ootsuka Y, Blessing WW, Nalivaiko E (2008) Selective blockade of 5-HT2A receptors attenuates the increased temperature response in brown adipose tissue to restraint stress in rats. Stress 11:125–133
271. Pehek EA, Nocjar C, Roth BL et al (2006) Evidence for the preferential involvement of 5-HT2A serotonin receptors in stress- and drug-induced dopamine release in the rat medial prefrontal cortex. Neuropsychopharmacology 31:265–277
272. Grønli J, Bramham C, Murison R et al (2006) Chronic mild stress inhibits BDNF protein expression and CREB activation in the dentate gyrus but not in the hippocampus proper. Pharmacol Biochem Behav 85:842–849
273. Krishnan V, Nestler EJ (2008) The molecular neurobiology of depression. Nature 455:894–902
274. Lee B-H, Kim Y-K (2010) The roles of BDNF in the pathophysiology of major depression and in antidepressant treatment. Psychiatry Investig 7:231–235
275. Nowacka M, Obuchowicz E (2013) BDNF and VEGF in the pathogenesis of stress-induced affective diseases: an insight from experimental studies. Pharmacol Rep 65:535–546
276. Marmigère F, Givalois L, Rage F et al (2003) Rapid induction of BDNF expression in the hippocampus during immobilization stress challenge in adult rats. Hippocampus 13:646–655
277. Molteni R, Calabrese F, Cattaneo A et al (2009) Acute stress responsiveness of the neurotrophin BDNF in the rat hippocampus is modulated by chronic treatment with the antidepressant duloxetine. Neuropsychopharmacology 34:1523–1532
278. Dranovsky A, Hen R (2006) Hippocampal neurogenesis: regulation by stress and antidepressants. Biol Psychiatry 59:1136–1143
279. Lu B, Chang JH (2004) Regulation of neurogenesis by neurotrophins: implications in hippocampus-dependent memory. Neuron Glia Biol 1:377–384
280. McEwen BS (2000) Effects of adverse experiences for brain structure and function. Biol Psychiatry 48:721–731
281. Pittenger C, Duman RS (2008) Stress, depression, and neuroplasticity: a convergence of mechanisms. Neuropsychopharmacology 33:88–109
282. Kim JJ, Diamond DM (2002) The stressed hippocampus, synaptic plasticity and lost memories. Nat Rev Neurosci 3:453–462
283. Pavlides C, Nivón LG, McEwen BS (2002) Effects of chronic stress on hippocampal long-term potentiation. Hippocampus 12:245–257
284. Bennett MR, Lagopoulos J (2014) Stress and trauma: BDNF control of dendritic-spine formation and regression. Prog Neurobiol 112:80–99
285. Govindarajan A, Rao BSS, Nair D et al (2006) Transgenic brain-derived neurotrophic factor expression causes both anxiogenic and antidepressant effects. Proc Natl Acad Sci U S A 103:13208–13213
286. Liu R-J, Aghajanian GK (2008) Stress blunts serotonin- and hypocretin-evoked EPSCs in prefrontal cortex: role of corticosterone-mediated apical dendritic atrophy. Proc Natl Acad Sci U S A 105:359–364

287. Watanabe Y, Gould E, Daniels DC et al (1992) Tianeptine attenuates stress-induced morphological changes in the hippocampus. Eur J Pharmacol 222:157–162
288. Czéh B, Michaelis T, Watanabe T et al (2001) Stress-induced changes in cerebral metabolites, hippocampal volume, and cell proliferation are prevented by antidepressant treatment with tianeptine. Proc Natl Acad Sci U S A 98:12796–12801
289. Jiang X, Xing G, Yang C et al (2009) Stress impairs 5-HT2A receptor-mediated serotonergic facilitation of GABA release in juvenile rat basolateral amygdala. Neuropsychopharmacology 34:410–423
290. Popoli M, Yan Z, McEwen BS, Sanacora G (2012) The stressed synapse: the impact of stress and glucocorticoids on glutamate transmission. Nat Rev Neurosci 13:22–37
291. Treccani G, Musazzi L, Perego C et al (2014) Stress and corticosterone increase the readily releasable pool of glutamate vesicles in synaptic terminals of prefrontal and frontal cortex. Mol Psychiatry 19:433–443
292. Bland ST, Schmid MJ, Der-Avakian A et al (2005) Expression of c-fos and BDNF mRNA in subregions of the prefrontal cortex of male and female rats after acute uncontrollable stress. Brain Res 1051:90–99
293. Pandey DK, Bhatt S, Jindal A, Gautam B (2014) Effect of combination of ketanserin and escitalopram on behavioral anomalies after olfactory bulbectomy: prediction of quick onset of antidepressant action. Indian J Pharmacol 46:639–643
294. Quesseveur G, Repérant C, David DJ et al (2013) 5-HT_{2A} receptor inactivation potentiates the acute antidepressant-like activity of escitalopram: involvement of the noradrenergic system. Exp Brain Res 226:285–295
295. Sibille E, Sarnyai Z, Benjamin D et al (1997) Antisense inhibition of 5-hydroxytryptamine2a receptor induces an antidepressant-like effect in mice. Mol Pharmacol 52:1056–1063
296. Choi M-J, Lee H-J, Lee H-J et al (2004) Association between major depressive disorder and the -1438A/G polymorphism of the serotonin 2A receptor gene. Neuropsychobiology 49:38–41
297. Jin C, Xu W, Yuan J et al (2013) Meta-analysis of association between the -1438A/G (rs6311) polymorphism of the serotonin 2A receptor gene and major depressive disorder. Neurol Res 35:7–14
298. Horstmann S, Lucae S, Menke A et al (2010) Polymorphisms in GRIK4, HTR2A, and FKBP5 show interactive effects in predicting remission to antidepressant treatment. Neuropsychopharmacology 35:727–740
299. Lucae S, Ising M, Horstmann S et al (2010) HTR2A gene variation is involved in antidepressant treatment response. Eur Neuropsychopharmacol 20:65–68
300. McMahon FJ, Buervenich S, Charney D et al (2006) Variation in the gene encoding the serotonin 2A receptor is associated with outcome of antidepressant treatment. Am J Hum Genet 78:804–814
301. Viikki M, Huuhka K, Leinonen E et al (2011) Interaction between two HTR2A polymorphisms and gender is associated with treatment response in MDD. Neurosci Lett 501:20–24
302. Blier P (2003) The pharmacology of putative early-onset antidepressant strategies. Eur Neuropsychopharmacol 13:57–66
303. Lopes Rocha F, Fuzikawa C, Riera R et al (2013) Antidepressant combination for major depression in incomplete responders--a systematic review. J Affect Disord 144:1–6
304. Rocha FL, Fuzikawa C, Riera R, Hara C (2012) Combination of antidepressants in the treatment of major depressive disorder: a systematic review and meta-analysis. J Clin Psychopharmacol 32:278–281
305. Frokjaer VG, Mortensen EL, Nielsen FA et al (2008) Frontolimbic serotonin 2A receptor binding in healthy subjects is associated with personality risk factors for affective disorder. Biol Psychiatry 63:569–576
306. McKeith IG, Marshall EF, Ferrier IN et al (1987) 5-HT receptor binding in post-mortem brain from patients with affective disorder. J Affect Disord 13:67–74
307. Oquendo MA, Russo SA, Underwood MD et al (2006) Higher postmortem prefrontal 5-HT2A receptor binding correlates with lifetime aggression in suicide. Biol Psychiatry 59:235–243

308. Serres F, Azorin JM, Valli M, Jeanningros R (1999) Evidence for an increase in functional platelet 5-HT2A receptors in depressed patients using the new ligand [125I]-DOI. Eur Psychiatry 14:451–457
309. Li Y, Luikart BW, Birnbaum S et al (2008) TrkB regulates hippocampal neurogenesis and governs sensitivity to antidepressive treatment. Neuron 59:399–412
310. Castrén E, Rantamäki T (2010) The role of BDNF and its receptors in depression and antidepressant drug action: reactivation of developmental plasticity. Dev Neurobiol 70:289–297
311. Kavalali ET, Monteggia LM (2012) Synaptic mechanisms underlying rapid antidepressant action of ketamine. Am J Psychiatry 169:1150–1156
312. Adachi M, Barrot M, Autry AE et al (2008) Selective loss of brain-derived neurotrophic factor in the dentate gyrus attenuates antidepressant efficacy. Biol Psychiatry 63:642–649
313. Eisch AJ, Bolaños CA, de Wit J et al (2003) Brain-derived neurotrophic factor in the ventral midbrain-nucleus accumbens pathway: a role in depression. Biol Psychiatry 54:994–1005
314. Koponen E, Rantamäki T, Voikar V et al (2005) Enhanced BDNF signaling is associated with an antidepressant-like behavioral response and changes in brain monoamines. Cell Mol Neurobiol 25:973–980
315. Lepack AE, Fuchikami M, Dwyer JM et al (2014) BDNF release is required for the behavioral actions of ketamine. Int J Neuropsychopharmacol 18(1):pyu033
316. Monteggia LM, Barrot M, Powell CM et al (2004) Essential role of brain-derived neurotrophic factor in adult hippocampal function. Proc Natl Acad Sci U S A 101:10827–10832
317. Saarelainen T, Hendolin P, Lucas G et al (2003) Activation of the TrkB neurotrophin receptor is induced by antidepressant drugs and is required for antidepressant-induced behavioral effects. J Neurosci 23:349–357
318. Shirayama Y, Chen AC-H, Nakagawa S et al (2002) Brain-derived neurotrophic factor produces antidepressant effects in behavioral models of depression. J Neurosci 22:3251–3261
319. Taliaz D, Nagaraj V, Haramati S et al (2013) Altered brain-derived neurotrophic factor expression in the ventral tegmental area, but not in the hippocampus, is essential for antidepressant-like effects of electroconvulsive therapy. Biol Psychiatry 74:305–312
320. Björkholm C, Monteggia LM (2016) BDNF- a key transducer of antidepressant effects. Neuropharmacology 102:72–79
321. Duman RS, Li N (2012) A neurotrophic hypothesis of depression: role of synaptogenesis in the actions of NMDA receptor antagonists. Philos Trans R Soc Lond Ser B Biol Sci 367:2475–2484
322. Schmidt HD, Duman RS (2007) The role of neurotrophic factors in adult hippocampal neurogenesis, antidepressant treatments and animal models of depressive-like behavior. Behav Pharmacol 18:391–418
323. Wang J-W, Dranovsky A, Hen R (2008) The when and where of BDNF and the antidepressant response. Biol Psychiatry 63:640–641
324. Chiaruttini C, Vicario A, Li Z et al (2009) Dendritic trafficking of BDNF mRNA is mediated by translin and blocked by the G196A (Val66Met) mutation. Proc Natl Acad Sci U S A 106:16481–16486
325. Colle R, Deflesselle E, Martin S et al (2015) BDNF/TRKB/P75NTR polymorphisms and their consequences on antidepressant efficacy in depressed patients. Pharmacogenomics 16:997–1013
326. Hosang GM, Shiles C, Tansey KE et al (2014) Interaction between stress and the BDNF Val66Met polymorphism in depression: a systematic review and meta-analysis. BMC Med 12:7
327. Schumacher J, Jamra RA, Becker T et al (2005) Evidence for a relationship between genetic variants at the brain-derived neurotrophic factor (BDNF) locus and major depression. Biol Psychiatry 58:307–314
328. Bath KG, Jing DQ, Dincheva I et al (2012) BDNF Val66Met impairs fluoxetine-induced enhancement of adult hippocampus plasticity. Neuropsychopharmacology 37:1297–1304
329. Chen Z-Y, Jing D, Bath KG et al (2006) Genetic variant BDNF (Val66Met) polymorphism alters anxiety-related behavior. Science 314:140–143

330. Liu R-J, Lee FS, Li X-Y et al (2012) Brain-derived Neurotrophic factor Val66Met allele impairs basal and ketamine-stimulated synaptogenesis in prefrontal cortex. Biol Psychiatry 71:996–1005
331. Yu H, Wang D-D, Wang Y et al (2012) Variant brain-derived neurotrophic factor Val66Met polymorphism alters vulnerability to stress and response to antidepressants. J Neurosci 32:4092–4101
332. Masi G, Brovedani P (2011) The hippocampus, neurotrophic factors and depression: possible implications for the pharmacotherapy of depression. CNS Drugs 25:913–931
333. Marek GJ (2008) Regulation of rat cortical 5-hydroxytryptamine2A receptor-mediated electrophysiological responses by repeated daily treatment with electroconvulsive shock or imipramine. Eur Neuropsychopharmacol 18:498–507
334. Muguruza C, Miranda-Azpiazu P, Díez-Alarcia R et al (2014) Evaluation of 5-HT2A and mGlu2/3 receptors in postmortem prefrontal cortex of subjects with major depressive disorder: effect of antidepressant treatment. Neuropharmacology 86:311–318
335. Kusumi I, Boku S, Takahashi Y (2015) Psychopharmacology of atypical antipsychotic drugs: from the receptor binding profile to neuroprotection and neurogenesis. Psychiatry Clin Neurosci 69:243–258
336. Nandra KS, Agius M (2012) The differences between typical and atypical antipsychotics: the effects on neurogenesis. Psychiatr Danub 24(Suppl 1):S95–S99
337. Park SW, Phuong VT, Lee CH et al (2011b) Effects of antipsychotic drugs on BDNF, GSK-3β, and β-catenin expression in rats subjected to immobilization stress. Neurosci Res 71:335–340
338. Buckley PF, Pillai A, Howell KR (2011) Brain-derived neurotrophic factor: findings in schizophrenia. Curr Opin Psychiatry 24:122–127
339. Pandya CD, Kutiyanawalla A, Pillai A (2013) BDNF-TrkB signaling and neuroprotection in schizophrenia. Asian J Psychiatr 6:22–28
340. Ray MT, Shannon Weickert C, Webster MJ (2014) Decreased BDNF and TrkB mRNA expression in multiple cortical areas of patients with schizophrenia and mood disorders. Transl Psychiatry 4:e389
341. Hashimoto T, Bergen SE, Nguyen QL et al (2005) Relationship of brain-derived neurotrophic factor and its receptor TrkB to altered inhibitory prefrontal circuitry in schizophrenia. J Neurosci 25:372–383
342. Pillai A (2008) Brain-derived neurotropic factor/TrkB signaling in the pathogenesis and novel pharmacotherapy of schizophrenia. Neurosignals 16:183–193
343. Gonzalez-Burgos G, Cho RY, Lewis DA (2015) Alterations in cortical network oscillations and parvalbumin neurons in schizophrenia. Biol Psychiatry 77:1031–1040
344. Inan M, Petros TJ, Anderson SA (2013) Losing your inhibition: linking cortical GABAergic interneurons to schizophrenia. Neurobiol Dis 53:36–48
345. Nakazawa K, Zsiros V, Jiang Z et al (2012) GABAergic interneuron origin of schizophrenia pathophysiology. Neuropharmacology 62:1574–1583
346. Brown JA, Ramikie TS, Schmidt MJ et al (2015) Inhibition of parvalbumin-expressing interneurons results in complex behavioral changes. Mol Psychiatry 20(12):1499–1507
347. Nakamura T, Matsumoto J, Takamura Y et al (2015) Relationships among parvalbumin-immunoreactive neuron density, phase-locked gamma oscillations, and autistic/schizophrenic symptoms in PDGFR-β knock-out and control mice. PLoS One 10:e0119258
348. Shen S, Lang B, Nakamoto C et al (2008) Schizophrenia-related neural and behavioral phenotypes in transgenic mice expressing truncated Disc1. J Neurosci 28:10893–10904
349. Mellios N, Huang H-S, Baker SP et al (2009) Molecular determinants of dysregulated GABAergic gene expression in the prefrontal cortex of subjects with schizophrenia. Biol Psychiatry 65:1006–1014
350. Weber ET, Andrade R (2010) Htr2a gene and 5-HT(2A) receptor expression in the cerebral cortex studied using genetically modified mice. Front Neurosci 4:36
351. Seo MK, Lee CH, Cho HY et al (2015) Effects of antipsychotic drugs on the expression of synapse-associated proteins in the frontal cortex of rats subjected to immobilization stress. Psychiatry Res 229:968–974

352. Angelucci F, Aloe L, Iannitelli A et al (2005) Effect of chronic olanzapine treatment on nerve growth factor and brain-derived neurotrophic factor in the rat brain. Eur Neuropsychopharmacol 15:311–317
353. Park SW, Lee CH, Cho HY et al (2013) Effects of antipsychotic drugs on the expression of synaptic proteins and dendritic outgrowth in hippocampal neuronal cultures. Synapse 67:224–234
354. Newton SS, Duman RS (2007) Neurogenic actions of atypical antipsychotic drugs and therapeutic implications. CNS Drugs 21(9):715–725
355. Halberstadt AL, Nichols DE (2010) Handbook of the Behavioral neurobiology of serotonin. Elsevier 4(7):621
356. Nichols DE (2004) Hallucinogens. Pharmacol Ther 101:131–181
357. Halpern JH, Pope HG (2003) Hallucinogen persisting perception disorder: what do we know after 50 years? Drug Alcohol Depend 69:109–119
358. Guirado R, Perez-Rando M, Sanchez-Matarredona D et al (2014) Chronic fluoxetine treatment alters the structure, connectivity and plasticity of cortical interneurons. Int J Neuropsychopharmacol 17:1635–1646
359. Kepser LJ, Homberg JR (2015) The neurodevelopmental effects of serotonin: a behavioural perspective. Behav Brain Res 277:3–13
360. Maya Vetencourt JF, Sale A, Viegi A, Baroncelli L, De Pasquale R, O'Leary OF, Castrén E, Maffei L (2008) The antidepressant fluoxetine restores plasticity in the adult visual cortex. Science 320(5874):385–388
361. Cabelli RJ, Hohn A, Shatz CJ (1995) Inhibition of ocular dominance column formation by infusion of NT-4/5 or BDNF. Science 267:1662–1666
362. Hata Y, Ohshima M, Ichisaka S et al (2000) Brain-derived neurotrophic factor expands ocular dominance columns in visual cortex in monocularly deprived and nondeprived kittens but does not in adult cats. J Neurosci 20:RC57
363. Huang ZJ, Kirkwood A, Pizzorusso T et al (1999) BDNF regulates the maturation of inhibition and the critical period of plasticity in mouse visual cortex. Cell 98:739–755
364. Roth TL, Lubin FD, Funk AJ, Sweatt JD (2009) Lasting epigenetic influence of early-life adversity on the BDNF gene. Biol Psychiatry 65:760–769
365. Cargnin S, Massarotti A, Terrazzino S (2016) BDNF Val66Met and clinical response to antipsychotic drugs: a systematic review and meta-analysis. Eur Psychiatry Mar:45–53
366. Notaras M, Hill R, van den Buuse M (2015b) The BDNF gene Val66Met polymorphism as a modifier of psychiatric disorder susceptibility: progress and controversy. Mol Psychiatry 20:916–930
367. Serretti A, Drago A, De Ronchi D (2007) HTR2A gene variants and psychiatric disorders: a review of current literature and selection of SNPs for future studies. 14: 2053–2069
368. Shinozaki G, Romanowicz M, Marek DA, Kung S (2013) HTR2A gene-child abuse interaction and association with a history of suicide attempt among Caucasian depressed psychiatric inpatients. J Affect Disord 150:1200–1203

Index

B.P. Guiard, G. Di Giovanni (eds.), *5-HT_{2A} Receptors in the Central Nervous System*, The Receptors 32, https://doi.org/10.1007/978-3-319-70474-6

C

D

I

K

T

The manufacturer's authorised representative in the EU is Springer Nature Customer Service Centre GmbH, Europaplatz 3, 69115 Heidelberg, Germany. If you have any concerns regarding our products, please contact ProductSafety@springernature.com

Printed and bound by CPI Group (UK) Ltd, Croydon, CR0 4YY
30/09/2026
02225229-0015